Autonomous Robots

Intelligent Robotics and Autonomous Agents
Ronald C. Arkin, editor

Behavior-Based Robotics, Ronald C. Arkin, 1998

Robot Shaping: An Experiment in Behavior Engineering, Marco Dorigo and Marco Colombetti, 1998

Layered Learning in Multiagent Systems: A Winning Approach to Robotic Soccer, Peter Stone, 2000

Evolutionary Robotics: The Biology, Intelligence, and Technology of Self-Organizing Machines, Stefano Nolfi and Dario Floreano, 2000

Reasoning about Rational Agents, Michael Wooldridge, 2000

Introduction to AI Robotics, Robin R. Murphy, 2000

Strategic Negotiation in Multiagent Environments, Sarit Kraus, 2001

Mechanics of Robotic Manipulation, Matthew T. Mason, 2001

Designing Sociable Robots, Cynthia L. Breazeal, 2001

Introduction to Autonomous Mobile Robots, Roland Siegwart and Illah R. Nourbakhsh, 2004

Autonomous Robots: From Biological Inspiration to Implementation and Control, George A. Bekey, 2005

Autonomous Robots
From Biological Inspiration to Implementation and Control

George A. Bekey

The MIT Press
Cambridge, Massachusetts
London, England

This book was set in Times New Roman on 3B2 by Asco Typesetters, Hong Kong.

Library of Congress Cataloging-in-Publication Data

Bekey, George A., 1928–
Autonomous robots : from biological inspiration to implementation and control / George A. Bekey.
p. cm. — (Intelligent robotics and autonomous agents)
Includes bibliographical references and index.
ISBN 978-0-262-02578-2 (hc. : alk. paper)—978-0-262-53418-5 (pb.)
1. Autonomous robots—Research. I. Title. II. Series.
TJ211.495.B45 2005
629.8′92—dc22 2004066198

To Shirley, Ron, and Michelle

Contents

Preface

Autonomous robots are intelligent machines capable of performing tasks in the world *by themselves*, without explicit human control over their movements. This book is an introduction to these remarkable systems, which have proliferated in recent years and promise to play a major role in our lives in the future. The book has several objectives:

- to provide a "guided tour" of the field of autonomous robots, in two ways: first, by reviewing the hardware implementations of several hundred current systems and some of their application areas (such as entertainment, industry, the military, and personal service); and second, by introducing some of the technology underlying these robots and their uses (including control, architectures, learning, manipulation, grasping, navigation, and mapping);
- to review the biological inspiration that forms the basis of many current and recent developments in robotics;
- to discuss some of the fundamental issues associated with robot control.

The breadth of the field can be seen from the fact that the book includes discussions of wheeled robots, legged robots (with two, four, six, and eight legs), flying robots, underwater robots, snakelike robots, climbing robots, jumping robots, and other kinds of robots.

We frequently define a robot as a machine that senses, thinks, and acts. In artificial intelligence such systems are known as "agents." Robots are distinguished from software agents in that they are *embodied* and situated in the real world. They receive information from the world through their sensors. They can be touched and seen and heard (sometimes even smelled!), they have physical dimensions, and they can exert forces on other objects. These objects can be balls to be kicked, parts to be assembled, airplanes to be washed, carpets to be vacuumed, terrain to be traversed, or cameras to be aimed. Robots are also subject to the world's physical laws, they have mass and inertia, their moving parts encounter friction and hence produce heat, no two parts are precisely alike, measurements are corrupted by noise, and, alas, parts break. Robots also contain computers, which provide them with ever-increasing speed and power for both signal processing and cognitive functions. The world into which we place these robots keeps changing; it is nonstationary and unstructured, so we cannot predict their behavior accurately in advance.

These are some of the features of autonomous robots. They suffer from all the limitations of the real world, but because they are physical, they also fascinate us. This is particularly true of humanoid robots, but there is some intrigue in all moving robots. They are an imitation of life, and we are drawn to watching them. It is not only the fact that they move that beguiles us, since many things move in the world, but that they appear to move intelligently, they avoid obstacles, they interact with one another, and they accomplish tasks. For those of us who design and build them, enabling them to perform these and other actions is precisely our goal.

This book is an introduction to both the science and practice of autonomous robots. It can be used as a textbook in senior-level or first-year graduate courses. It is also a book of reference readings for practitioners in industry. Although there is some mathematics in the book, it appears primarily in connection with issues of control and localization. The book does not, however, offer a rigorous treatment of robot control. Rather, it attempts to stimulate, to pose questions, to review the way in which robots are designed, constructed, and used, and to provide some perspective on a rapidly changing field. Some chapters, like those providing overviews of mobile robots (chapter 7), legged locomotion (chapters 8 and 9), and humanoid robots (chapter 13) should be accessible to anyone with some engineering or computer science background, or even an intelligent layperson with no formal technical training.

Autonomous robots are increasingly evident in many aspects of industry and everyday life. As Rodney Brooks of MIT stated in a recent article, "The robots are here!" They are accepted by military organizations, since they are capable of reconnaissance and other missions. They are very evident in the entertainment industry, where they appear as pets or even as soccer players. In the service industries, robots are being used or considered for use in such tasks as vacuuming carpets, washing airplanes, filling gasoline tanks of automobiles, and delivering meals in hospitals and mail to offices.

I believe that during the next twenty years we will see autonomous robots appear in many aspects of our personal and professional lives. We may not recognize some of them, since they may be embedded in our cars or kitchen appliances or innumerable other objects with which we interact. Mobile autonomous robots will also become increasingly evident, not only in the exploration of distant planets or undersea environments, but also in the performance of numerous services for people, in health care, industry, the environment, and our homes. These will indeed be exciting years for roboticists.

There are several hundred pictures of robots from many laboratories throughout the world in this book. Even so, the field is growing so rapidly that it was impossible to include all of the robots that have been developed. So I apologize to all my colleagues whose robots are not mentioned in the book and whose articles are not cited. I am grateful to all of you for helping to create this dynamic field.

This book would not have been possible without the help of numerous colleagues and friends. I would like to acknowledge my intellectual debt of gratitude to Rodney Brooks of MIT (from whose work I learned to look at the intelligence and control of robots in a totally new way), to my USC colleague Michael Arbib (whose astounding breadth of knowledge about both the brain and robotics continues to be a source of inspiration for me), and to my late colleague Rajko Tomović from the former Yugoslavia (from whom I learned many analogies between human control and robot control). I am also immensely grateful to my other USC colleagues, particularly Maja Matarić and Gaurav Sukhatme, for their support, for sharing some of their deep knowledge of the field with me, and for continuing and vastly enhancing the robotics program I started many years earlier. I also want to thank all my former doctoral students, particularly those who worked in robotics and related areas, including Andrew Frank, John Coggshall, Jim Chang, Fred Hadaegh, Dan Antonelli, Howard Olsen, Tasos Chassiakos, Huan Liu, Dit-Yan Yeung, Danilo Bassi, Gerard Kim, Patti Koenig, Arvin Agah, Tony Lewis, John Kim, Gaurav Sukhatme, Alberto Behar, Michael McHenry, Ayanna Howard, Jim Montgomery, and Stergios Roumeliotis. Over the years it has become clear that I have learned more from them than they have from me.

The book would have never been finished without the help of Arun Bhadoria and Catherine Hrabar. Most of the line drawings in the book were made by Arun, a former student whose friendship I value immensely. Catherine wrote most of the letters requesting permission to use figures, organized all the captions for the figures, and made sure that no details were omitted. Were it not for her amazing skills at organization, the details associated with the book would have been a nightmare. Thank you both, from the bottom of my heart.

Finally, without the support, love, and patience of my wife, I could not have finished the book. Thank you, dear Shirley, for putting up with me when my head was somewhere else and I was buried in the book, instead of replacing burned-out light bulbs the way a husband should.

George A. Bekey
Arroyo Grande, California

1 Autonomy and Control in Animals and Robots

Summary

This chapter introduces the main theme of the book: control of autonomous robots based on biological principles. Numerous mobile robots (with various degrees of autonomy) are presented and discussed to provide a context for the rest of the book. There are overviews of control issues in robotic systems and the overall architecture of mobile robots, including sensors, actuators, and intelligent processors, illustrated by multiple examples.

1.1 What Is Autonomy?

Autonomy refers to systems capable of operating in the real-world environment without any form of external control for extended periods of time. Thus, living systems are the prototypes of autonomous systems: They can survive in a dynamic environment for extended periods, maintain their internal structures and processes, use the environment to locate and obtain materials for sustenance, and exhibit a variety of behaviors (such as feeding, foraging, and mating). They are also, within limits, capable of adapting to environmental change.

The emphasis on behaviors makes it clear that we do not consider a rock an autonomous system. Clearly, it exists in the world without external control, but it is capable neither of operating in the world nor of exhibiting any behaviors.

The emphasis in this book is on autonomous systems created by humans. Frequently, these systems draw inspiration from biology, but not always. For example, many autonomous systems use wheels for locomotion, and no wheels exist in nature. "Capable of operating" implies that these systems perform some function or task. This function may be that intended by their human creator, or it may be an unexpected, emergent behavior. As these systems become more complex, they are likely to exhibit more and more unexpected behaviors.

It should be clear that at the present time, most robots are not fully autonomous, within the scope of the preceding definition. They are not capable of surviving and performing useful tasks in the real world for extended periods, except under highly structured situations. However, if the environment is sufficiently stable and the disturbances to it are not too severe, robots can indeed survive and perform useful tasks for extended periods. Furthermore, the field of robotics is a very active research area at this time, and we can expect robots to exhibit increasing levels of autonomy and intelligence in the near future. In certain structured situations, for example, the international RoboCup competitions, small teams of robots already exhibit full autonomy while playing "robot soccer" (Asada and Kitano 1999a).

1.2 What Is a Robot?

In this book we define a robot as *a machine that senses, thinks, and acts.* Thus, a robot must have sensors, processing ability that emulates some aspects of cognition, and actuators. Sensors are needed to obtain information from the environment. Reactive behaviors (like the stretch reflex in humans) do not require any deep cognitive ability, but on-board intelligence is necessary if the robot is to perform significant tasks autonomously, and actuation is needed to enable the robot to exert forces upon the environment. Generally, these forces will result in motion of the entire robot or one of its elements (such as an arm, a leg, or a wheel).

This definition of a robot is very broad. It includes industrial robot manipulators, such as those used for pick-and-place, painting, or welding operations, provided they incorporate all these three elements. Early industrial manipulators had neither sensing nor reasoning ability; they were preprogrammed to execute specific tasks. Currently most industrial robots are being equipped with computer vision and other sensors and include on-board processors to allow for some autonomy. The definition also encompasses a wide range of mobile robots, from the smallest (currently about 1 cm^3) to robot planes, helicopters or submersibles, humanoids and household robots. Figure 1.1 illustrates some of the robots discussed in this book. It is evident from the pictures in the figure that robots come in a wide variety of shapes and sizes, with varying degrees of autonomy, intelligence, and mobility. We describe each of the robots depicted in figure 1.1 briefly in section 1.8. In later chapters of this book we will encounter them again, with considerably more detail on their anatomy and function.

1.3 Problems of Robot Control

Given the view of autonomy outlined in section 1.1, then what is "robot control"? There appears to be a contradiction between *autonomy*, which implies that a robot is capable of taking care of itself, and *control,* which appears to imply some sort of human intervention. To be sure, some form of "high-level control" is required to ensure that the robots do not harm any humans or equipment or other robots. In effect, this high level of control implies an implementation of Asimov's laws, which can be paraphrased as follows:

1. A robot should never harm a human being.
2. A robot should obey a human being, unless this contradicts the first law.
3. A robot should not harm another robot, unless this contradicts the first or second law.

However, there are other levels of control. At the "lowest" level, we want to be sure that the motors driving robots' wheels or moving their legs are used in stable configurations and do not begin to oscillate when activated. At the next level of control, we need to design robots so that they do not collide with one another or with obstacles, while at the same time maintaining stability at the lowest level. We also expect the robots to be able to perform a number of behaviors, such as "foraging" (gathering prespecified objects from the environment) or "flocking" or "following" (e.g., Matarić 1994), while at the same time avoiding obstacles and maintaining stability. Software architectures allowing for such control processes to proceed in parallel are known as *subsumption* architectures (Brooks 1986) or *behavior-based* architectures (Arkin 1998).

The various levels of control discussed in the foregoing are shown in figure 1.2. The software organization associated with these multiple levels is often termed the *control architecture* of a robot. We examine these various aspects of control in detail in succeeding chapters of this book, but some of the basic issues are discussed in the following paragraphs. Clearly, the higher levels of control provide inputs to the lower levels, but there is also feedback from the lower levels to the upper levels. Sensors provide inputs to the lowest (and sometimes the intermediate level); actions upon the world are exerted from the lowest level.

Note that the upper box in figure 1.2 indicates human input is involved in high-level robot control. Low-level control is clearly autonomous, whereas intermediate-level control is generally autonomous in contemporary mobile robots but may still involve some human input. As indicated previously, this is an extremely active area of research, and we can expect increasing autonomy even at the highest level. "Structure shift," referred to in the figure in the context of high-level control, implies an ability on the part of a robot to reconfigure its physical structure; some robots are already capable of some autonomous reconfiguration (see, e.g., Rus and Chirikjian 2001; Shen, Salemi, and Will 2002).

Low-level control systems encountered in other, more common venues are frequently taken for granted, without recognition that they were in fact designed using

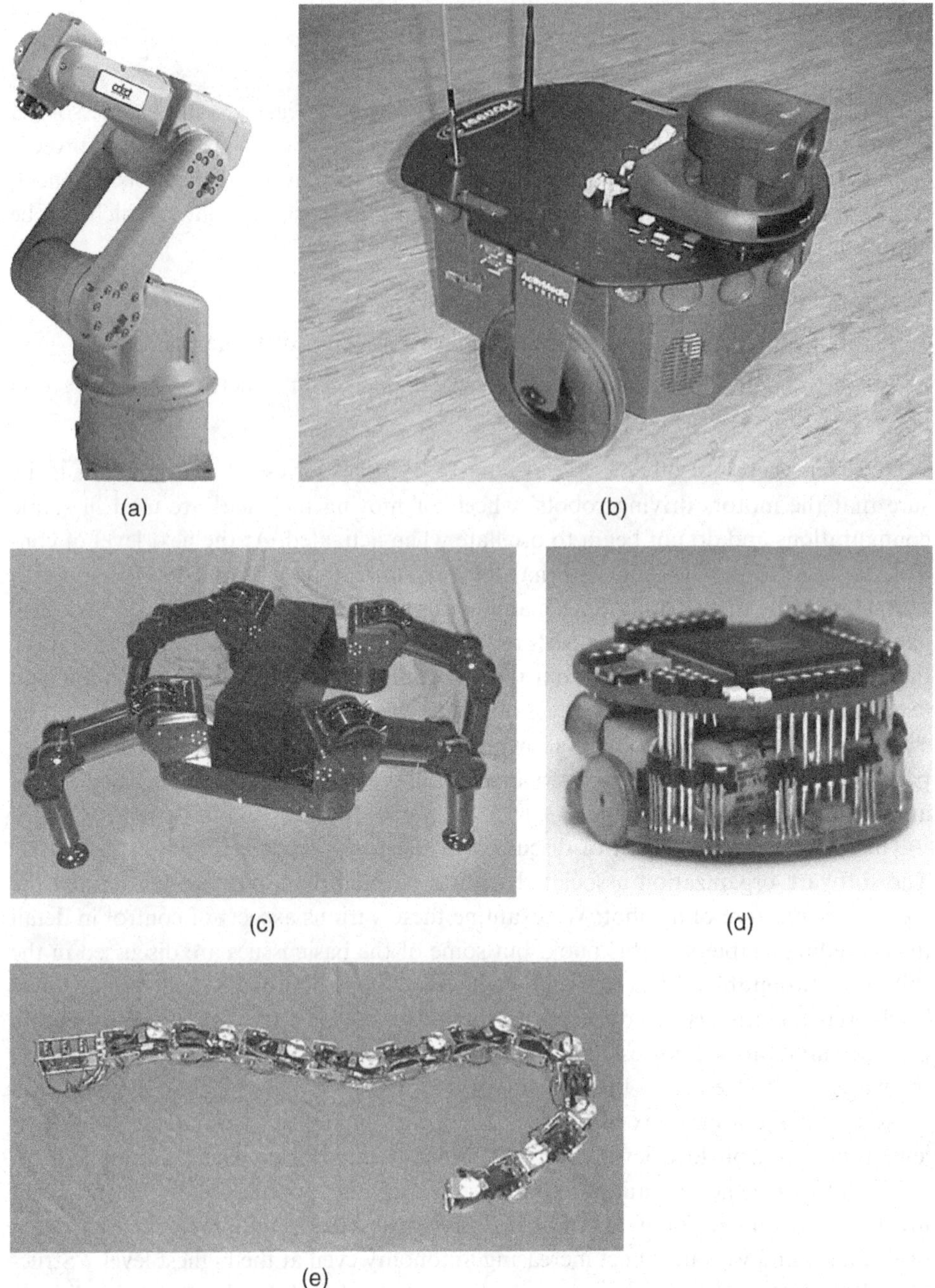

Figure 1.1
(a) a typical industrial manipulator (photograph courtesy of Adept Technology, Inc.); (b) a Pioneer mobile robot, commonly used in research laboratories (photograph courtesy of ActivMedia Robotics); (c) a large quadruped robot (TITAN IX) developed by the Hirose-Yoneda Laboratory at the Tokyo Institute of Technology (photograph courtesy of Shigeo Hirose); (d) a small Khepera robot, about 7 cm in diameter, originally developed at the Swiss Federal Institute of Technology and available commercially from

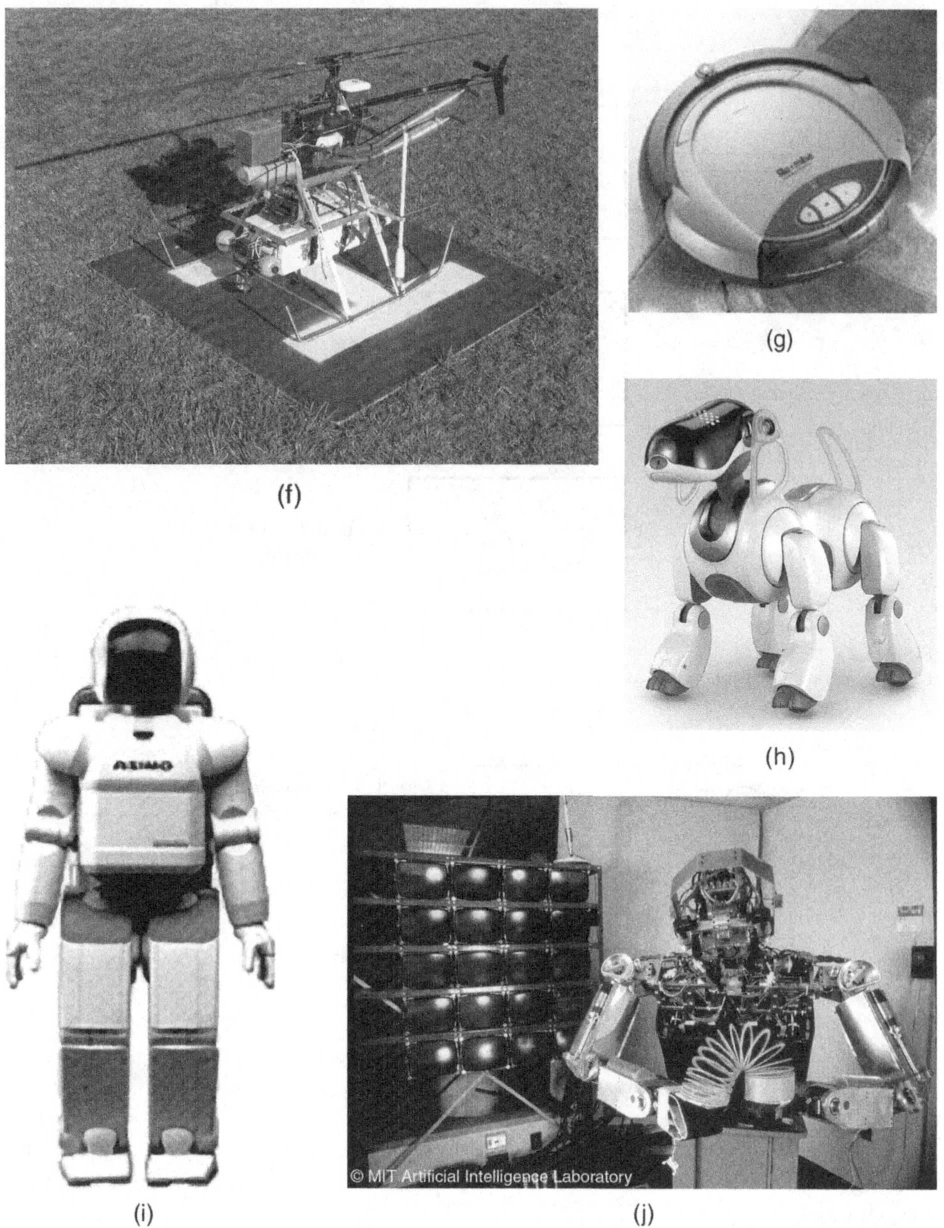

(f) (g) (h) (i) (j)

Figure 1.1 (continued)
K-Team S.A. in Lausanne, Switzerland (photograph courtesy of K-Team); (e) articulated snakelike robot constructed by Kevin Dowling at Carnegie Mellon University (courtesy of Kevin Dowling); (f) the AVATAR robot helicopter developed at the University of Southern California courtesy of Gavrav Sukhatine; (g) Roomba, a household vacuum-cleaning robot from iRobot Corporation (photograph courtesy of iRobot Corporation); (h) AIBO, a pet robot from Sony Corporation (photograph courtesy of Sony); (i) ASIMO, a biped walking robot from Honda Motor Company Ltd. (photograph courtesy of Honda); (j) Cog, a humanoid torso with significant cognitive abilities as well as arm and head movements, developed at the Artificial Intelligence Laboratory at the Massachusetts Institute of Technology (photograph courtesy of Rodney Brooks and Annika Pfluger)

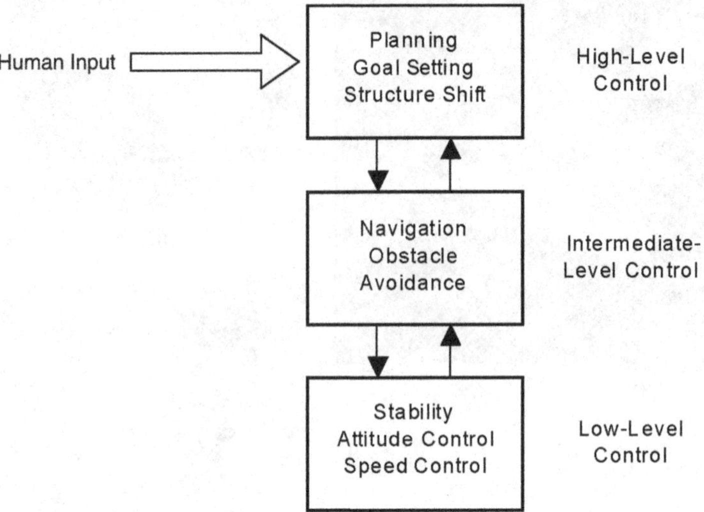

Figure 1.2
Levels of control in autonomous robots

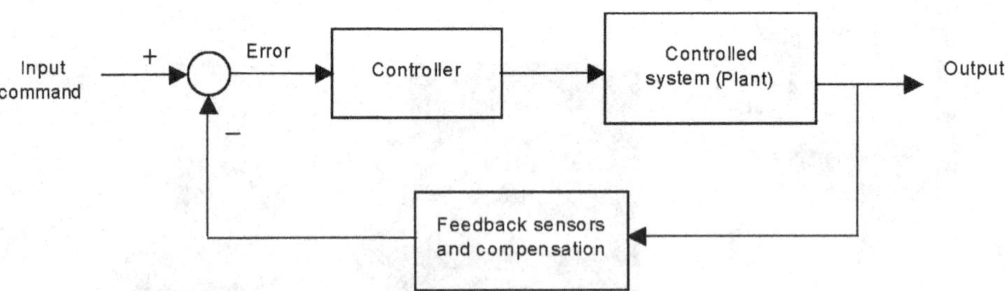

Figure 1.3
Basic control system

the techniques of control theory. For example, such automobile systems as power steering or power brakes are in fact feedback control systems, with the general structure shown in figure 1.3.

The *input command* in the figure may represent, for example, in the case of power steering, the desired orientation of the front wheels (as commanded by the steering wheel). The error is the difference between the commanded direction and the actual direction. This error signal is the input into a controller (frequently a microprocessor) that generates an input signal to the actual motor that moves the wheels. The resulting wheel direction is now measured using sensors and compared with the input command. Systems of this type are known as negative feedback control systems, since the feedback signal has the opposite sign from the input command. It is important to note that the system illustrated in this figure is a dynamic system, not a static one. This means that it is described by differential equations to represent the variables and their rates of change (derivatives with respect to time); static systems are described by algebraic equations, since they do not depend on time.

Note, however, that nothing is as simple as it seems in a feedback control system. If the driver applies a clockwise motion to the steering wheel, followed a fraction of a second later by a counterclockwise motion (as may happen when avoiding an object in the roadway), and these motions are repeated, it is important that there be no delay in comparing the feedback signal with the input. Assume that the driver produces the second (counterclockwise) command 0.5 seconds after the first command. Assume that the controller, controlled system, and feedback boxes do not alter the shape of the signal they receive, but simply delay it in time by 0.5 seconds. Then the input command, actual output position, and error signals will have the form shown in figure 1.4.

If the feedback signal is delayed by 0.5 seconds, it is evident that when it arrives at the comparator and is subtracted from the input signal, it will in fact add to the input, thus producing an error signal that grows in time, as shown in the lower waveform of figure 1.4. Improperly designed control systems may display such increasing oscillations even when the input signal is returned to zero. This phenomenon is known as *instability*. Many robots move so slowly that such unstable behaviors are highly unlikely. However, future generations of autonomous robots are expected to exhibit much higher response speeds, and hence their low-level controllers will have to be designed carefully to avoid instabilities.

Although the foregoing example concerned the steering control system in an automobile, similar systems provide the speed or orientation control of a wheeled robot, for example, or the leg position control of walking robots. Of course, in order to control these or any other variables, we need to be able to measure their values and then exert correcting forces. Hence, the issues of control system design are intimately related to selection and design of sensors and actuators. We consider these various aspects of robot control systems in later chapters. Various approaches to the design of engineering control systems are discussed in chapter 4.

1.4 Biologically Inspired Robot Control

In a general sense, engineering and biological control systems have similar structures, as illustrated in figure 1.5. Panel (a) shows a prototypical biological control system, in which the command signal is provided by the central nervous system, the computations required by the controller are performed either locally or by the brain, and the "plant" refers to the dynamics of the controlled system. Panel (b) shows the

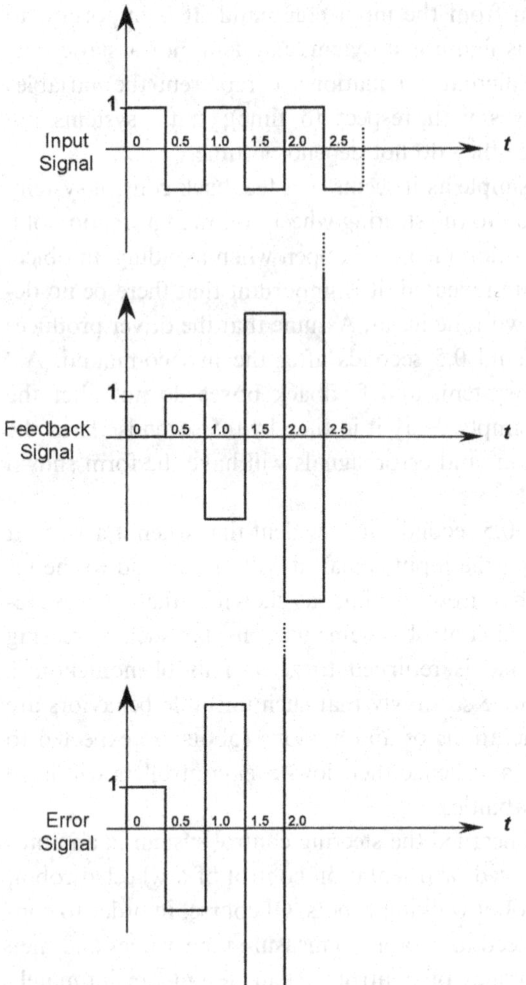

Figure 1.4
Emergence of oscillations in a feedback system

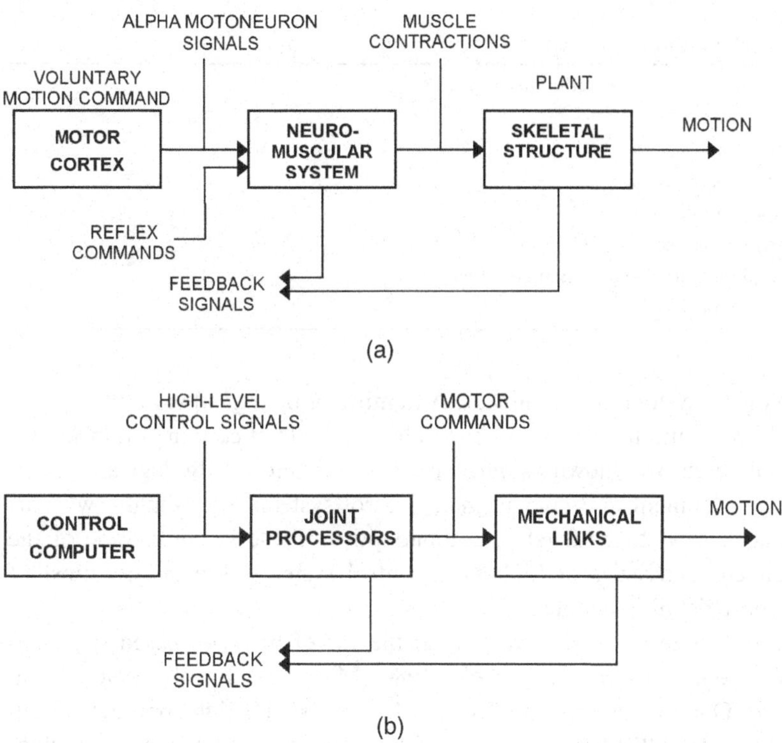

Figure 1.5
Control systems: (a) biological and (b) analogous engineering (robot)

engineering counterpart to the biological system in panel (a). It is equivalent to the control system shown in figure 1.3 but shows some of the details associated with controlling the legs of a walking robot.

It is important to note that one should not take analogies of the type shown in figure 1.5 too seriously, since biological systems are more complex than and may behave in ways quite distinct from human-designed systems. Consider, for example, the control of body temperature. It is well known that the body core temperature in humans is maintained at approximately 37°C. However, the system that maintains this constant temperature (and a number of other homeostatic systems) does not behave like the engineering systems in figure 1.2, since there is no reference temperature input. In other words, whereas an engineering system may have a "reference" value (e.g., the voltage from a battery), the body's temperature control system has no such standard value. Furthermore, the diagram depicted in figure 1.2 is used primarily (but not exclusively) to describe linear control systems, that is, systems in which both the plant and the controller can be described by linear differential equations.

Table 1.1
Some characteristics of biological control systems

Adapt to changes in internal and external environment
Usually nonlinear
Hierarchical organization
Include redundancy
May involve multiple control loops
Control is frequently distributed
Control may be based on multiple performance criteria
May display limit cycle oscillations

In general, biological systems are nonlinear. A number of biological variables appear to oscillate with small amplitudes in a manner characteristic of certain nonlinear systems. These oscillations are known as *limit cycles* (see chapter 4). Biological systems adapt to their environment and change their control systems accordingly, whereas engineering systems tend to be fixed and nonadaptive. Table 1.1 lists some of the more important characteristics of biological control systems. We discuss most of these issues in more detail in chapter 2.

Recent research has increasingly emphasized the use of behavior-based strategies for control of autonomous robots (Brooks 1986; Maes and Brooks 1990; Arkin 1998; Beer 1990). One of the major motivating factors behind this approach to autonomy arises from the difficulties associated with traditional methods, which require accurate knowledge of the robot's dynamics and kinematics, as well as carefully constructed maps of the environment in which they operate. Such approaches are not well suited to time-varying and unpredictable, unstructured situations. As a solution to this problem, Brooks and others have proposed reactive strategies: The robot senses the environment and reacts with appropriate behaviors as required. As we show in chapter 5, current behavior-based architectures augment reactive behaviors with planning and reasoning; the latter are sometimes referred to as *deliberative* components of the robot's control architecture.

1.5 Sensors

Robots need sensors both to receive information from the outside world and to monitor their internal environment. Many (but not all) robot sensors are devices that attempt to imitate some of the properties of animal senses. In this section we provide a brief introduction to the major sensory systems in animals, indicate how some of the features of these sensors are incorporated into sensing devices used with mobile robots, and list the major limitations of these devices. Sensors are discussed in greater detail in chapter 3.

Exteroceptive sensors are used to obtain information from the external environment. We frequently speak of the "five senses" (vision, hearing, olfaction, touch, and taste), but each of these major categories in living systems encompasses an exquisite and complex array of sensory dimensions. Furthermore, perception of the outside world by animals is based on an interaction of the sensory apparatus proper with corresponding processing centers in the brain. Hence, design of an artificial "eye" for a robot requires not only a light-sensitive receptor, but some aspects of the visual cortex. Robot vision systems usually include a camera (with resolution comparable to or better than standard television cameras) and software designed for such tasks as detecting edges, enhancing contrast, or recognizing objects. Clearly, some of these features are based on living prototypes. The compound eyes of certain insects have other properties, not yet imitated by robot sensors.

The human visual system includes neural circuits tuned to the perception of lines in particular directions. The frog visual system is highly tuned to the detection of moving insects (Lettvin et al. 1959). The design of robot sensory systems may include some aspects of both of these features. This would be highly desirable if one wished, for example, to design a robot helicopter to recognize particular landmarks on the ground or a robot frog to catch and digest flies. The situation is similar with the other senses. In general, then, robot sensors are very limited compared to their living equivalents. For example, some animal sensory systems, such as the olfactory system of certain insects (capable of detecting a few molecules of pheromones) or the visual system of raptors, such as eagles or hawks, are incredibly sensitive to particular signals. Engineering approximations to some of these animal sensory systems are used on robots. Most robots are equipped with obstacle detectors that operate using ultrasound or lasers. These detectors emit a signal and pick up the echo from an object, in a manner analogous to the navigation system of bats. On the other hand, it is possible to design robot sensors capable of detecting physical phenomena not detectable by sensory systems of living animals. For example, a robot equipped with a Geiger counter can detect ionizing radiation, which vertebrates cannot do. Similarly, robot sensors can be designed to detect ultraviolet or microwave radiation. Other exteroceptive sensors in robots include those able to detect sound, object texture (touch), certain odors, temperature, and slippage. These and other sensors are discussed in chapter 3.

Proprioceptive sensors monitor the organism's or robot's internal environment. In view of the differences between engineering systems and living systems, proprioceptive robot sensors may not have living models. For example, a robot may need to monitor its wheel rotations and its battery voltage level, which have no human or animal counterparts. Monitoring the joint angles and leg motor currents in a legged robot corresponds to human systems' obtaining of information from Golgi tendon organs and muscle spindles, but the analogy is gross at best. In robots employing artificial muscles (which contract when stimulated), it may be possible to design sensors with properties that mimic those in living muscle, but it is not clear that such a design would be desirable in view of their large number and complexity. The goal of proprioceptive sensors in robots is to provide signals indicative of the robot's internal states, in order to improve control, to identify and correct faults, or to provide feedback to humans.

1.6 Actuators

A robot must be able to interact physically with the environment in which it is operating. In fact, the key difference between a robot and a "softbot" or software agent lies primarily in a robot's having actuators that permit it to affect the environment, say, by exerting forces upon it or moving through it, which a softbot lacks. Various actuators are discussed in detail in chapter 3. Some of the most common actuators are

- artificial muscles of various types, none of which are very good approximations of living muscles;
- electric motors, the most common actuators in mobile robots, used both to provide locomotion by powering wheels or legs, and for manipulation by actuating robot arms (special-purpose motors, such as stepper motors, are used for precision movement);
- pneumatic and hydraulic actuators, used in industry for large manipulation tasks, but seldom for mobile robots.

1.7 Intelligence

As noted in section 1.2, a robot is a machine that senses, thinks, and acts. Computers are the brains of robots and are essential elements of these systems. The continuing decreases in size and weight of microprocessors, coupled with increases in speed and memory, have had major effects on the development of mobile robots. Since at present a single chip can have the processing power of a mainframe computer only twenty years ago, small mobile robots are now being equipped with extremely powerful on-board computers. Hence, the limited cognitive abilities of these robots (and thus their ability to perform increasingly complex tasks) are due to software rather than hardware. "Intelligence" appears in these systems in a number of ways:

1. *Sensor processing* Raw outputs from sensors are not very useful for controlling behavior. Thus, robot vision systems frequently include special-purpose software for locating areas in the environment that differ in some way from the surround ("blob

detection"), for edge detection, for contrast enhancement, and so on. Biological sensors also include such preprocessing (e.g., "What the frog's eye tells the frog's brain" [Lettvin et al. 1959]). Placing some of this processing in the sensors reduces the load on the robot's central processor.

2. *Reflex behaviors* In living systems there are rapid reaction paths from sensing to actuation (reflexes) that do not involve higher centers in the nervous system. Withdrawing the hand upon touching a hot stove or the knee-jerk reflex are examples. In the latter, when the physician taps the tendon below the patella with her hammer, specialized sense organs within the muscle fibers (*muscle spindles*) sense the resulting lengthening of the fiber and send a signal to the spinal cord, resulting in contraction of the muscle. When the leg of an insect contacts an obstacle, such as a rock, while walking, the leg is withdrawn and moved higher in an attempt to clear the obstacle. When included in the behavior of a walking machine, such behavior appears "intelligent" (Brooks 1989; Sukhatme 1997).

3. *Special-purpose programs* Such applications as navigation, localization, and obstacle avoidance may be included in robot software.

4. *Cognitive functions* Artificial-intelligence research is providing robot computers with continual improvements in a number of cognitive functions, including reasoning, learning, and planning.

The organization of the above components of robot control software is known as the robot's *software architecture*. It is generally hierarchical in structure, with the reactive components appearing at the lowest level and those components involving planning and learning at the highest level, as illustrated in the "three-level architecture" of figure 1.6. Human control inputs are applied at the highest level. Robot architectures are discussed in detail in chapter 5.

All of these aspects of robot design are discussed in later chapters of the book. Suffice it to say at this point that the field is moving very rapidly, so that we can expect fairly dramatic improvements in robot intelligence in the coming decades. But will the robots then be so intelligent that they may refuse to obey us (see chapter 15)?

1.8 A Brief Survey of Current Robots and Associated Control Issues

We now consider the robots illustrated in figure 1.1. Robotics developed along two distinct paths, one concerned primarily with manipulation of objects, and a second with mobility. Although both types of robots are discussed in later chapters, the book's emphasis is on mobile robots. Here we introduce only some of the major aspects of the robots in this figure and indicate some of the control problems that appear with each major design.

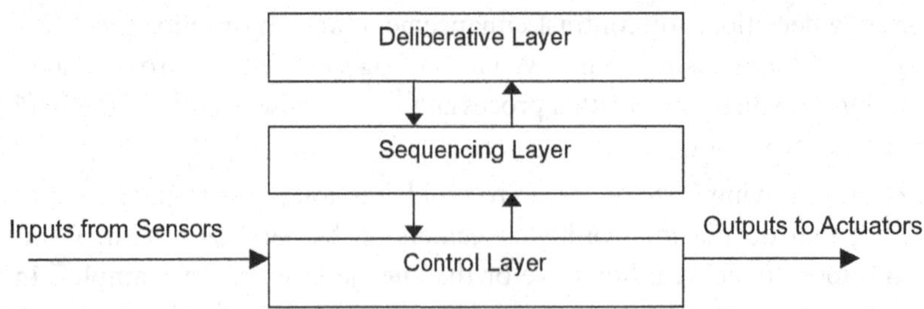

Figure 1.6
Typical three-level architecture for mobile robots

1.8.1 Industrial Manipulators

The development of manipulators began with devices to facilitate remote handling of radioactive materials, during and shortly after World War II. Human operators could move remote arms and grippers by means of joysticks, instrumented gloves, and so on. Such systems were known as "tele-manipulators," since they involved manipulation at a distance. These systems led to manipulators for industry to facilitate such tasks as material handling, welding, and spray painting, all of which present some hazards to human operators. Typical industrial manipulators are illustrated in figure 1.7 (see Engelberger 1980; Niku 2001). Manipulation is discussed in chapter 10.

Note that the robots in this figure display a superficial resemblance to a human arm. The "shoulder" joint allows for motion of the entire structure, as well as rotation about a vertical axis. The "elbow" joint allows for motion in a single plane. The wrist, which is not clearly shown in either illustration in this figure, may allow for 3 degrees of freedom (dof) of rotation (pitch, roll, and yaw). In addition, the end effector or gripper may allow for an additional degree of freedom, resulting in a total of 7 dof for the robot. Practical robots may have fewer dof.

It should also be noted that early manipulators had no sensors. They were entirely preprogrammed to follow desired trajectories. A master painter could move the end effector (holding a spray gun) of a painting robot to follow the desired spray path; this motion was memorized by the system and then repeated exactly, time after time. Contemporary industrial manipulators, in contrast, are equipped with vision, touch, and other sensors as well as on- or off-board computers, thus endowing these systems with the possibility of some autonomous and adaptive behavior.

The control issues in manipulator design center on problems of coordinate transformation. Control of the end effector ("hand") position of the robot is obtained through motors, some of which may be located at the shoulder or elbow of the device. Hence, it is necessary to transform desired end effector positions and orienta-

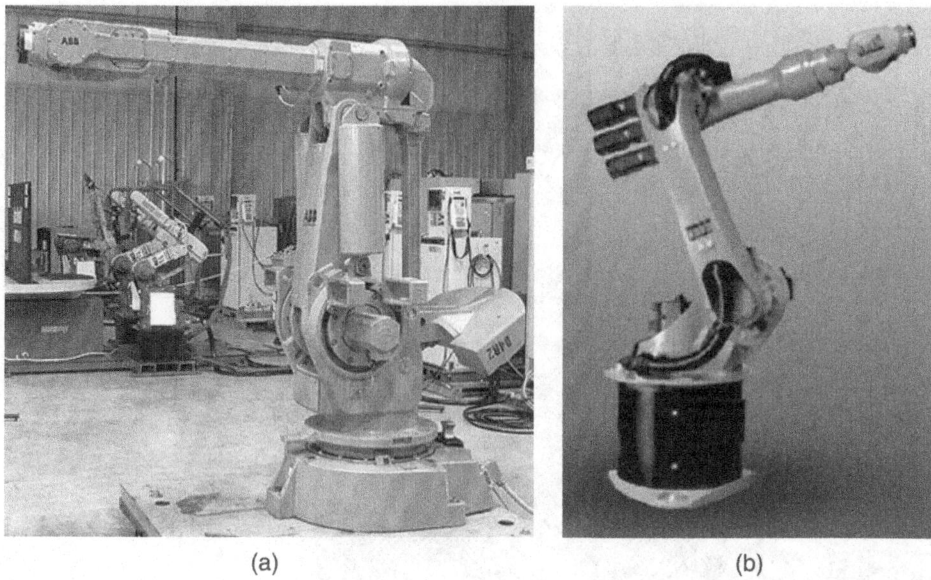

(a) (b)

Figure 1.7
Typical industrial robots: (a) IRB 6400 spot welding robot (photograph courtesy of ABB Ltd.) and (b) KR15 loading and palletizing robot (photograph courtesy of KUKA Roboter GmbH)

tions into appropriate motor rotations at these other joints. Typically, the transformation requires the inversion of coordinate transformation matrices (see chapter 10).

1.8.2 A Pioneer Mobile Robot

As noted previously, the second path of robot development was concerned with mobile robots, devices capable of moving in their environment by means of legs or wheels. (More recent mobile robots are also capable of mobility in the air and under water.) Figure 1.8 shows a commonly used research robot, the Pioneer 3-AT, made by ActivMedia Robotics, in Amherst, New Hampshire. Note that this robot has four wheels, a number of sonar proximity sensors along the front surface, and a complex of communication equipment on its top surface. Some Pioneers are equipped with laser range sensors, global positioning satellite (GPS) receivers, and other sensing devices.

Control of the position, orientation and, velocity of the Pioneer (and other wheeled robots) is obtained through electric motors driving the wheels. Generally, the control systems for the driving motors make use of feedback; careful design is needed to avoid the possibility of instability (see chapter 4). Differential control of a pair of wheels may allow the robot to turn in place, or the robot may have a minimum turning radius (like an automobile). Further control problems will arise when

Figure 1.8
Pioneer 3-AT robots (photograph courtesy of ActivMedia Robotics)

the robot is attempting to navigate from a starting to a goal position. The simplest way to control a trajectory is by counting wheel revolutions. Unfortunately, this method, known as *odometry*, can lead to very large trajectory errors, since the robot wheels may slip. Hence, more complex methods of trajectory control are needed (possibly involving Kalman filters or other statistical methods), or additional sensors (such as vision) may be used to identify landmarks for navigation. Wheeled robots are discussed in more detail in chapter 7.

1.8.3 TITAN Quadrupeds

By contrast with the wheeled Pioneer robot, a number of laboratories have designed and built legged robots with two, four, six and even eight legs. Clearly, the broad

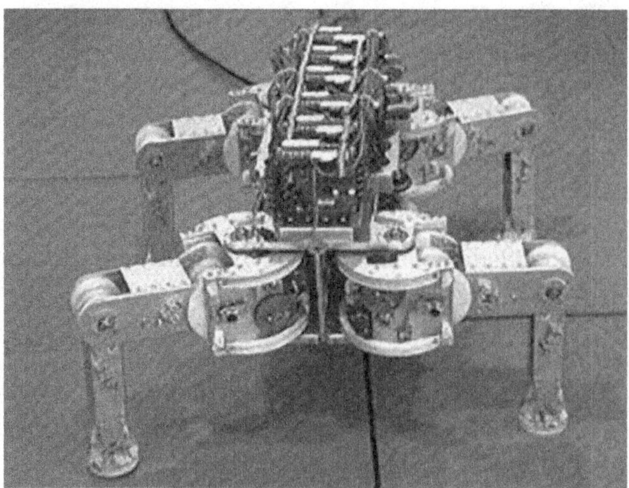

Figure 1.9
TITAN VIII quadruped robot (photograph courtesy of Shigeo Hirose)

architecture of these machines is based on biological prototypes, but the robots generally have fewer degrees of freedom and hence are much less complex than the animals on which they are based. The four-legged machine shown in figure 1.9 (known as TITAN VIII, with TITAN standing for Tokyo Institute of Technology Aruku Norimono [walking vehicle]), was constructed in the Hirose-Yoneda Laboratory of the Tokyo Institute of Technology. Shigeo Hirose is one of the best robot designers in the world at the present time. He developed the theory of stability for legged locomotion machines and has built a series of increasingly sophisticated quadruped robots. The major features of TITAN VIII include

- 12 degrees of freedom (each leg has 3 dof);
- a weight of 19 kg, not including battery and computer, and ability to carry a payload of 5 to 7 kg;
- a potentiometer in each joint for position feedback for control;
- unique control systems known as Titech, designed in Hirose's laboratory, as motor drivers;
- ability to walk at a maximum speed of 0.9 m/sec, which is remarkably fast for a machine of this weight.

TITAN VIII is commercially available for a price (at the time of this writing) of 1.5 million yen.

The latest in the TITAN series is TITAN IX, shown in figure 1.1(c). This robot is equipped with adaptive feet, enabling it to walk on uneven terrain; it also has the ability to recover from falls and can climb stairs autonomously. One of the most interesting features of this robot is that it can stand on three legs and use the fourth leg, equipped with an adjustable gripper, as a manipulator. Four-legged robots are discussed in more detail in chapter 9.

Two types of control problems appear in connection with quadruped robots, involving static and dynamic stability. Static stability refers to the ability of the robot to maintain an upright posture while standing (or while moving very slowly, so that dynamic effects are negligible). With quadrupeds (also known as tetrapods), this means that when one leg is off the ground, the projection of the system's center of gravity must lie within the triangle formed by the three legs on the ground. It is evident that if the center of gravity is outside of this triangle, it will exert torque and cause the animal or robot to fall. Animals adjust their center-of-gravity position as they walk to ensure static stability. In robots this may require an active control system. The situation is more complex when the animal moves rapidly, for example, a horse in gallop. In this case all four legs may be off the ground for brief periods, and dynamic stability requires the consideration of inertial forces due to the motion as well as gravity.

1.8.4 Khepera Mobile Robot

The robot illustrated in figure 1.1(d), the Khepera mobile robot, is designed for "tabletop robotics" experiments. It was designed by Jean-Daniel Nicoud at École Polytechnique Fédérale de Lausanne (EPFL), usually referred to in English as the Swiss Federal Institute of Technology, and is now manufactured by K-Team S.A. in Lausanne. Its cylindrical body is approximately 7 cm in diameter. It has two active wheels and a third support point. The basic robot includes the drive motors and on-board processor, but many additional modules, compatible with the size and shape of the platform, namely, with the same form factor, can be added. These include vision and other sensors as well as a gripper.

A wide variety of software is available for the Kheperas, ranging from that for navigation and obstacle avoidance to that for simulators. Kheperas are being used extensively for research in such areas as robot learning and group interaction. Good control system design ensures that the wheels of Kheperas do not break into oscillations even with rapid changes of velocity.

1.8.5 Snake Robots

The robots discussed in the foregoing consist of a body and either wheels or legs to provide mobility. Several investigators have also constructed articulated, segmented robots whose motion approximates that of snakes. Figure 1.10 shows a snake robot developed at the Fraunhofer Institute for Autonomous Intelligent Systems in Germany. See also figure 1.1(e) that shows a snake robot designed and constructed by Kevin Dowling at Carnegie Mellon University.

The undulating movement of a snake (or, equivalently, an eel, such as the lamprey) requires coordination and sequencing of muscle contractions along the spinal cord, enabling the successive segments of the animal to move in turn (Grillner and Dubuc 1988). Thus, building a robot snake requires the construction of a multisegmented body with the appropriate control sequence to allow smooth, undulating movements, as in Ostrowski and Burdick 1996 and Dowling 1997. The movements of a snake are assumed to be controlled by central pattern generators in the spinal cord (Grillner and Dubuc 1988). The Dowling snake in figure 1.1e consists of ten segments; each link has a 2-dof servo, thus allowing movement in three dimensions. The front segment holds a television camera, thus providing for the robot a "snake's eye" view of the world. Snake robots are discussed further in chapter 7.

1.8.6 A Robot Helicopter

The wheeled and legged robots we have discussed in the previous sections were designed to work on land, but there are also robots that fly through the air (e.g., Montgomery, Fagg, and Bekey 1995) or swim under water (e.g., Yuh, Ura, and Bekey 1997). Figure 1.11 shows an autonomous robot helicopter, AVATAR (Autonomous Vehicle Aerial Tracking and Reconnaissance), developed at the University of Southern California (USC) by the author's colleagues and their students.

Clearly, the problems inherent in the control of a robotic air vehicle are quite different from those involved in the control of land vehicles. First, to remain airborne, the vehicle must generate sufficient lift to overcome both drag and gravitational forces. This implies a need for sufficient forward speed for an aircraft and sufficient rotational velocity for the rotor (effectively, a rotary wing) on a helicopter. Speed, in turn, means that dynamic effects cannot be neglected, as they often can be and are with relatively slow land vehicles.[1]

The dynamics of helicopters are quite complex, since they include aerodynamics, blade bending (and possible oscillations), and the interaction among various control modes. Much of the control in helicopters is obtained by adjustment of the pitch of the rotor blades, once every revolution (hence referred to as "cyclic"). This cyclic adjustment increases lift on one blade while decreasing it on the other, affecting both the vehicle's pitch and its roll. A control mode called the "collective" changes the pitch of the rotor blades by the same amount (collectively); this change affects the

1. In my opinion, some control of dynamic effects will need to be included in most mobile robot models in the near future, as these vehicles become faster and are used in increasingly unstructured domains. Dynamic models are scarce partly because a great deal of robotics work is done by computer scientists, and many computer science curricula do not include differential equations.

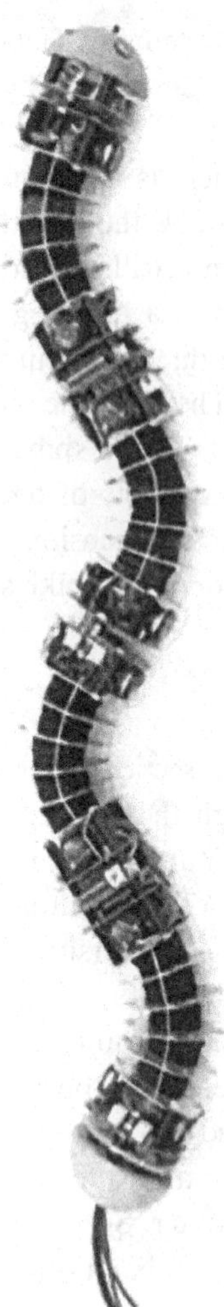

Figure 1.10
Snake robot developed at the Fraunhofer Institute for Autonomous Intelligent Systems in Germany (photograph courtesy of the Fraunhofer Institute)

Figure 1.11
The AVATAR robot helicopter (in flight) developed at the University of Southern California (photograph courtesy of Stefan Hrabar and Gaurav Sukhatme)

thrust, thus increasing or decreasing the helicopter's lift. Tail rotor pitch affects the vehicle's yaw. There is a great deal of cross-coupling between control modes. For example, changes in the thrust level (from the throttle or the collective) produce torques about the yaw axis, which need to be counteracted by the tail rotor to ensure that the helicopter's heading does not change. Hence, the resulting differential equations governing helicopter dynamics are highly nonlinear. For this reason many investigators in the area of robot helicopters (e.g., Montgomery [1999]) have not attempted to solve the complete set of equations governing the dynamics, relying rather on heuristics, simulation, and trial-and-error methods to find proper values for control system gains.

The control of robot helicopters is greatly facilitated by using a hierarchical, behavior-based architecture of the type employed in AVATAR. The architecture of this robot helicopter is discussed in chapter 5 as a case study.

Interest in robot air vehicles of various types is likely to increase in the future, for both military and civilian applications. These may include airplanes, helicopters, and even flying vehicles with flapping wings, modeled on birds and insects.

1.8.7 Roomba, a Household Robot

The use of robots as household helpers has been in the public imagination for generations. The popular 1960s television program *The Jetsons* featured a household robot, dressed like a maid with an apron, who pushed a vacuum cleaner. In 2002, the U.S. robotics company iRobot began selling a completely autonomous vacuum-cleaning robot, Roomba (figure 1.1g), at a base price of $199. This remarkable little robot is turned on, a room size (small, medium, or large) is selected, and the machine is fully autonomous thereafter. It moves about the room in an increasing spiral motion until it reaches a wall, where it shifts to a wall-following mode using sonar sensing. The exterior of the round machine is equipped with touch sensors, so that when it comes in contact with an object (say, a chair leg), it shifts direction. With a battery life of about one and a half hours, it is capable of cleaning even a large room, although not as well as a handheld vacuum cleaner.

Roomba has a number of interesting features. For example, it senses the edge of a surface on which it is traveling and stops if there is a sudden drop (e.g., at the top of a stairwell). As a result of its performance and modest cost, it is reported that some 1 million of these robots had been sold as of late 2004. A European vacuuming robot, manufactured by Electrolux, is named Trilobite. The Roomba and the Trilobite are among the first household robots; many more are expected in the coming decade.

1.8.8 AIBO, an Entertainment Robot

Sony introduced AIBO (Artificial Intelligence Robot), the first robot pet, in the late 1990s. The machine has since been modified and improved several times, as well as drastically reduced in cost. The third-generation AIBO ERS-7, released in 2004, is shown in figure 1.1(h).

AIBO is a remarkable robot in a number of ways:

1. It is capable a number of autonomous behaviors, such as chasing and playing with a ball, by virtue of its excellent vision system and the coupling of vision with leg movements (both for chasing and for "kicking" the ball).

2. If it falls down, it is capable of getting up.

3. It can receive input commands from human users by touch or by voice, which enables it perform other behaviors such as lying down, sitting, or waving.

4. The on-board computer is sufficiently sophisticated so that robot laboratories (if given access to the code by Sony) can program new behaviors.

5. AIBOs can interact with one another, so that they are now being used by a number of institutions in "robot soccer" competitions (RoboCup) (Asada and Kitano 1999a).

These behaviors make the robot surprisingly lifelike, so that it can be a playmate for children as well as a laboratory tool or a soccer player. The next decade is likely to bring many more semiautonomous toys. The potential market in entertainment robotics is very, very large, compared to, say, that in industrial robotics.

1.8.9 ASIMO, a Biped Walker

In the late 1990s Japan announced a national research effort to create humanoid robots, that is, machines with a physical resemblance to a human body structure and having an ability to walk on two legs. In response to this challenge, Honda Motor Company Ltd. designed and built a series of walking machines known as the P-1, P-2, and P-3 (figure 1.12). As we discuss in later chapters, biped walkers had previously been built, but none was capable of the degree of stable walking (and stair climbing!) that characterized Honda's machines. All the P-series humanoids were teleoperated from a complex remote control station, but once turned on to, say, forward walking, or climbing a set of stairs, the robot was capable of performing the

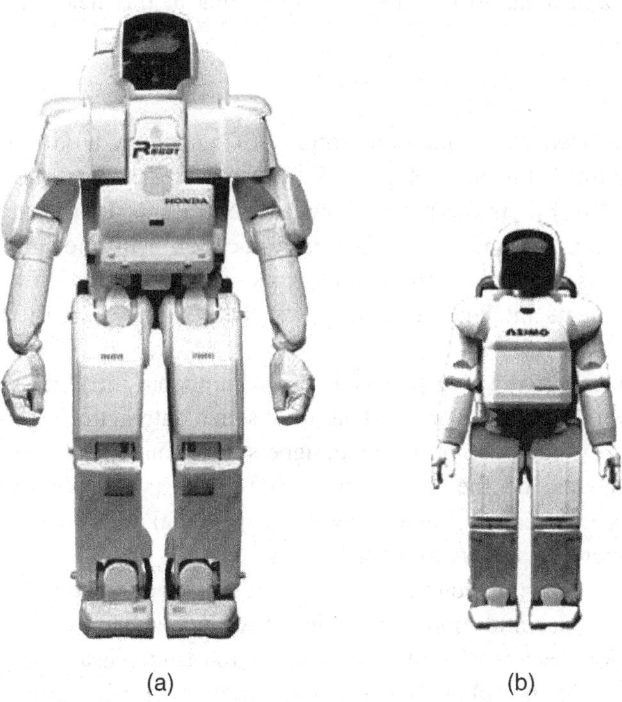

Figure 1.12
Humanoid robots from Honda: (a) P-3, (b) ASIMO (photographs courtesy of Honda Motor Company Ltd.)

repeated movements on its own. These machines were quite large (some two meters tall) and heavy, due in part to the on-board batteries the robots carried in a backpack. In 2001 Honda introduced ASIMO (Advanced Step in Innovative Mobility), a robot the size of child, also illustrated in figure 1.16. Biped robots are discussed in more detail in chapter 8.

Walking robots have always fascinated people. Although the P-series robots and ASIMO were neither very autonomous nor very "intelligent," their ability to walk stably on two legs made them a sensation wherever they were displayed. (Recent versions of ASIMO display a number of features associated with intelligence, such as speech recognition and synthesis.) When a machine of approximately human size is capable of walking, it immediately assumes a humanlike character, and people tend to assume that it may also have other humanlike characteristics. Clearly, our homes and places of business are built for access to erect bipeds. The popular imagination leads directly to humanlike robots capable of assisting us in our homes. Although this type of autonomy is not likely to be achieved in the near future, these walking robots have spurred an international interest in humanoid robots, leading to international conferences and a number of research programs in this field (see chapter 13 and section 1.8.10).

1.8.10 Cog

The final example in this overview of autonomous robots is Cog (figure 1.1(j)), a humanoid torso developed in the Artificial Intelligence Laboratory at the Massachusetts Institute of Technology (MIT). As the figure shows, in contrast to the mobile robots described above, Cog is a humanoid torso in a fixed location, with movable arms, head, and eyes. The arms are sophisticated subsystems, as is the head. However, the power of Cog lies in its intelligence, that is, its ability to interact with humans and to learn.

In fact, the fundamental thesis of the Cog project is that "humanoid intelligence requires humanoid interactions with the world." This means that interaction with humans forms the basis of the robot's design. Cog's designers have found that even a few simple humanlike behaviors on the part of the robot (such as following a human's position with its camera eyes, or gesticulating while speaking) are sufficient to make humans treat the robot as if it were another human.

Cog can be considered to be a "brain" coupled with a set of sensors and actuators that try to approximate the sensory and motor dynamics of a human torso, except for a flexible spine. The major degrees of motor freedom in the trunk, head, and arms of a human are all there in the robot. It has vision (from cameras), hearing was being developed as this book was being written, and some proprioception was possible via joint position sensors. As noted, the uniqueness of this robot arises from its intelligence, so that it can also be viewed as a hardware platform for artificial-intelligence research. Cog was retired from active research in late 2004. Humanoid robots are discussed in detail in chapter 13.

1.9 Concluding Remarks and Organization of the Book

This chapter has provided an overview of the entire field of mobile robots. We have introduced some of the fundamental concepts (including biological models), discussed sensors and actuators for robots and the basic control architectures used in robot software, and touched on robot intelligence. Since one of the themes of the book concerns control of robots, we have dealt with both low-level and high-level control concepts. Finally, the breadth of the field has been illustrated with a large number of examples from current research in and applications of mobile robots.

The rest of the book is organized as follows: Chapters 2–6 deal with background material, including robot hardware, control from both a biological and an engineering perspective, software architectures, and robot intelligence. Specifically:

• Chapter 2 introduces biological control systems, to provide a basis for evaluating robot control systems.

• Chapter 3 surveys the major hardware components of robots, including structure, sensors, and actuators, as well as cognitive architectures and control methods.

• Chapter 4 is an engineering counterpart to chapter 2, with an overview of low-level robot control.

• Chapter 5 deals with software architectures for mobile autonomous robots.

• Chapter 6 is devoted to issues surrounding intelligence and learning.

Chapters 7–13 discuss various implementations and applications of robots. Specifically:

• Chapter 7 provides an overview of robot locomotion, including a discussion of wheeled, legged, flying, swimming, and crawling robots.

• Chapters 8 and 9 are devoted to legged locomotion. Chapter 8 deals with biped locomotion, and Chapter 9 considers locomotion with four, six, and eight legs.

• Chapters 10 and 11 survey robot manipulation, including both arms and hands. Although the focus of the book is on robot mobility, many mobile robots have (or will have) arms. To grasp objects, these arms must terminate in hands. Hence, chapters 10 and 11 provide a brief overview of these topics.

• Chapter 12 considers the control and coordination issues in multiple-robot systems.

• Chapter 13 concerns humanoid robots, thus building on the material on biped locomotion in chapter 8 as well as that on arm movements and grasping in chapters 10 and 11.

Chapters 14–15 cover current and future research and are followed by an appendix. Specifically:

• Chapter 14 focuses on issues of localization ("Where am I?"), navigation, and mapping, all currently subjects of intensive research efforts.

• Chapter 15 presents the author's view of the future of robotics, including both the potential increased usefulness of robots to humanity and the possible dangers that may arise from large numbers of increasingly intelligent and autonomous robots.

• For readers unfamiliar with the basic concepts of linear feedback control and the use of Laplace transforms, an intuitive introduction to these concepts is provided in the book's appendix.

2 Control and Regulation in Biological Systems

Summary

The purpose of this chapter is to present some aspects of the ways in which animals regulate their internal environments and control their movements in the world. We then compare and contrast these control and regulation methods with those used in engineering control systems, in order to provide a framework for the design of robot controllers. In the concluding section of the chapter we outline some common approaches to robot control, indicating the way in which inspiration from biology affects both the design and the implementation of these controllers.

2.1 Homeostasis

The automatic regulation of the internal environment of living things is frequently described by the word "homeostasis," coined by the physiologist W. B. Cannon (1939). (Actually, the first discussion of the constancy of the internal environment as being essential to life probably came from Claude Bernard in 1865.)[2] The "internal environment" in mammals refers to such variables as body core temperature, blood pressure, concentration of carbon dioxide (CO_2) in tissues, or concentrations of certain biochemical substances in the blood. These and other variables tend to stay within fairly narrow ranges during the normal behavior of the animal but may change dramatically during emotional states or heavy exercise or in the presence of disease. From an engineering point of view, we consider these quantities to be "regulated."

To illustrate the distinctive features of a homeostatic control system, consider the control of blood volume, as illustrated in figure 2.1. The body has a number of

2. "La fixité du milieu intérieur est la condition de la vie libre."

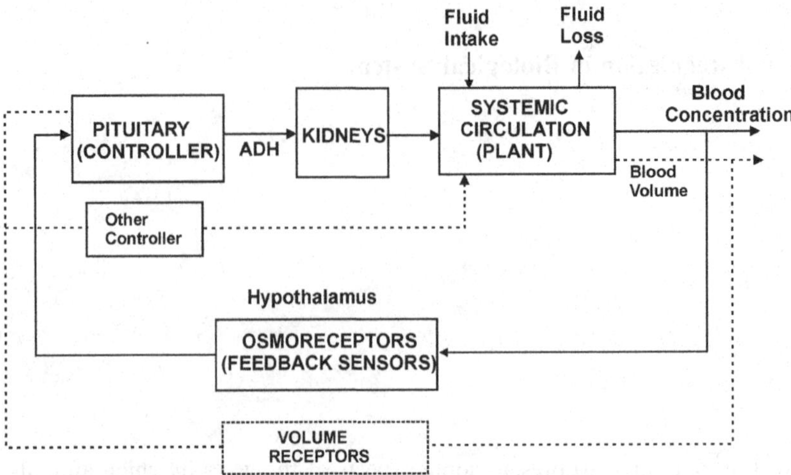

Figure 2.1
Blood volume control system

mechanisms for regulating blood volume, including control of urine volume, perspiration, or loss of water during breathing. The last, known as *insensible water loss,* can be substantial. These mechanisms of regulating the loss of fluid from the body are balanced by fluid intake from drinking. Let us concentrate for the moment on the control of urine excretion as a method of regulating blood volume. Assume that a person (or vertebrate animal) spends time in the desert with no drinking water available. Water will be lost through evaporation from perspiration and convection. The loss of fluid will increase the density of the blood and hence change its osmotic pressure across the numerous semipermeable membranes in the body. This change in osmotic pressure is sensed by *osmoreceptors* in the thalamus, which then sends signals to the pituitary gland to increase the production of a hormone known as *anti-diuretic hormone* (ADH). When it reaches the kidneys, this hormone changes the filtering properties of the kidney tubules so as to decrease the amount of fluid removed from the blood and reaching the bladder. Hence, urination is decreased and blood volume is conserved, while at the same time thirst increases in order to urge the intake of water.

Several aspects of this system are apparent. It is complex, involving sensing, neural control signals, chemical control signals, and actuation. Note also that there are multiple control loops, as indicated by the dotted lines in the figure. The body also has blood volume sensors, which provide yet another set of feedback signals. This form of redundant control is typical of biological control systems. Let us examine the differences between engineering and biological control systems in more detail.

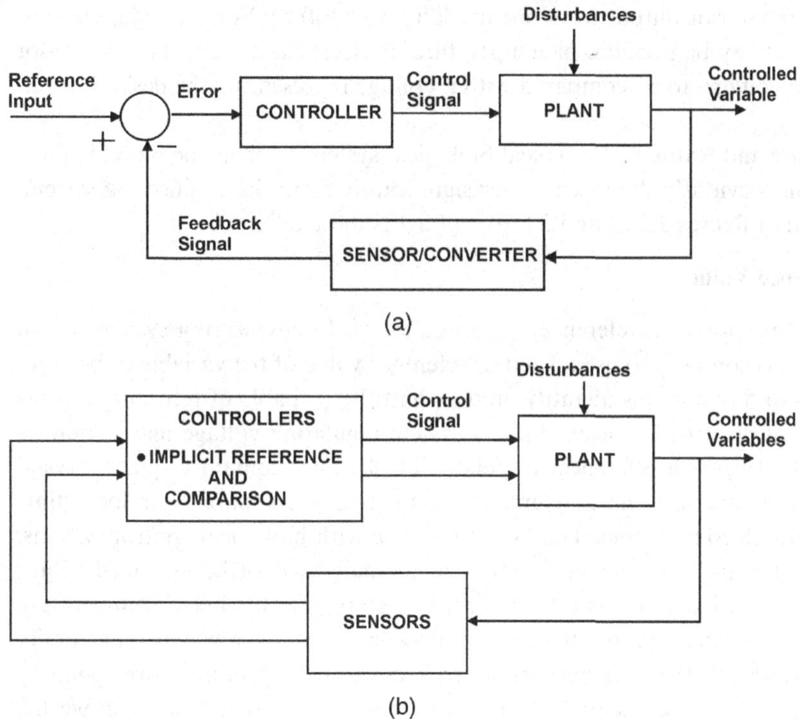

Figure 2.2
Comparison of (a) engineering and (b) biological feedback control systems

2.2 Engineering and Biological Control Systems

Let us now compare the homeostatic system of figure 2.1 with an engineering feedback control system, as illustrated in figure 1.3 and reproduced in figure 2.2(a). As we discussed in chapter 1, a feedback control system requires

- a desired or *reference value* of the variable being controlled;
- a *sensor* to measure the controlled variable;
- a *comparator* to determine the difference between the desired value of the variable and its actual current value;
- a *controller* to provide appropriate commands to *actuators* or *effectors* to change the variable in the desired direction.

These elements of a closed-loop control system are illustrated in figure 2.2(a). Note that the figure explicitly indicates that the feedback sensor may also be a converter that changes the sensor output from one modality to another. For example, the controlled variable may be pressure or temperature or electrical current, but the sensor output will be voltage, to be compared with a voltage representing the desired or reference value.

The structure and features of a typical biological system, such as the blood volume control system previously discussed, differ significantly from the engineering system. A comparison of figures 2.2(a) and 2.2(b) highlights these differences.

2.2.1 Reference Values

Consider the question of a reference or desired value. In engineering systems (as in figure 2.2(a)) it is common to use a fixed or reference value of the variable to be regulated, sensors to monitor this quantity, and a controller capable of returning it to its desired values if disturbed. For example, a system regulating voltage may contain a high-quality battery as a reference. Regulated clocks may contain a quartz crystal whose vibrations are used as a reference, or they may compare their local time against Greenwich Mean Time. This is not the case with biological control systems, in which a "set point" or reference input signal usually cannot be identified. Thus, there is no block of ice to act as a temperature reference for the body's temperature control system, no fixed number to act as a blood pressure reference or an osmotic-pressure reference. Rather, changes in controlled variables produce corresponding changes in control signals (neural, chemical, or mechanical) that tend to move the variable back to its regulated value by means of negative feedback. It is important to note that as a result these "reference values" are also variable and influenced by other factors. For example, one of the mechanisms used in the control of blood pressure (as well as volume) is a change in the volume of blood filtered by the kidneys and the amount of water extracted and passed to the bladder for excretion as urine. This "reference value" is not fixed; so-called renal hypertension raises the reference value in this system.

An analogous situation arises in certain engineering control systems. For example, a thermostat-controlled temperature control system does not compare room temperature to a fixed reference. Rather, it uses a strip of two dissimilar metals with different temperature expansion coefficients. As the room temperature changes, the strip bends in one direction or another, turning on a heater or cooler as required. Clearly, such controllers are not as precise as those using a fixed reference, but they tend to be simpler to design and implement.

2.2.2 Comparator and Negative Sign

There is no obvious "comparator" in biological control systems that subtracts a number representing the actual value of a variable from a reference value to produce an error signal. This does not mean that biological control systems do not use negative feedback. Quite the contrary, most are negative feedback systems. Rather, the negative feedback is built into the combination of sensor and controller. Thus, the increased stimulation of the pituitary that produces an increase in ADH secretion results in a decrease in fluid loss, hence the minus sign associated with negative feedback.

2.2.3 Sensors

Both biological and engineering systems require sensors. Thus, homeostasis requires biological sensors for all the variables being regulated. (In biological systems the sensors are known as *receptors*.) Indeed, the living organism has a variety of such sensors. Baroreceptors (or simply baroceptors) located in the arch of the aorta and elsewhere in the circulation are sensitive to changes in blood pressure, osmoceptors in the hypothalamus measure the osmotic pressure of the blood, muscle spindles produce outputs that are sensitive to muscle contraction force and velocity, and so forth. Sensory systems contain many receptors, including not only the well-known receptors in the retina of the eye or in the inner ear, but thousands of receptors in the olfactory system and the tactile system. Many of the receptors for homeostatic variables are *stretch receptors*. As these sensors are stretched, say by an increase in pressure, they produce a change in the firing frequency along nerve fibers leading from the receptor to the central nervous system (CNS). (Fibers leading to the CNS are known as *afferent* fibers, whereas those bringing command signals from the CNS are knows as *efferent* fibers.) For example, stretch receptors in the large veins and the atria of the heart function as blood volume sensors.

2.2.4 Actuators

Actuators or effectors are present in both biological and engineering control systems. In biological systems they may be mechanical (e.g., muscles). For example, blood pressure is regulated in part by contractions of the small arterial vessels (arterioroles). A reduction in cross-section of these vessels increases their resistance to blood flow and hence increases the blood pressure upstream. In some cases the actuators are chemical in nature. Explicitly identifiable sensors do not exist for all homeostatic variables. In some cases the specific variable may not be "measured" by a sensor, but deviations from the normal values may increase or decrease hormonal secretions. In such situations, the role of sensor and actuator are combined.

2.2.5 Adaptivity

Although not explicitly shown in figure 2.2(b), biological systems differ from standard engineering systems in being adaptive. This means that a change in the values of the controlled variables may result not only in feedback control, but also in changes in the characteristics of the controller or the controlled system.

2.2.6 Control Redundancy

Figure 2.2(b) illustrates another aspect of biological systems, namely, control redundancy, as has already been mentioned in connection with the blood volume control system. Even with this system, the story is more complex than previously indicated. We have discussed primarily the control of volume on the basis of the osmotic pressure sensed by osmoreceptors. Blood pressure, sensed by baroceptors in arteries, also affects ADH secretion. Further, stretch receptors in the large veins and in the aorta respond to changes in cross-section resulting from variations in blood volume and also affect ADH secretion. But that is not all. The kidneys also regulate the concentration of sodium in the blood through adjustment of excretion via the kidneys. Here the control signal is provided by another hormone, aldosterone, secreted by the adrenal cortex. Ultimately, the control of blood volume involves the combined effect of all these control loops.

2.2.7 Multipurpose Controllers

Biological control systems are frequently multipurpose. Thus, perspiration affects both body temperature and fluid volume. The respiratory system is concerned primarily with gas exchange, but it also eliminates heat and water from the body. Changes in heart rate and peripheral vasoconstriction affect ADH secretion (and hence blood volume) as well as cardiac output and blood pressure. This last system is illustrated in figure 2.3. A consequence of the multipurpose nature of many biological controllers is the fact there is cross-coupling between them. In other words, control of one variable may affect another one, sometimes in an undesirable way.

It should be evident from the preceding discussion that engineering and biological systems differ in a number of ways and that homeostatic systems are generally more

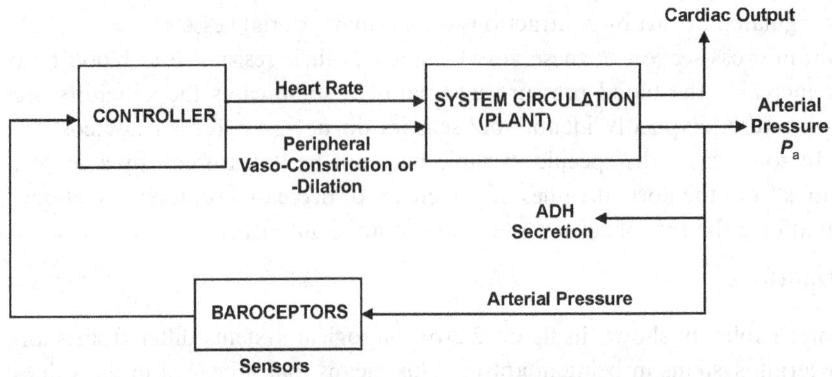

Figure 2.3
Blood pressure control system

complex. As we show later in the chapter, simple robot control systems basically follow the model of figure 2.2(a). On the other hand, as robots become more complex (by the addition of sensors and more sophisticated architectures) and interact with humans in a greater variety of ways, they will require control systems with more complex properties that increasingly resemble those of living systems. This will certainly be true of humanoid robots.

2.3 Multiple Levels of Control: Control Architecture

Another characteristic of biological control systems is that they operate on multiple levels. Consider the control of leg movements in walking, for example. As indicated in chapters 8 and 9, legged locomotion on smooth, level terrain is controlled by local "reflex" loops. In humans this means that the rhythmical pattern of leg movements during constant-speed walking is controlled primarily by local pattern generators involving the spinal cord, without involvement of the motor cortex in the brain. Thus, soldiers can fall asleep while marching. However, adaptation to changes in the environment (such as changes in the slope of the terrain or the slipperiness of the surface) may involve control inputs from the cerebellum. Requirements for precise foot placement (say, in rocky terrain) may require inputs from higher centers in the brain.

It appears that the body encourages local control whenever possible. Control of water reabsorption in the kidneys is governed by local mechanisms. These mechanisms may then affect the whole organism. For example, the kidneys may require increased blood pressure for proper water regulation; the net effect may be central high blood pressure (renal hypertension). Hand and arm movements may be controlled by reflexes (such as withdrawal from hot surfaces) or they may be controlled from higher centers. We can visualize these multiple centers of control as represented in figure 2.4.

As shown in the figure, the highest level of control involves planning of actions (e.g., for voluntary movements), and the lowest level consists of reflex control. Multilevel, hierarchical control structures of this type are commonly used in robotics, as we showed in the previous chapter and discuss in chapter 5. It is particularly important to note that control at each level requires feedback information. Thus, voluntary control of peripheral muscles is obtained by means of signals transmitted by alpha motoneurons (see chapters 8 and 9). Nerve fibers carrying information from the brain to the periphery are known as efferent fibers; those bringing feedback information (e.g., from stretch receptors in the muscles) are known as afferent fibers. The importance of feedback is highlighted by the fact that rehabilitation of patients with certain neuromuscular diseases is much easier if the patient's afferent fibers are intact (see Perry 1992). Experiments with monkeys have shown that if the afferent feedback

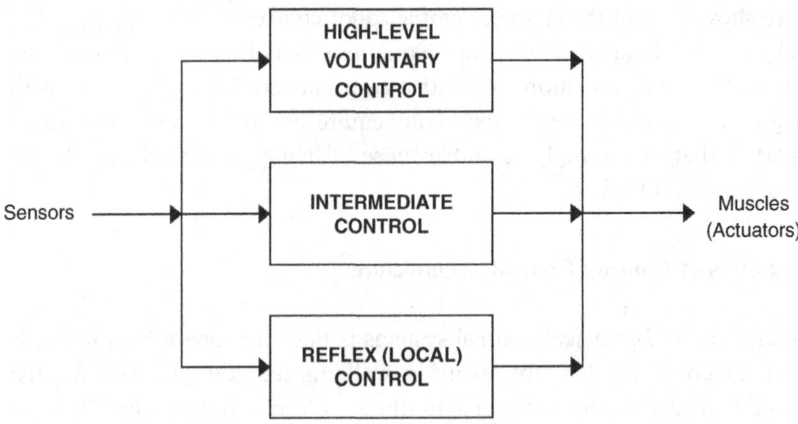

Figure 2.4
Multiple levels of control in living organisms

fibers in a limb are severed, the monkey will not use the affected limb even when the voluntary efferent fibers are intact.

As another illustration of the multiple levels of control, consider again the blood volume control system discussed in section 2.1. Fluid control based on ADH secretion is an intermediate control mechanism, whereas drinking is a voluntary, high-level control mechanism. Water loss through respiration appears to be a local phenomenon.

Clearly, biological control systems are feedback control systems, but they differ from engineering control systems in a number of ways. The following section analyzes two additional homeostatic control systems to further illustrate and clarify the differences between these systems and engineering systems.

2.4 Other Biological Control Systems

2.4.1 Control of Body Temperature

To further explore the multiplicity of interacting control loops, consider the human body's temperature control system, as illustrated in figure 2.5. It is well known that in humans the body temperature is maintained at approximately 37°C. Significant deviations from this temperature are indications of disease. The figure does not show the sensors activating the body's temperature control system, but they are present both centrally and on the periphery of the body. The central receptors are in the brain (the thalamus); the peripheral receptors are in the skin. The figure illustrates several control mechanisms present under cold conditions (shown by the solid lines) and others present under heat stimulus (shown with a dashed line).

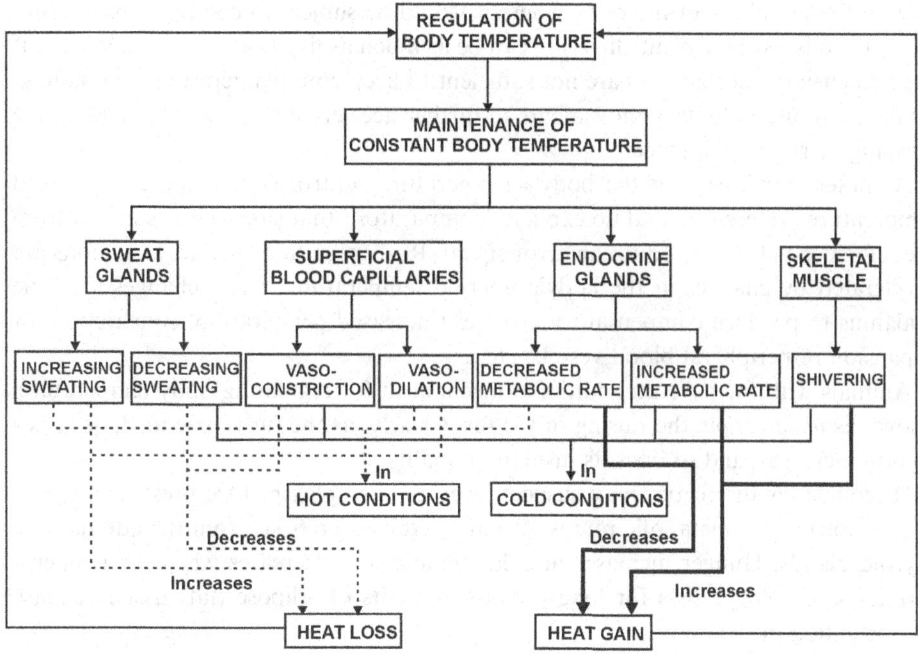

Figure 2.5
Major elements of temperature control system in humans

In high-temperature environments maintenance of the body core temperature at 37°C makes use of two major mechanisms: perspiration and convection. Under normal temperature conditions some heat is lost through convection, both from the surface of the skin and from the moist surfaces of the alveoli in the lungs. In fact, under basal conditions, human metabolism produces about 70 Calories per hour; 16 of these Calories are lost through convection. This is *insensible perspiration.* As the body's heat receptors indicate an increase in ambient temperature, several mechanisms come into play. The sweat glands are activated and secrete a fluid consisting mainly of water, but some salt as well. As this water evaporates, it cools the skin. In hot, dry climates, sweating can result in the loss of 3–4 liters of water per hour. Clearly, this will produce rapid dehydration, if the water is not replaced by drinking. In addition, the peripheral blood vessels (arterioroles) dilate, removing heat from the body core, allowing greatly increased blood flow through the surface capillaries to provide the fluid needed for sweating and to cool the skin through convection. In addition, hunger decreases, since an intake of food would result in increasing metabolism and hence more central heat production.

In very cold environments, the body uses other mechanisms to protect the core temperature, including shivering, increases in metabolic rate, and restriction of blood flow to the periphery of the body where it would be subject to cooling. The amount of heat produced by violent shivering can be as much as five times the resting level. If the homeostatic mechanisms are not sufficient to keep core temperature from falling, humans can use voluntary movements to further accelerate their metabolic rate, like jumping or running in place.

As noted previously, in the body's temperature control system there is no fixed temperature "reference" and no explicit "comparator" that subtracts the actual from the reference value to produce an error signal. Rather, certain chemical reactions are accelerated by changes in the body's internal temperature. These changes spur the thalamus to produce compensatory changes (increased perspiration, constriction, or expansion of peripheral blood vessels, etc.).

Animals with fur have an additional mechanism for regulating body temperature, known as *piloerection*, the raising of the fur. Clearly, as the fur is extended, its cross-section increases, and so does its insulating quality.

Of course, as the reader may suspect, even the complex picture presented here is not complete. The metabolic rate is also affected by secretions from the adrenal and thyroid glands. Hunger increases in cold climates and decreases when the temperature rises. If cold persists for long periods, deposits of adipose (fat) tissue increase the insulation of the body.

2.4.2 Control of Skeletal Muscles and the Stretch Reflex

An additional example of a simple negative feedback system is the knee-jerk reflex in humans. Although not specifically a homeostatic mechanism, it illustrates other aspects of biological control. This reflex is invoked when a person sits on the edge of a table and his patellar tendon (just below the kneecap) is struck rapidly, resulting in the stretching (lengthening) of the thigh muscle to which this tendon is attached.

Living muscles are equipped with a number of sensors that are essential in the control of the force the muscle exerts. Chief among these sensors are the *muscle spindles* and the *Golgi tendon organs*. Muscle spindles are specialized receptor cells located in specialized fibers that are parallel with the force-producing muscle fibers in a skeletal muscle. Thus, they contract when the muscle contracts and send signals to the brain along nerve fibers known as *Ia afferent fibers*. However, the strength of these signals is not simply proportional to the contraction force. Rather, it depends on both the amplitude and the velocity of contraction. It should be noted that the muscle spindle is a *mechanoreceptor*, a structure that senses mechanical movement and produces an electrical signal. In engineering control systems sensors are frequently transducers, in that they translate one quantity into another. For example, a potentiometer translates rotational motion into a voltage. Since control computers nearly always use electrical signals (generally binary in form), such translating sensors are essential for control.

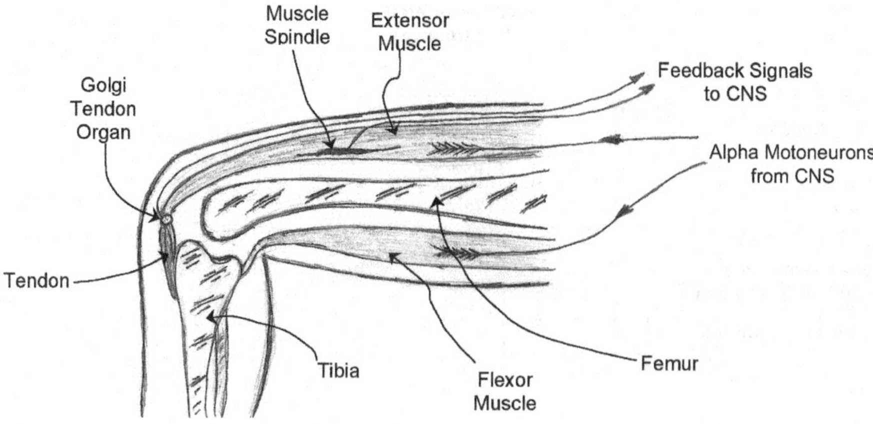

Figure 2.6
Major elements of human skeletal muscle control systems

In addition to muscle spindles, the skeletal muscles of mammals are also equipped with specialized sensor structures located in the tendons, known as Golgi tendon organs. (Tendons are specialized structures that serve as attachments between muscle and bone.) Figure 2.6 provides a schematic illustration of these major structures. The Golgi tendon organs appear to bias the muscle when contraction signals arrive along alpha motoneurons, the fibers responsible for initiating contractions. Thus, they may play a safety role, ensuring that the exerted forces are not so large as to cause possible damage to the supporting structure.

Note that figure 2.6 refers to skeletal (*striated*) muscles, used to move arm or leg segments with respect to the body (the skeleton). These muscles should be distinguished from *smooth* muscles, such as those contracting the heart or the peripheral arterial vessels. Figure 2.7 places the muscle in a feedback control loop (see chapter 4), to indicate that muscle force is an actively controlled function. Similarly, the force produced by a robot actuator is an actively controlled quantity.

Since the muscle spindles are attached in parallel with muscle fibers, stretching the muscle will also stretch the spindles and cause them to increase the firing frequency along nerve fibers leading to the spinal cord. These incoming (afferent) fibers make connections with outgoing (efferent) fibers, which in turn carry stimulation messages back to the same thigh muscle, causing it to contract in the well-known knee-jerk reflex. Note that this contraction shortens the fibers, so that it is actually a result of negative feedback, as illustrated in figure 2.7.

Several observations can be made regarding this control system:

• The knee jerk is an entirely involuntary (reflex) response, at the lowest level of the control responses in figure 2.4.

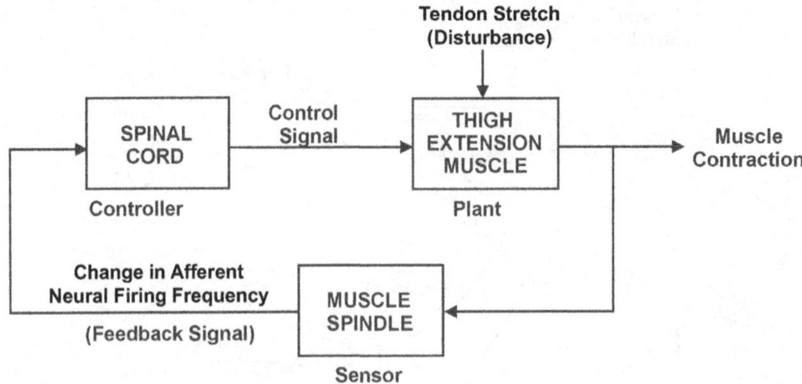

Figure 2.7
Muscle stretch reflex (adapted from Khoo 2000)

- The system corresponds to the canonical biological feedback control system of figure 2.2(b). The muscle spindles are the sensors, the thigh muscle is the "plant," the hammer blow that stretches the tendon is the disturbance, and the reflex center in the spinal cord is the "controller."
- Note that this is indeed a negative feedback system, but there is no comparator and no error signal as in engineering systems. The "minus sign" of the negative feedback is embedded in the response of the components of the system.

We will discuss this control system in more detail in connection with the study of human and biped robot locomotion in chapter 8.

2.5 Nonlinearities in Biological Control Systems

As we show in chapter 4, engineering control systems commonly are designed to behave linearly. This implies that they obey superposition; that is, twice as large an input signal will produce twice as large a response.[3] By contrast, biological control systems frequently involve nonlinearities. Some nonlinear behavior is to be expected. For example, since biological variables cannot exceed certain values, they exhibit upper limits that may show up in mathematical models as saturation nonlinearities.

Consider, for example, the firing frequency of certain sensory receptors, as a function of the sensed variable (the stimulus). An ideal linear receptor would have a response proportional to the input stimulus over the full range if inputs, as shown in figure 2.8.

3. This is an oversimplified definition of the principle of superposition. See chapter 4 and the appendix for a more careful discussion of this important principle.

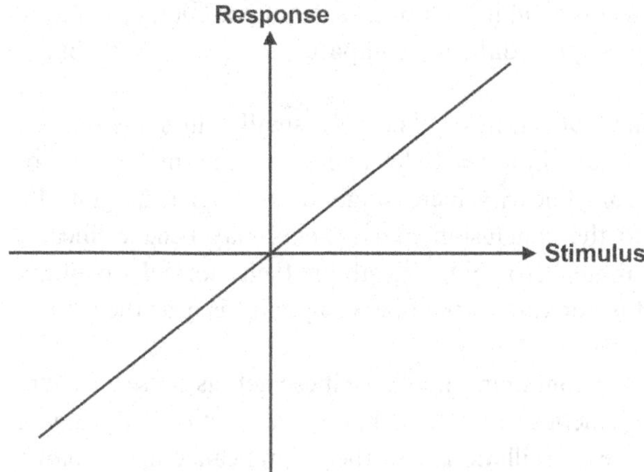

Figure 2.8
Ideal linear relation between stimulus and response

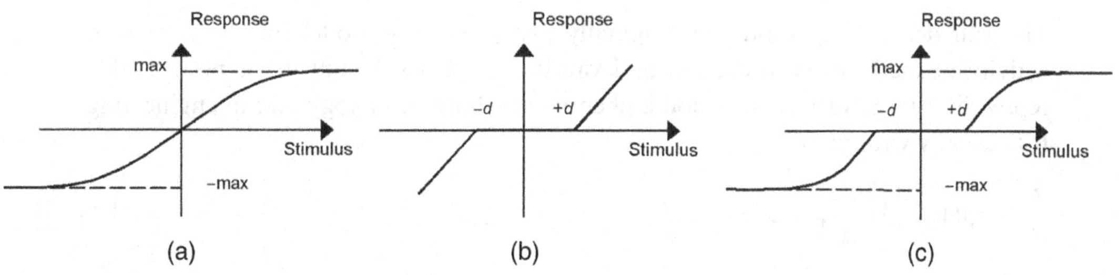

Figure 2.9
Nonlinear relation between a stimulus and response: (a) saturation, (b) dead zone, (c) both saturation and dead zone

On the other hand, an actual biological receptor might have a nonlinear response characteristic similar to one of those illustrated in figure 2.9. Figure 2.9(a) shows a *saturation* characteristic. There is range of stimulus values over which the input-output relationship is nearly linear. Beyond this range, it takes a larger and larger input to obtain a given increment of response, until the response reaches its maximum possible value. Since receptors are always limited to some maximum output (say a maximum neural firing frequency), it is evident that all biological receptors display some form of saturation.

Another common nonlinearity is known as the *dead zone*. Typically, a biological system will not respond to an input stimulus until the stimulus exceeds some minimum value. (Such a property has adaptive value, since it may conserve energy.) A typical input-output relationship exhibiting dead zone nonlinearity is shown in figure 2.9(b).

Of course, many biological systems exhibit both of these nonlinearities, as in figure 2.9(c), which makes their analysis quite difficult, compared to the study of linear models, as we show in chapter 4.

Note that the behavior of figure 2.9(a) is nearly linear for small values of stimulus. Hence, biological systems may behave in a nearly linear manner for small values of input signals, but will deviate from linearity increasingly as the signal magnitudes grow. Similar reasoning leads to the conclusion that systems may behave linearly for small excursions not only about zero, but also about their normal operating point. (Of course, this statement is not valid if the operating point is near the saturation limit of a variable.)

One of the properties of systems containing nonlinearities such as those in figure 2.9 is that they may exhibit spontaneous oscillations known as *limit cycles*. To gain some insight into the nature of these oscillations and their significance in a broader view of nonlinear systems, we consider a particular model of biological systems known as the van der Pol equation.

2.5.1 The van der Pol Equation

The van der Pol equation was originally proposed as a model for the oscillatory activity of the heart (van der Pol and van der Mark 1928), but it has been used to represent a large variety of periodic phenomena, both in biology and in engineering. It is usually written as

$$\frac{d^2y}{dt^2} - \beta(1 - y^2)\frac{dy}{dt} + y = 0 \tag{2.1}$$

Consider the coefficient of the first-order term. If we think of equation (2.1) as describing a mechanical system, then the first-order term corresponds to velocity, and its coefficient $\beta(1 - y^2)$ represents damping. If response $y(t)$ is small, so that $y^2 < 1$, the system exhibits negative damping, and hence the response will tend to grow. If $y(t)$ is large, the damping term will be positive and large, thus attenuating the response. The net effect of the second term in the equation is to produce a steady-state oscillation of $y(t)$ regardless of its initial value. The waveshape of the oscillation (the limit cycle) depends on the magnitude of the parameter β. Typical waveshapes are shown in figure 2.10.

A plot of the derivative dy/dt versus $y(t)$ for any second-order system is known as the phase plane plot (see chapter 4). For the van der Pol equation in particular, the phase plane plot is shown in figure 2.11, which also shows the trajectories of the solution of the equation for various initial conditions. Note that all the solutions converge on the same trajectory; those that start out with values larger than the final steady-state solution gradually attenuate, whereas those that begin with small values gradually grow until they converge on the same solution.

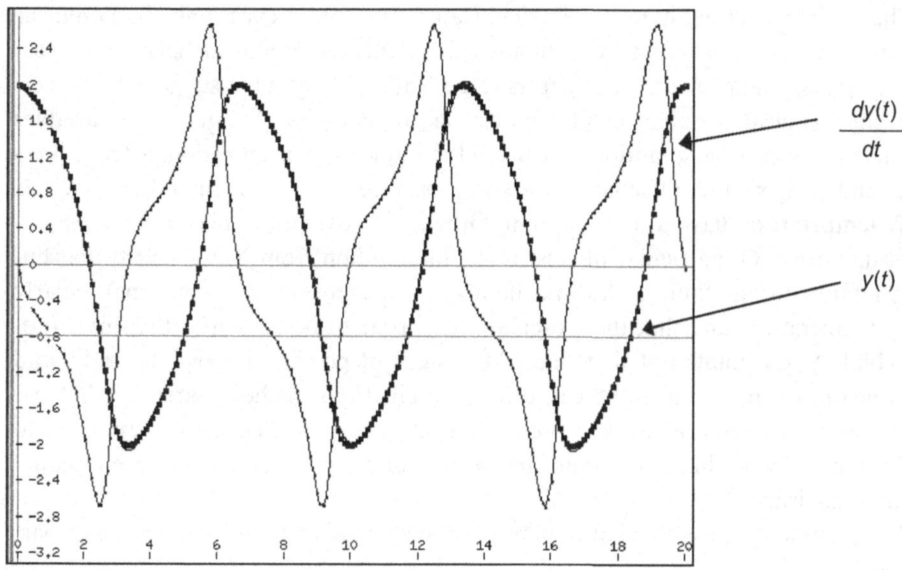

Figure 2.10
Oscillatory solutions to the van der Pol equation. The solid line represents the magnitude of $y(t)$; the broken line represents its derivative dy/dt

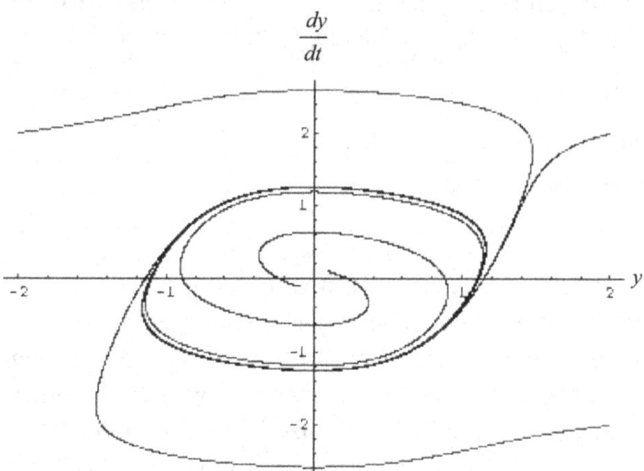

Figure 2.11
Phase plane diagram of the van der Pol equation, showing convergence to steady-state oscillation

The existence of limit cycles is significant, since they exist only in nonlinear systems, and as we have observed, biological systems *are* nonlinear. During the past twenty years a number of investigators (e.g., Yates [1988]) have suggested that certain physiological variables exhibit limit cycle oscillations. Among these variables are many homeostatic quantities, such as blood glucose concentration, arterial pressure, and temperature. Of course, we know that many of these quantities (such as body temperature) have a daily rhythm. Others, like ovulation, may have a twenty-eight-day cycle. Of particular interest is the observation from Yates (1988) that human infants display limit cycles in a number of homeostatic variables, particularly body temperature, and that these oscillations appear to be critical to the survival of the child. Yates points out that one of the signs of possible impending crisis (and even death) of an infant is the cessation of oscillations in these variables. If these observations are confirmed, they present an interesting challenge to physiology: to understand why nonlinearities (and the ensuing limit cycle oscillations) are essential to the organism.

We examine the properties of nonlinear systems in general and limit cycles in particular in chapter 4.

2.6 Cost Functions

Engineering control systems are frequently designed in such a way that they perform optimally with respect to some specific criterion. Typically, such criteria are

- minimum mean square error;
- minimum energy consumption for some specific tasks;
- minimum overshoot in response to some inputs;
- maximum velocity in reaching some target.

Since biological systems are "designed" by evolution, it is not clear what criterion they optimize in their performance. Many biological systems appear to operate on the basis of minimum energy expenditure for some range of tasks. Such a criterion seems intuitively reasonable for an organism under normal conditions. On the other hand, when threatened, the organism may need to maximize its ability to flee from danger. The problem is that we do not know the criteria used by biological control systems to optimize their performance. However, some investigators have suggested possible optimization criteria that *may be used* by a given biological system and then solved mathematical models of the system to test these hypotheses. For example, Otis, Fenn, and Rahn (1950) assumed that the respiratory control system chooses the frequency of breathing so as to minimize the work of the respiratory muscles in each cycle of breathing. They perfomed a mathematical analysis of the system to find the optimal breathing frequency based on this hypothesis. Using system parameters from several normal subjects, they arrived at the "optimal" frequency of fourteen breaths per minute, which is within the normal range for resting subjects. In this study the breathing waveform was assumed to be sinusoidal. Later work by Yamashiro and Grodins (1971) relaxed this assumption and arrived at an optimal value of about sixteen breaths per minute, also within the normal range. Such analyses are difficult, since they attempt to take the nonlinearities of the system into account.

2.7 Control of Functional Motions in Humans

The general principles of biological control discussed in the foregoing apply to the physiological systems of humans as well as those of animals. However, human beings exert enormous influences on their external environment, both through the action of their own muscular system and through the use of tools and machines that can greatly amplify their own strength and reach, increase their speed of response, and enable them to manipulate the world at microscopic and even molecular scales. Even so, the functional motions of humans in their ordinary interaction with the world appear to be based on a number of unique factors, related to both to human sensory systems and to human cognitive abilities. (The phrase "functional motion" refers to a motion that is intentional and purposeful, aimed at a specific function, such as walking, stepping over a rock, reaching for an object, or grasping it with an appropriate hand posture.) We consider some of the simplifying principles of the control of functional motions in chapter 4.

2.8 Relevance to Robot Control

The purpose of this chapter was to present some of the unique aspects of biological control systems and to compare and contrast them with engineering control systems. We discuss the latter in chapter 4. Biological control systems appear in most of the book's chapters as models or inspirations for the design of robot controllers.

Controllers for robot manipulators (chapter 10) began as simple linear feedback control systems. However, since these systems were modeled on the human arm, it soon became apparent that more complex controllers were required in order to obtain some of the versatility of that arm. Adaptive control enabled the gains of the controller to self-adjust, depending on whether the manipulator motion was in a vertical plane (and thus affected by gravity) or in a horizontal plane. Further improvements, again modeled on the human arm, were needed to obtain satisfactory performance in both large reaching movements and fine positioning movements. It has also become essential to integrate vision and manipulation systems, again following biological inspiration.

The situation is even more interesting with respect to mobile robots. Although most small mobile robots use very simple linear controllers at the lowest (reflex) level, they also perform reasoning and planning at high levels. Many mobile robots use a multitude of sensors. Hence, in common with living systems, they must integrate the readings from these sensors in order to make movement decisions. As we have shown in this chapter, redundancy in receptors is common in animals, and such integrative decisions are essential to the survival of the organism. For example, is a visual danger signal more important than an auditory signal from a different direction? These issues, referred to a *sensor fusion* in the engineering literature, continue to be of interest to robotics researchers. As we move toward increasingly autonomous humanoid robots, issues surrounding biological models for their control will become more and more important. Humanoids will be required to monitor their internal environments (robot homeostasis?) as well as to have a multitude of sensors to provide information from the exterior environment. Some understanding of the basic principles of biological control will be essential for the design of these new systems.

2.9 Historical Background

The systematic analysis of biological control systems began with the work of Norbert Wiener (1961), the father of cybernetics, who was the first to model control phenomena in both animals and machines. Fred Grodins (1963), a physiologist, published a pioneering book on mathematical analysis of the cardiovascular and respiratory systems shortly thereafter. Several other books on biological control systems followed in the next few years (Bayliss 1966; Milhorn 1966; Milsum 1966; Riggs 1970; Albergoni, Cobelli, and Francini 1974). Except for Bayliss, which is largely conceptual, the other books are quite mathematical in their presentation of biological control. Following this remarkable series of books, no major books on the subject were published until the excellent textbook by Khoo (2000), some thirty years later, which includes MATLAB programs for many of the biological models it presents.

The actual application of biological control principles to robotics is still in its infancy. One of the earliest books is Beer, Ritzmann, and McKenna 1993, which concerns primarily insect models. Other recent books on engineering approaches to biology include Webb and Consi 2001 and Mange and Tomassini 1998, neither of which is concerned primarily with control issues. Most papers in this field are presented at conferences like "Animals and Animats," or "Simulation of Adaptive Behavior" or published in journals like *Robotics and Autonomous Systems, Autonomous Robots*, or *IEEE Transactions on Robotics and Automation.*

3 Fundamental Structural Elements

Summary

This chapter summarizes the properties of the major structural elements of mobile robots, including the components that make up their "bodies" (structural links and joints), sensors (for receiving information about the world and their own internal environment), and actuators (for exerting influences upon the world). On-board computers and other elements, such as communication devices, are also reviewed briefly.

3.1 The Structural Elements

This chapter summarizes some of the major issues associated with structure, sensing, actuation, and control of autonomous robots. The goal is not to provide an encyclopedic treatment of all these elements. Rather, we discuss some of their features and highlight some of the problems associated with their use.

The bodies of mobile robots are constructed of a variety of materials: plastics, metals, ceramics, etc. However, they have certain elements in common. Portions of the structure provide the rigidity required for the robot to maintain its posture and structural integrity. Other elements, sometimes known as the *links*, are connected by *joints*, forming a *kinematic chain*. Thus, the kinematic chain in a leg would include the hip joint, the upper leg link, the knee joint, the lower leg link, the ankle joint, and the foot, as indicated in figure 3.1.

The links in a robot generally consist of a single structural element and do not attempt to imitate the exact structure of animal legs. For example, the lower leg in humans includes two bones, the tibia and the fibula, in contrast to the single element in the typical robot leg. Similarly, the joints of robot legs and arms are much less complex than human or animal joints. They are frequently made simply of hinge joints, whereas joints like the human knee are much more complex structures in which the instantaneous center of rotation moves as a function of the joint angle.

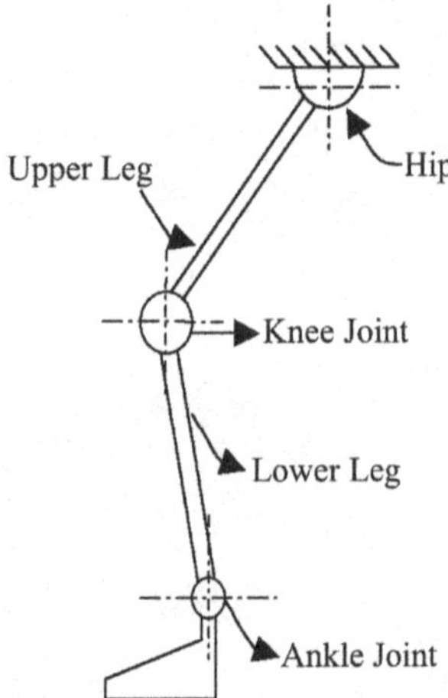

Figure 3.1
Links and joints in a robot leg

Similarly, the main body structure (the "back" or "spine") is frequently rigid in robots, or it may include two or three joints, whereas it may be multijointed and quite flexible in animals. Nevertheless, the links and joints in a robot form a kinematic chain. As we show in chapter 8, the mathematical analysis of such chains is not easy. The problem is basically the following: Under normal conditions, in human (or animal) motion, the central nervous system commands that an end point (foot or hand) move to a desired position. However, the system can control the chain only from its attachment point to the body, by contracting muscles that position a given link with respect to its attachment point to a joint. Such control is known as *forward kinematic control.* To move, say, the hand from a given starting position to a desired end position, it is necessary to perform a backward calculation, to determine the necessary muscle contractions about the shoulder joint, the elbow joint, and the wrist. In addition, the trunk may twist to move the arm forward or backward. The animal brain performs such coordinate transformations constantly and with apparent ease. For robots, it is necessary to perform a series of matrix operations in what is known as the *inverse kinematic problem.*

Joints in living systems may allow for motion in more than one direction, for example, the proximal joint in fingers, the wrist or the ankle. Each such motion direction indicates a *degree of freedom* for the joint. Mechanical joints may possess a single degree of freedom, such as a hinge, or multiple dof, such as universal joints, which allow movement in two orthogonal directions. We can describe the movement properties of a robot component in terms of the number of dof it possesses. In order to reach an arbitrary point in space (x, y, z), it is evident that a robot arm requires 3 dof. If, in addition, it is required to place the end effector ("hand") in an arbitrary orientation, it will require 3 additional dof (pitch, roll, and yaw). Thus, a general-purpose manipulator will have 6 dof, whereas special-purpose industrial manipulators may only have 4 or 5. It should be noted that the human arm has more than 6 dof. This is evident since it is possible to move one's elbow without disturbing the position or orientation of the hand. Manipulators with more than 6 dof are termed *redundant.*

It should also be noted that a complete robotic system, such as a walking machine, may have large numbers of dof. If each of four legs has 3 dof, the body has 3 dof, and the neck 2 dof, the complete robot will have 17 dof. Some of the robots designed for emotional expressiveness have large numbers of dof in the face. For example, the robot Leonardo discussed in chapter 13 has 66 dof in the face, head, and arms.

We have used the word "muscle" in the above paragraph to indicate the actuator that moves a given link with respect to a particular joint. Unfortunately, the artificial muscles in existence (e.g., Klute and Hannaford 2000) are not adequate substitutes for mammalian muscle. Some, like the nonlinear actuators developed by Pratt and Williamson (1995), possess remarkably lifelike properties, but they are still much simpler than their organic prototypes. Living muscles are remarkable structures. They are highly nonlinear, so that the force they exert depends both on the amount of stretch of the muscle and its velocity of contraction. Furthermore, muscles exert force only when they shorten, whereas an electromechanical actuator may be able to exert force in two directions. In order to exert torques on bones, muscles are attached to them by means of tendons. Tendons are not simply attachment devices but have dynamics of their own. Some muscles exert force around a single joint, whereas others may cross two joints before attaching, thus exerting force in a more complex manner. Some aspects of the control of skeletal muscles were discussed in chapter 2; they are considered in more detail in connection with human walking in chapter 8.

3.2 Actuators for Robots

As described in chapter 2, a mammalian skeletal muscle is an actuator that moves a particular skeletal link around a joint, thus changing its position with respect to another link. It receives control inputs along alpha motoneurons and sends back feedback signals along afferent fibers, as indicated in figure 2.6. Robot actuators are analogous structures that move robot links with respect to other links. However, robot structures need not be imitations of animal structures. Thus, robot joints can be either rotary (also known as revolute), as shown in figure 3.2, or prismatic

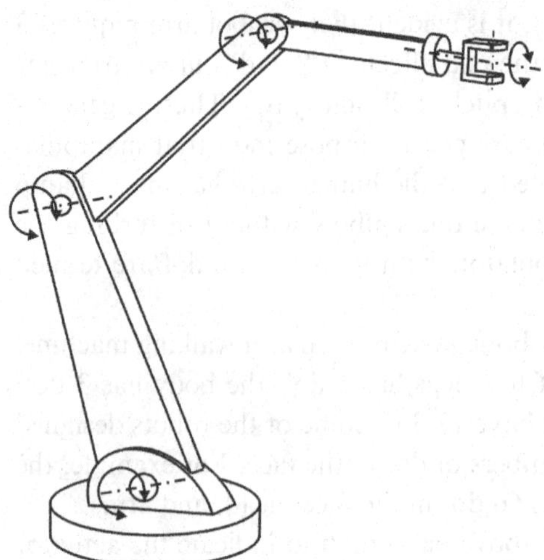

Figure 3.2
Robot with only rotary joints

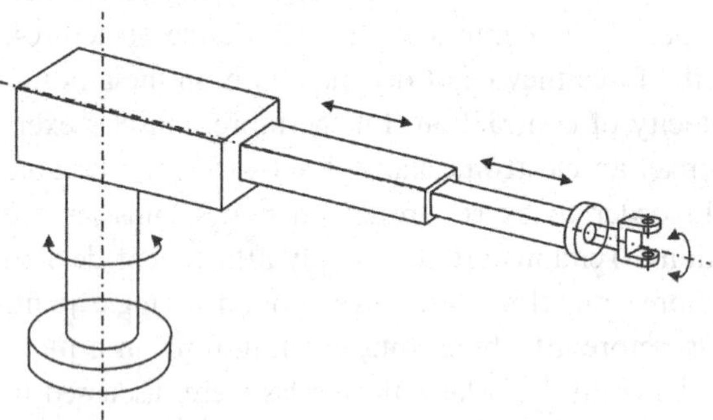

Figure 3.3
Robot with both rotary and prismatic joints

(telescoping), as shown in figure 3.3. Furthermore, whereas muscles can only exert force in one direction (while they shorten), robot actuators may be able to exert forces both while shortening and while lengthening. Actuators are needed to exert forces on the environment and to produce locomotion (through wheels or legs or other mechanisms), as well as to close and open grippers and other end effectors.

The most common actuators used with mobile robots are electric motors, usually direct-current (DC) servomotors. Other actuator devices include artificial muscles, electromagnetic actuators, and shape memory alloys; there are a number of others. Pneumatic and hydraulic motors are used with industrial manipulators, but generally not with mobile autonomous robots because of the weight penalty associated with carrying pressure pumps, fluid reservoirs, and the like.

3.2.1 Electric Motors

Electric motors are the most common source of torque for mobility and/or manipulation in robotics. Most mobile robots use electric motors to generate wheel movement. Walking machines use electric motors to produce joint rotations at the hip, knee, or ankle; manipulators use electric motors to produce joint rotations at the shoulder, elbow, and wrist, as well as closure of the "hands" or end-effectors. These motors are frequently direct-current servomotors (i.e., they are used in a control loop analogous to that in figure 1.3). They may also be stepping motors, devices capable of responding directly to digital signals with a fixed angular rotation in response to a one-bit change in the input command. Stepping motors may be employed in open-loop as well as closed-loop modes.

When used with rotary joint systems, motors can produce torque by being mounted directly on the joints (Asada, Kanade, and Takeyama 1982) or by pulling on cables. The cables can be thought of as tendons that connect the actuator (muscle) to the link being moved. Since cables can apply force only when pulled, it is necessary to use a pair of cables to obtain bidirectional motion around a joint. The mechanical complexity that may arise as a result of cable drive systems is perhaps most apparent in the control of the Utah-MIT robot hand (see chapter 11) (Jacobsen et al. 1986). This hand has four fingers, each with 4 degrees of freedom. In order to provide bidirectional motion about all these sixteen joints requires sixteen motors and thirty-two cables! Both direct and cable drive systems are used in wheeled robots. Clearly, mounting motors directly on joints allows for bidirectional rotation, but such mounting may increase the physical size and weight of the joint, and this may be undesirable in some applications.

The fact that by their very nature, electric motors produce rotational motion raises a second issue in regard to their use in robots. When linear translation is required, it is necessary to translate rotational to linear motion. Prismatic joints, for example, require linear translation rather than rotation from the motor. Leadscrews, belt-and-pulley systems, rack-and-pinion systems, or gears and chains are typically used to transform rotational to translational motion. (When belts are used, they are commonly toothed to avoid slippage.) Figure 3.4 illustrates these common forms of rotary to linear conversion.

Another common (but nonlinear) form of conversion from rotational to translational motion involves the use of cams. A cam is an irregularly shaped contour driven by a rotational motion, on which rides a *follower*, typically a rod with a small wheel, that is constrained to move in a given translational direction. Cams are used

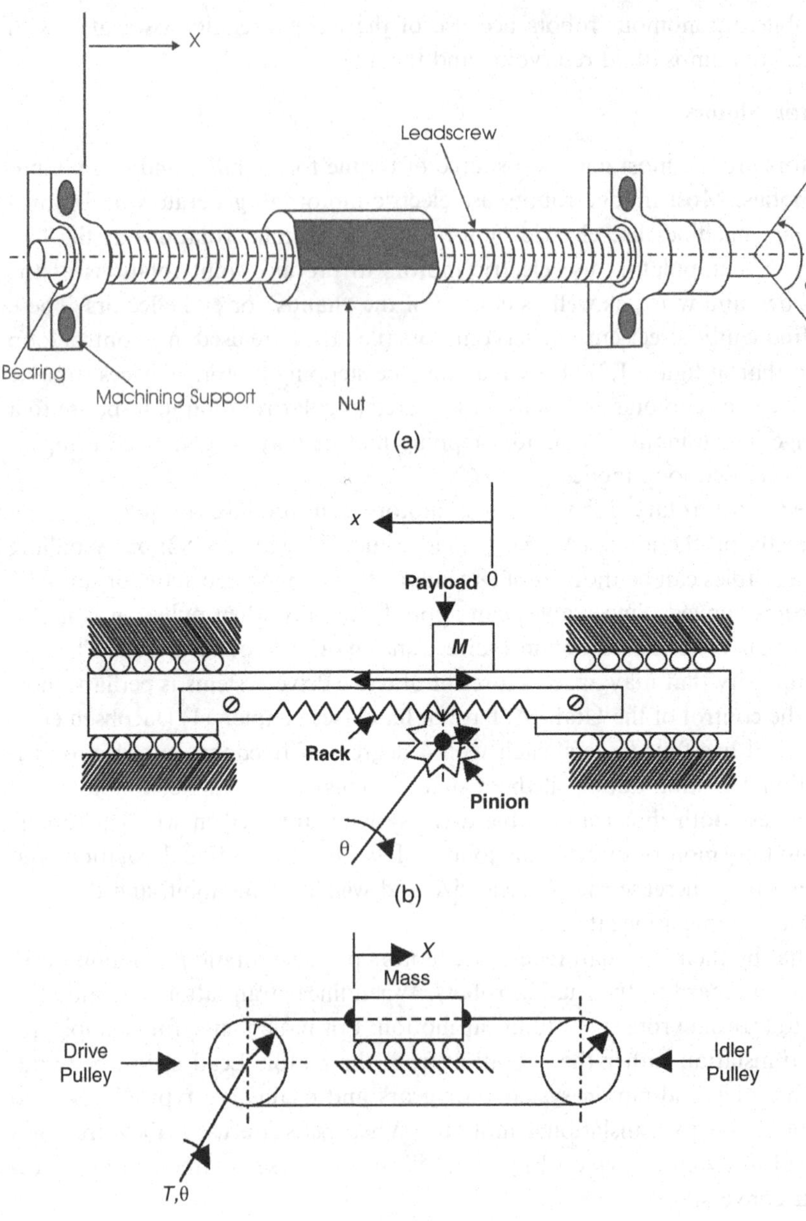

Figure 3.4
Linear actuation using electric motors: (a) leadscrew, (b) rack and pinion, (c) belt and pulleys (adapted from Klafter, Chmielewski, and Negin 1989)

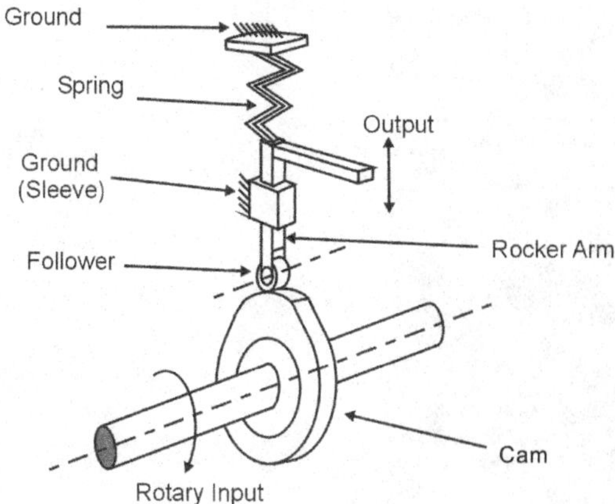

Figure 3.5
Cam and follower

to turn switches on and off, open or close valves once each revolution, and so on. A typical cam is shown in figure 3.5.

3.2.2 Artificial Muscles: McKibben Type

In living organisms physical movement is produced by muscles. As we have seen, muscles in vertebrates are surprisingly complex structures. They consist of assemblages of individual fibers of various types, with nerve fibers providing command inputs and a number of internal sensors to provide feedback signals. Muscles contract (shorten) when activated. Since they are attached to bones on two sides of a joint, the longitudinal shortening produces joint rotation. Because muscles shorten actively and lengthen passively, pairs of muscles attached on opposite sides of a joint are required to produce bilateral motion.

During the past forty years a number of attempts have been made to build artificial muscles, that is, actuators with properties analogous to those of living muscle. The earliest of these was the so-called McKibben muscle. The earliest implementations of this device consisted of a rubber bladder surrounded by a sleeve made of nylon fibers in a helical weave. When activated by pressurized air, the sleeve prevented it from expanding lengthwise, and the device shortened like living muscle. In the 1960s there were attempts to use McKibben muscles to produce movement in mechanical structures strapped to nonfunctional arms of quadriplegics. The required compressed air was carried in a tank mounted on the person's wheelchair. These experiments were never completely successful.

Contemporary versions of the McKibben muscle, illustrated in figure 3.6, have been developed by Hannaford and his colleagues at the University of Washington

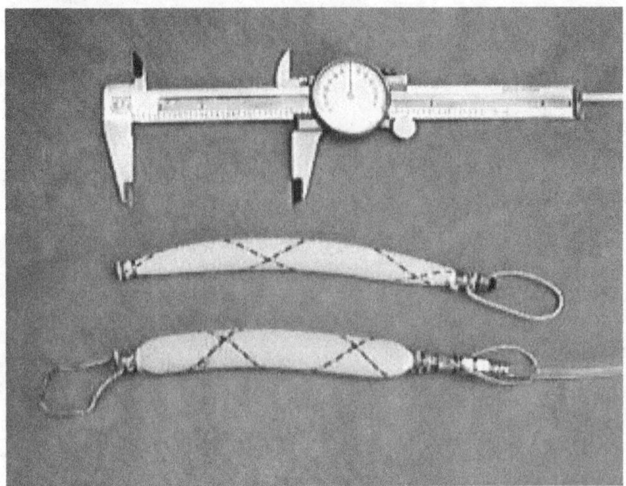

Figure 3.6
McKibben-type artificial muscle (courtesy of Blake Hannaford)

(Klute and Hannaford 2000). Brooks (1997) developed an artificial muscle for control of the arms of the humanoid torso Cog, discussed briefly in chapter 1, and Pratt for control of leg movements in a biped walking robot (Pratt and Williamson 1995). Pratt's muscles are interesting because of their nonlinear properties, which emulate certain aspects of living muscle. Even with recent improvements, however, it is fair to say that no artificial muscles developed to date can match the properties of animal muscles.

3.2.3 Artificial Muscles: Shape Memory Alloys

Shape memory alloys (SMAs) have unusual mechanical properties. Typically, they contract when heated, which is the opposite of what standard metals do when heated (expand). Not only does an SMA contract when heated, but it also produces a thermal movement (contraction) one hundred times greater than that produced by standard metals.

Since they contract when heated, SMAs provide yet another source of actuation for robots. After contraction, the material gradually returns to its original length when the source of activation is removed and it is allowed to cool. One material with this property is Nitinol, an alloy of nickel and titanium. Nitinol wire can be as thin as 0.025 mm (0.001″). It is evident that SMAs, when formed into cables, can be used as artificial muscles.

SMAs have two major problems when used as artificial muscles, however: They cannot generate very large forces, and they recover their original length slowly.

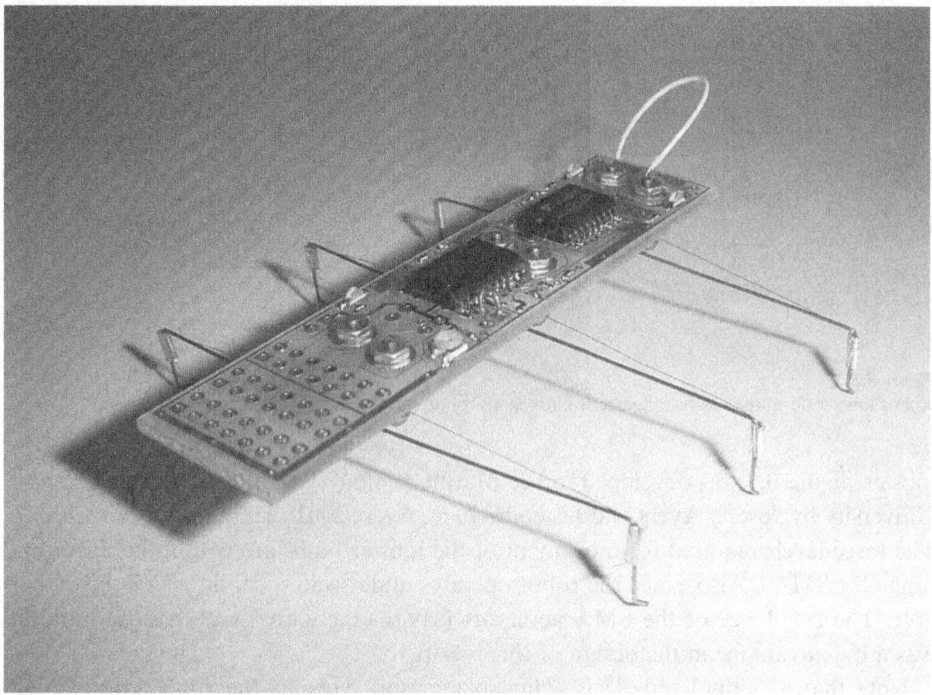

Figure 3.7
Stiquito with attached microcontroller (photograph courtesy of James Conrad and Jonathan Mills)

When heated (say, by electrical stimulation) and shortened, SMAs must be cooled to recover their original length. Depending on the environment in which they are used, this cooling may be slow, thus reducing the frequency response of any artificial muscle in which they are employed. Both of these problems have seriously limited their applications in robotics, though they have been used for motion control in a variety of small legged robots, such as the Stiquito, a small, inexpensive, hexapod, insect-like robot developed by James Conrad and Jonathan Mills (Conrad and Mills 1997) and used extensively in high schools and colleges to introduce students to the basic ideas of robotics. The books by Conrad and Mills are unique in that they contain all the supplies needed to build the robot. The latest version of Stiquito, built by James Conrad and incorporating a piggyback control board using a peripheral interface controller (PIC) processor, is illustrated in figure 3.7 (Conrad and Mills 2003). Each leg has an attached Nitinol wire.

A fascinating application of SMAs that completely overcomes the above difficulties involves their use in artificial muscles to produce leg movements in a robot

Figure 3.8
Robot lobster developed at Northeastern University (photograph courtesy of Joseph Ayers)

lobster (figure 3.8), as developed in the Marine Biology Laboratory at Northeastern University by Joseph Ayers and his colleagues (Ayers 2001; Safak and Adams 2002). The force levels required for movement of the lobster limbs are within the operating range for SMAs. Also, since the robot operates under water, cooling is readily available. The small size of the SMA actuators (say, as compared with electric motors) was a big advantage in the design of this robot.

Note that a Nitinol muscle is a linear actuator, whereas the robot lobster's leg segments rotate about the respective joints. Hence, it is necessary to convert linear motion into rotary motion in this robot by attaching one end of the Nitinol muscle modules to a point on the leg segment and the other to a system of pulleys using Kevlar strings. A typical leg of the lobster robot, showing the SMA actuators and the pulleys, is depicted in figure 3.9.

3.2.4 Artificial Muscles: Electroactive Polymers

A third class of artificial muscles, which has gained considerable prominence in recent years, involves the use of electroactive polymers (EAPs). Like SMAs, these materials also change their shape when electrically stimulated, which makes them usable as artificial muscles (Bar-Cohen 2000, 2001; Shahinpoor and Kim 2001). The most attractive feature of EAPs is their ability to emulate biological muscles with a high degree of toughness, large actuation strain, and inherent vibration damping. Both SMA-actuated and EAP-actuated robots can be made highly maneuverable, noiseless, and agile. Unfortunately, as with SMAs, the force actuation and mechanical energy density of EAPs are relatively low, which limits the applications for which they can currently be considered.

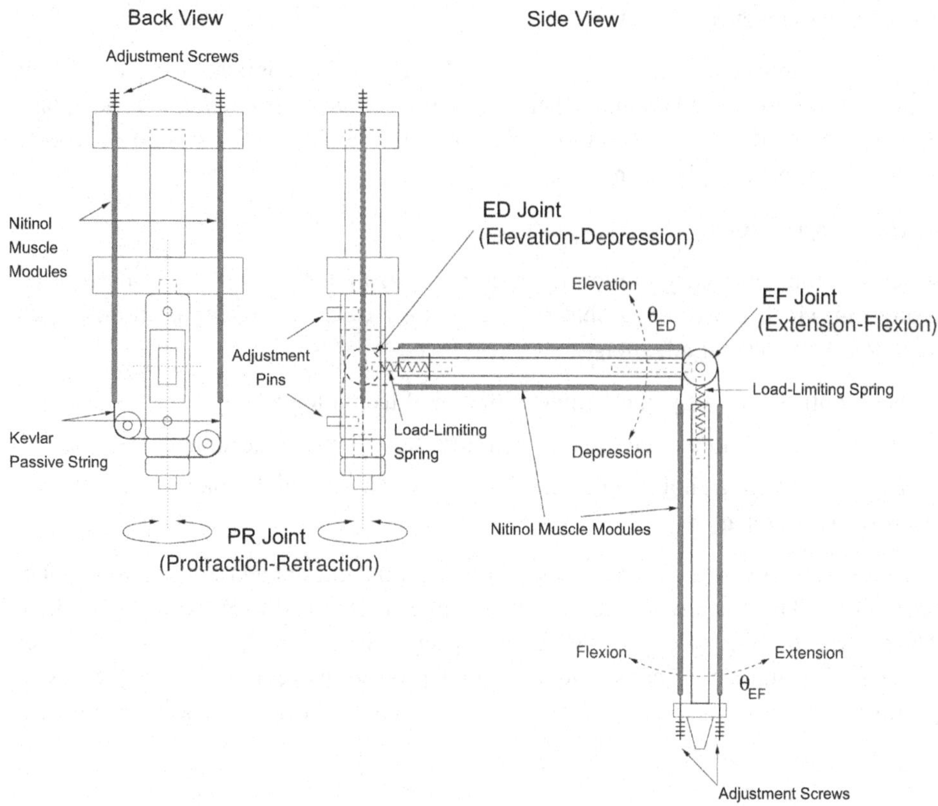

Figure 3.9
Typical leg of robot lobster (courtesy of Koray Safak and Joseph Ayers)

3.2.5 Pneumatic and Hydraulic Actuators

Large manipulators in industry frequently employ hydraulic drives, since such drives provide a higher torque-to-weight ratio than electric motors. Hence, they are used for actuation in robots that need to move very large loads, say, in industrial manipulators that carry entire vehicles. Because of the maintenance problems associated with pressurized oil (including leaks), hydraulic motors are not used in smaller mobile robots. Of course, hydraulic actuators are common in such applications as automobile brakes, but not in robotics.

Pneumatic drives have been used as actuators in robots in the past but are not currently popular, since air is compressible, resulting in nonlinear behavior of the actuator.

3.2.6 Electromagnetic Actuators

Where only small linear motions are required—for closing contacts, for example—it is possible to use solenoids and other forms of *electromagnetic actuation*. Solenoids may also be used to deploy portions of a robot that are normally stowed or folded, as in certain space applications.

3.2.7 Stepper Motors

When incremental rotary motion is required in a robot, it is possible to use special motors in which the rotor is able to assume only discrete angular positions. Such actuators have several advantages:

- Their control is directly compatible with digital technology.
- They can be operated open loop by counting steps, with an accuracy of ± 1 step.
- They can be used as holding devices, since they exhibit a high holding torque when the rotor is stationary.

There are two basic types of stepper motor: permanent magnet and variable reluctance (VR). The basic structure of a VR stepper motor is illustrated in figure 3.10. The central structure (rotor) rotates with respect to fixed (stator) structure. The latter consists of a number of coils wound around projections representing sequences of positive and negative poles, respectively. Note that in the figure, the rotor and the

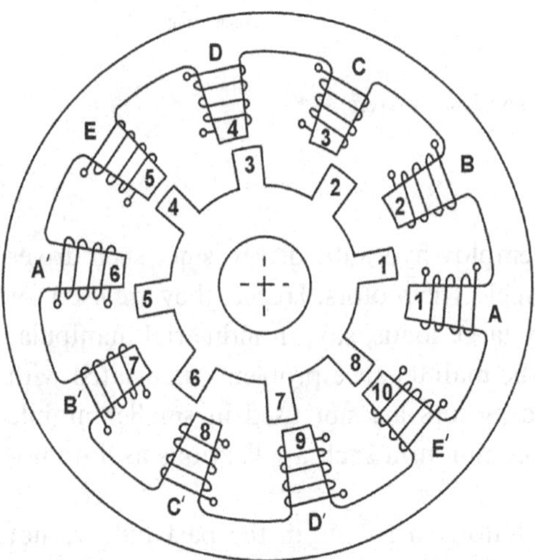

Figure 3.10
Structure of a VR stepper motor

stator of the motor have an unequal number of teeth: The rotor shown has eight teeth, whereas the stator has ten. As illustrated in the figure, when poles E-E′ are energized, teeth 4 and 8 of the rotor line up with these poles. If we now energize poles D-D′, the closest rotor teeth to these poles will be 3 and 7, so that the rotor will jump by 9°. (The spacing between the stator teeth is $360/10 = 36°$, whereas the spacing between rotor teeth is $360/8 = 45°$, hence the "step" of the motor will be $45 - 36 = 9°$.) Stepper motors are very popular in such applications as printer drives and certain industrial robots.

3.2.8 Other Actuators

The previous sections do not discuss all available actuators, but they give an indication of the great variety of ways in which robot motion can be produced. It should be noted that actuation is still a major area of research in robotics. When one considers that an ant can carry an object several times its body weight, it is clear that robot actuators leave a great deal to be desired.

3.2.9 Linkages

Actuator outputs may connect to the robot through various linkages, some of which convert rotational into translational motion, others into oscillating motion. So-called four-bar linkages include pantographs, used to reproduce motion at a different scale, namely, larger or smaller than the original. Many types of linkages are described in standard books on engineering mechanics.

3.3 Sensors for Robots

Animals are equipped with an astonishing variety of specialized sensors that enable them to monitor both their external and internal environments. As shown in chapter 2, homeostasis (maintenance of a nearly constant internal environment) depends on proprioceptive sensors that provide indications of deviations from desired values in such internal body variables as arterial blood pressure, body core temperature, degree of muscular contraction, and concentration of hormones and other biochemical substances in the bloodstream. Exteroceptive sensors provide indications of such variables as air temperature, illumination, color, sound intensity and frequency, and surface roughness. Migrating birds are able to detect changes in their orientation with respect to the earth's magnetic field and use this information for navigation. Fish can detect movement of water. A recent book (Hughes 1999) summarizes some of the more exotic senses animals possess. It should be noted that many of the sensors in the animal kingdom are used in control loops. For example, if blood pressure (as measured by the baroceptors in the aorta) decreases, the blood pressure control system causes peripheral vasoconstriction, thus increasing the downstream impedance to blood flow and causing the central blood pressure to rise. The output of the sensor is not measured and recorded, but used only for closed-loop control.

In some cases animals are able to determine the value of a particular variable by processing of sensory inputs. For example, observation of the "waggle dance" of the honeybee allows other bees to determine the vector direction and approximate distance to a source of pollen. Such a determination implies a form of computation, but not necessarily any associated intelligence.

Like their counterparts in the animal kingdom, robot sensors are also numerous. They are used to monitor such phenomena as motion, visual scenes, sounds, smells, electromagnetic radiation, and others. We review some of these sensors in the next section.

3.3.1 Proprioceptive Sensing

As indicated previously, both robots and animals need to sense both their internal and external environments. Sensing the internal environment (proprioception) makes homeostasis possible, since the information from the sensors is used in feedback control mechanisms. Sensing the external environment (exteroception) enables humans and animals to avoid danger, to satisfy hunger and thirst, and to accomplish voluntary goals. Robots need external sensing to avoid danger and to accomplish goals set by their designers. In this section we summarize the properties of the most commonly used sensors in robotics. Readers interested in further detail on these or other sensors are urged to consult the literature (e.g., Everett 1995) or the available journals in this field, such as the *IEEE Journal of Sensing.*

3.3.1.1 Position Sensing: Wheeled Robots

Most mobile robots use wheels for locomotion. Hence, position sensing is simply the sensing of wheel rotation, usually accomplished with a potentiometer (pot). In principle, one should be able to estimate the distance traveled simply by counting wheel revolutions. This cannot be done with a pot, since it is usually limited to a small number of revolutions, but it can be done with a simple electromagnetic device. For example, it is possible to count the number of times that a coil passes a magnet. Special-purpose generators, such as synchros or resolvers, can also be used to measure rotations. Unfortunately, this is not a practical way of estimating distance, even though this is the way in which an automobile odometer works. The wheels of a robot do not have identical diameters; there is some slippage (depending on the nature of the ground surface); and a robot does not travel in an exactly straight line. Hence, odometry distance estimates gradually deviate further and further over time from the actual distance traveled. The problems of odometry have been studied extensively (see, e.g., Borenstein and Koren 1987). Reliable estimates of the distance

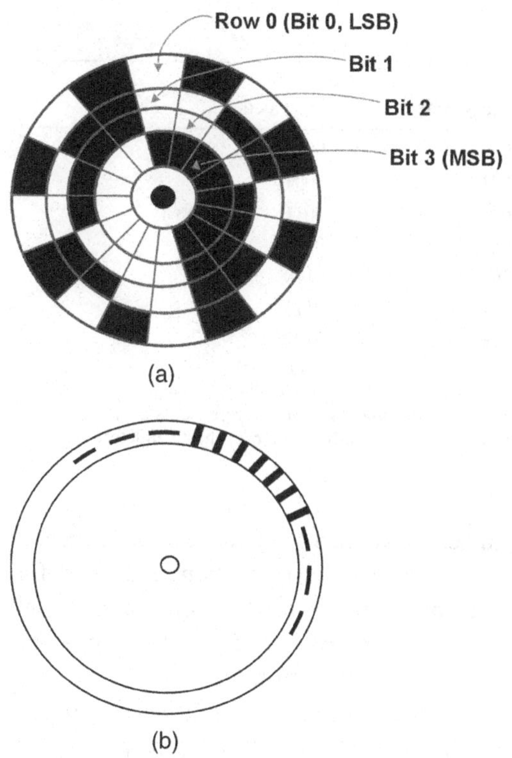

Figure 3.11
Encoders for sensing angular position: (a) absolute encoder, (b) incremental encoder

traveled by a robot may require external references. It is also possible to improve the accuracy of estimation by means of statistical procedures, such as Kalman filtering (see chapter 14).

A more practical way of measuring rotation is based on the use of encoders, which have the advantage of providing outputs directly in digital form. There are two common types of encoders, shown schematically in figure 3.11. A typical *absolute encoder* is shown schematically in figure 3.11(a), and an *incremental encoder* is shown in figure 3.11(b). An absolute encoder provides a digital output corresponding to an angular position, whereas an incremental encoder provides only a relative position with respect to a given reference. Consider the absolute encoder first.

It can be seen from figure 3.11(a) that the innermost ring of the absolute encoder depicted consists of a black stripe and a similar white stripe, each covering 180°. A light source, an optical sensor, and appropriate coding can be used to provide a 1 and a 0 from the sensor for each 180° of rotation, as the corresponding color passes

Table 3.1
Coding of encoder output

Decimal code	Rotation range (degrees)	Binary code	Gray code
0	0–22.5⁻	0000	0000
1	22.5–45⁻	0001	0001
2	45–67.5⁻	0010	0011
3	67.5–90⁻	0011	0010
4	90–112.5⁻	0100	0110
5	112.5–135⁻	0101	0111
6	135–157.5⁻	0110	0101
7	157.5–180⁻	0111	0100

Note: The superscript minus (⁻) indicates that the corresponding digital code represents an angle in the given range to an increment below the maximum value. Other methods of preventing indeterminacy at the boundary values are possible.

by the sensor, representing the most significant bit (MSB) of the output sequence. The next ring outward produces a 1 and a 0 each quarter revolution, and so on for successive rings toward the exterior. A four-bit encoder, in which the increment of rotation corresponding to one least significant bit (LSB) is 22.5°, willl produce the output sequence shown in table 3.1 during the first 180° of rotation.

Such an encoder wheel can be used to provide a measurement of wheel rotation to any desired degree of accuracy. Note that the wheel in this figure provides a binary-code output; it is also possible to use other representations, such as the Gray code.

The incremental encoder shown in figure 3.11(b) simply produces a sequence of pulses starting from an arbitrary position. Clearly, if this is an optical encoder, there will be light sources and sensors. The systems are more complex than indicated here, but we wish only to summarize the principles upon which they operate. It should also be noted that that such encoders can provide a digital output corresponding to linear (as well as rotational) displacement, by simply "unrolling" the stripes in the figures and presenting them along a linear scale.

3.3.1.2 Position Sensing: Legged Robots

We have already noted that human muscles include sensors that provide feedback information to the central nervous system regarding muscle contractions. Hence, it is possible to deduce limb movement from such afferent information. In walking robots, limb motion may be sensed by potentiometers located at the joints; this method is also used to sense displacement in robot manipulators. It is important to note that while these pots can be used to infer the location of the limb segments with respect to the robot body, they are not useful of determining the distance traveled on the ground. In principle it should be possible to estimate distance by knowing the step length and then counting the number of steps. However, since there will be some slippage (depending on surface material, friction, and other parameters) and the step length will not be identical from leg to leg, such distance estimates will be highly imprecise. This lack of precision is equivalent to that presented in the odometry problem we encountered in the previous section with wheeled robots. More precise distance estimates require external references and exteroceptive sensors.

In some walking machines a combination of limb displacement and contact sensing is used to trigger specific behaviors. For example, if a foot encounters an obstacle when moving, displacement sensors and leg motion actuators are used to raise the leg higher in an attempt to avoid the obstacle (Sukhatme 1997).

3.3.1.3 Velocity

In principle, the speed of rotation of a wheel or leg segment can be estimated through differentiation of the position output. However, since the position estimate is inherently noisy, differentiation will accentuate the noise and is generally not practical (unless accompanied by filtering, which then produces a delayed estimate). In wheeled robots (and automobiles or rotating machinery) rotational velocity measurements are usually obtained from *tachometers*, special-purpose electromagnetic devices that produce an output proportional to rate of rotation. The speedometer in an automobile provides an estimate of linear velocity based on the output of a tachometer and knowledge of the wheel diameter. Clearly, this reading will vary as the tire wears. Note that in an automobile, the odometer reading is obtained through integration of the tachometer output. This reading has all the problems described previously.

3.3.1.4 Acceleration

Estimates of acceleration can be obtained through differentiation of the velocity estimate, but this is usually undesirable since, as pointed out previously, differentiation is a noise-amplifying process. Under certain conditions, some robots also estimate *jerk*, the derivative of acceleration. Devices known as accelerometers provide a direct estimate of linear acceleration, by measuring the force exerted by a known, moving mass on a restraint, and using Newton's law ($F = ma$).

3.3.1.5 Load

An important internal measure concerns the load being carried by the robot. Several methods are commonly used for obtaining this measure. For manipulators equipped with grippers, it is possible to mount force sensors directly on the gripper. A commonly used indirect measure of load is the current drawn by the actuator motor. Driving motor currents provide an indication of torque requirements; these would be expected to increase with an increased load (e.g., grasping and lifting a heavy object, moving up a hill, pushing an object).

3.3.1.6 Internal Temperature

Since the electronics on board a robot may fail if the internal temperature becomes either too high or too low, some applications require thermometers linked to appropriate control mechanisms. For example, a spacecraft or a planetary rover may be able to change its orientation with respect to the sun if its internal temperature is too high or too low. Other robotic vehicles may be able to turn internal heaters on or off as required to keep the temperature within acceptable limits. Since the operation of motors generates heat, it may be necessary for a robot to cease activity until the temperature is reduced, say by radiation.

3.3.1.7 Battery Charge

Autonomous robots commonly use batteries to supply the electrical power required for mobility, communication, computation, and even sensing. Hence, it is essential that they monitor the level of charge of the batteries if recharging mechanisms are available. (Of course, in some applications, the robot's useful life will be assumed to end when the batteries are discharged. This may be true of many military applications, where recharging is not possible.) When recharging of batteries is possible, it can take place in a number of ways, depending on the application. For example:

- A robot may simply signal to human operators or to larger recharging robots that its power level is becoming low. (In effect, this is a call like "Daddy, please plug me in!")
- When autonomous robots operate within a structured environment, say, the interior of a factory or a home, they may just locate an electrical outlet or charging station and plug themselves in. The AIBO entertainment robot (see chapter 9) uses such a charging station.
- On planetary environments, which change from light to dark, it may be necessary for a robot to cease all action until its solar panels can again face the sun and recharge the batteries.
- Robots that generate electricity on board through biochemical or similar processes may have to cease all action until these relatively slow processes can recharge the batteries.

3.3.1.8 Failures

One of the most critical needs for internal sensing in robots is to detect failures of components or subsystems. Such sensing will be highly specific to the particular components for which failure detection is needed. For example, actual or incipient motor failure may be detected through excessive current flow, which simply requires an ammeter. Inability of the robot to maintain a desired posture may be indicated by inclinometers. Sensing the current to the batteries, when the panels are deployed, could reveal failure of the solar charging system. Detecting the failure of a leg to move requires sensing the joint angles, whereas sensing of the failure of a wheel to turn requires shaft encoders or similar devices that measure rotation. Of course, such sensing will be of little use if corrective measures are not possible. Recently, the use of a bank of Kalman filters has been suggested as a way of monitoring (and possibly overcoming) certain failures (Goel et al. 2000). We believe that future deployment of groups of distributed robots will require the development of "maintenance robots" charged with receiving and identifying failure signals from members of the group and then performing repairs. In space vehicles this will mean replacement of repair functions currently performed by astronauts in space suits with such teleoperated devices as the NASA Robonaut (for "robotic astronaut") described in chapter 13 (Ambrose et al. 2000; Bluethmann et al. 2000) and eventually by autonomous robots.

3.3.2 Exteroceptive Sensing

A variety of sensors are used in robotics to monitor the external environment. This section discusses the most important ones.

3.3.2.1 Vision

A vision sensor is one of the most common sensors included in contemporary robots. It usually includes a camera (with resolution comparable to standard television cameras) and software designed for such tasks as edge detection, contrast enhancement, and recognition of particular objects. Some of these features are based on living prototypes. For example, some animals' visual systems have mechanisms for contrast enhancement.

Some current vision systems are equipped with circular or conical mirrors that enable them to "see" in all directions, that is, in 360° of arc. A typical device of this type is illustrated in figure 3.12 (Baker and Nayer 2001).

3.3.2.2 Proximity Sensing

Perhaps the most common sensors on mobile robots are devices that send and receive ultrasonic signals. The transmitted signals are bounced from an object and received by the sensor; the time interval provides a measure of the distance from the robot. Although not very precise, such sensors are inexpensive and readily available (e.g., being used for range measurements on certain cameras). Some robots are surrounded by a ring of ultrasonic sensors, as illustrated in figure 3.13.

Lasers are also commonly used to enable robots to sense the presence of objects, other robots, walls, doorways, and so on. They provide more precision in certain ranges than ultrasonic sensors, but at significantly higher cost. One of the most

(a)

(b)

Figure 3.12
Omnicam panoramic vision system: (a) camera showing parabolic mirror under the lens, (b) typical view (photographs courtesy of Amin Atrash)

(a)

(b)

Figure 3.13
Mobile robots equipped with a number of ultrasonic proximity sensors: (a) Pioneer 3 robot (photograph courtesy of ActivMedia Robotics), (b) Magellan Pro Mobile Robot (photograph courtesy of iRobot Corporation)

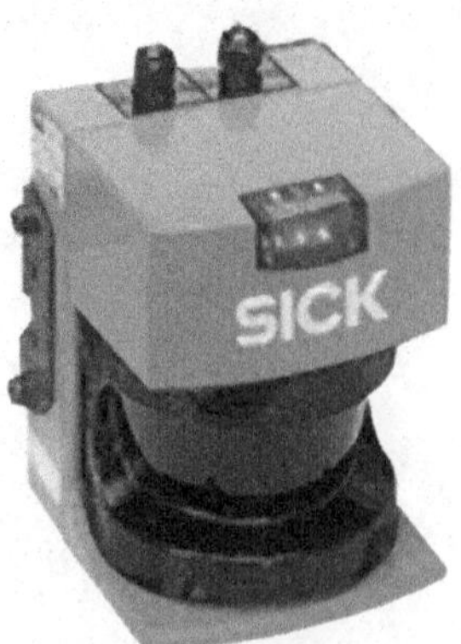

Figure 3.14
Laser range finder Model LMS 200 (photograph courtesy of SICK, Inc.)

widely used high-quality laser range finders, made by Sick, Inc., is illustrated in figure 3.14. The Pioneer 3 robot depicted in figure 3.13(b) includes a camera and a gripper. It can also be equipped with a laser range finder.

3.3.2.3 Audition

Hearing is provided to robots by means of microphones, transducers for converting acoustic energy into electrical signals. In general, robot hearing does not have many of the properties of animal hearing systems, such as sound location from binaural reception or highly sensitive time and frequency detection. Many current robot hearing systems are used only for detection of alarm signals, though increasingly one finds robots designed to respond to voice commands. The processors of such robots require a speech recognition module.

3.3.2.4 Olfaction

Currently some robots are being equipped with olfactory systems. An olfactory sensor enables a robot, for example, to detect the presence of nitrogen-based compounds, such as those that may be used in explosives, or to trace the path of a person or a load of chemicals. Generally, robot "noses" are tuned to particular smells and do not yet have the versatility or sensitivity of a dog's nose. A source for commercial odor sensors for robots is Applied Sensor, Inc., a Swedish company with branches in the United States and Germany (Aiken 2004).

3.3.2.5 Touch

Robots must be able to sense the physical world through direct contact. The simplest sensors just close a switch on contact with an external object; for example, bumpers on the robot can be equipped with microswitches. Some robots have been equipped with artificial fingertips, containing a dense array of solid-state devices to provide an indication of contact force and/or pressure and even lateral (slippage) forces. Such devices are particularly useful in the control of grasping. Other touch-sensing devices make use of a collection of small rods that move until they make contact. Such an array can also provide an indication of the shape of the object being touched. Photodetector devices are also used. In systems employing photodetectors, when the contact point on the robot comes in contact with a surface, a rubber or elastomeric material is compressed and a rod moves, interrupting a light beam. Touch can also be used to compress materials whose resistive or capacitive properties then change, thus providing an indication of contact. Clearly, there is a great variety of touch-sensing systems.

3.3.2.6 Slippage

Slip is closely related to touch. Measurement of slip is particularly important in grasping (see chapter 11), to ensure that a grasped object does not fall from the robot's "hand." In humans, some of the same sensory mechanisms in the fingertips are responsible for measurement of both slippage and pressure. In robotic hands, slip has been measured by means of small spheres (analogous to the ball in a computer mouse) and by special semiconductor elements embedded in the fingertip of a robot hand. Since this a rather specialized sensory mechanism, primarily of interest in grasping, it is not discussed further in this chapter.

3.3.2.7 Taste

The author is not aware of any robots equipped with a sense of taste as such. Some robots are specifically equipped with sensors to detect concentrations of particular chemicals, but not to classify them along a taste dimension such as sour or sweet or bitter.

3.3.2.8 Temperature

Human skin is equipped with temperature, pressure, and slippage sensors. All these capabilities can be provided to robots.

3.3.2.9 Other Sensors

The sensors discussed in the foregoing have counterparts in biology. In addition, robots are frequently equipped with other sensors that have no such counterparts, such as the following:

- Wheel rotation measurement devices: We have already indicated that the discernment of motion about a joint can be considered a proprioceptive sense. On the other hand, measurement of distance traveled using vision is probably an exteroceptive sense.

• Ionizing radiation: Whereas humans are not able to detect X rays and other forms of radiation (although some animals apparently can), robots can be equipped with Geiger counters and other radiation detectors. These sensors are particularly useful in robots working in nuclear power plants or performing cleanup activities in radioactive environments.

• Vibration: It is known that vibrations at certain frequencies can cause serious harm to humans and animals, yet we do not have the proper sensors to detect these vibrations. Such sensors can be provided easily to robots.

• Chemical concentration gradients: Robots have been built to detect and follow concentration gradients of specific chemicals, thus enabling them to follow chemical trails in the environment, in a way analagous to that in which insects may follow pheromone trails.

In principle, robots can be equipped with any of the myriad sensors developed for industrial and engineering applications, provided that the sensors are compatible with size, weight, and energy constraints.

3.4 Localization

External sensing includes the need for the robot to know where it is (known as *localization*) and what else is present in its surroundings. Being able to answer the question "Where am I?" may also enable the robot to take corrective action if the location and or direction are not the intended ones. This important issue is discussed in chapter 14.

3.5 Computation and Communication

Clearly, robots need computers to process sensory inputs, to generate commands to the actuators, and to perform such cognitive functions as reasoning and planning. Most contemporary mobile robots use on-board processors, ranging from simple 8- or 16-bit single-chip microprocessors using one of several real-time operating systems to PC boards using Windows or Linux operating systems. Since this field changes very rapidly, any description of computers or associated software employed in robots would be rapidly obsolete, so computing in robots is not discussed further in this book. Specific issues associated with particular applications are discussed in connection with those applications.

Mobile robots need to communicate with their base station, with one another, and with humans. Most robot electronic communication systems employed in the past used radio frequency (RF) transmitters and receivers. Currently, many mobile robots working in both indoor and outdoor environments use wireless Ethernet communication. This technology is based on IEEE Standard 802.11 and generally uses the trade name Wi-Fi. Communication with distant robots, such as Mars rovers, requires specialized hardware and software systems designed to work with very low power and narrow bandwidths. This is another area with implications beyond robotics that is not discussed further in this book. On the other hand, communication with humans by means of voice, gestures, and other means is considered in connection with humanoid research in chapter 13.

4 Low-Level Robot Control

Summary

This chapter summarizes the main issues involved in the analysis and design of low-level control systems for robots. The ideas, including the notions of stability and sensitivity, are presented in an intuitive way. The chapter's goal is to introduce the reader to the fundamental concepts of both linear and nonlinear control theory and the applicability of these ideas to the control of both robot manipulators and mobile robots. We then introduce some principles for the design of controllers for feedback systems, with an emphasis on proportional-integral-derivative (PID) controllers commonly used in robotics. The mathematical treatment of linear feedback control systems using frequency domain methods is presented in the appendix.

A later section of the chapter discusses some of the basic ideas involved in more advanced control system issues, such as adaptive, nonlinear, and optimal control systems, as well as systems in the presence of uncertainty. We also present a view of biologically inspired control systems, using the concepts of control theory to provide additional perspective on the issues raised in chapter 2 and to highlight some of the differences between living and engineering systems.

4.1 Engineering Control: An Intuitive Introduction to Its Advantages and Limitations

Control is defined as the ability to cause the state of a system to move to a desired value or set of values. For example, we use the accelerator and brake pedals of an automobile to control its speed, that is, to reach or maintain a desired speed or to follow a desired velocity profile. In a robot, we may desire to control linear velocity, turning rate, position of an arm, or other variables.

An engineering control system consists of at least two parts: the *controller* and the object to be controlled, the *plant*. These elements may be configured with or without feedback, as shown in figure 4.1.

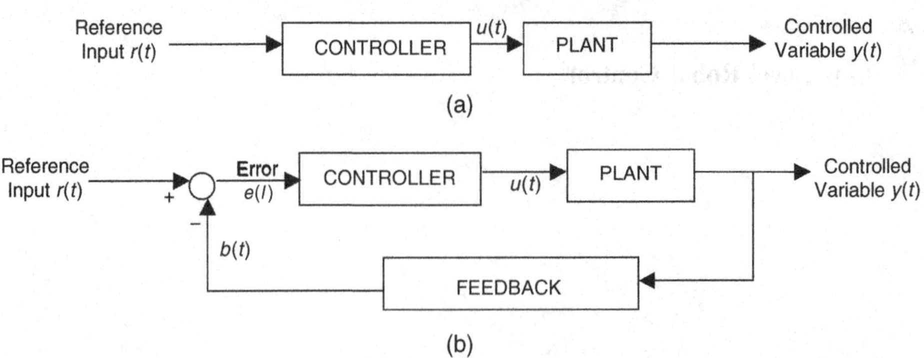

Figure 4.1
(a) Open-loop and (b) closed-loop control systems

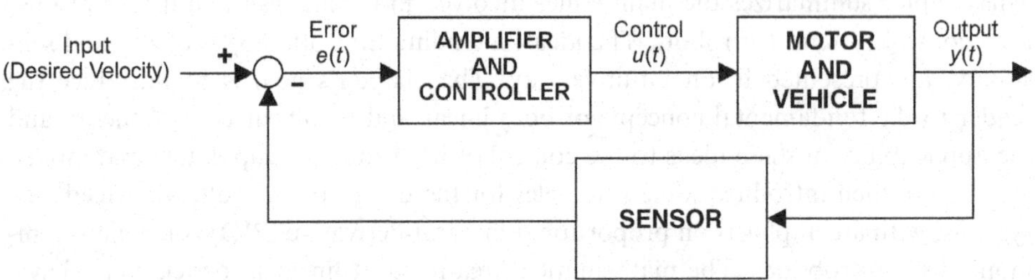

Figure 4.2
Feedback control system for robot velocity

Assume that we wish to control the velocity of the driving wheels on a mobile robot. The desired velocity (or, in some cases, the position) $r(t)$ is the input to the system. To find out how rapidly the wheels are actually turning, we use sensors (such as tachometers). These sensors are the feedback elements of figure 4.1. The actual velocity $y(t)$ is compared to the desired velocity using a comparator, thus producing an error signal $e(t)$. Note the minus sign on the comparator. To find the error we need to subtract the measurement of the actual velocity from the desired value (hence the term "negative feedback"). The error signal drives a controller (to be discussed in the following section) and amplifier. The resulting control signal $u(t)$ drives the motor and hence moves the entire robot, the plant. Such systems are also known as servomechanisms,[4] or simply *servos*; the diagram in figure 4.2 represents a velocity servo.

4. Some authors reserve the word "servomechanism" for control systems in which the reference input is zero; others restrict it to systems in which the output or controlled variable is mechanical movement. No such distinctions are made in this book.

Let us now return to figure 4.1. Figure 4.1(a) shows an open-loop configuration, whereas figure 4.1(b) illustrates a closed-loop control system. Some older home heating systems are open loop; they work as follows: Assume that we know that during each 15 minutes, the furnace produces a 1°F rise in room temperature. If the room is at 56°F, and we wish to raise the temperature to 60°F, we need to keep the furnace burning for 1 hour. Note that there is no sensor to tell us what the *actual* temperature is after 3 hours; we depend on an accurate model of furnace and room. As the furnace gets older, it will probably be less efficient, and after 1 hour of being on, it may raise the room temperature by only 3°. The solution to this inaccurate furnace problem is to use a sensor on the output or controlled variable (temperature in this case) and compare it with the desired values for this variable; the difference is the *error*, as shown in figure 4.1(b). After suitable processing by the controller (as is discussed later), the error signal controls the furnace. Now, the furnace will burn as long as needed until the error is reduced to zero.

In typical control systems we need to control more than one variable; in that case each arrow in the diagrams of figure 4.1 represents a vector of quantities, and the system produces an measurement of error for each of the variables. A measure of the errors, the nature of the control task, and the plant dynamics are the elements needed for the synthesis of the controller. In order to apply mathematical methods to the implementation of the controller, it is necessary to have a model of the plant. ("Non-numerical" control methods may use a symbolic [rather than a mathematical] model of the plant, as described below.)

Let us examine the differences between these two approaches to control more carefully. The controller "gain" K_c changes units, say, from degrees of dial rotation to time duration. Then, in the open-loop case, the controller output is given by

$$u(t) = K_c r(t), \tag{4.1}$$

where K_c represents the controller gain. Let the plant be represented by

$$y(t) = K_p u(t), \tag{4.2}$$

where K_p represents the gain of the plant, $u(t)$ is the controller output (in this case, the number of minutes of operation), and $y(t)$ is the output variable (the increase in room temperature). For the furnace under discussion, K_p is the change in room temperature per unit time of operation. Combining equations (4.1) and (4.2):

$$y(t) = K_c K_p r(t) = K r(t). \tag{4.3}$$

Thus, the controlled output is proportional to the desired value $r(t)$. Now, if the furnace properties have changed by 1%, we have $K' = 1.01K$ and

$$y'(t) = K'u(t) = 1.01y(t). \tag{4.4}$$

Thus, it is clear that in this system, a 1% error in the plant transfer characteristics produces a 1% error in the controlled variable. In other words, open-loop control requires precise knowledge (an accurate mathematical model) of the plant.

Let us now look at the feedback case. Referring to figure 4.1(b) and assuming the controller and plant properties to be the same as before, the controller output is now proportional to the error signal $e(t)$:

$$u(t) = K_c e(t) = K_c[r(t) - b(t)], \tag{4.5}$$

where the feedback signal is related to the output by

$$b(t) = K_f y(t). \tag{4.6}$$

The feedback gain K_f may represent the properties of the sensor used to measure the output variable and the change of units, say, from degrees to volts. We can now eliminate the controller output $u(t)$ by combining equations (4.5) and (4.6) with (4.2):

$$y(t) = K_p K_c[r(t) - K_f y(t)]. \tag{4.7}$$

Solving for the output or controlled variable $y(t)$:

$$y(t) = [K_c K_p/(1 + K_c K_p K_f)]r(t). \tag{4.8}$$

This expression indicates that for the feedback control system, the output is obtained by multiplying the reference input by a ratio in which the numerator is the product of the forward-path gains in figure 4.1(b) and the denominator is one plus the product of all the gains in the loop. Comparing this expression to the corresponding open loop (equation (4.3)), we can see that the effect of a change in the plant gain on the output is now influenced also by the controller gain K_c and the feedback gain K_f. Let us make the same assumption as before, namely, that the plant gain changes by 1%, so that now $K_p' = 1.01K_p$. By substituting in equation (4.8) and dividing both numerator and denominator by K_c we obtain

$$y(t) = [1.01K_p/(1/K_c + 1.01K_f K_p)]r(t). \tag{4.9}$$

Note the term $1/K_c$ in the denominator. If the controller gain K_c is large, this term becomes negligibly small and the output becomes

$$y(t) \approx [1.01K_p/1.01K_f K_p]r(t) = (1/K_f)r(t). \tag{4.10}$$

This is quite a remarkable result, since it indicates that in the presence of feedback, the output is nearly independent of the plant gain K_p, with the degree of approximation depending on the controller gain. In typical feedback amplifiers, for example, the controller gain may be as large as 10^6 or even 10^8. In our example, this means that if the feedback scale factor is one, changes in the furnace properties will have a negligible effect on the final room temperature, which, in the steady state, will reach the desired value $r(t)$.

The phenomenon we have just described is known as *sensitivity*; that is, we can say that the closed-loop system is *less sensitive* to plant parameter variations than the open-loop system. It can be shown that the closed-loop system is also less sensitive to disturbances (noise) than the open-loop one.

Of course, this improvement in performance has come at a price.[5] The feedback control system requires a controller with much higher gain than that in the open-loop system; it requires sensors to measure the output variables, and it requires comparators. An even higher price is the fact that feedback systems may encounter instabilities not possible with open-loop systems, as we discuss in section 4.3.

Given these advantages and difficulties, why would we prefer feedback systems? In practical engineering terms, if system response is nearly independent of plant characteristics, this means we can buy less expensive components. Using a feedback audio amplifier requires a higher gain controller, but the basic amplifier itself can be made of less precise components, saving the manufacturer some money. No two amplifiers will have exactly the same open-loop properties, but they may behave nearly identically when the loop is closed.

This insensitivity to parameter variations may be the reason for the prevalence of feedback control in biological systems. Mammals maintain nearly constant core temperature, arterial blood pressure, and other so-called homeostatic variables (see chapter 2) by using feedback control, even as their internal and external environments change. As we have seen, this implies the need for sensors, such as the baroceptors used for pressure sensing, which are located in the aorta and elsewhere in the arterial system. On the other hand, biological systems display other peculiarities we shall consider later.

The furnace control system discussed in the foregoing is highly simplified. The analysis was based on the assumption that the controller, plant, and feedback elements respond instantaneously to an input change, so that their outputs are equal to their inputs, modified only by a scale factor or gain. In the real world, this is almost never the case, since physical elements have dynamics (their responses are not instantaneous and may depend on the frequencies present in the inputs) and further, they may be nonlinear and depend on the magnitude of the input as well. We will consider these effects later in the chapter. Further, we have assumed that the control system had a single input and a single output signal. In practice, a complex system may have multiple inputs and outputs. The preceding analysis is still valid if we consider each of the variables to represent a vector of quantities.

5. This comment is based on the fundamental principle in engineering design that all improvements have a cost. In other words, there is no free lunch.

The analysis and design of linear feedback control systems is usually carried out in the frequency domain. A brief introduction to methods for such analysis and design is given in the appendix. Readers not familiar with the notions of transfer functions and the stability of feedback control systems should review the material in the appendix or in a standard introductory text on feedback control systems. Without this background, it will be difficult to understand the remainder of this chapter. Basic knowledge of these concepts is required to understand the design of the so-called PID controllers that are used in nearly all low-level control systems for mobile robots.

4.2 Robot Controller Design Principles

Now we turn our attention to the principles that govern the design of robot controllers. We assume that the controller will be linear and described by a transfer function $G_c(s)$ (see the appendix for a discussion of transfer functions). This is clearly an important issue for robot control, particularly for low-level control of drive motors, manipulators, and so on. Rather than discuss controller design in the abstract, we will apply it to a specific case of a robot position servo.

4.2.1 Design of a Closed-Loop Position Control System for a Robot

We have seen a block diagram of a typical robot control system in figure 4.2. Figure 4.3 depicts basically the same system, with a little more detail in each of the boxes.

We assume that the reference input $r(t)$ is a desired angular position, either of a manipulator segment, or of the wheel in a mobile robot. The system output is the actual angle $c(t)$. The output of comparator 1 is then the position error e_p, since it compares the desired and the actual angular (or linear) position. Note that the posi-

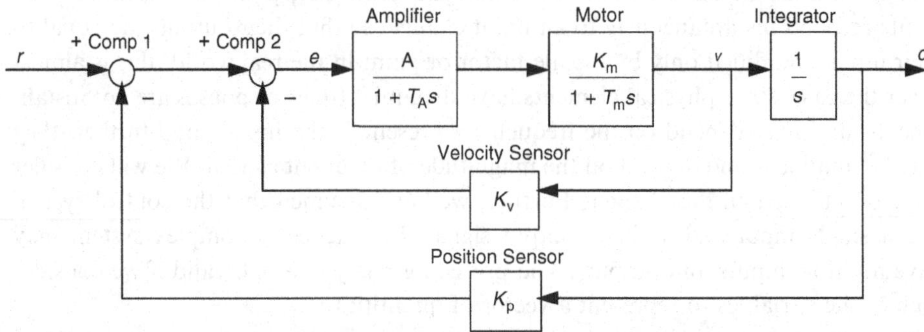

Figure 4.3
Typical robot servo loop

tion $c(t)$ is obtained by integrating the velocity $v(t)$, obtained at the output of the motor block, with respect to time. This block may actually represent both the motor and the load of the robot structure. If the robot's structural load is appreciable, it may be necessary to add another lag term (such as the denominators of the motor and amplifier transfer functions) to the denominator of this transfer function. The velocity is sensed by an appropriate sensor (perhaps a tachometer). It should be noted that this velocity feedback term is compared to the position error (in comparator 2), which plays the role of a velocity command or desired velocity. The output of comparator 2 is a combined error signal $e(t)$. It should be evident that it was not necessary to show the two comparators separately in the figure; both feedback signals could be fed into a single comparator (also known as a *summing junction*). Both the amplifier and motor are represented as first-order expressions of the type discussed in the appendix. Note that for mobile robots, it is common to have only a velocity control system, rather than one for position control; that is, the on-board processor commands vehicle speed rather than position. In that case the outside loop of figure 4.3, including the integrator, position sensor, and comparator 1, would be omitted, and the diagram would look more like that in figure 4.2.

Now consider what happens if the velocity loop is open (say, as the result of a broken wire, or simply because the velocity sensor gain has been set to zero). We can see that the loop consists of three first-order elements (two lags and an integrator), with the open-loop transfer function

$$G_{\mathrm{OL}}(s) = \frac{AK_{\mathrm{m}}K_{\mathrm{p}}}{s(1+T_{\mathrm{A}}s)(1+T_{\mathrm{m}}s)}, \tag{4.11}$$

so the maximum phase shift around the loop is 270°. Thus, the system is conditionally stable, depending on the value of loop gain at the point at which the phase shift is $-180°$ (see the appendix for a discussion of stability of the feedback system). When the velocity feedback loop is activated, by combining the various blocks, we can find the new composite open-loop transfer function to be

$$G_{\mathrm{OL}}(s) = \frac{AK_{\mathrm{m}}(K_{\mathrm{p}}+K_{\mathrm{v}}s)}{s(1+T_{\mathrm{A}}s)(1+T_{\mathrm{m}}s)}. \tag{4.12}$$

It is evident that the effect of the velocity feedback is equivalent to adding a numerator term to the transfer function (4.11). The new system is more stable than the pure position control system, and it becomes more and more stable as the velocity gain is increased.

The result we have just obtained is common in motion control systems, including those for mobile robots. In systems in which position control is desired, one frequently uses tachometers to add velocity feedback, just to improve stability.

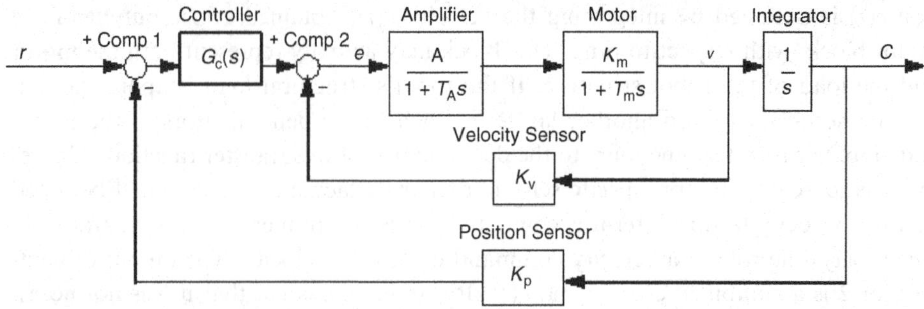

Figure 4.4
Typical robot servo loop with controller

4.2.2 PID Controllers

Consider now a slightly more complex version of figure 4.3, in which the motor/load block is of second order (as will be the case when the inertia of the moving vehicle is not negligible). Let us also include the effect of friction. Without going into the details of what happens to the transfer function (it becomes more complex, as would be expected), the important point is that even with a fixed reference input, the system will be unable to achieve zero position error; that is, we will find that in the steady state

$$c - r \neq 0. \tag{4.13}$$

This is true in nearly all mechanical or electromechanical systems: The effect of friction in the components will create a position error. Whether such a position error is important depends on the application. In some manipulator applications, highly precise control of end position is critical, and one must find ways of eliminating such errors. One way of accomplishing this is to use a more complex controller.

Consider figure 4.4, in which a controller has been inserted between the first and second comparator. Specifically, we will use a proportional-integral-derivative (or PID) controller, defined by

$$U(s) = \left(K_P + \frac{K_I}{s} + K_D s\right) E_p(s), \tag{4.14}$$

where $e_p(t)$ is the position error $r(t) - K_p c(t)$, $u(t)$ is the controller output, and velocity and position sensors are represented simply by constants. (Do not confuse the feedback gain K_p with the controller gain K_P; the subscript in the latter is a capital *P*.) This equation can be represented as in figure 4.5.

What are the advantages of this controller, and how can it solve the problem of steady-state position error? Clearly, the proportional gain term K_P simply multiplies

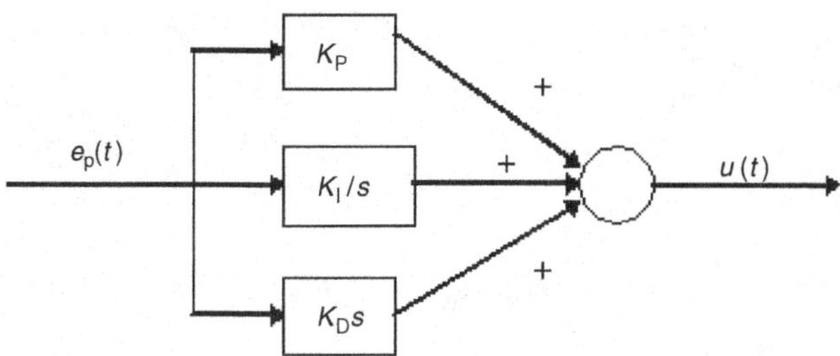

Figure 4.5
PID controller

the amplifier gain A. Consider the second term in (4.14). The transfer function $1/s$ corresponds to the operation of integration (see the appendix), so that the second term indicates that the error will be integrated, multiplied by a constant K_I, and added to the proportional term. This means that any small position errors will be integrated and grow with time, thus providing a larger input to the motor and causing it to move so as to decrease the error.

What about the third term in (4.14)? Consider a situation in which the error in response to some input is large, but decreasing. Using a proportional term will produce too large a correction, since it does not take into account the decreasing error. The integral term takes time to produce an effect, so a proportional term alone will result in an overshoot in the correction. The solution to this problem is to add a term to the controller that produces an output proportional to the slope (derivative with respect to time) of the error. Now, if the error is positive but decreasing, the net output of the controller will be smaller than that due to the proportional term alone. Evidently, this term gives the controller some anticipation. Adjustment of the three gains K_P, K_I, and K_D makes it possible to obtain desirable performance characteristics, including zero steady-state error and diminished overshoot.

Most mobile robots use PID controllers for low-level control of their drive motors. Sometimes the integral term or the derivative term is omitted, resulting in a PD (proportional-derivative) controller or a PI (proportional-integral) controller, respectively.

4.3 Control of Multilink Structures

A robot manipulator or an articulated leg of a walking robot consists of rigid links or segments connected by joints, as in figure 4.6. The manipulator structure differs from the leg structure mainly in that the end effector on a manipulator may have additional degrees of freedom, whereas robot feet are frequently passive. There are two approaches to the control of such structures: position control and force (or torque)

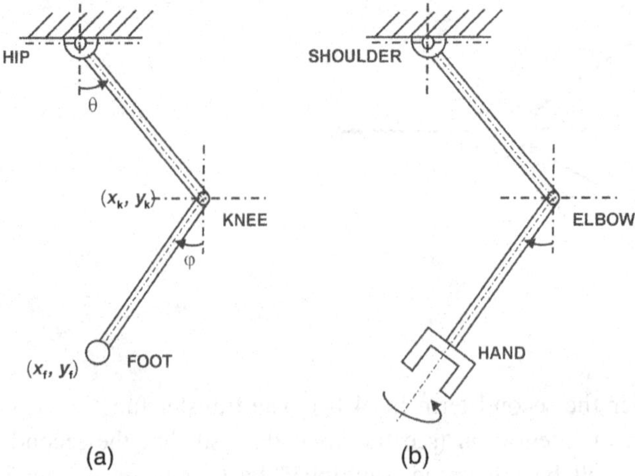

Figure 4.6
Kinematics of robot (a) leg and (b) arm

control. The first approach implies that the goal is to control the position of the end point (e.g., the foot in a leg controller). The second approach implies that in addition, for example, to controlling the position of a gripper, it may be important to control the force it applies to the grasped object. This is clearly important if the objects to be grasped can be crushed or broken by excessive force or can slip as a result of the application of inadequate force.

The major problem in position control is that final position of the end point is affected by both motors, the shoulder and elbow in the arm, and the hip and knee in the leg. If we rotate these two joints, say, by angles θ_1, θ_2, it is easy to compute the resulting position change $\Delta x, \Delta y$. Such computation of the Cartesian coordinates of the end point of a chained multilink structure, given the link lengths and their angular orientation, is known as the *forward-kinematics* problem. However, the goal of control is the opposite, that is, to move the joints so as to achieve a specific end-point position in both axes. This is known as the *inverse-kinematics* problem. Thus, in order to control the position of the end point of a multilink structure, it is necessary to solve an inverse problem. Specifically, for the leg model in figure 4.6(a), the knee coordinates are given by

$$x_k = l_1 \cos\theta, \qquad y_k = l_1 \sin\theta, \tag{4.15}$$

where l_1 is the length of the upper leg segment and the foot coordinates are given by

$$x_f = x_k + l_2 \cos \varphi,$$
$$y_f = y_k + l_2 \sin \varphi. \tag{4.16}$$

where l_2 is the length of the lower leg segment. Substituting (4.15) into (4.16) yields the foot coordinates:

$$x_f = l_1 \cos \theta + l_2 \cos \varphi,$$
$$y_f = l_1 \sin \theta + l_2 \sin \varphi. \tag{4.17}$$

These equations represent the forward kinematics. To apply control signals to the motors at the joints, these two transcendental equations must be solved for the angles θ and φ. Unfortunately, there is not a unique solution to the inverse problem; an alternate solution yielding the same foot coordinates can be found at $-\theta$ and $-\phi$. Although this is not a significant issue in this simple case, it can indeed cause major difficulties in some cases.

The foregoing transformation from Cartesian to rotational coordinates can be generalized to an arbitrary number of joints. Assume that there are three joints and that we need to define the foot location in x, y, and z. Let us denote the joint angles by θ_i, $i = 1, 2, 3$, and the Cartesian coordinates by $x = x_1,\ y = x_2,\ z = x_3$. Then we can define the vectors

$$\boldsymbol{\theta} = \begin{pmatrix} \theta_1 \\ \theta_2 \\ \theta_3 \end{pmatrix}, \quad \mathbf{x} = \begin{pmatrix} x_1 \\ x_2 \\ x_3 \end{pmatrix},$$

and replace (4.17) with the general relationship

$$\mathbf{x} = \mathbf{F}(\boldsymbol{\theta}). \tag{4.18}$$

As indicated previously, this expression cannot be inverted, since the solution may not be unique (or may not exist at all). However, although in general the position expression (4.18) cannot be used to find the inverse kinematics, the corresponding rate expression is invertible. Hence, we differentiate (4.18) to obtain

$$\dot{\mathbf{x}} = \mathbf{F}'(\boldsymbol{\theta})\dot{\boldsymbol{\theta}}. \tag{4.19}$$

The terms in the matrix $\mathbf{F}'(\boldsymbol{\theta})$ are given by $\partial F_j/\partial x_k,\ j, k = 1, 2, 3$. This matrix is known as the *Jacobian* and denoted by $\mathbf{J}(\boldsymbol{\theta})$; the forward kinematics can thus be represented as

$$\dot{\mathbf{x}} = \mathbf{J}(\boldsymbol{\theta})\dot{\boldsymbol{\theta}} \tag{4.20}$$

and the inverse kinematics by

$$\dot{\boldsymbol{\theta}} = \mathbf{J}^{-1}(\boldsymbol{\theta})\dot{\mathbf{x}}. \tag{4.21}$$

The solution of (4.21) is an important issue in the control of complex industrial manipulators and multijointed legs in walking machines.

4.4 State Space Approach: Theory, Advantages, and Limitations

The preceding sections of this chapter have concentrated on frequency domain models and transform methods for controller design. An alternative approach is based entirely in the time domain and makes use of the concept of *state.* The state space approach relies on two elements: the state vector and state equations. The state of the system is a minimum number of variables required to describe the behavior of a system, given its initial conditions and inputs. The idea of state stems from mechanics, since it is well known that the motion of a dynamical system is determined by the evolution of its position and velocity vectors. Thus, only two variables (position and velocity) are required to describe the state of a mechanical system. The vector-matrix description of forward and inverse kinematics in the previous section is an example of a state space formulation; the compactness and generality of such a description is evident.

The idea that complete system dynamics can be represented by a moving point in a plane can be extended conceptually to an n-dimensional vector space, but without the corresponding visualization. Essential in the concept of the state is the nonredundant feature of the state vector. No matter what set of variables is selected to describe the behavior of the dynamical system, the completeness of the description implies a minimum number of components of the state vector, known as the state variables. It is natural to extend the state concept from mechanics to any process (electrical, chemical, etc.) whose evolution can be described by a set of variables directly related to physical quantities or derived from them. The notion of state can be further generalized to apply to abstract general dynamical systems described by differential equations, without specific reference to any specific physical domain. Such a broad interpretation of the state has proved to be very useful in control theory as it concerns general issues of control synthesis and optimization, regardless of the specific nature of the plant. It must not be forgotten, however, that the concept of state implies that a mathematical model or description of the system is available.

Formally, a state space model of a system (including control) can be developed as follows. Consider a linear system described by the transfer function

$$\frac{X(s)}{U(s)} = \frac{K}{s^2 + 3s + 2}. \tag{4.22}$$

The equivalent differential equation (see the appendix) is

$$\ddot{x}(t) + 3\dot{x}(t) + 2x(t) = Ku(t). \tag{4.23}$$

Now let $x = x_1$ and $\dot{x}_1 = x_2$. Then $\dot{x}_2 = -2x_1 - 3x_2 + Ku$. Combining the latter two first-order equations, we obtain the vector-matrix expression

$$\begin{bmatrix} \dot{x}_1 \\ \dot{x}_2 \end{bmatrix} = \begin{bmatrix} 0 & 1 \\ -2 & -3 \end{bmatrix} \begin{bmatrix} x_1 \\ x_2 \end{bmatrix} + \begin{bmatrix} 0 \\ K \end{bmatrix} u(t). \tag{4.24}$$

This expression can be written in vector-matrix notation as

$$\frac{d\mathbf{x}(t)}{dt} = \mathbf{A}\mathbf{x}(t) + \mathbf{B}\mathbf{u}(t). \tag{4.25}$$

This is state equation for the linear system of equations (4.22) and (4.23), where $\mathbf{x}(t)$ is the state of the system and $\mathbf{A}$ and $\mathbf{B}$ are constant matrices. To solve this equation, it is necessary to have initial conditions for the state variables (the components of the state vector).

For a general nonlinear system the evolution of the system state is described by the dynamical equation

$$\begin{aligned} \frac{d\mathbf{x}(t)}{dt} &= \mathbf{f}(\mathbf{x}(t), \mathbf{u}(t), t), \\ \mathbf{x}(t_0) &= \mathbf{x}_0, \\ \mathbf{y}(t) &= \mathbf{g}(\mathbf{x}(t), \mathbf{y}(t), t), \end{aligned} \tag{4.26}$$

where $\mathbf{x}(t)$ is the n-dimensional state vector, $\mathbf{u}(t)$ is the m-dimensional control vector ($m \le n$), $\mathbf{x}_0$ is the initial state of the system, and $\mathbf{y}(t)$ represents the output (of dimension k, $k \le n$). The variables $\mathbf{f}$ and $\mathbf{g}$ indicate (possibly nonlinear) vector functions of the state, control and time. Thus, this expression is quite general, since it allows for nonlinearity and time dependence. For the linear, time-invariant systems described earlier, equation (4.26) reduces to

$$\begin{aligned} \frac{d\mathbf{x}(t)}{dt} &= \mathbf{A}\mathbf{x}(t) + \mathbf{B}\mathbf{u}(t), \\ \mathbf{x}(t_0) &= \mathbf{x}_0, \\ \mathbf{y}(t) &= \mathbf{C}\mathbf{x}(t) + \mathbf{D}\mathbf{u}(t), \end{aligned} \tag{4.27}$$

where $\mathbf{A}, \mathbf{B}, \mathbf{C}$, and $\mathbf{D}$ are constant matrices. Clearly, this expression could be represented by Laplace transforms. Note that for a simple mechanical system, the state $\mathbf{x}(t)$ would be two-dimensional, the elements being position and velocity. The first and second lines of equations (4.26) and (4.27) are known as the *state equations*, and the third line of each is known as the *output equation*. The set of all states, $\mathbf{X} = \{\mathbf{x}(t)\}$, is called the state space. In general, the output equation includes a term representing additive noise.

Using the above terminology, the behavior of the dynamic system can be described as motion along a trajectory in the abstract state space. State space representation has a number of advantages:

• The equations have the same form regardless of the order of the system (i.e., compactness).

• Powerful mathematical and computational tools exist for handling such equations. Note that the form of the operators $\mathbf{f}(.)$ and $\mathbf{g}(.)$ is unimportant if the equations are solved on a computer. Furthermore, the discrete form of the first line of (4.26) can be written as

$$\mathbf{x}_{n+1} = \mathbf{f}(\mathbf{x}_n, \mathbf{u}_n, n). \tag{4.28}$$

It is evident that this discrete-state equation is also a recursion formula for computing the state, give the initial condition and the control input.

• The order and form of the equations is independent of choice of state variables. In many systems there are many possible choices of state variables.

• Consistent with the concept of state, it is clear from these equations that the evolution of the state vector is completely determined by the state equations, the initial conditions, and the control input $\mathbf{u}(t)$.

• The control signal $\mathbf{u}(t)$ can be synthesized using mathematical tools and analytical representation of the performance criterion.

Note that the state equations allow the time variable t to proceed in either the positive or the negative direction. Physically, this is equivalent to saying that the dynamical system may have memory whose contents can be erased without affecting the system's existence. This property is typical of mechanical, electrical, and other engineering systems, whereas living and "soft" systems cannot be driven to past states without destruction.

It should be evident that the state space approach can be used efficiently to model a wide range of dynamical processes, including robotic systems. On the other hand, it should also be clear that by its very definition, the state space model may not be suitable for representing sensor-driven dynamic processes, whose analytical forms are not well known. Functional motions in biological systems rely heavily on sensor-driven control mechanisms such as reflexes, pattern generators, and skill-based motor acts. Advanced robotic systems, equipped with tactile and vision sensors, may therefore require more complex control structures consisting of state space models, artificial-intelligence-based models, and neural networks.

4.5 Nonlinear Robot Control

Most of the preceding sections of this chapter have concerned *linear control systems*, that is, systems in which the controlled plant can be described by linear differential equations. (As we have seen, such systems can also be described in the frequency domain by Laplace transforms.) A linear system is one that satisfies the *superposition principle*, which states that if the response to an input $x_1(t)$ is $y_1(t)$, and the response to an input $x_2(t)$ is $y_2(t)$, then the response to an input $ax_1(t) + bx_2(t)$ will be $ay_1(t) + by_2(t)$, where a and b are constants.

For better or for worse, however, nature is not linear. For example, in mechanical system, at low levels of force, a system may not move at all as a result of sticking friction. At high levels of force, an increase in force may not, because of saturation, result in an increase in the velocity of a system. This is certainly true of the force exerted by human muscles. Similarly, an increase in light intensity beyond a certain point may not increase the response of visual receptors. A decrease in stimulus may not produce the exact reverse response of an increase of stimulus if a system has hysteresis. These effects are examples of *nonlinearities*.

In robotics, nonlinear effects show up in other ways as well. Consider a robot manipulator. Clearly, the robot's control system will behave differently if the robot arm is moved in a horizontal plane than if it moves up (against gravity) or down (with gravity). A robot aircraft will face different air densities as it moves up from sea level to altitudes at which the air is thinner. Similarly, a robot submersible will face increasing pressures as it descends. We can then ask: If the system to be controlled is nonlinear, then can we use linear controllers, like the PID controllers discussed previously?

The answer to that question is "sometimes yes, sometimes no." It depends partly on the nature of the particular nonlinearity. This section considers several approaches to control of nonlinear systems.

4.5.1 Linearization

If real-world systems are not linear, how do we justify using linear methods for control? The justification arises from the fact that many systems are *approximately linear*, at least over a portion of their operating range. Consider an automobile speed control system. Over some range of speeds, there will be an approximate linear relation between the movement of the accelerator pedal and the vehicle velocity. This will not be true at very low speeds, and certainly not at very high speeds that approach the upper limit of velocity. The situation is more complex yet, because wind resistance is proportional to the square of the velocity. In other words, doubling the velocity will result in a fourfold increase in wind resistance. Even so, over some range of speeds, the system can be approximated by a linear relation between control input and output velocity. Note that the approximation will be different around 20 mph than around 40 mph. Nevertheless, over a range of speeds, linear tools can be used successfully. The key concept is that, in general, linearity is approximate, and the approximation is *local*, in the neighborhood of some operating point.

As an example, consider the equation for an ordinary pendulum with no damping:

$$\frac{d^2\theta}{dt^2} + \frac{mgl}{I} \sin\theta = 0. \tag{4.29}$$

In this equation θ is the angle of the pendulum with the local vertical; m, g, l, and I are constants representing the mass, gravity, length and inertia, respectively, of the pendulum. If the pendulum is displaced to an initial angle θ_0, it will oscillate with constant amplitude. If this equation were used to represent the leg of a walking machine, we would have to add a damping term and a control input on the right-hand side. Note that this equation is nonlinear, so that we cannot apply linear-analysis tools to it.

To linearize the equation, we recall that the nonlinear term $\sin\theta$ can be represented by a Taylor's series expansion:

$$\sin\theta = \theta - \frac{\theta^3}{3!} + \frac{\theta^5}{5!} - \cdots \tag{4.30}$$

For small values of the angle, we can approximate $\sin\theta$ with the first term in the series. Then, if we let $mgl/I = c$, equation (4.29) becomes

$$\ddot{\theta} + c\theta = 0, \tag{4.31}$$

which is a linear equation that can be solved using Laplace transforms or other methods. The quality of the linearization can be assessed by comparing the sine wave with the straight line approximation used in the foregoing, as in figure 4.7. Note that the quality of the approximation is reasonable near zero, and our goal was to study the pendulum for small oscillations around zero. (The approximation would not, however, be very good around $\theta = \pi/2$.)

In general, if we have a nonlinear plant represented by equation (4.26), we can rewrite it as follows:

$$\begin{aligned} \dot{\mathbf{x}} &= \mathbf{f}(\mathbf{x}, \mathbf{u}), \\ \mathbf{y} &= \mathbf{g}(\mathbf{x}, \mathbf{u}). \end{aligned} \tag{4.32}$$

To linearize this equation around an operating point $\mathbf{x}_0, \mathbf{u}_0$, we assume that the deviations from this point (also known as perturbations) are small, and we represent the state vector, the control vector, and the output vector, respectively, by

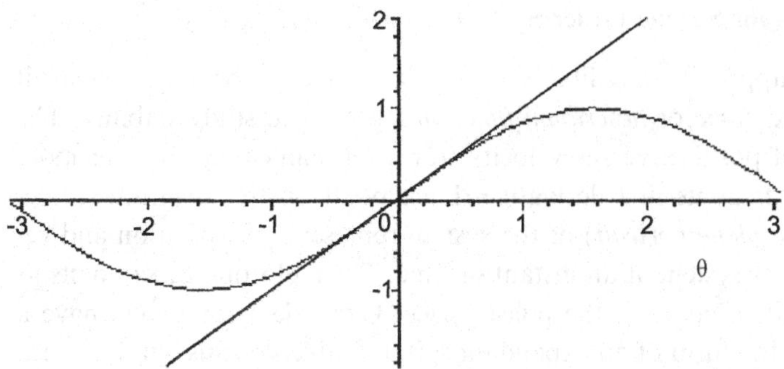

Figure 4.7
Linear approximation of a sine wave

$$\mathbf{x}(t) = \mathbf{x}_0 + \mathbf{x}'(t),$$
$$\mathbf{u}(t) = \mathbf{u}_0 + \mathbf{u}'(t), \tag{4.33}$$
$$\mathbf{y}(t) = \mathbf{y}_0 + \mathbf{y}'(t).$$

Equation (4.33) makes the assumption that the perturbation quantities, denoted by the prime symbol, are small. Hence, we can expand the functions $\mathbf{f}(\mathbf{x}, \mathbf{u})$ and $\mathbf{g}(\mathbf{x}, \mathbf{u})$ in equation (4.32) in Taylor's series, as we did with the sine function for the pendulum, and keep only the first-order terms, as follows:

$$\dot{\mathbf{x}}' = \left.\frac{\partial \mathbf{f}}{\partial \mathbf{x}}\right|_{\mathbf{x}_0, \mathbf{u}_0} \mathbf{x}' + \left.\frac{\partial \mathbf{f}}{\partial \mathbf{u}}\right|_{\mathbf{x}_0, \mathbf{u}_0} \mathbf{u}',$$
$$\mathbf{y}' = \left.\frac{\partial \mathbf{g}}{\partial \mathbf{x}}\right|_{\mathbf{x}_0, \mathbf{u}_0} \mathbf{x}' + \left.\frac{\partial \mathbf{g}}{\partial \mathbf{u}}\right|_{\mathbf{x}_0, \mathbf{u}_0} \mathbf{u}'. \tag{4.34}$$

These equations can be written in the form

$$\dot{\mathbf{x}}' = \mathbf{A}\mathbf{x}' + \mathbf{B}\mathbf{u}',$$
$$\mathbf{y}' = \mathbf{C}\mathbf{x}' + \mathbf{D}\mathbf{u}'. \tag{4.35}$$

These are linear equations, like those in equation (4.27). It should be noted that the matrix coefficients in (4.35) contain derivatives of the original nonlinear functions, evaluated at the linearization point $\mathbf{x}_0, \mathbf{u}_0$.

In practice some robot designers simply ignore the nonlinearities in the controlled system and adjust controller parameters by trial and error until they get acceptable performance. This approach is not recommended, since the controller gains it yields may not be best for *any* operating point.

4.5.2 Special Methods for Second-Order Systems

Many systems behave approximately like second-order systems. For these systems it is common to use *phase plane* or *describing-function* methods to study stability. The phase plane is a plot of position versus velocity (for a mechanical system), or more generally, the output versus its first derivative. Each point on this diagram (sometimes known as a *phase plane portrait*) of the system represents the position and velocity of the second-order system at an instant of time. Thus, plotting these points as a function of time yields a curve in the phase space. Unstable systems will have a phase plane portrait in the form of an expanding spiral. Stable systems will have trajectories that spiral toward the origin or another equilibrium point. Certain nonlinearities give rise to particular phase plane trajectories (D'Azzo and Houpis 1995; Kuo 1995).

As an example, consider the van der Pol equation discussed in chapter 2, whose phase plane portrait is given in figure 2.11.

Describing-function methods are used to study the stability of nonlinear control systems in the frequency domain. Interested readers are referred to the literature (e.g., Kuo 1995).

The most common way of designing controllers for nonlinear systems is to select the form of a controller and then use computer simulation to allow rapid searching of parameter spaces for values that yield satisfactory performance. We caution again that nonlinear systems do not behave like linear ones. Thus, a nonlinear system may behave much like a linear system for a particular magnitude or frequency spectrum of inputs, but its performance may be drastically different from that of a linear system with a smaller or larger input, or one with a different spectrum.

We have only, in this section, scratched the surface of a large and complex subject. Detailed analysis of nonlinear control systems, including issues of stability, the effects of disturbances, feedback design, and so on, is performed in a number of books. A strong background in linear systems theory is recommended for those pursuing further reading in this area, as the books are highly mathematical (e.g., Isidori 1997, 1999; Khalil 2001).

4.6 Adaptive Control and Other Approaches

Control system design is a large and complex field. The academic discipline includes several approaches to *optimal control*, that is, the design of controllers that perform in the best manner possible with respect to given performance criteria. Most of these methods are particularly suited to linear systems and use quadratic criteria.

Several ideas from artificial intelligence have been used as a basis for design of robot controllers, including knowledge-based systems, neural networks, and evolutionary algorithms. Basically, controllers employing these approaches use some form of learning to accumulate the knowledge required for control. Hence, these methods are reviewed in the context of robot learning in chapter 6.

This section concentrates on *adaptive control.* As indicated previously, one of the problems involved in designing controllers for certain robots is that they may be operating in widely varying environments, so that a single control design may not be satisfactory. Let us say that we have designed a PID controller for a robot aircraft that is quite satisfactory at sea level but leads to very poor, highly oscillatory responses at higher altitudes. An obvious solution to this dilemma is to design several controllers, each appropriate to a given altitude range, and then switch among them using an altimeter to provide the information needed to determine which controller to use. For robot manipulators, it may be necessary to vary the gain as a function of the load being carried, and even based on the direction of movement with respect to the gravity vector. In this way, each controller, although still *suboptimal,* is designed for a particular operating range. The basic structure of such a controller is shown in figure 4.8.

In applications in which the movements of the robot and the environment in which it will operate are known in advance, it is possible to program controller gains as a function of time. Such controllers are known as *programmed gain controllers.*

The gain-switching approach is a way of compensating for changes in robot dynamics as a result of changes in the robot's environment or load. This suggests that if it were possible to track the changes in the robot dynamics, one could then introduce compensatory changes in the controller. Such an approach, known as the

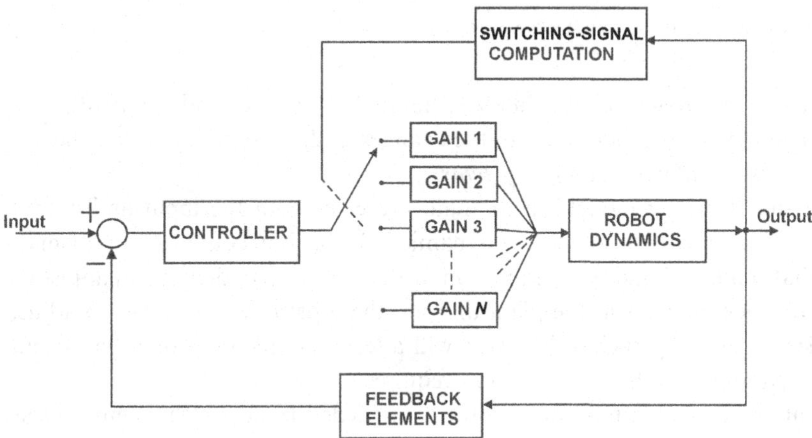

Figure 4.8
Variable-gain controller

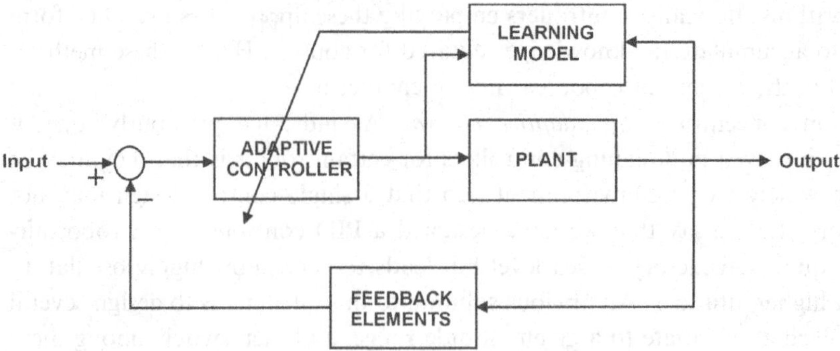

Figure 4.9
Adaptive controller with a learning model

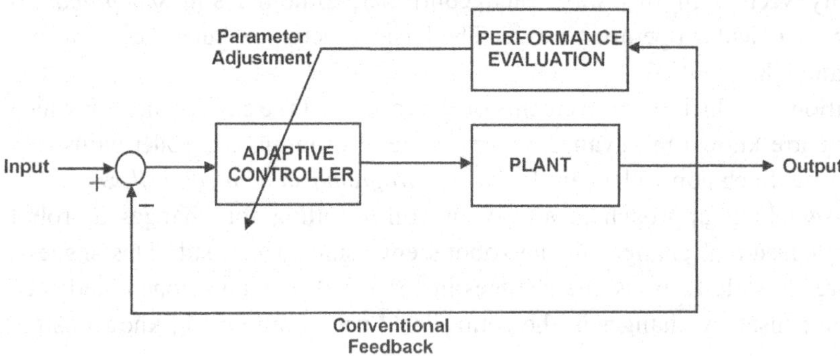

Figure 4.10
Adaptive controller based on performance evaluation

"learning model" approach, is illustrated in figure 4.9 (Mendel and Fu 1970). It is one way of making a controller adapt to the changes in the system it controls; hence, it is one of a class of *adaptive control systems.*

Note that the "learning model" in the figure receives both the input and output signals from the plant (the robot system dynamics). If this box contains a mathematical model that represents the system or plant at the initial time, then it can adjust its parameters to track changes in the plant and use these parameter changes to adjust the controller. Frequently such adjustment will affect only the controller gains, but all controller parameters can be adjusted as required.

An alternate way of obtaining the information needed to adjust the controller as the environment in which the system is operating changes is based on evaluation of a performance measure, as shown in figure 4.10. The idea here is that changes in the environment will affect the way the system performs, as indicated by an appropriate

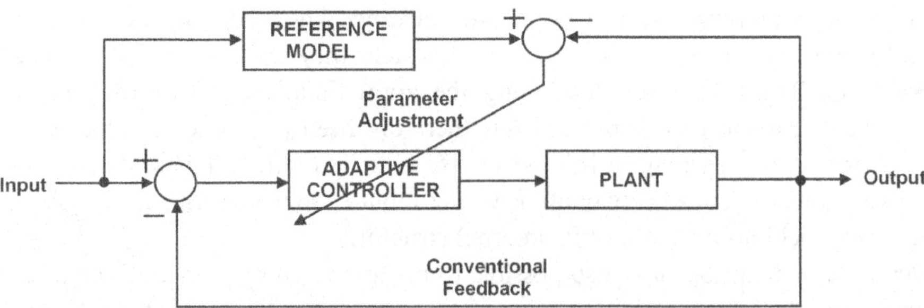

Figure 4.11
Reference model adaptive control

performance measure. It may be argued that this is in fact the proper way to effect adaptation, since if the environment does not affect the system performance, why adjust the controller? Such performance measures as mean square deviation from a steady-state value or response to a sudden input are commonly used in this context. A system with a controller incorporating this type of adaptation does not require an accurate model of the plant.

Yet a third method of adaptation is shown in figure 4.11. This approach is based on the idea that we often expect our systems to behave in a certain way, and deviations from such behaviors are unacceptable. For example, we may wish our robot submarine to behave as if it were a second-order system with a certain damping ratio. The submarine may in actuality be a twentieth-order system, but in response to step inputs, it could behave approximately as a second-order system. We term a mathematical model that produces desirable behavior of this type the *reference model* for the system. Note that the same inputs are applied to the closed-loop system as to the reference model. If the two responses differ, the information is used to adjust the controller.

4.7 Model-Free Approaches to Control

The control methods we have presented in the foregoing depend on the availability of mathematical models of the system being controlled. As we saw earlier, for simple mobile robots the structure of such models can be constructed without much difficulty. It may necessary to use parameter estimation methods to find the values of the parameters, but at least in principle, the required mathematical models can be developed. However, as systems become more complex, such as the underwater robots or flying robots discussed in chapter 7, even the structure may be too complex to describe in simple mathematical terms. When this is the case, it may be appropriate to use a method of representation that can approximate the input-output behavior of an arbitrary system without any advance knowledge of its structure. Artificial neural networks provide such an approach; they are considered *universal approximators.* The process of developing the approximation is basically one of *learning*; the discussion of this method has therefore been postponed to chapter 6, which discusses robot learning. It is important to note that although neural networks may provide an excellent approximation to the input-output behavior of a system, they generally yield no insight into its internal structure.

Under certain conditions it is also possible to use a method of simulated evolution to approximate the input-output behavior of a system. Methods of this type, known as genetic or evolutionary algorithms, are also discussed in chapter 6.

4.8 Uncertainty in Control System Design

The presentation in this chapter has tacitly assumed that all the signals used for control are measurable and noise free. Of course, this is not the case. The measurement of wheel position and velocity of a mobile robot will be noisy. The robot may be moving over rough terrain, so that the sensor outputs incorporate randomness as well as the desired signal. There may be environmental disturbances, say, in the vicinity of high-voltage power lines. The sensors themselves may be noisy and produce an output including some randomness, even in the absence of external disturbances. This type of uncertainty gives rise to additional terms in the model equations. Thus, the linear model of equation (4.27) is replaced with

$$\frac{d\mathbf{x}(t)}{dt} = \mathbf{A}\mathbf{x}(t) + \mathbf{B}\mathbf{u}(t) + \mathbf{w}(t),$$

$$\mathbf{x}(t_0) = \mathbf{x}_0, \tag{4.36}$$

$$\mathbf{y}(t) = \mathbf{C}\mathbf{x}(t) + \mathbf{D}\mathbf{u}(t) + \mathbf{v}(t),$$

where the term $\mathbf{v}(t)$ represents additive noise in the system output, and the term $\mathbf{w}(t)$, known as *system noise,* is used to represent our lack of complete knowledge of the system. For example, we may have ignored some nonlinearities in writing the equations for the system model; they are now included in the system noise term. Both noise terms are described in statistical terms. If they are scalar, we can describe them in terms of their mean and variance; if they are vector quantities as in (4.36), we describe them in terms of their expected values and covariance matrices.

The presence of disturbances changes the representation of a system from a deterministic one to a random one. The simplest approach to disturbances is to use low-pass filters to reduce the effect of the noise. This is particularly effective if the noise contains significantly higher frequencies than the signal. More complex methods of estimating the signal values to be used in control make use of such methods as Wiener or Kalman filtering. We discuss Kalman filters briefly in chapter 14 in connection with localization and navigation, in which the robot needs to determine its position in the world using noisy data.

Another approach to dealing with uncertainty arises from linguistic descriptions of systems. For example, in describing a particular mobile robot, an engineer may say things like: "This robot moves faster when the outside temperature is high, and the humidity is low. To keep the speed constant, we need to reduce the controller output a little under these conditions." Note that linguistic statements of this sort are highly imprecise. Statements involving terms like "high," "low," "faster," or "a little" are known as *fuzzy statements*. One can attempt to quantify them exactly, say, by measuring the increase in speed for every 5° of temperature change, or one might decide to design the controller using *fuzzy logic* (Kosko 1994; Wang 1994).

A fuzzy controller usually consists of a set of rules and methods for quantifying the inferences made with these rules. A typical set of rules for motion control of a mobile robot moving on rough terrain might look something like the following (adapted from Niku 2001, 324):

If the slope is Very-Steep Up and the terrain is Very-Rough, the speed should be Very-Slow;

If the slope is Steep-Up and the terrain is Rough, the speed should be Slow;

If the slope is Level and the terrain is Moderate, the speed should be Fast.

The cited reference suggests some twenty rules to describe this controller. Ultimately, once decisions are made based on the fuzzy rules, a deterministic (or *crisp*) set of values needs to be selected for the actuators; this is accomplished by a process known as *defuzzification.*

The development of fuzzy logic is due to Zadeh (1965), who is known as the "father of fuzzy logic," and fuzzy logic has achieved increasing importance in robot control in recent years. The use of fuzzy rules in conjunction with neural networks leads to so-called fuzzy controllers, which have been applied successfully to the control of mobile robots (Godjevac 1997).

4.9 Biologically Inspired Control: Basic Principles

The control of robots by means of the methods described in the foregoing is based on the availability of a mathematical model of the plant. Further, most of the methods assume that the plant, that is, the robot mechanism, is linear. This approach is consistent with developments in control since its beginnings in the 1940s and lead to a numerical control algorithm, generally implemented with microprocessors. For mobile robots, these control algorithms are incorporated in chips and carried on board.

Industrial manipulators may use dedicated processors, located on the factory floor. Since multiple-linkage structures are highly nonlinear, the solution of the model equations for such structures would require a major computing effort, and current robot controllers in industry instead generally use linearized models and relatively simple control laws (such as PID control), which can be implemented on available microcomputers. A similar situation exists with respect to the control of multifingered robot hands, where the computational task is even more difficult to carry out in real time, since each finger is in effect a multiple-degree-of-freedom manipulator.

These remarks are not intended to imply that the control equations of manipulators or mobile robots cannot be solved in real time. Quite the contrary, faster computers and novel computational architectures have made real-time control of robots relatively easy. The nature of the computing problem has been indicated here only to provide a basis for comparison with the control of functional motions in animals and humans. (By the phrase "functional motion," we mean a motion that is intentional and purposeful, aimed at a specific function, such as walking, stepping over a rock, reaching for an object, or grasping it with an appropriate hand posture.) In living systems, the equations of motion are obviously not represented explicitly in the form of mathematical equations that must be solved in the brain before a motion can be executed. Repetitive motions, such as those involved in walking on level terrain or flying in smooth air, may involve the use of "central pattern generators" that trigger a sequence of muscle contractions to produce the motion. Purposeful single motions, such as grasping a given object or stepping over an obstacle, may be based on simplified models of the world, large stored knowledge bases and learned mappings that are used to translate world coordinates into appropriate motor neuron commands. Such mappings may be stored in the form of connection patterns in neural networks. However, they are not solved as equations. Human beings do not solve the differential equations of motion of their legs before taking a step, or of finger motion before grasping an object.

Clearly, there are both similarities and fundamental differences between control in living systems and in human-made systems. Some of these differences were discussed in chapter 2. For a number of years the author and his colleagues have been concerned with developing a new class of biologically inspired robot controllers. To distinguish these controllers from those based on standard numerical algorithms, we term such biologically inspired controllers *nonalgorithmic*. They have been used as a basis for control of artificial limbs and of grasping in robot hands (Bekey and Tomović 1986; Bekey, Tomović, and Zeljković 1990; Bekey et al. 1993).

As described in chapter 2, the control of leg, arm, and hand movements in humans is extremely complex. Hence, the design of controllers for robot hands, arms, and legs must be based on a number of simplifying assumptions. We have found that the following principles are applicable to the control of human extremities:

1. *Hierarchical organization* Functional motions of human extremities are planned and executed at several levels. At the highest level are decisions concerned with task or mission planning and organization. At lower levels the execution of a task (such as walking on level ground or grasping a cylindrical object) is based on accumulated knowledge, experience, training, and sensory feedback.

2. *Simplification of the problem domain* We postulate that humans simplify sensory input to apply learned strategies from one situation to related ones. For example, hand approach trajectories for grasping of objects are largely independent of the details of object shape. Such simplification makes it possible to deal with the world without saturating the cognitive and processing structures of the brain with details irrelevant to the task at hand.

3. *Clustering* Sensory inputs appear to be clustered into relatively small numbers of equivalence classes. Thus, we believe that the relatively small number of grasp postures used by humans (e.g., power grasp, hook grasp, pulp pinch) is due to the clustering of the infinity of object shapes in the world into a small number of geometric primitives, such as cylinder, parallelepiped, and sphere. Clustering may be equivalent to the idea that geometric shapes are represented in the brain by *Geons* (Biederman 1987).

4. *Reflex control* Reflex control refers to motions that are triggered by a sensory input and then proceed automatically until an additional sensory input is received. This principle can be used to explain the closure of all the fingers until an object of arbitrary shape is enclosed, or the forward motion of the leg during the swing phase of a stride to full extension. Such reflexes appear at the lowest level of control, as discussed in chapters 2 and 5.

5. *Heuristics of motion control* We also believe that the control of functional motions in humans is frequently based on heuristics derived from experience. Such heuristics are used to select the grasping point on a tool or the foot orientation when stepping over a log.

6. *Minimum-effort motion* It appears that a number of motions are carried out in such a way as to minimize the effort required for control. This principle may account for the way in which stored energy is used in leg motion or for the trajectory of the arm during a reaching task.

These six principles have been used as a basis for the control philosophy of the Belgrade-USC robot hand, a highly autonomous, shape-adaptive grasping device described in detail in chapter 11. We do not imply that these principles constitute a complete theory of robot hand control. Rather, they are presented here as an alternative to conventional approaches to control that may be applicable to robot control as well as the control of functional motions in humans.

5 Software Architectures for Autonomous Robots

Summary

The organization or architecture of the software in an autonomous robot has a major influence on the specific approaches to the robot's control. This chapter summarizes several major architectures used in mobile robots. First, we review traditional vertically oriented hierarchical software structures (including NASREM). We then review three-level architectures, in which speed control, attitude control, and the like are in the lowest level of the hierarchy, with higher levels being reserved for execution and planning functions. We then look at behavior-based architectures, in which sensing and action are tightly coupled, with units of "behavior" being horizontally rather than vertically organized (but with a possible planning layer above the behaviors). Finally, we examine hybrid architectures that blend the reactivity of behavior-based systems with deliberative components. Two case studies are presented to illustrate some of the concepts.

5.1 What Is a Robot Architecture?

In this book, "robot architecture" is used to refer to the software organization of a robotic system. Thus, architecture implies *software* architecture. In computer science, on the other hand, the word usually refers to the *design* process used in the development of the software. Thus, a standard textbook on computer architecture (Stone 1980) defines it as follows:

> Computer architecture is the discipline devoted to the design of highly specific and individual computers from a collection of common building blocks.

Arkin (1998) uses a similar definition for robotic architecture, again emphasizing that it is a design discipline. Our view is that the architecture represents the structure of the software, the way in which the robot processes sensory inputs, performs cognitive functions, and provides signals to output actuators, independently of how it was designed. In this sense we follow the definition given in a current textbook on artificial intelligence (Russell and Norvig 2002, 786):

The architecture of a robot defines how the job of generating actions from percepts is organized.

Thus, the definition of architecture in this book concerns the practical structure of a robot's software. It may begin with an abstract design, but its goal is to define the way in which sensing, reasoning, and action are represented, organized, and interconnected.

5.2 Where Does Control Fit into Robot Software?

The word "control" has several different meanings in robotics. To some designers it refers exclusively to ensuring that a robot's movement remains stable and performs according to control system design criteria. This means that the motor and/or wheel speeds are maintained as closely as possible to the desired set points, that the orientation of the vehicle stays close to the desired path, or that the attitude of the vehicle remains horizontal in spite of irregularities in the surface over which it travels. Other criteria may consider steady-state error, recovery from disturbances in minimum time, and the like. This aspect of control is frequently denoted as "low-level control." The design of control subsystems follows the principles introduced in chapter 4; for example, PID controllers are commonly used. Other robot designers use the word "control" to indicate the ability of the robot to follow directions toward a goal. From this point of view, path planning and a robot's ability to follow the desired path, as well as the ability of a robot to follow verbal commands, are part of "high-level control."

In the design of aircraft or missiles one frequently separates these two aspects of control by referring to (a) control *of* the trajectory and (b) control *about* the trajectory, which correspond to high-level and low-level control, respectively. Control of the trajectory is more commonly known as *guidance*; its function is to maintain the desired trajectory to the goal. Sensory inputs needed to accomplish this goal may come from global positioning satellites (GPS) or vision, perhaps augmented by internal maps of the terrain over which the aircraft or missile must pass. While the vehicle is flying toward the intended goal or target, it is essential that it maintain a stable orientation and not oscillate about any of its axes. This is control *about* the trajectory; it is low-level control. Inputs are obtained from gyroscopes or inertial platforms or integrating accelerometers; deviations from the desired attitude result in immediate adjustment of control surfaces. In robotics the word "control" is used for both aspects. Some robot designers attempt to distinguish low-level control by calling it the servomechanism (or simply servo) level. In traditional control theory (James, Nichols, and Phillips 1947), the word "servomechanism" referred to feedback control systems in which the reference input is zero. The word is also used (e.g., in the *Microsoft Press Computer Dictionary*) to denote feedback control systems in which the final output is mechanical movement of some sort. Although this definition is applicable to most low-level control systems in robotics, we consider it to be an unfortunate use of the word, since it separates control of wheel position from control of voltage or light intensity; all these control systems could be described by the same mathematical expressions.

As we show in later sections of this chapter, low-level control is frequently reactive; that is, there is very close coupling of sensing and action. As soon as a sensor detects a change in the attitude of a vehicle (say that it begins to roll), signals are sent to the actuators to perform a corrective movement. On the other hand, high-level control frequently requires reasoning, so sensory information must be processed by a deliberative processor that in turn may send signals to actuators. Thus, Matarić (2002) defines robot control as follows:

> Robot control is the process of taking information about the environment, through the robot's sensors, processing it as necessary in order to make decisions about how to act, and then executing those actions in the environment.

It should be clear from the preceding discussion that the software architecture of a robot should encompass the entire range of control issues, from low-level control to high-level planning, decision making, and reasoning required to achieve a given goal. Furthermore, as Matarić (1992a) notes,

> an architecture provides a principled way of organizing a control system. However, in addition to providing structure, it imposes constraints on the way the control problem can be solved.

5.3 A Brief History

Perhaps the earliest robot architecture was that used by W. Grey Walter, a physiologist, in the design of his "tortoise" in 1953 (figure 5.1). This remarkable little machine had many of the features of contemporary robots: sensors (photocells for seeking light and bumpers for obstacle detection), a motor drive, and built-in behaviors that enabled it to seek (or avoid) light sources, wander, avoid obstacles, and recharge its battery. Although the behaviors were prioritized (e.g., avoiding obstacles had higher priority than seeking light), it lacked any explicit control system design. The friction in the motors and relatively slow response resulting from the use of relays for switching provided sufficient damping to prevent unstable oscillations from developing. Since the sensors were directly coupled with the actuators (through

Figure 5.1
Walter's tortoise (photograph courtesy of Owen Holland)

two vacuum tubes Walter referred to as "nerve cells"), the architecture was basically behavior-based, as described in section 5.5.[6]

A quite different architecture characterized a robot constructed at the Stanford Research Institute (SRI) in 1969. This robot, Shakey (figure 5.2), was also equipped with a vision system (significantly more complex than Walter's photocell) and bump sensors. It was capable of steering by means of differential control of its two drive motors, and it could control tilt and focus of its video camera. However, the sensor outputs were not directly coupled with the drive motors. Rather, they formed inputs to a "thinking" layer that used an artificial-intelligence planner known as the Stanford Research Institute Problem Solver (STRIPS) (Fikes and Nilsson 1971). The planner used pictures obtained from the robot's vision system to construct a map of the robot's surroundings (an office area populated with large colored blocks) and a goal position entered by the user to compute a navigation plan, which was then executed. Thus, the operation of the robot consisted of the "sense-think-act" paradigm (also referred to as "sense-plan-act") illustrated in figure 5.3. The robot was known as Shakey because the television camera mount was not rigid and the camera shook as the robot moved.

5.4 Hierarchical and Deliberative Architectures

The SRI robot Shakey was the prototype of the deliberative approach to robot architectures. This approach is characterized by an architecture with a hierarchical structure in which each layer provides subgoals (or explicit instructions) to the layer

6. Owen Holland of the University of the West of England discovered the original 1951 version of Walter's tortoise, replaced a capacitor and found it in perfect working order. It is now on exhibit at the London Science Museum.

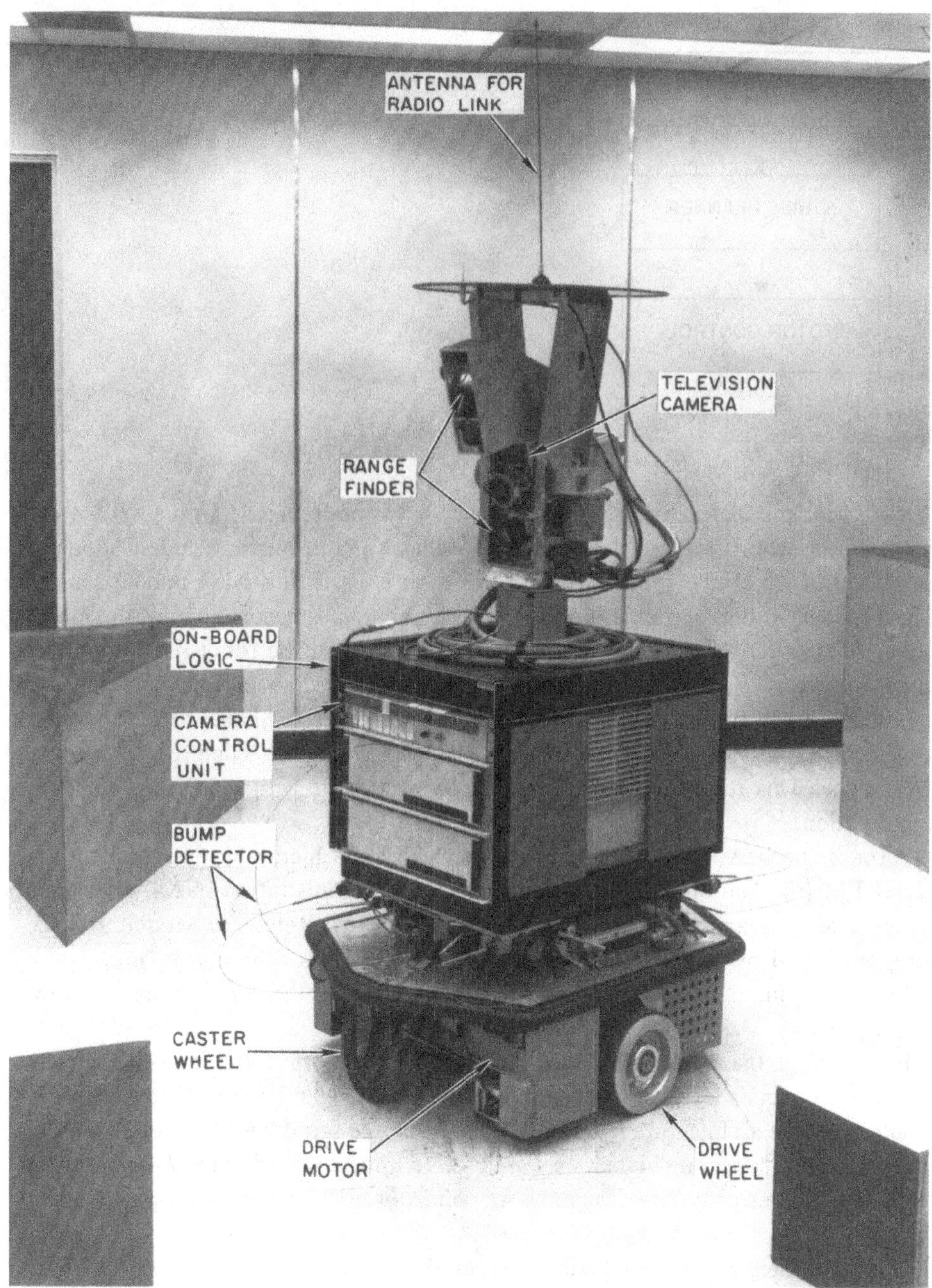

Figure 5.2
Shakey, from the Stanford Research Institute (photograph courtesy of SRI International)

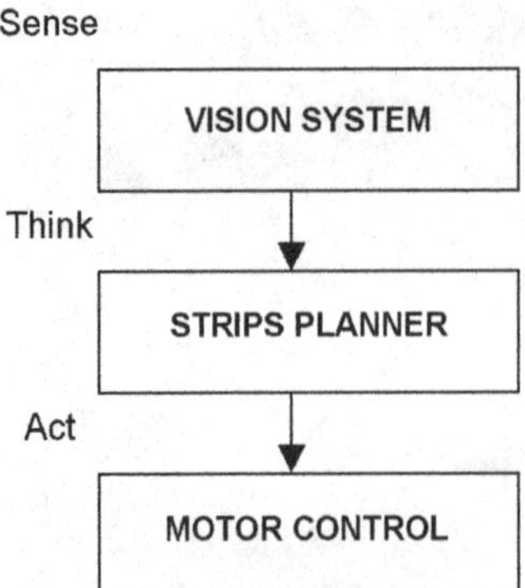

Figure 5.3
Sense-think-act architecture of Shakey

below. (In some hierarchical designs, a layer may communicate with several lower layers in the hierarchy.) Furthermore, hierarchical architectures include a model of the world in which the robot moves. Perception is used to modify and update the world model, so that action is produced by planning and reasoning from the model, rather than directly from the perception. In other words, the robot senses and then thinks before it can move. Because of the time required to perform the planning operations, early robots based on this architecture (such as Shakey) were restricted to stationary worlds in which nothing changed while the cognitive procedures took place. This restriction has been relaxed somewhat with increasing speeds of computation.

Perhaps the best known (and most extreme) of the hierarchical architectures is NASREM (the National Aeronautics and Space Administration [NASA]/National Institute of Standards and Technology [NIST] Standard Reference Model) architecture, developed by James Albus at the National Bureau of Standards (now NIST) (Albus, McCain, and Lumia 1987). Six levels characterize this architecture, as shown in figure 5.4.

Each layer in the architecture is assigned certain functions, ranging from control at the lowest or servo level to strategic planning at the highest level. Note that sensor inputs enter the architecture from the lowest level, the servo level. Processed sensory inputs at each succeeding higher level are used to update a world model. Thus, in this architecture, control system design, as we defined it in chapter 4, takes place only at the servo level.

Global, strategic planning takes place at the highest level. Since such planning concerns long-range goals, the time required to complete it is not a major concern. Visualize, for example, the strategic-planning process of a worldwide agricultural enterprise, which may include making decisions on multiyear planting in different time

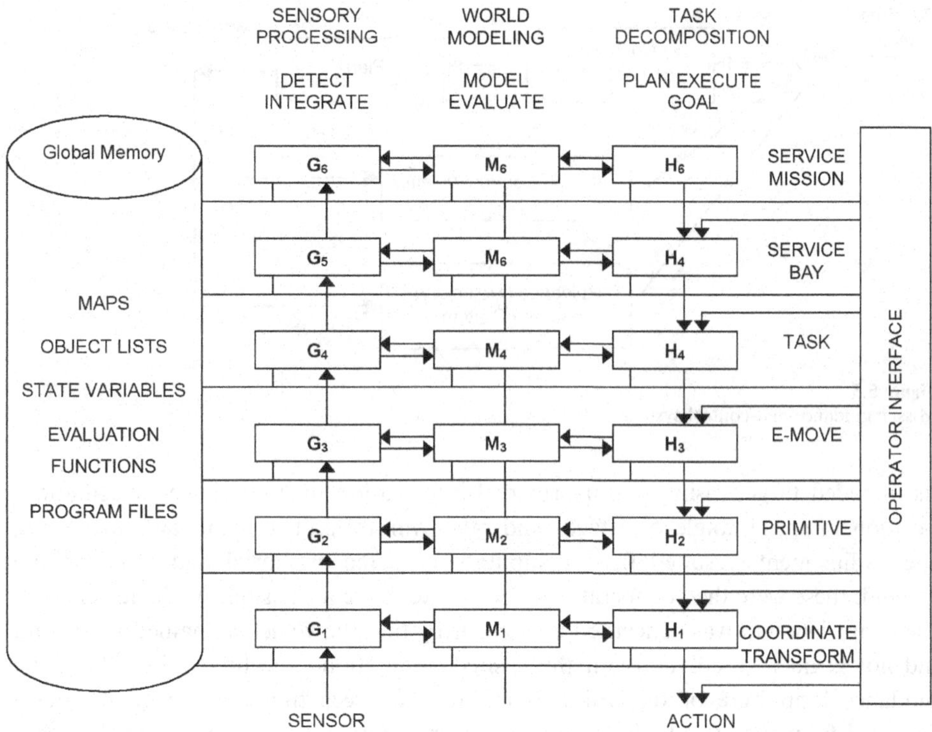

Figure 5.4
NASREM architecture (illustration courtesy of James Albus)

zones. Further down the hierarchy, where local short-term planning is involved, it may be necessary to allocate harvesting equipment to different fields, while ensuring that all crops are harvested when ready. Clearly, this planning task has a much more constrained time horizon than the strategic plan. At the lowest level is the design of the speed and force control systems for the individual tractors or harvesting machines.

When planning and control are viewed from a global perspective, as in the preceding paragraph, it is evident that a hierarchical structure of the type in figure 5.4 is both appropriate and necessary. The situation is not as clear when a single mobile robot is concerned, as we discuss later. The author was involved in the early design of guidance and control systems for ballistic missiles. In these systems, the hierarchy included only two levels, as discussed briefly in section 5.2 and illustrated below in figure 5.5.

The upper or guidance level of the system in the figure was concerned with ensuring that the missile's trajectory was adjusted as required to enable the missile to reach

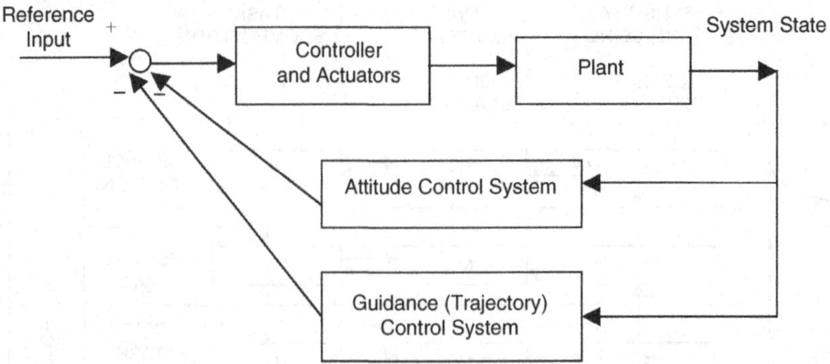

Figure 5.5
Missile guidance-and-control system

its intended target, using sensor-provided information on the missile's orientation, position in space, angle of attack, and other variables. The commands producing such adjustments resulted from computations, some on board and some on the ground; these were the "deliberations" needed to make a decision. At the lower level, the control system was concerned with ensuring that the vehicle remained stable and did not begin to oscillate about the desired orientation. The time constants of the guidance loop were of the order of minutes, whereas those of the control loop involved frequencies as high as several hundred Hertz. It is interesting to note that flight experiments demonstrated that in fact the coupling between the two layers was essentially negligible. Thus, control system design flaws that produced some low-level oscillations had a negligible effect on the missile's trajectory. Similarly, changes in the commanded orientation of the missile had a negligible effect on control system behavior. The two subsystems could be designed separately.

The preceding comments are not intended to diminish the importance of deliberative, hierarchical architectures, which are clearly essential in the control of many systems, both large and small. Rather, they are intended to indicate that there is also a place for systems in which there is a close coupling between sensing and action, with a minimum of deliberation, as in the control layer of figure 5.5. We shall study such systems in the following section. The theory of hierarchical systems has been greatly amplified by Meystel (1986).

5.5 Reactive and Behavior-Based Architectures

By contrast with the architectures described in the previous section, we now examine a different approach originally proposed by Brooks (1986). Architectures resulting from this approach are characterized by a close coupling of perception and action,

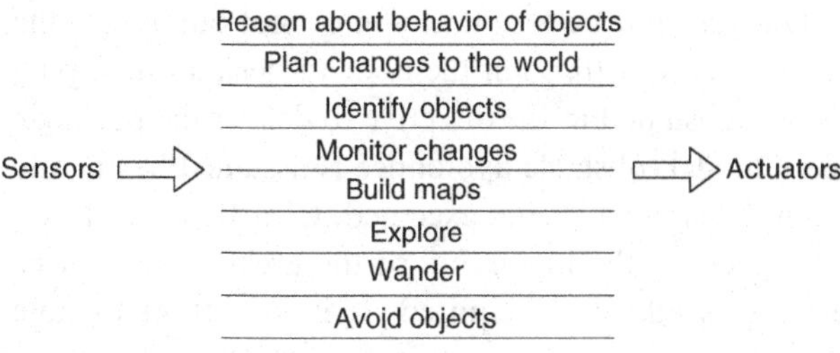

Figure 5.6
Direct coupling of perception and action

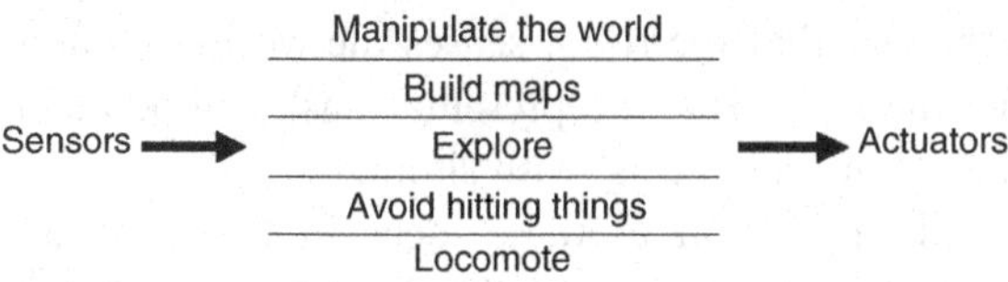

Figure 5.7
Reactive, behavior-based simulation model

without an intermediate cognitive layer. Figure 5.6, taken from an early drawing of Brooks, illustrates this principle by indicating that "cognition" is in the eye of the beholder, but that a system that couples action directly with sensing can display behaviors that appear to be "intelligent."

Brooks (1986), in a now-classic paper, advocated a layered architecture for a mobile robot. His approach was a radical departure from the sense-think-act paradigm used in Shakey (and illustrated in figure 5.3). By contrast, Brooks argued that building complex world models and using symbolic reasoning were hindrances to successful development of mobile robots, and that the environment in which a robot moved was in fact the only "model" the robot needed to be able to act. Rather than using the "vertical" structure of the sense-think-act model, he suggested a horizontal decomposition in terms of behaviors, as shown in figure 5.7.

In architectures based on Brooks's model, all the behaviors work in parallel, concurrently, and in general, asynchronously. In early versions of such architectures, the robot's behaviors were implemented using finite-state machines that were difficult to explain to non–computer scientists. More recent versions describe the robot's behaviors in the form of rules that state the robot's response to a given input from sensors (e.g., "In the event of a collision, turn motor to reverse and back up"). The actual physical implementation of the rules is left to the designer.

In Brooks's model, the higher, more complex behaviors *subsume* those beneath them. Thus, exploring subsumes obstacle avoidance by the robot. To illustrate this principle, assume that you give someone a package to deliver to the office across the hall. It is not necessary to say, "Oh, yes, and be sure that you do not bump into the furniture on the way. Also, be sure to go out through the door rather than bumping into the walls." Those behaviors are subsumed in the directive to deliver the package. In architectures based on Brooks's model, obstacle avoidance is assumed to operate autonomously while the navigation behavior is being executed. Clearly, lower levels have no awareness of levels above them in the hierarchy, so the architecture can be designed incrementally and new layers added as required. It is important to note that in this type of architecture, sensors provide inputs to these behaviors, but there is no specific deliberative layer. For the reasons given, this type of architecture is also known as a *subsumption architecture.* Brooks emphasized two basic aspects of behavior-based robots: They had to be *situated* in the world and be *embodied* in hardware. Situatedness emphasizes the fact that the robot senses the world and acts upon the outputs of its sensors (rather than on abstract representations); embodiment emphasizes the physical nature of the robot (in contrast with simulations).

Brooks encountered a great deal of opposition to these ideas in the artificial-intelligence (AI) community. As a result he began to call the traditional approach GOFAI ("good old-fashioned AI") and his approach "nouveau AI." The history of these ideas, as published in a series of papers over a number of years, is summarized in *Cambrian Intelligence* (Brooks 1999).

The design of subsumption-type behavior-based robots relies on a number of heuristics, as described by Matarić (1992a). A design of this type must do the following:

1. Qualitatively specify the behavior needed for the task—that is, describe the way the robot should respond to the world.

2. Decompose the robot behavior into observable, disjoint actions. These actions serve as subgoals as the difference between the robot's current state and the goal is reduced.

3. Connect these actions in terms applicable to the robot's actuators.

In addition, a subsumption architecture must carefully restrict communication between levels, to ensure proper coordination so that in fact higher, more complex levels subsume lower, simpler behaviors. For example, an early foraging study (Matarić 1994) used the following behaviors, in order of increasing complexity:

- wandering
- avoiding
- pickup
- homing

The behaviors were programmed using priority-based arbitration as the coordination mechanism. The lower levels in the hierarchy have no knowledge of the higher levels.

It should be evident that purely reactive control does not require any reasoning. Yet the behavior of a robot with a tight coupling between perception and action can appear surprisingly "intelligent." For example, the six-legged robot Genghis from the MIT Artificial Intelligence Laboratory (Brooks 1989) and the four-legged robot built in the USC Robotics Research Laboratory (Sukhatme 1997) included a touch sensor on the front of the legs. Both robots had a built-in behavior such that if the sensor contacted an obstacle, the leg on which it was mounted would be withdrawn and raised higher for another try at overcoming the obstacle. Observation of such robots gives an impression of purposefulness.

It is important to note that the introduction of reactive, behavior-based architectures by Brooks was a revolutionary change in the field. Unfortunately, purely reactive architectures did not perform satisfactorily on some complex tasks.

We do not discuss behavior-based architectures in detail in this book, since they are exhaustively treated in the outstanding book by Arkin (1998) on this subject.

5.6 Hybrid Reactive-Deliberative Architectures

The two approaches to high-level robot control discussed in the previous sections developed independently. The deliberative sense-map-plan-act approaches came earlier, with the Shakey robot being among the first of this species; reactive, subsumption architectures came later. As would be expected, it was only a matter of time before researchers began to investigate hybrid versions of the two types of architecture. Among the first was Matarić (1992b), one of Brooks's students, who added a planning layer on top of the multiple reactive layers. From this point of view, the simple reactive structure of figure 5.7 can be modified to include higher-level layers, as shown in figure 5.8.

Note that in figure 5.8 the first five layers correspond to the reactive architectures described earlier, whereas the top two layers are concerned with planning and deliberation. (Note also that the figure includes a map-building layer. As pointed out

Level	Behavior
7	Reason about the world; modify plans
6	Formulate and execute plans
5	Identify objects in the world
4	Observe changes in the world; modify behavior
3	Build a map of the environment
2	Explore the world
1	Wander about the world
0	Avoid hitting obstacles

Figure 5.8
Hybrid reactive-deliberative structure

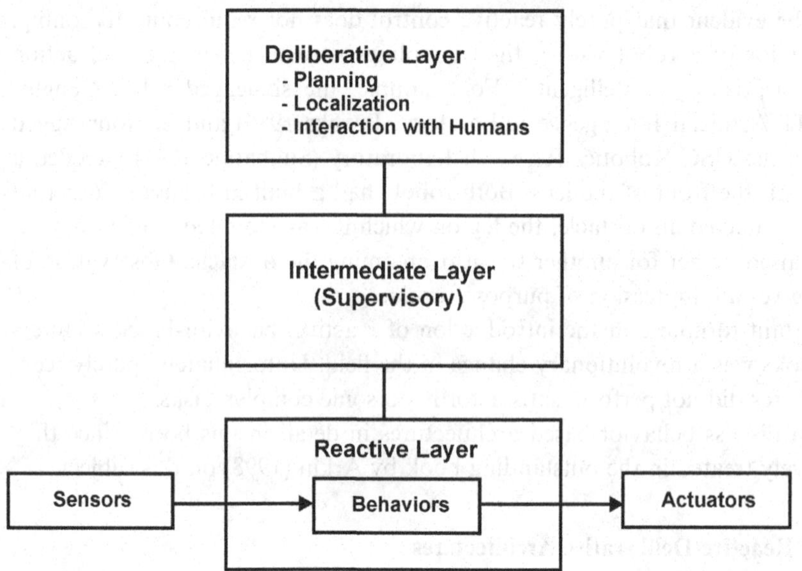

Figure 5.9
Typical three-layer hybrid reactive-deliberative architecture

previously, a number of advocates of reactive architectures dispute the need for map building, suggesting that the robot's environment is the only map it needs.) The structure of figure 5.8 is also a subsumption architecture, since each layer subsumes all lower layers, and the lower layers are assumed to have priority over the higher ones. Thus, avoiding an obstacle will take priority over executing a plan to reach some distant goal.

A typical hybrid architecture consists of three layers, as shown in figure 5.9. The bottom layer is the reactive layer discussed above, in which sensors and actuators are closely coupled. The upper layer provides the deliberative component, including interaction with humans, planning, localization ("where am I?"), and other cognitive functions. The so-called supervisory layer is intermediate between the other two. Its function will become clearer in some of the examples that follow.

A further refinement of this three-layer model involves assigning a hierarchical structure to the deliberative portion of the hierarchy. So called *hierarchical-deliberative* planning architectures (see figure 5.10) subdivide the planning component into sections depending on the length of the planning horizon, both in time and in space. Short-term planning is largely local in nature and concerns only goals in the vicinity of the robot and those reachable within a "short" time interval (where "short" depends on the particular task, the speed of the robot, and the nature of the environment). Clearly, the short-term planner must react rapidly and requires precise

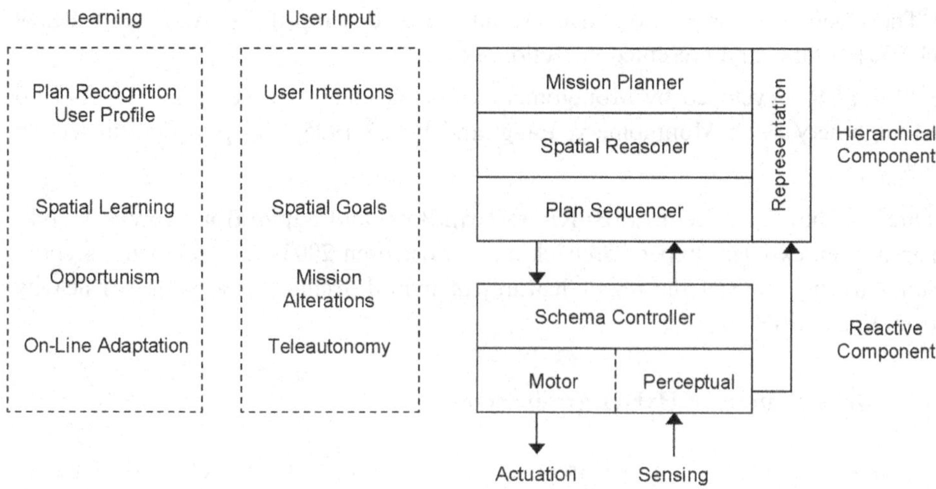

Figure 5.10
Typical hierarchical-deliberative structure (adapted from Arkin 1998)

knowledge of the robot's environment. On the other hand, at the highest level, the global planner is concerned with the robot's overall action strategy; hence, it has more time to obtain its plans and requires less precision. Hierarchical planners are strongly dependent on good models of the world (maps) to perform planning. On the other hand, behavior-based, reactive approaches depend on close coupling of perception and action and thus are well-suited to rapidly changing environments.

A number of investigators have constructed robots with hybrid deliberative-reactive architectures. An excellent summary of such architectures up to 1990 is given by Arkin (1998), who is also the developer of one of the first (and best) hybrid architectures, known as AuRA (Autonomous Robot Architecture) (Arkin 1986, 1987) (see figure 5.10). Other representative hybrid architectures include the following:

- Atlantis, developed by Gat (1992) at the NASA Jet Propulsion Laboratory
- Generic Robot Architecture, developed by Noreils (1993) in France
- Xavier, developed by Simmons (1994) and his colleagues at Carnegie Mellon University
- 3T, developed by Bonasso and his colleagues (1997)
- Saphira, developed by Konolige and Myers (1996) at SRI
- TeamBots, developed by Balch (2000)
- BERRA (Behavior-based Robot Research Architecture), developed in Sweden by Lindström, Orebäck, and Christensen (2000)

• The Tropism System Cognitive Architecture, developed by Agah and Bekey (1997a) at USC and presented in section 5.8

• AVATAR, developed by Montgomery and colleagues at USC (Fagg, Lewis, and Montgomery 1993; Montgomery, Fagg, and Bekey 1995) and presented in section 5.9

Three of these architectures (BERRA, TeamBots, and Saphira) are carefully compared in an excellent paper (Orebäck and Christensen 2003). The following section, summarizing some of the major features of hybrid architectures, is drawn heavily from this paper.

5.7 Major Features of Hybrid Architectures

The major features of a robot architecture, according to Orebäck and Christensen (2003), include the following:

• robot hardware abstraction

• extensibility and scalability

• limited run time overhead

• actuator control model

• software characteristics

• tools and methods

• documentation

We discuss each of these items briefly in the following sections. Readers interested in more detail are urged to consult Orebäck and Christensen's paper.

5.7.1 Robot Hardware Abstraction

The purpose of an abstract hardware description is to make changes in software and control as easy and convenient as possible when hardware is changed. Such items as sensors, actuators, motor drivers, and bumpers can be described abstractly. The architecture should encapsulate hardware-specific commands into a single file in the source code, so that this is the only part of the code that needs to be changed when hardware is altered. Abstractions can be made on different levels of the hierarchy of control, as illustrated in figure 5.11.

5.7.2 Extensibility and Scalability

Extensibility refers to the capability to add new software and hardware to the robotic system. This is a very important feature, since robots in research laboratories are constantly being changed and improved through the addition of new features. In

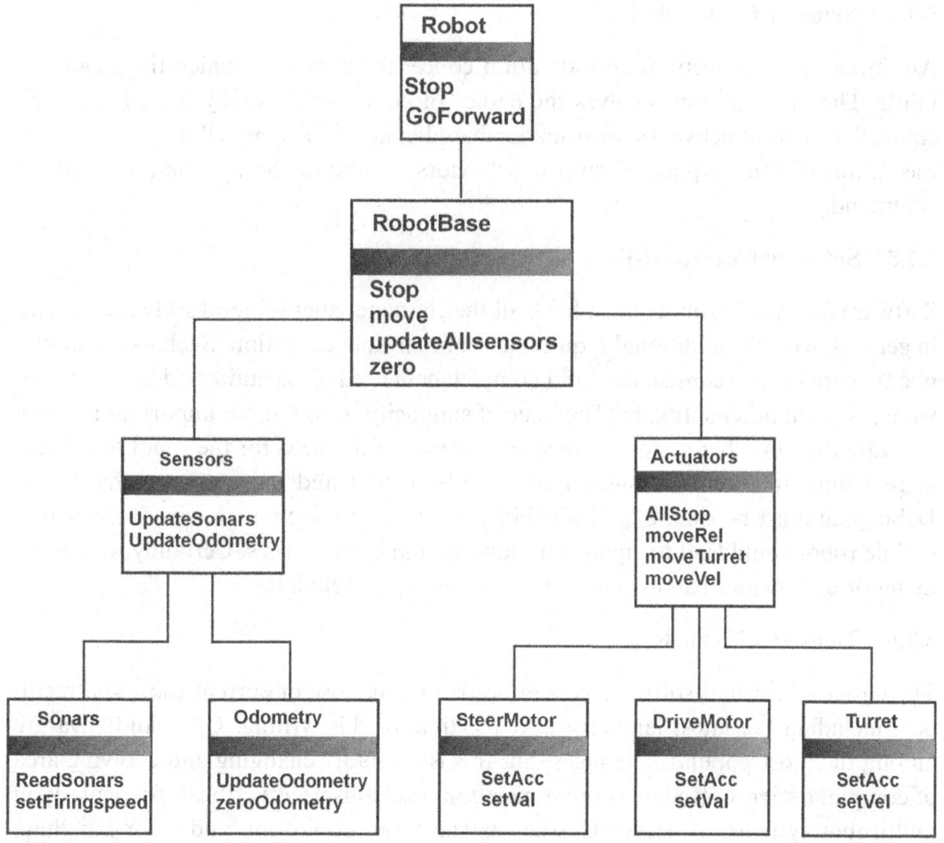

Figure 5.11
Example of hierarchical hardware abstraction (adapted from Orebäck and Christensen 2003)

hardware, this could mean simply the addition of new sensors; in software it could mean the addition of new behaviors to the robot's repertoire. Scalability has several possible meanings, but it usually refers to expanding a system from a single robot to multiple robots. The architecture on which the robots are based needs to provide for the communication and control links associated with such a distributed system, as well as avoiding bottlenecks.

5.7.3 Run Time Overhead

Overhead consists of such factors as memory and central processing unit (CPU) requirements, frequency (the number of control loops at the bottom level being updated each second), and end-to-end latency (the maximum time required for a sensor reading to have an effect on an actuator command).

5.7.4 Actuator Control Model

An important architectural consideration concerns the way in which the robot will move. The first decision involves the choice of control philosophy (e.g., fuzzy-logic controllers, neural networks, or behavior-based control). The model may also require the fusion of the outputs of several behaviors to obtain the appropriate actuator command.

5.7.5 Software Characteristics

Software for mobile robots must have all the characteristics of good software systems in general, with the additional requirement of real-time operation. Such issues as simplicity, correctness, consistency, and completeness need to be addressed in robot software just as in other software. The issue of simplicity is even more important in robot software than in other systems, however, because of the need for the robot to operate in real time and for its designers to be able to add and integrate new hardware. Debugging must be simple, and reliability must be extremely high, since failure of a mobile robot could lead to significant damage and even injuries. Certainly, the use of flying or underwater robots puts a great premium on reliability.

5.7.6 Tools and Methods

The design of a robot software system needs to make use of current tools and methods, including the latest languages. At the time of this writing, C++ and Java are among the most popular languages, but this is a rapidly changing and evolving area of computer science. Reliable communication mechanisms are critical, particularly in multirobot systems, as are such issues as run time monitoring and error handling. These software tools will become more and more important as processes are distributed over a large number of systems and interact with a large number of robots.

5.7.7 Documentation

A robot architecture needs to be sufficiently documented that it survives the departure of its key developers from the laboratory where the work was done. This is particularly true at universities, where novel architectures are frequently developed by graduate students. When these students graduate and leave, they may have many of the architectural details in their heads, having never committed them to paper or computer. Orebäck and Christensen (2003) state that proper documentation of robot architecture should include the following components:

- philosophy of the architecture
- programmer's guide
- user's guide

• reference manual that describes the relevant application program interface (API) at each level

• code documentation

This is an ambitious list, even without including on it the most difficult problem, namely, keeping the documentation current as systems evolve and change.

5.8 Case Study 5.1: The Tropism-Based Architecture

As an example of an architecture for adaptive control of robots in changing environments, we now present the tropism system architecture developed at the University of Southern California for control of multiple robots (see chapter 12). As will be shown in what follows, the tropism-based architecture is another example of a hybrid reactive-deliberative architecture; it emphasizes learning: reinforcement learning for short-term learning and evolutionary algorithms for long-term learning (Agah 1994; Agah and Bekey 1997d). Learning is discussed in the following chapter.

The tropism-based architecture is based on the following basic principles:

1. The robot must be able to translate perception into action. Each robot must have sensors that enable it to perceive those aspects of the world that will influence behaviors. Since this architecture is designed for use with multiple robots, they must have the ability to detect other robots and ascertain whether they are friendly or dangerous; they must be able to detect objects to be gathered; they must be able to avoid obstacles and barriers; and they must able to detect their home base. The objects in the robots' world are of three kinds: simple objects, which can be carried by a individual robot; dual objects, which must be decomposed before they can be carried; and heavy objects, which can be carried only by two robots acting cooperatively.

2. The robots must be able to improve their task performance with practice; that is, they must be able to learn. On successive repetitions of the same task, they are expected to gather more objects and use less energy per gathering action.

3. The robots should be able to accomplish a global goal using only local information.

4. The colony of robots must improve its ability to perform its task over successive generations.

5. The action of a robot in response to a given stimulus should not be precisely predictable, any more than is the action of a living creature, but the probability of a robot's performing an appropriate action should increase with learning.

These principles were implemented at USC using the Tropism System Cognitive Architecture, illustrated in figure 5.12. This architecture is based on the tropisms, that

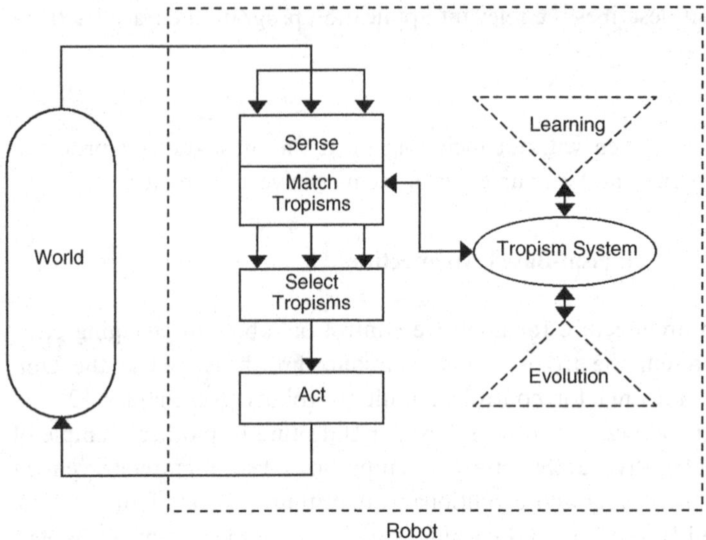

Figure 5.12
Tropism System Cognitive Architecture (from Agah 1994, reproduced with permission of author)

is, the *likes* and *dislikes* of the robot, which are termed positive and negative tropisms, respectively. Such an architecture transforms the robot's sensing of the world into appropriate actions and enables it to survive and function in an uncertain world. The concept of positive and negative tropisms as principal mechanisms of intelligent creatures was first discussed in Walter 1953.

Figure 5.12 presents the architecture in block diagram form. As shown in the figure, the robot perceives an entity in the world. The entity could be an object to be gathered, another robot, a predator, or an obstacle. This perception triggers a tropism, if one is available, and the robot performs an action. The likelihood that a particular action is performed in response to the same percept will then either strengthen or weaken the tropism. Let us now examine this process in more detail.

Robots with the Tropism System Cognitive Architecture can sense both the *type* and *state* of the entities they encounter. For instance, the entity that is sensed could be a predator, and its state could be "active." Denoting the set of entities by $\{\varepsilon\}$, the set of entity states by $\{\sigma\}$, the set of possible robot actions by $\{\alpha\}$, and the tropism values by τ_i, with $0 \leq \tau_i \leq \tau_{\max}$, a *tropism element* can be represented as a set of relations. In each of these relations, an entity and the state of that entity are associated with an action by the robot and an associated tropism value:

$$\{(\varepsilon, \sigma)\} \rightarrow \{(\alpha, \tau)\} \tag{5.1}$$

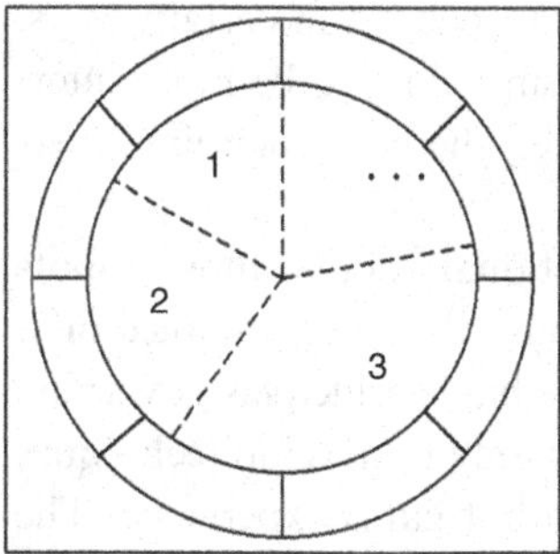

Figure 5.13
Roulette wheel model of random action selection (from Agah 1994, reproduced with permission of author)

The larger the magnitude of the tropism value, the more likely it is that the robot will perform the action. The entity, its state, the robot's action, and the tropism value can then be represented by a four-tuple, the tropism element:

$$(\varepsilon, \sigma, \alpha, \tau). \tag{5.2}$$

The tropism system of a robot can be represented by Σ, the set of tropism elements. As the tropism system can dynamically change, Σ is a function of time, $\Sigma(t)$.

Once a robot performs a sensory sweep of its surroundings (the available sensory area), the set of the tropism elements is checked for any matching entity and entity state. For all the matched entities and states, the associated action and the corresponding tropism value are marked for possible selection. One action is selected from the chosen set through the use of a biased roulette wheel, as shown in figure 5.13. Each potential action is allocated a space (slot) on the wheel, and the size of the space allotted is proportional to the associated tropism value. For example, action 3 in the figure is associated with a larger tropism value than is action 2, which in turn is associated with a larger tropism value than action 1. Thus, an action with a large tropism value is more likely to be selected, and thus to be performed by the robot, than one with a low value. Consequently, a random selection is made on the roulette wheel (i.e., the wheel is "spun," determining the robot's action).

When the tropism values remain fixed throughout a set of experiments, we refer to tropism architecture as static:

$$\Sigma(t+1) = \Sigma(t). \tag{5.3}$$

The repeated loop of perception, invocation of the tropism system, and action produces the behavior of the robot. The tropism values can be set by the experimenter in a variety of styles, ranging from random selection, to trial-and-error setting.

Calculated settings can be developed to allow the robot to behave in certain ways. For example, the robot can be programmed via its tropism values to be more interested in attacking predators than in gathering objects (i.e., the corresponding areas on the robot's roulette wheels are adjusted accordingly).

Although a static tropism architecture produces functional robots, those robots have no ability to adapt to changes in the environment, and their performance on a given task does not improve with time, as required by the basic principles governing our design. To allow learning to take place, it is necessary only to develop techniques for automatic adjustment of the tropism values as the robot gains experience. The processes of addition, modification, and deletion of tropism elements yields a dynamic tropism system. The methodologies for accomplishing these adjustments to the robot's tropism values are described in chapter 6.

As an example of the change that can occur in a tropism element, consider figure 5.14. In this figure, the specific tropism representing a robot's preference for gathering objects increases with successive generations. In effect, this increase represents the reward the robot receives for successful gathering, which changes the sizes of the spaces allocated to the various actions on the robot's roulette wheel and increases the probability that it will perform successful gathers in the future. Note also that the curve in the figure includes positive and negative standard deviation bars. The process is not deterministic, as indicated by the random aspect of the roulette wheel discussion in the foregoing.

We discuss the subject of improvement in performance with generations at greater length in connection with robot learning in the next chapter. Here the goal was only to illustrate an approach to an architecture that facilitates learning in a multiple-robot system.

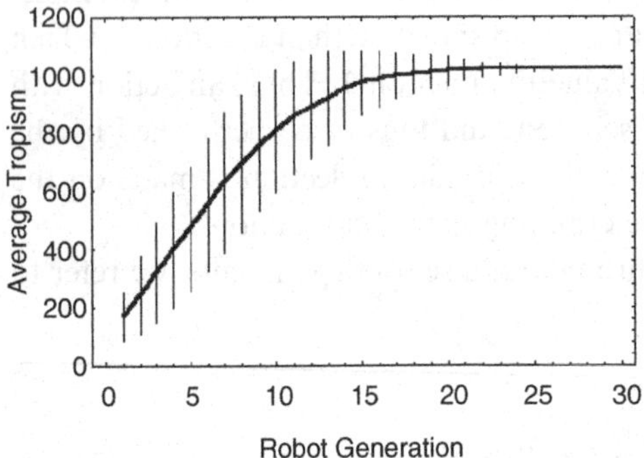

Figure 5.14
Variation of tropism for object gathering during simulated evolution (from Agah 1994, reproduced with permission of author)

5.9 Case Study 5.2: The USC AVATAR Architecture for Autonomous Helicopter Control

Several teams of graduate students in the Robotics Research Laboratory at the University of Southern California have developed a hybrid architecture for control of an autonomous helicopter, AVATAR, discussed briefly in chapter 1. Although it is primarily behavior-based, this architecture also includes a planner at the highest level (Fagg, Lewis, and Montgomery 1993; Montgomery, Fagg, and Bekey 1995; Montgomery 1999).

A helicopter is a highly complex flying machine that is controlled and powered by two rotors (as shown in figure 5.15). The main rotor provides lift as well as rolling and pitching moments that can affect the altitude, angular orientation (pitch and roll), and lateral motion of the helicopter. Thrust from the smaller tail rotor is used to counteract the torque created by the main rotor and to control the desired heading (yaw).

A helicopter pilot has access to four control variables: three for the main rotor and one for the tail rotor. The main rotor control inputs are

• *collective rotor blade pitch* adjusting the pitch of both rotor blades by the same amount increases or decreases the lift;

• *cyclic rotor blade pitch* increasing the pitch of one blade while decreasing the pitch of the other once per revolution (cycle) (depending on where in the cycle such an adjustment occurs, it will influence either the pitch (longitudinal cyclic) or roll (lateral cyclic) of the vehicle);

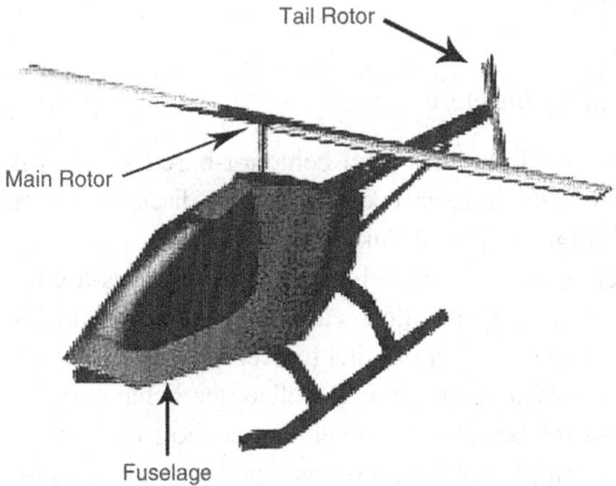

Figure 5.15
Schematic of helicopter, showing main and tail rotors

• *tail rotor collective pitch* adjustment creates a change in torque about the yaw axis.

Helicopter control is extremely difficult for a number of reasons. First, the vehicle is inherently unstable and cannot maintain a desired attitude and altitude in the absence of control inputs. Second, all the control variables are coupled, so that adjustment of one control input will produce a variety of effects. This is partially due to the large gyroscopic effect created by the main rotor (e.g., a pitch up motion will result in right roll as well). Since the main rotor is used to provide lift as well as pitching and rolling moments, the longitudinal, lateral, and normal motions are coupled, and adjustment of any one may cause changes in the others as well. Third, the helicopter operates in a time-varying environment, since it is subject to such disturbances as wind currents and temperature changes. Hence, a human helicopter pilot requires extensive training. Similarly, learning to fly a radio-controlled model helicopter (using joysticks) is difficult and time consuming; it may take a year of training to learn to fly a model helicopter reliably.

This section summarizes the architecture of the control system used for AVATAR and examines in detail the architecture of one of the controllers making up the overall complex control architecture. A considerably more detailed discussion of robot helicopters, with a number of other examples, can be found in chapter 7. The current version of this gasoline-powered helicopter is equipped with the following sensors:

• a Novatel differential GPS system, providing positional accuracy of 20 cm

• a Boeing inertial measurement unit (IMU), including three axes of linear acceleration measurement and a three-axis gyroscope, providing state information

• a downward pointing color CCD camera

• ultrasonic sonar to assist in landing

5.9.1 Behavior-Based Control System Architecture

The control architecture of AVATAR is hierarchical and behavior-based, as shown in figure 5.16. Several versions of this vehicle have been built. The discussion that follows is adapted from Saripalli, Montgomery, and Sukhatme 2003.

As discussed earlier in this chapter, a behavior-based controller partitions a control problem into a set of loosely coupled subcontrollers, each responsible for a particular task or behavior. These subcontrollers act in parallel to accomplish the overall performance goal of the system. As shown in the figure, helicopter behaviors are grouped into three categories. Low-level behaviors, shown in the bottom row of controllers, are those responsible for rapid, reflex responses, so they feature close coupling of sensors to actuators. These behaviors include those responsible for maintaining stability, thus enabling the craft to hover. The midlevel behaviors are also

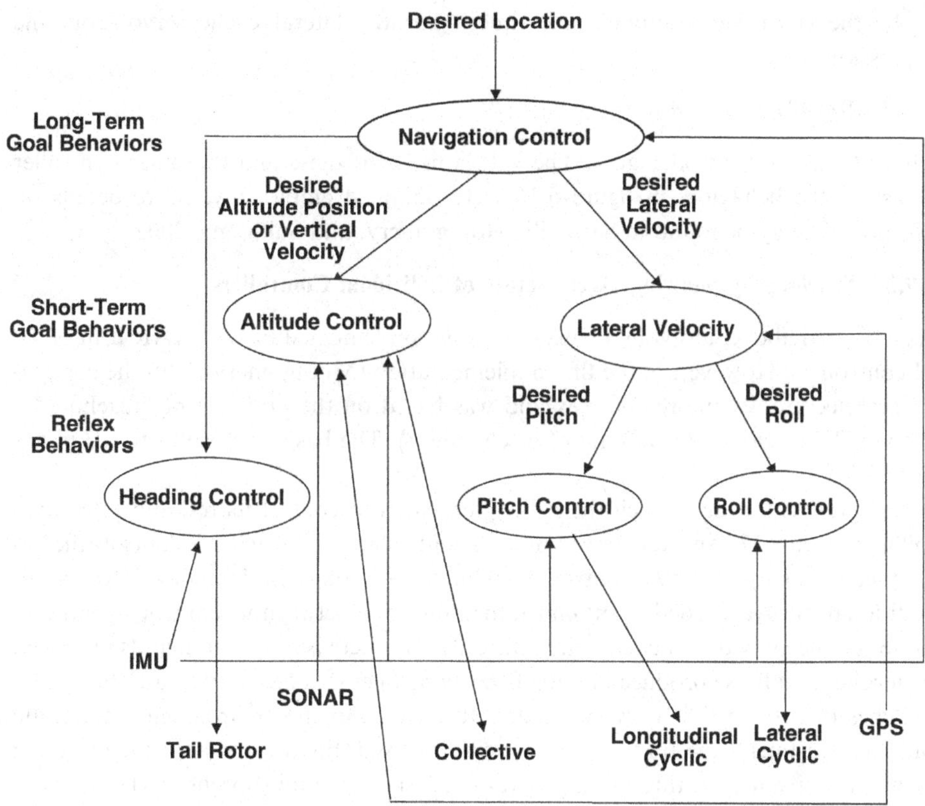

Figure 5.16
Behavior-based architecture for robot helicopter control (illustration courtesy of Gaurav Sukhatme)

short-term but require somewhat more complex processing. The top-level controller is responsible for the long-term planning of the vehicle flight. We refer to the behaviors as controllers, interchangeably.

More specifically, the *heading controller* attempts to hold the desired heading using inputs from the inertial measurement unit (IMU) to actuate the tail rotor; the *altitude controller* uses the sonar, GPS, and IMU to control the collective and throttle as required to maintain the desired altitude above ground. The *lateral velocity controller* provides inputs to the *roll* and *pitch controllers*. The *navigation controller* provides a desired heading to the heading controller, desired altitude to the altitude controller, and a desired velocity to the lateral velocity controller.

The low-level behaviors are implemented using simple proportional (P) controllers, whereas the altitude behavior makes use of a proportional-integral (PI) controller (see chapter 4). For example, if the actual roll angle is θ and the desired roll angle is θ_d, the controller computes a torque τ for the lateral cyclic servo from the expression

$$\tau = K_p(\theta_d - \theta),$$

where K_p is the controller gain. The values used for these and the other controller gains for the behaviors in figure 5.16 were obtained empirically. More details on this process may be found in Saripalli, Montgomery, and Sukhatme 2003.

5.9.2 Teaching by Showing: Architecture of Individual Controllers

As indicated above, individual behaviors were implemented in AVATAR using P or PI controllers. However, in the first implementation (Montgomery 1999) the controller architecture was more complex and was based on the principle of "teaching by showing," or *learning by imitation* (see chapter 6). The basic idea is illustrated in figure 5.17.

The process through which the helicopter learns to control its roll and pitch axes follows figure 5.17. An experienced human controller (the "teacher") controls the helicopter manually. A fuzzy controller attempts to imitate the human and generates additional rules as needed to minimize the difference between its control signals and those produced by the teacher. When the difference between the commands produced by teacher and those produced by the fuzzy controller is sufficiently small, the system switches control to the fuzzy controller. If changes in the environment necessitate further adjustment, the fuzzy rule base remains fixed, but a rapid-adjustment neural network, implemented through a general-regression method (Specht 1991), provides fine tuning as needed. In this way, the combination of fuzzy controller and neural

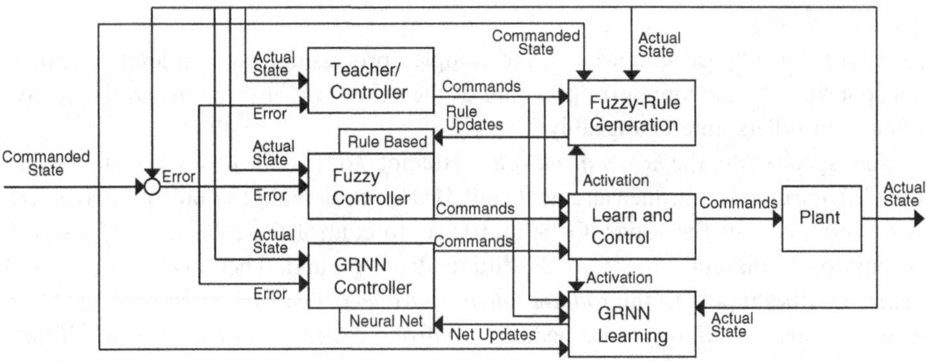

Figure 5.17
The teaching-by-showing architecture used in case study 5.2 (GRNN: general-regression neural network) (illustration courtesy of James Montgomery)

network provide a level of control for the system that matches that of the teacher as closely as desired.

As indicated by Montgomery (1999), the teaching-by-showing method was quite successful in simulation studies, in which it was used to synthesize controllers for pitch and roll. The teacher was either a human pilot or a nonlinear mathematical controller derived by Maharaj (1994) for control of the same plant. However, the method was not successful in producing stable controllers for helicopters in actual (rather than simulated) flight. It is possible that the learning mechanisms were not able to adjust to the degree of variability experienced by the helicopter in the real world. Nevertheless, teaching by showing is included here as an example of an architecture for learning control gains that may be successful in other, less demanding applications.

5.10 Open Architectures in Robotics

As with other areas of rapid change and evolution in technology, much of the work in mobile robots in the commercial sector is proprietary. Sony Corporation has taken a step toward enabling researchers working with entertainment robots, such as AIBO (see chapter 9), to modify aspects of the robot software and develop new applications. The underlying software and hardware architecture of AIBO was described in 1997 under the name OPEN-R (Fujita and Kageyama 1997). Unfortunately, OPEN-R itself is not "open-source" as in Linux, Emacs, or other open-source projects. However, Sony is attempting to make OPEN-R a standard for entertainment robotics. It is not clear to what extent the actual AIBO architecture accomplishes the goals set out for it by Fujita and Kageyama in 1997. Nevertheless, the architecture laid out in their paper is still clearly relevant for the development of new applications of mobile robots, including AIBO. The goals of OPEN-R are to provide

- a common interface for such components as sensors and actuators, so as to make them interchangeable among robots, analogous to the plug-and-play capabilities of personal-computer components;
- a method for determining the functions and specifications of various components;
- a layered architecture, making it easier to develop software and hardware applications.

The layered architecture includes an application programming layer (APL), a hardware adaptation layer (HAL), and a system service layer (SSL). The software platform is fully object oriented, so all hardware and software components can be identified as objects. The prototype of this architecture is illustrated in figure 5.18.

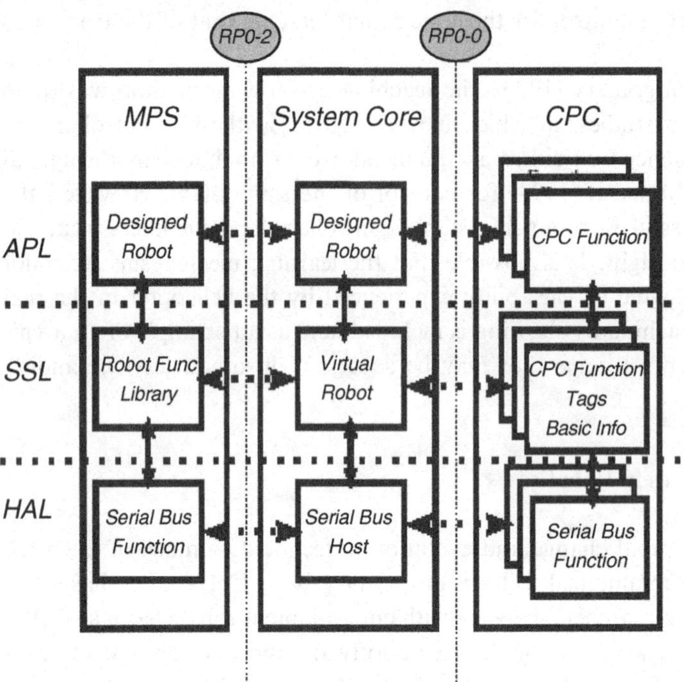

Figure 5.18
Prototype OPEN-R layered reference model (illustration courtesy of Masahiro Fujita)

In this figure, MPS represents Media for Program Storage and CPC represents the Configurable Physical Components (including sensors and actuators in all addressable components of the robot, such as the head or legs). The System Core contains a CPU as well as digital signal processing (DSP) chips for such activities as sound processing. In 2002 Sony released a software development kit (SDK) implementing many of the ideas proposed in Fujita and Kageyama's paper.

Although OPEN-R is a major step toward open architectures, we believe it will be a long time before true interchangeability of either hardware or software components is possible. Even software modules for basic applications, like obstacle avoidance and visual feature recognition, are frequently tied to specific sensors and/or actuators and are protected by their manufacturers.

5.11 Concluding Remarks

It should be evident from the discussion in this chapter that robot architectures are an active area of research. In particular, there is an increasing need for robust architectures suitable for multiple-robot systems. We have discussed one particular simple example of such an architecture; others are presented in chapter 12, as part of the discussion on multiple-robot systems and their control. The second case study in the chapter was intended to illustrate that multiple modes of control of a robotic system are frequently needed, so the system's architecture must be designed to accommodate them.

In many cases, current architectures are developed in different laboratories, using different assumptions and approaches, which make portability extremely difficult. At this time, there is no standardization, and there are no common frameworks for specifying behaviors and higher-level aspects of robot functionality. There is also no common approach to hardware abstraction. These issues make it clear that portability of different architectures is still a goal in the distant future.

6 Robot Learning

Summary

In this chapter we review various approaches to learning in robots, that is, endowing them with an ability to improve their performance on given tasks with practice. Learning requires some evaluation of the performance and feedback to the learner, enabling it to modify its strategy for performing the task in question. In view of the feedback involved, the study of learning is highly relevant to a study of robot control. In fact, this aspect of learning is frequently referred to as a control policy. Following an overview of the major issues in robot learning and the distinction between it and machine learning in artificial intelligence, we discuss reinforcement learning, neural-network-based learning, and learning using evolutionary algorithms. Other approaches to learning are also considered briefly. The concluding section discusses some of the limitations of cognitive processes in robots and directions for future development.

6.1 The Nature of Robot Learning

The question of robot intelligence is a complex one, since there is no commonly accepted definition of what constitutes intelligence in humans, to say nothing of robots. On the first day of classes on artificial intelligence taught by the author, students are asked to list the major attributes of intelligence. In addition to such attributes as problem solving, reasoning, or abstract thinking, the list the students come up with *always* includes learning. In other words, the students believe that an improvement in skill or performance with experience (associated with learning) is an essential component of intelligence.

Learning is clearly important in robotics, both from a positive and from a negative point of view. From a positive viewpoint, we look for an improvement in a robot's task performance. This improvement could be in the form of *increased speed* in a given task, *improved accuracy* in performance of the task, or an *expansion* in the robot's ability to handle a variety of tasks in the world. From a negative viewpoint, we would like to ensure that our robots make fewer mistakes with increasing experience. The old adage "Practice makes perfect" does not apply to robots exactly, but "Practice makes better" implies that learning is taking place. Learning is particularly important as robots move out of the laboratory and into unstructured real-world environments.

One of the best introductions to robot-learning known to the author is Connell and Mahadevan 1993a. In response to the question "What should robots learn?" Connell and Mahadevan give three answers:

1. *Hard-to-program knowledge* Robots should be able to learn (preferably directly from humans, say, by imitation) skills that may be difficult to program. For example, any activity involving manipulation requires programs that can deal with arm kinematics and dynamics as well as specification of the desired locations of target objects and the motions to be performed.
2. *Unknown information* The use of robots to explore underground caves or the interior of collapsed buildings would be helped greatly if robots so employed have the ability to learn a map of their surroundings as they move.
3. *Changing environments* A robot programmed for a particular environment may have great difficulties in the same environment at a future time (when it may have changed), or in a different but similar environment. The world is dynamic and changes constantly; robots must be able to keep up with the changes by learning.

In this chapter we concentrate on the ability of robots to learn. We begin with traditional views of robot learning, primarily reinforcement learning. In later sections of the chapter we consider neural networks, evolutionary algorithms, and the ability of robots to learn by imitating humans (or other robots). These last three methods are all biologically inspired and hence deserve their place in this book.

6.2 Learning and Control

Recall that in chapter 4 we discussed the notion of adaptive control, that is, the design of controllers that adjust their parameters in the face of changing environments. Consider, for example, the control of a rocket from the time it leaves the launch pad until it reaches the altitude of twenty thousand miles, where communication satellites are in orbit about the earth. Clearly, the dynamics of the vehicle will change drastically as it moves from a dense atmosphere at sea level to the rarified one at the edges of space. The same gains that provide good performance (e.g., stability, fuel economy, etc.) at sea level will not work at higher elevations. Consider again the diagram of such a rocket in two different configurations, shown in figures 6.1(a) and 6.1(b).

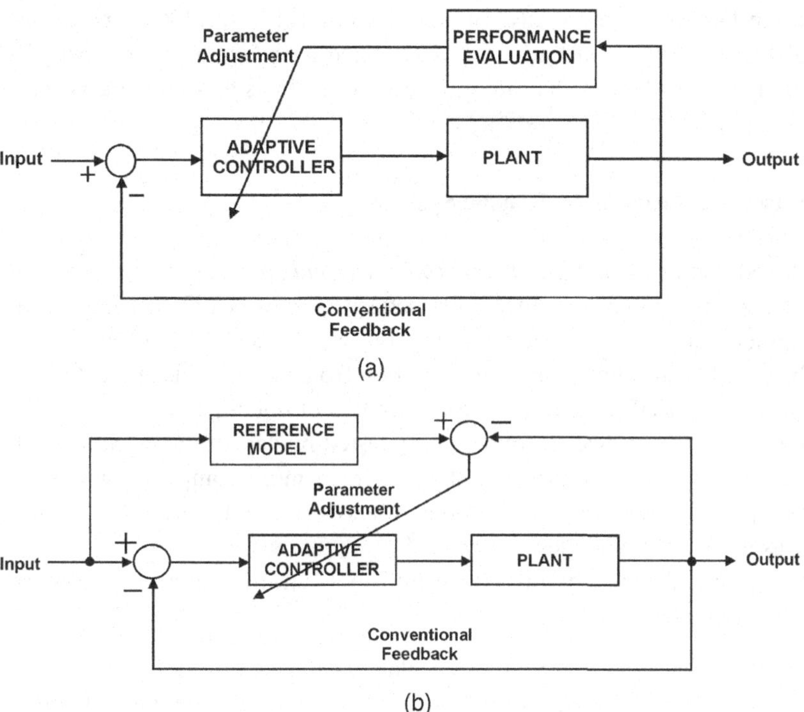

Figure 6.1
Adaptive ("learning") controllers: (a) adaptation based on performance evaluation, (b) adaptation using a reference model

Figure 6.1(a) includes a "performance evaluation" element that compares the system response to some criterion. If there are deviations from the desired response values, the parameters of the controller are adjusted to drive the system's performance back toward optimum. Figure 6.1(b) incorporates a "reference model" that receives the same input as the control system and produces the desired response. For example, we might want the system to behave at all times and in all environments as a second-order system with critical damping; that becomes the reference model. Deviations from this response are used to adjust the controller parameters to produce the desired response.

Two observations can be made about figures 6.1(a) and 6.1(b): First, the control system is nonlinear, because its parameters (and not just its inputs) are being adjusted. Second, we can think of the controllers as *learning* the proper parameter settings to give the desired response. This is exactly what we wanted our learning robot controller to do, as defined in point 3 of Connell and Mahadevan's discussion of what robots should learn: to be able to function well (if not optimally) in new environments. In fact, in the early days of adaptive control in the 1960s and 1970s, controllers of the type shown in figure 6.1 were called *learning controllers* (Mendel and Fu 1970). The point here is that adaptive control systems and learning systems have a great deal in common.

6.3 General Issues in Learning by Robotic Systems

It should be noted that *machine learning* and *robot learning* are not synonymous. Machine learning takes place in a computer. For such learning to qualify as robot learning, the computer must have interactions with the world, since the robot itself is situated in the world. The robot receives information from the world using its sensors and acts upon it using its effectors or actuators. Many, but not all, approaches to machine learning have been applied in robotics. Interested readers are urged to consult some of the many books on artificial intelligence for a more comprehensive treatment of this subject (e.g., Winston 1992; Poole, Mackworth, and Goebel 1998; Hopgood 2000; Russell and Norvig 2002).

Consider the generic diagram of an "intelligent robot" shown in figure 6.2. Several items should be noted with respect to this figure:

- The events or experiences of the robot take place *in the world*, outside of its own skin, and it reacts to them by means of actions on the world. Teaching a robot anything basically involves a mapping from perceptions of the world to actions in the world.
- In order to learn, the robot must have some *internal model* of the tasks, circumstances, and/or environments in which it works. This model may be very simple, such

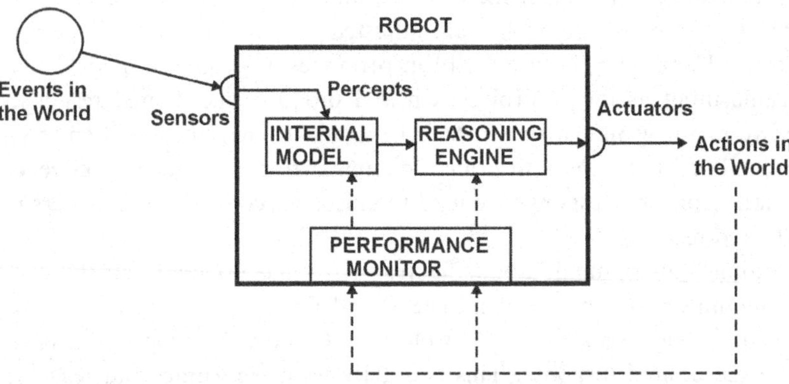

Figure 6.2
Generic architecture of robot learning

as assuming that all objects in the world are either red or blue, thus enabling it to classify sensory inputs. If the model is incorrect, in the sense of not corresponding to reality, the robot's actions may not be appropriate to the task at hand. As we shall see, if the robot does not have a internal model or representation of the task it is expected to perform, it can learn one through repeated experiments.

• The *reasoning engine* in the figure represents the specific learning algorithm being used by the robot.

• In order to learn, the robot must have some way of *evaluating* or *measuring its performance*. In other words, it must have a way of evaluating whether it has learned something and how well.

• Typically, the learning algorithm is tested on a number of *training data sets*. In robotics (as contrasted with standard AI programs) this means performing a series of tasks involving data from the world and actions upon the world. If successful, the method is then tried again on a set of *test data*. Separating the training from the test data avoids the possibility of circular reasoning, in which the robot is trained on some example tasks and then demonstrates its expertise on the same tasks.

• Note that the diagram shows *feedback* from the world back to the robot. In *supervised learning* there is a teacher or critic that gives the robot instant feedback on the success of each action it takes. In *unsupervised learning* there is no instant feedback, and the robot must discover the categories and patterns in the data by itself.

• Since the robot acquires data from the world, its percepts are subject to *disturbances* or simply *noise*. In other words, sensors are not perfect, and the data they provide may be corrupted by noise. Although this subject will not be discussed at length in this book, it is clearly of great importance, and considerable research is being conducted on it.

We consider in this chapter only a few approaches to learning: reinforcement learning, neural-network learning, and learning through the use of evolutionary algorithms.

6.4 Reinforcement Learning

Learning through reinforcement is probably the best-known method of learning in common experience. Training a dog with biscuits as rewards for good behavior ("good dog!") and harsh words or a hit with a folded newspaper ("bad dog!") is an example of psychological reinforcement learning. The situation is similar in robot learning. The problem is that this "good robot–bad robot"[7] feedback is generally

7. A wise roboticist once advised me, as I was assigning humanlike feelings to a robot: "Do not anthropomorphize robots. They don't like it."

provided only after a sequence of actions, and the correct solution for a given trial is not provided. This means that the learning takes place by trial and error, followed by performance evaluation, but without indicating the correct behavior. For example, a robot may be tasked with finding a target and given a reward (*positive reinforcement*) only when it finds it. Learning under such a system is indeed possible, but it tends to be very slow, requiring very large numbers of trials. By contrast, supervised learning would take place if a teacher provided immediate feedback following each action ("you should not have turned left here," "this was the right turn"). However, the application of supervised learning requires that the teacher prepare a series of training examples in advance that cover all possible situations the robot might encounter. This is clearly not possible in unstructured, outdoor environments. Excellent discussions of reinforcement learning may be found in the literature (e.g., Arkin 1998; Connell and Mahadevan 1993c; Barto 1992).

In reinforcement learning (frequently abbreviated RL), then, the goal of the robot program is to construct a strategy for maximizing the reward (Sutton 1990). This strategy, which involves a mapping from the robot's perceived states to actions, is known as a *control policy*. Some researchers define reinforcement learning as a problem of optimization, that is, inferring a control policy that optimizes a scalar cost or evaluation function. As discussed in case study 6.1, in our laboratory at USC we used RL to teach a mobile robot to avoid collisions with obstacles as it moved about the lab (Fagg, Lotspeich, and Bekey 1994). We chose a *negative-reinforcement* policy; that is, the robot was "punished" by appropriate parameter adjustments when it actually collided with an object. From this very sparse information, the robot's learning-control system had to develop a control policy that kept it from being punished. The problem of how to determine a control policy is diagrammed in figure 6.3.

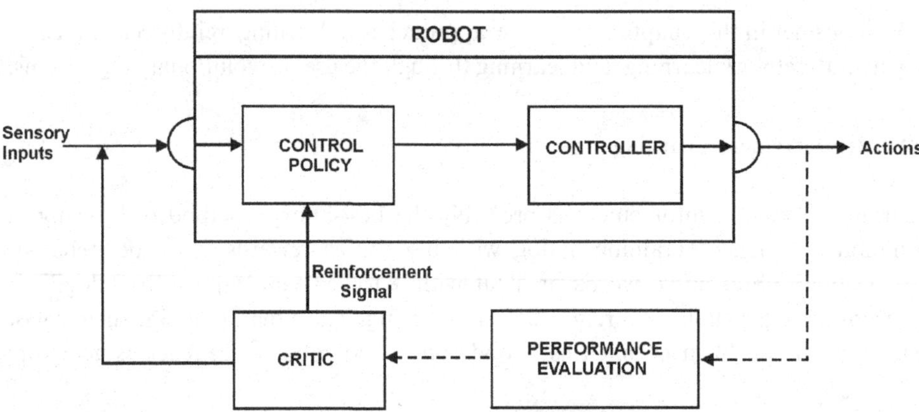

Figure 6.3
Generic view of reinforcement learning showing the control policy

The box labeled "critic" in the figure looks at the sensory inputs and the robot's performance on the task at hand and provides reinforcement signals. Note that the critic can be simple indeed, as in the example from our laboratory, in which the critic simply assigned a value of -1 to each collision.

Actually, the situation for our robot was more complex yet, since it was equipped with two sensors: a sonar proximity sensor and a bumper switch. Clearly, the bumper provided an actual indication of collision, but the sonar gave an advance indication of an upcoming collision. Hence, the robot needed to decide which of these two sensory inputs was more important in avoiding negative reinforcements. This is known as the *credit assignment problem.* This problem has two primary aspects: *structural credit assignment*, that is, figuring out which sensory data were associated with a particular reward or punishment received, and *temporal credit assignment*, that is, determining which actions in a sequence of actions performed by the robot were responsible for the reward or punishment. For our collision-avoiding robot, the determination of which of its two sensors was primarily responsible for a particular punishment was a structural credit assignment problem.

One of the best studied problems in the reinforcement approach to learning is the box-pushing problem illustrated in figure 6.4 (Connell and Mahadevan 1993b). The task of the robot in the figure (named Obelix) was to locate boxes and push them across the room without getting stuck in corners. The task was difficult, because the only sensors available to the robot were

Figure 6.4
Obelix: A box-pushing robot (from Connell and Mahadevan 1993c, reproduced with permission of Kluwer Academic Publishers)

```
Monolithic_reward (old_state, action, new_state)
    begin
            IF      action = forward
                    and BUMP(old_state)
                    and BUMP(new_state)
                    and ¬STUCK(new_state)
            THEN    return 1
            ELSE    return 0
    end
```

Figure 6.5
Reward function for monolithic learner (adapted from Connell and Mahadevan 1993c)

- a ring of eight sonars (four facing forward and two each left and right) that gave the approximate (NEAR or FAR) locations in only sixteen bits of information;
- an infrared detector facing forward (in the box on top of the robot in the figure) that provided a binary response function at a distance of 4 inches;
- a motor current monitor.

When the IR sensor was activated, it indicated that there was something against the front of the robot, and it turned on an output bit called BUMP. The motor current exceeded a threshold (and produced a bit called STUCK) when the robot was attempting to push a wall or was stuck. A major problem was that the lack of vision made it difficult for the robot to distinguish boxes from walls. From the eighteen bits of information (over 256,000 perceptual states) the robot had to choose one of five actions: move forward or turn in place right or left by 22° or 45°.

Connell and Mahadevan first attempted to build a single (monolithic) controller using reinforcement learning, with the reward function given in figure 6.5. Basically, the learner's reward function was set to 1 when it actually pushed a box (i.e., BUMP was activated in two successive times while moving forward, and the second state was not STUCK) and to 0 otherwise.

This attempt at inducing learning in a robot was unsuccessful for a number of reasons (Connell and Mahadevan 1993b):

- As indicated previously, the size of the sensor input space was very large.
- Use of the simple reward function in figure 6.4 resulted in very infrequent rewards, and hence the robot learned very slowly.
- Attempts to construct a more complex reward function to provide the robot with more frequent rewards for successful pushing also failed. For example, putting greater emphasis on avoiding getting stuck resulted in Obelix's avoiding all contact with objects; putting more emphasis on contact produced the undesirable effect of Obelix's attempting to push not only boxes but walls as well. In effect, the robot got stuck in local minima.

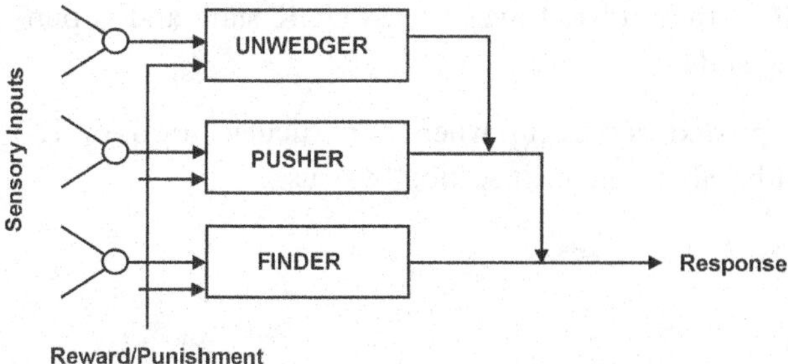

Figure 6.6
Behavior-based architecture for box-pushing task

• Connell and Mahadevan were unable to find a way to include the robot's performance history information in the reward function, and the robot tended to repeat its mistakes (e.g., forgetting that it had just pushed a wall and attempting to do it again).[8]

• The relatively simple sensors used on the robot also produced problems, since in sonar images, walls looked like boxes and vice versa.

The solution to these problems involved decomposing the box-pushing problem into three separate subtasks and using a subsumption (behavioral) architecture (chapter 5), as illustrated in figure 6.6. Note that the robot must find a box, it must be able to push it, and it must avoid getting stuck. Each of these behaviors has its own sensory inputs to which it reacts, and hence each behavior has its own rewards.

Let us consider these three behaviors in turn:

1. The *finder* is rewarded whenever the robot is close to a box, that is, when it moves forward and there are NEAR bits in the sensor input vector. If the robot's movement leads it away from a box, so that NEAR bits are turned off, it is punished.

2. The *pusher* is rewarded when the BUMP sensor is activated, and the reward continues as long as BUMP remains active. The behavior is punished if the robot loses the box for some period of time (the time lapse allowing for possible temporary loss resulting from rotation of the box) and if the robot stops when the box is wedged against a wall.

8. I have a pet desert tortoise named Touché, a creature not known either for its swift intelligence or for its ability to learn new tricks. For the past forty years, Touché has been attempting to push its way through a wrought iron fence where the spacing between the pickets is about half the width of its shell. It walks away, and moments later it tries again. Apparently, neither Obelix nor Touché is able to learn from its mistakes.

3. The *unwedger* is rewarded if a stuck robot leaves the STUCK state and is punished if it persists in the STUCK state.

The preceding decomposition proved successful when a particular form of RL known as Q-learning, described briefly in the next section, was used.

6.5 Q-Learning

Assume that the robot is in a state s, where s represents a specific set of values of all the robot's parameters, such as controller gains, physical location, sensor readings, and battery charge. The set of all possible states is denoted by S. The robot is capable of performing an action $a \in A$, where A is the set of all possible actions. The basic idea of Q-learning is to enable the robot to learn the most appropriate behaviors, that is, to take those actions that will yield the highest possible combination of immediate reward and expected future reward for a given state s. Q-learning denotes the value of taking a given action in a given state by the function $Q(s, a)$. For a robot to obtain a complete control policy, it must learn the values of Q over the entire state-action space $S \times A$ (Watkins and Dayan 1992).

Let us make this abstract definition more operational by looking at the way the robot's actions are updated. The *value* of performing an action a in state s, $Q(s, a)$, is defined as the sum of the immediate reward r plus the utility $E(y)$ of the state y resulting from the action, where

$$E(y) = \max_a(y, a). \tag{6.1}$$

Note that equation (6.1) requires that we search over all possible states with the action a to find the maximum. Then, the function $Q(s, a)$ is updated as

$$Q_{t+1}(s, a) \leftarrow Q_t(s, a) + \alpha(r + \lambda E(y) - Q_t(s, a)) \tag{6.2}$$

Note that the utility $E(y)$ is multiplied by a coefficient λ, a discount factor that reduces the influence of expected future rewards, where $0 < \lambda \leq 1$. During learning, the sum of current and expected rewards, $r + \lambda E(y)$, is not yet equal to the current value of Q. Hence, the update equation (6.2) states that the next value of Q will be the current value plus an error term, multiplied by a coefficient α that determines how rapidly the error is corrected. Hence, this coefficient is sometimes called the *learning rate*.

Essentially, this method produces a table of state-action pairs from which the robot can select the action that maximizes $Q(s, a)$. Using Q-learning and the behavior-based architecture shown in figure 6.5, Connell and Mahadevan (1993b) were successful in enabling their robot to learn the box-pushing task.

6.6 Case Study 6.1: Learning to Avoid Obstacles Using Reinforcement Learning

The purpose of the study discussed in this section was to apply reinforcement learning to the design of a behavior-based controller for a mobile robot (Fagg, Lotspeich, and Bekey 1994). As with other RL approaches, the teacher provided the robot only with an evaluation of its behavior. The task of the robot's learning system was to infer a control policy from this feedback information. This section summarizes the paper cited, with an emphasis on the RL approach.

6.6.1 The Robot and Its Task

The robot used in this study was a converted radio-controlled model truck named Marvin (figure 6.7). It was approximately 45 cm long, 25 cm wide, and 25 cm high. Four wheels (rubber tires about 9 cm in diameter) gave it the appearance of a dune buggy. An electric motor provided power to the drive shaft, which actuated the rear wheels through a differential gear box.

6.6.1.1 Control

Low-level control was performed by a Motorola 68332 processor, installed on a custom made printed circuit (PC) board designed and fabricated by students in the USC Robotics Research Laboratory. Joystick control was available for robot teleoperation and for providing high-level, real-time feedback to the robot. The low-level controller communicated via an RS-232 link to a Sun SPARCstation, where high-level control decisions were made.

Figure 6.7
The mobile robot Marvin, used in the RL experiment discussed in case study 6.1 (photograph from author's files)

6.6.1.2 Sensors

The robot was equipped with two sets of sensors: (1) Microswitches attached to the bumpers of the car detected front and rear collisions with the walls; and (2) five ultrasound (sonar) range sensors (oriented to the front, rear, up, left, and right of the robot) provided indications of decreasing distance to the walls or obstacles in the laboratory. Reinforcement signals were derived directly from sensory feedback (e.g., punishment when the vehicle collided with a wall), or from the radio control (RC) joystick.

6.6.1.3 Actuators

The vehicle controls were very simple. The robot was capable of moving forward or backward, and the steering wheel could be turned to its full extension either right or left or pointed straight forward. A electric motor provided power to the rear wheels, and a servomotor controlled the steering of the front wheels. The maximum possible speed was approximately 30 kph; however, the programmed speed in use was about 10 cm/sec.

We experimented with several tasks for Marvin, including wall following, obstacle avoidance, and exploration of the environment. Here we report only on the obstacle avoidance task. The robot moved in the laboratory environment, in which it faced common obstacles like chair and table legs. The reinforcement policy consisted of punishment for running into an obstacle (or wall) (i.e., reinforcement = −1) and either a reward for not running into a wall or no reward at all, as described in section 6.6.3.

6.6.2 The Network Model for Action Selection

It should be evident that Marvin needed a method for mapping sensory inputs to appropriate outputs. The method adopted in this study was based on our previous work with neural networks (see section 6.7) for modeling the primate visual system. Basically, the idea is to enable an active sensor unit in the network to activate a corresponding unit in a *feature detector layer*. The activated feature layer units will form a pattern. A competition among the activated units will determine which is the "strongest" using a winner-take-all approach. The remaining units vote for one of the possible actions.

The architecture of the network implementing this process is shown in figure 6.8. Note that the sensor inputs **I** (five sonars and two bumpers) activate a particular pattern of activity at level **F**. (The raw inputs can be weighted by level **W**). The feature detector units then interact with one another, in a manner similar to the way in which visual sensor units in animals interact, by increasing contrast with their neighbors, thus producing the pattern at level **G**. This effect, which we term "local winner-

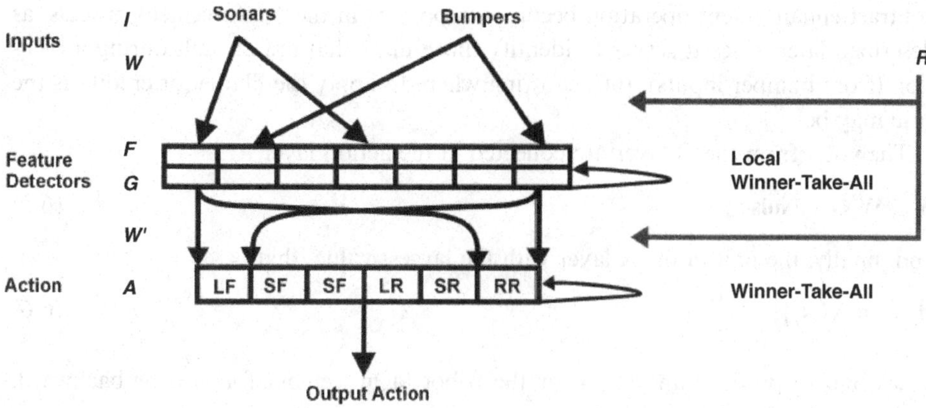

Figure 6.8
Network architecture for selecting robot action

take-all," enhances near neighbors and suppresses those units that are further away. Those that continue to be active then vote for an action to be taken at the *action selection layer* **A**. Although this is not shown in the figure, each feature detector unit provides an input to each action unit, multiplied by weights denoted as **W**′. There are six possible actions—LF-left forward, SF-straight forward, RF-right forward, LR-left reverse, SR-straight reverse, and RR-right reverse—as indicated in the figure. The votes are summed at each action unit, and the unit with the highest activity is selected to produce the action at the next time step. Note also that the reinforcement input R affects the weights **W** and **W**′.

The mapping from sensors to feature detectors is implemented as

$$\mathbf{F} = \mathbf{WI} + \mathbf{Noise}, \tag{6.3}$$

where **F** and **I** are vectors representing the feature detectors and inputs respectively, while **W** is a matrix of weights. **Noise** is a vector of disturbances added to the **F** layer.

The interaction among **F** units to produce the **G** vector is implemented by the rule

$$\begin{aligned} \text{Winner}_i &= \max\{F_i\} = 1, \\ G_i &= \text{Winner} * F_i. \end{aligned} \tag{6.4}$$

This rule can be interpreted as follows: At the **F** layer, the most active unit (based on sensor inputs) in a given size neighborhood is selected, declared the local winner, and assigned the value one. The local winner then acts as a selector to pick out those values from the **F** layer that are transferred to the **G** layer to increase contrast. This contrast enhancement operation becomes important in the reinforcement process, as described later, since it serves to identify those units that are at fault during a collision (from bumper inputs), thus showing where to apply the blame (or credit, as the case may be).

The votes from the **G** layer are collected in the action layer **A**,

$$\mathbf{A} = \mathbf{W}'\mathbf{G} + \mathbf{Noise}', \tag{6.5}$$

and, finally, the unit in the **A** layer with the largest value, that is,

$$A_i = \max_j \{A_j\}, \tag{6.6}$$

is selected to produce an action for the robot (a movement forward or backward, with or without steering).

6.6.3 Reinforcement Learning

The previous section described the mapping from sensor inputs to motor commands in the robot using a neural network. Now we consider the learning process, which makes adjustments to the matrices **W** and **W**$'$ based on the reinforcement information received from the teacher. The signal R can be positive or negative, depending on the nature of the reinforcement, and its magnitude is related to the degree of appropriateness or inappropriateness of the robot's behavior. We concentrate here on the structural credit assignment problem, that is, how to assign blame to the individual weights in **W** and **W**$'$ given a particular state of the network of figure 6.8 and a reinforcement signal from the teacher. Recall that the reinforcement signal will simply tell the robot whether it performed well or not; it is the learning system that has to figure out where to place the blame for a signal R.

Consider first a positive reinforcement signal (i.e., a signal to recognize the fact that Marvin did not collide with an obstacle in the immediate past). We want to increase the probability that Marvin takes the same action when the identical situation arises again. To accomplish this, we want to strengthen the weights from the currently active inputs to the feature detectors at level **F**, in order to recognize the same features. Also, we want to make sure that the same action is taken; this is accomplished by strengthening the weights **W**$'$ from the active features to the selected action.

The situation is similar with $R < 0$. We interpret this as an error in selecting the right action. Hence, we need to decrease the strength of the connections from the feature detector units to the selected action (**W**$'$), so that it is less likely to be selected again under the same circumstances. To insure that we recognize the same situation, we need to increase the strength of the connections from the current inputs to the feature detector layer (**W**).

The update rule just described (for both positive and negative reinforcements) can be expressed mathematically as

$$\Delta W_{ij} = \alpha |R| I_i G_j W_{ij}, \tag{6.7}$$

$$\Delta W'_{jk} = \alpha' R G_j A_k W'_{jk}. \tag{6.8}$$

Clearly, equation (6.7) concerns updates to $\mathbf{W}$, and (6.8) to $\mathbf{W}'$. In these equations α and α' are known as *learning rate constants*. The subscripts on I, G, A, W, and W' indicate a specific element in the given vector or matrix. Once these increments are computed, we update the weights by

$$W_{ij} \leftarrow W_{ij} + \Delta W_{ij}, \tag{6.9}$$

and similarly for each W'. This rule is very similar to the neural-network update rule described in section 6.7.

6.6.4 Results

We experimented with various reinforcement policies, including

1. punishment for running into walls or obstacles with no rewards;
2. the use of both punishment and reward, with the punishment signal being larger than the reward signal;
3. the use of both punishment and reward, with the reward signal being larger than the punishment signal.

The most successful policy was the second one, with the punishment signal being larger than the reward signal. Punishment alone resulted in extremely long learning times, since the system spent a great deal of time searching for an appropriate action. Table 6.1 shows the results of the actions selected by the system during learning in relation to the sensory inputs and the corresponding feature detectors. This table may be viewed as a set of if-then rules, indicating the appropriate action for a given set of inputs. Over a set of learning experiments these rules were consistently learned. Note that in these experiments the robot's sonars were enhanced to provide near-range, midrange, and far-range signals. It should be evident that although not perfect, the selected actions are certainly reasonable. In our laboratory, Marvin stopped hitting obstacles after some 20–30 minutes of learning (five to seven hundred movement steps).

6.6.5 Other Results

Fagg, Lotspeich, and Bekey 1994 also included extensive results from our experiments on temporal credit assignment. A problem with temporal credit assignment

Table 6.1
Actions learned by Marvin using reinforcement learning

Sensory inputs	Supported actions
Midrange left sonar Far-range left sonar Far-range up sonar	Straight forward
Front-bumper contact Near-range forward sonar Midrange up sonar Far-range up sonar	Straight reverse
Far-range up sonar	Straight forward
Rear-bumper contact	Right forward
Far-range up sonar	Left forward
Near-range forward sonar Midrange forward sonar Near-range right sonar	Left reverse

arises in these experiments because the movements of the robots depend on past as well as present actions. Hence, it is necessary to assign credit to a whole sequence of control decisions, and not a single one, as implied in the previous section. In the approach described in our paper, based on the work of Sutton (1988), the highest weight was given to the most recent decision, with exponentially decreasing weights for decisions made further back in time (Fagg, Lotspeich, and Bekey 1994).

6.7 Learning Using Neural Networks

6.7.1 Biological Basis

Neural networks (or, more properly, artificial neural networks) were originally conceived as models of the way information is processed in the brain. Figure 6.9 illustrates a simplified neuron, the fundamental processing element in the nervous system, including the brain. The human brain contains some 10^{11} neurons, which make some 10^{14} connections with one another. The processing power of the brain is thought to reside in the connections among neurons, which are reorganized during learning.

From figure 6.9, we note that the major components of a neuron are

• the *cell body*, or *soma*, which contains the nucleus of the cell;

• an *axon*, a long fiber stretching out from the cell body whose length may range from a centimeter to as much as a meter (the end of the axon divides into multiple branches, which is known as *axonal arborization*);

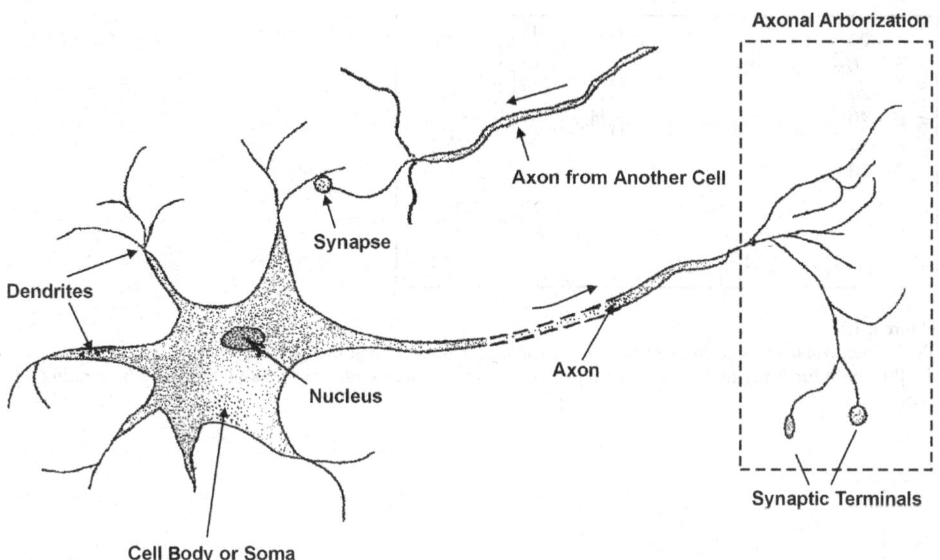

Figure 6.9
Simplified representation of a neuron

• *dendrites*, multiple fibers branching out from the cell body in a dense network of branches.

Neuronal signals are trains of electrical pulses (see figure 6.10). In effect, the pulses, known as *action potentials*, are regenerated along the axon, so that their shape does not change during transmission. Hence, information is carried in the frequency of firing. Signals are propagated along the axons, from the cell body to the axonal terminals, where they make connections (known as *synapses*) with the dendrites of other cells. A single neuron may form synapses with thousands, even hundreds of thousands, of other neurons.

The synapse is a complex electrochemical process. The arrival of a neural spike at the end of a dendrite causes the release of a neurotransmitter (such as acetylcholine), which moves across a gap between the transmitting and receiving cells. Its arrival at the receiving cell causes the generation of a new electrical spike, so that transmission can continue to the cell body, where all the incoming signals are integrated. The connection between an incoming signal and a neuron may be inhibitory or excitatory, so the integration of multiple incoming signals may be viewed as a weighted sum. If the sum of incoming signals at a given time exceeds a threshold, the neuron will fire and send a new spike out along its axon. This picture is highly simplified; in reality, the process is much more complex. Interested readers are urged to consult a text on

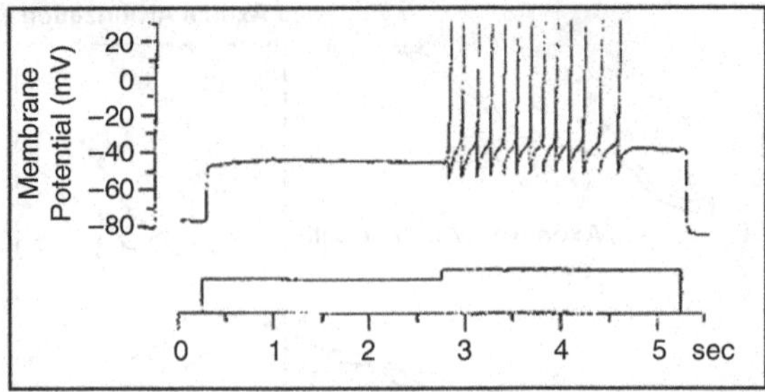

Figure 6.10
Typical neuronal spike train. The bottom of the figure shows the stimulation level, which increases above the threshold for firing at about 2.8 sec. The spike train is measured across the membrane surrounding an axon

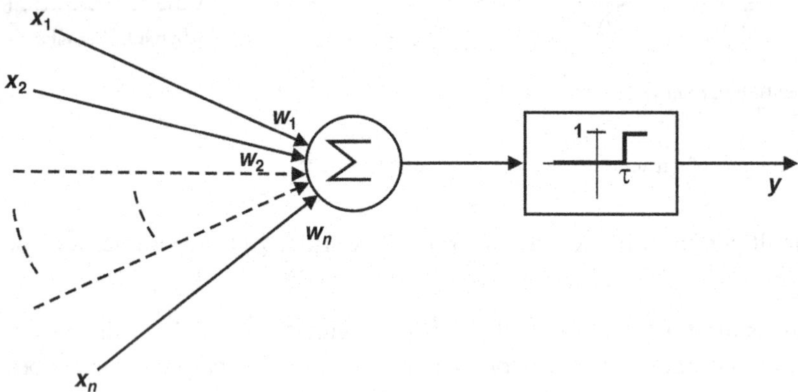

Figure 6.11
McCulloch-Pitts neuron

neurophysiology or the remarkable *Handbook of Brain Theory and Neural Networks* (Arbib 1995).

6.7.2 Computational Neurons and Perceptron Learning

The first simple computational model of a neuron, as proposed by McCulloch and Pitts (1943), is shown in figure 6.11. The model replaces the spike trains of actual neurons with continuous input signals. When the weighted sum of the inputs exceeds a threshold, the "neuron" produces an output that is transmitted to the next neuron:

$$y = T\left(\sum_{j=1}^{n} w_j x_j\right) + \tau, \tag{6.10}$$

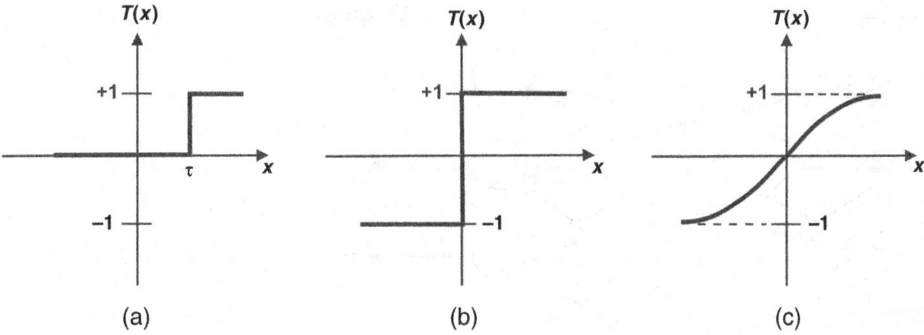

Figure 6.12
Common activation functions: (a) threshold, (b) step, and (c) sigmoid

where $T(.)$ is a nonlinearity known as the *activation function*, the weights are known as *synaptic weights*, and τ is the threshold. In practice it is common to replace the threshold term with an additional input of one with a weight equal to τ.

Three common forms of activation functions are shown in figure 6.12. They are defined as follows:

$$\text{Threshold} \quad T(x) = \begin{cases} 1 & \text{if } x \geq \tau, \\ 0 & \text{if } x < \tau, \end{cases} \tag{6.11}$$

$$\text{Step} \quad T(x) = \begin{cases} +1 & \text{if } x \geq 0, \\ -1 & \text{if } x < 0, \end{cases} \tag{6.12}$$

$$\text{Sigmoid} \quad T(x) = \frac{1}{1 + e^{-x}}. \tag{6.13}$$

Networks of simple neurons were introduced in the 1950s (Rosenblatt 1957) under the name "Perceptrons." At present the name "Perceptron" is restricted to a single layer of neurons with a threshold nonlinearity, as shown in the left-hand side of figure 6.13, where each neuron (denoted by a circle) performs the function of equation (6.10). Since each output is independent of the others, we restrict our attention to a single neuron, but we can "stack" as many as needed to represent a given function. Note that in figure 6.13 the neurons are termed *units*, to distinguish them from biological neurons. (They are also sometimes termed *artificial neurons.*)

Consider now how such a network might learn an input-output characteristic of some system. A single unit can be used to represent certain Boolean functions, such as AND, NOR, and NOT (but not the EXCLUSIVE-OR). Say that the inputs are obtained from sensors that provide binary values (e.g., a contact sensor that is either ON or OFF, or an infrared sensor that indicates that an object is within some

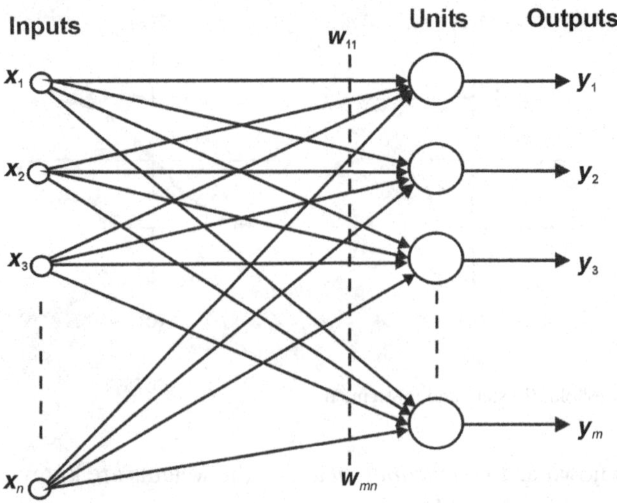

Figure 6.13
Perceptron network

distance or not), and the network has two outputs that turn motors on or off. Let us assume that the goal of the network is to enable the robot to avoid obstacles. Initially the weights $w_{j,i}$ (to the jth unit from the ith input) are unknown and set to arbitrary values. If for a given example x_i the output of a unit is y and the desired output is y_d, then the difference between these values is an error. The Perceptron learning rule is

$$\begin{aligned} \Delta w_i &= \alpha x_i(y_\mathrm{d} - y), \\ w_i &\leftarrow w_i + \Delta w_i, \end{aligned} \tag{6.14}$$

where the constant α is known as the *learning rate*. The effect of this rule is to adjust the weights such that the error is reduced. Rosenblatt (1962) proved that this rule (also known as the *delta rule*) will converge to a set of weights that correctly matches the examples given to the network for a large class of functions. It turns out that Perceptrons are limited to learning the input-output relationships of *linearly separable* functions, as proven by Minsky and Papert (1969). Hence, even such simple functions as the EXCLUSIVE-OR cannot be represented by a Perceptron.[9]

The delta rule is an example of a *gradient optimization* procedure. To understand the principles of this type of procedure as applied to the Perceptron of figure 6.13,

9. I was a Ph.D. student at the time when the Rosenblatt convergence theorem was published, and I remember the excitement among the students about the possibility of creating intelligent machines from Perceptrons. It was not until multilayer neural networks were explored in the 1980s that some of the potential of this technology became apparent.

assume that our goal is to minimize a performance index *PI* based on the error between the desired output y_d and the actual network output $y(t)$, such as

$$PI = \frac{1}{2}[y(t) - y_d(t)]^2. \tag{6.15}$$

We use a quadratic function for *PI* so that it has a clearly defined minimum whether the error is positive or negative. As we know, $y(t)$ depends on the weights, as given in equation (6.10). Let us assume that the activation function is simply unity and that we have only one input x and only one weight w, so that $y = wx$. Then the slope of the *PI* curve is given by the derivative

$$\frac{dPI}{dw} = (y - y_d)\frac{dy}{dw} = ex, \tag{6.16}$$

where the error is indicated by e. Since we desire to adjust the weight w in such a way that *PI* is minimized, we choose an increment with the sign opposite to that of the slope (6.16):

$$\Delta w = -\alpha\frac{dPI}{dw} = -\alpha ex, \tag{6.17}$$

where α represents the size of the step. Note that this expression is identical to the learning rule given in equation (6.14), for the case of a single parameter or weight. When there are multiple weights, as in the Perceptron network, we have to calculate the slope in the parameter space by differentiating with respect to all the weights to obtain a vector known as the *gradient* of the performance index *PI* and denoted as

$$\mathbf{grad\ PI} = \begin{vmatrix} \dfrac{\partial PI}{\partial w_1} \\ \dfrac{\partial PI}{w_2} \\ \vdots \\ \dfrac{\partial PI}{w_n} \end{vmatrix}. \tag{6.18}$$

Since the performance index now depends on more than one parameter, we use the notation for partial (rather than total) derivatives. Then the learning rule can be stated as

$$\mathbf{\Delta w} = -\alpha\ \mathbf{grad\ PI}. \tag{6.19}$$

This simply means that the change in the weight vector will be in the same direction as that in the local gradient vector. Since the gradient vector always points in the locally steepest direction, this method is also known as *steepest descent* (because of the minus sign). In general, algorithms in which small changes are made in a direction that improves a performance index are known as *hill-climbing algorithms.*

6.7.3 Learning in Multilayer Feedforward Networks

It is straightforward to imagine adding intermediate layers of units between input and output in figure 6.13. Apparently Rosenblatt and others suggested doing so in the 1950s but did not pursue the idea, since they were not able to find a way to adapt the learning rule to adjust the intermediate weights. As we saw in equation (6.14), the learning rule makes use of the error between the actual and desired output, but what is a corresponding "error" for intermediate units? A method for learning in multilayer networks known as *back-propagation* was developed by Bryson and Ho (1969) and formalized by Werbos (1974); see also Werbos 1995 for a discussion of more recent developments.

Consider the three-layer network in figure 6.14, which contains an input layer, an output layer, and an intermediate or "hidden" layer of units. (Clearly, it is possible to have more than one hidden layer.) Learning in multilayer networks proceeds as in Perceptrons: Example inputs are presented to the network, and the network outputs are compared to the actual or desired outputs corresponding to the example inputs. The problem now is to determine how to allocate the error not just to the weights in the output layer, but to those in the intermediate or hidden layer or layers as well. Note that in the figure there are many weights connecting each input to an output.

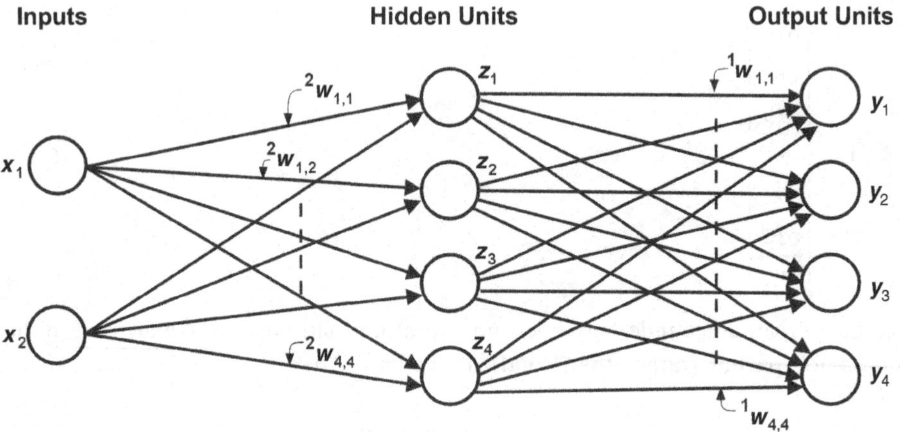

Figure 6.14
Three-layer feedfoward network showing hidden layer

The back-propagation algorithm begins with an update rule very similar to the Perceptron update rule, namely:

$$\Delta^1 w_{j,i} = \alpha z_j (y_{di} - y_i) T_i' = \alpha z_j e_i T_i', \tag{6.20}$$

where z_j is the output of the jth unit in the hidden layer, α is the learning rate as before, and T' is the derivative of the activation function $T(x)$. The last term is necessary because multilayer networks generally use sigmoid activation functions, as in figure 6.12(c). In contrast with the binary activation functions of figure 6.12(a) or 6.12(b), the output in a sigmoid activation function is never equal to one, so that differentiation of the performance index leads to the additional term T'. Since the sigmoid activation function has a simple analytical form (see equation (6.13)), it is not difficult to compute this derivative.

Now we need to determine how to propagate the errors from the output layer back to the weights in the hidden layer to write an update equation analogous to (6.20). This is accomplished by means of the chain rule of differentiation; interested readers can perform the exercise themselves by substituting the weighted input sums for the terms z_j in (6.20). The resulting expression is

$$\Delta^2 w_{k,j} = \alpha x_k T_j' \sum_{i=1}^{n} (w_{j,i} e_i T_i'). \tag{6.21}$$

Comparing this expression with (6.20) makes it evident that the sum over the output nodes plays the role of the output error for the hidden layer. In effect, it divides the output error according to the strength of the connection between the hidden layer and the output layer. Equation (6.21) is known as the *generalized delta rule*.

Although the mathematics involved may appear formidable, once these expressions are programmed, the learning process proceeds very directly. Multilayer networks are *universal approximators*; they can be used to learn (approximately) any input-output relationship, without knowing the analytical form of that relationship. This clearly makes them a very powerful learning tool. They have a number of properties, some good and some not so good:[10]

1. The degree of approximation that can be achieved is generally not known in advance. Hence, a multilayer network will learn the desired input-output relationship to the best of its ability, but that may not be very good. The approximation can sometimes be improved by adding more hidden layers or adding more units to existing hidden layers, at the expense of more complexity and longer convergence times.

10. A recent book quotes John Denker (an investigator of speech recognition using neural networks) as saying, "Neural networks are the second best way of doing just about anything" (Russell and Norvig 2002, 585).

2. Connectionist networks, such as figure 6.14, display an important property known as *generalization*. This means that when a network has converged to a stable set of weights, its outputs not only will match the example input-output data on which it was trained but will predict intermediate behaviors as well. In other words, we can apply other inputs not in the training set to the network and predict the corresponding outputs. The accuracy of prediction depends on the smoothness of the surfaces in weight space.

3. Convergence in such networks can be frustratingly slow and typically requires numerous examples. The usual procedure is to run through the entire set of examples with the initial set of weights (which may be randomly selected) before using the update equations. This run through the full set of examples is known as an *epoch*. Thousands of epochs may be required before convergence is obtained. Attempts to increase the convergence rate by increasing the learning rate α may lead to oscillations.

4. Yet another way to increase the speed of convergence is to use a technique called *momentum*. The idea is that nothing succeeds like success. When calculating the current increment in the weight from the ith input to the jth hidden layer, $\Delta w_{j,i}(t)$, we add a fraction of the *previous* increment. This tends to keep the weight changes going in a successful direction. The weight change equation (6.20) is now modified to

$$\Delta w_{j,i}(t) = \alpha z_j e_i T_i' + \beta \Delta w_{j,i}(t-1), \tag{6.22}$$

and similarly with equation (6.21). The quantity β is known as the *momentum parameter*.

It should be noted that the title of this section includes the word "feedforward." Another class of networks also allows for feedback terms from the output units and/or hidden units. Such networks are known as *recurrent networks*. Clearly, they are much more complex to analyze than feedforward networks.

6.7.4 Representation of Individual Neurons

Individual neurons have been represented as weighted sums of inputs to a nonlinearity, the activation function, as in equation (6.10). This representation is highly idealized, since actual physical neurons cannot respond instantaneously to the sum of the inputs they receive from incoming axons. A more realistic model is known as the "leaky integrator" (Arbib 1995):

$$\tau \frac{dy_j}{dt} + y_j = \sum_{i=1}^{n} w_{j,i} x_j + b_j, \quad j = 1, 2, \ldots, n,$$

$$FR_j = f(y_j), \tag{6.23a}$$

where τ is the time constant of the circuit; $w_{j,i}$ is the weight of the connection from the ith to the jth neuron; b represents a bias or threshold value; and FR is the firing rate of the neuron, a nonlinear function of the membrane potential y. Some investigators include the activation function directly in the definition:

$$\tau\frac{dy_i}{dt} + y_i = T\left(\sum_i w_i x_i\right) + b_i, \tag{6.23b}$$

where $T(\ )$ is typically the sigmoid function of figure 6.12(c). Note that in the steady state, when the derivative becomes zero, these expressions revert back to the previous definitions of computational neurons.

6.8 Case Study 6.2: Learning How to Grasp Objects of Different Shapes

This case study concerns the use of a neural network to enable a robot hand to learn how to grasp a cylinder successfully. Robot hands are discussed in detail in chapter 11. Here our emphasis is on the neural-network features. More details can be found in Liu, Iberall, and Bekey 1989.

Human hands are extremely versatile, capable not only of grasping, but of a wide variety of other functions ranging from pointing to punching (MacKenzie and Iberall 1994). The grasping of even simple objects is a surprisingly complex process, depending on the geometry of the object, the purpose of the grasp, and the angle of approach. For example, grasping a knife for cutting is very different than grasping it for the purpose of passing it to another person. Consider a circular cylinder, with diameter approximately that of a human finger, standing on its end. Clearly, it can be grasped by approaching it from the top or from the side; each of these approaches will result in a different placement of the fingers and palm. The specific arrangement of the fingers and palm in grasping is known as a *grasp mode*. The most common human grasp modes are the following, as illustrated in figure 6.15:

1. *Power or cylindrical grasp* The fingers are flexed fully, and the opposed thumb is flexed over the fingers to increase power, as in holding a hammer.

2. *Tip or precision pinch* The tips of the fingernails of the index finger and thumb are brought together, as in lifting a pin from a flat surface.

3. *Hook grip* The fingers are all flexed at the interphalangeal joints and extended at the metacarpo-phalangeal joints, as in carrying a suitcase.

4. *Pulp pinch* The pulps of the index finger and thumb are opposed, with the distal interphalangeal joints extended, as in gripping a sheet of paper.

5. *Spherical grasp* The fingers are flexed, as when holding a ball.

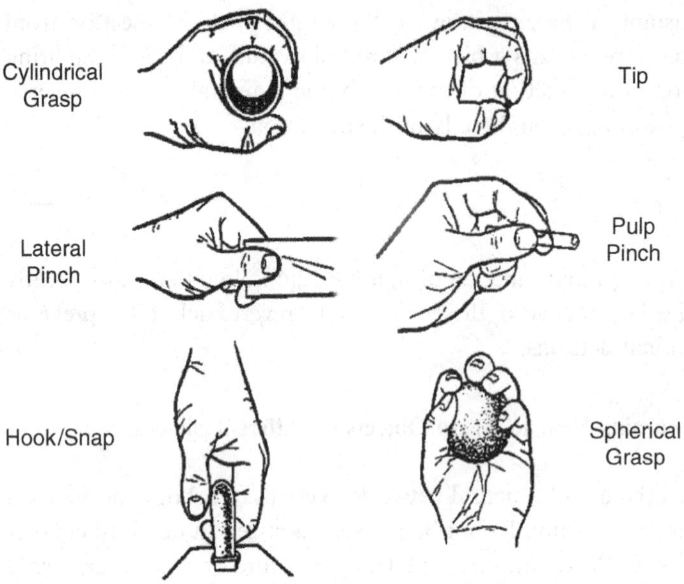

Figure 6.15
Six generic grasp modes (adapted from Mackenzie and Iberall 1994)

6. *Lateral pinch* The pulp of the thumb is opposed to the radial side of the middle phalanx of the middle finger, as in turning a key.

A robot hand may not be able to reproduce all these modes. In the experiment discussed in this case study, we used the Belgrade-USC hand, described in chapter 11 and illustrated in figure 6.16. This is a five-fingered hand capable of performing a power grasp (pg), pulp pinch (pp), lateral pinch (lp), and hook grip (hg).

Consider now a robot task of grasping an arbitrary object. We assume that a camera and vision system provide the robot with the geometric parameters of the object. The robot now must learn which of the grasp modes of which it is capable are appropriate for the given object. For simplicity, we reduced the problem to one of selecting the appropriate grasp mode for cylinders of various lengths and diameters, without considering the approach orientation. Clearly, if a cylinder to be grasped has a small diameter compared to the span of the hand, then pulp pinch may be appropriate. If the cylinder diameter is approximately equal to the width of the palm, a power grasp would be required. Thus, the task of the neural network was to learn the feasible grasp modes for any given cylinder diameter and length within the robot hand's physical limitations.

We selected a network with one hidden layer consisting of four neurons, two input units to represent the cylinder diameter and length, and three output units, as illus-

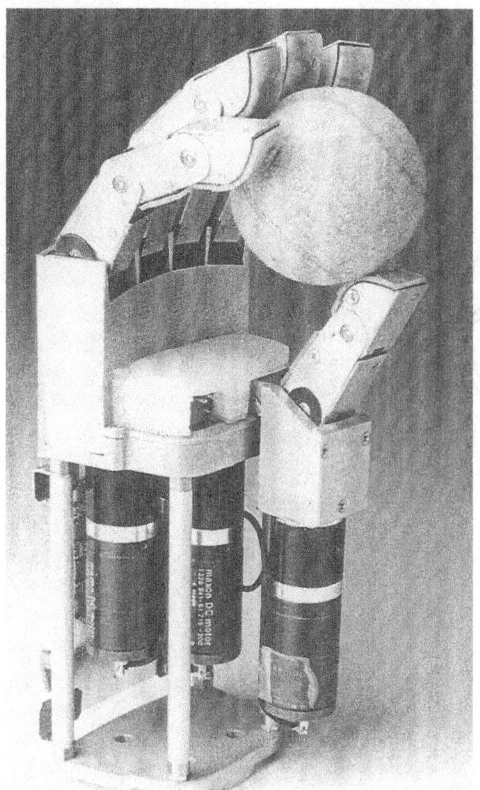

Figure 6.16
Belgrade-USC hand (see chapter 11 for detailed description) (photograph from author's files)

trated in figure 6.17. The number of output neurons was determined by the possible combinations of available grasp modes. The hook grip was not considered appropriate for the cylinders considered here, so there were only three modes available.

The possible combinations of these modes and their binary coding for a given cylinder were

$\{pg, pp, lp\} = 6 = 110$

$\{pg\} = 5 = 101$

$\{pg, pp\} = 4 = 100$

$\{pp\} = 3 = 011$

$\{pp, lp\} = 2 = 010$

$\{lp\} = 1 = 001$

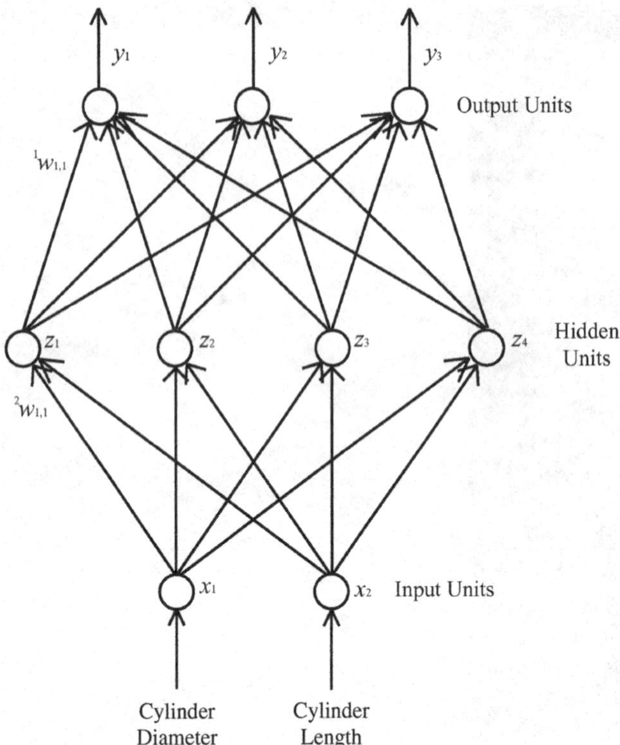

Figure 6.17
Neural network for learning cylinder grasp modes

Thus, if a given cylinder was suitable for grasping using either pulp pinch or lateral pinch, the output was encoded as the number 2. If it could be grasped only by a power grasp mode, it was encoded as 5. We used a binary representation of the numbers 1–6, thus requiring three output nodes. The coding scheme was designed specifically so that if two modes were possible, they fell between the number corresponding to the individual modes. Hence, $\{pp, pg\}$ was designed to fall between $\{pg\}$ and $\{pp\}$. Note that the combination $\{pg, lp\}$ does not appear because it would not make sense, since power grasp is appropriate for large-diameter cylinders for which lateral pinch would not be appropriate. A total of nine examples were given to the network, as listed in table 6.2.

After 4,595 iterations, the network converged on a set of weights for which the outputs correctly matched the training examples. Furthermore, the network generalized very well. We allowed the cylinder diameters to cover the range $\{0.0, 7.0\}$ cm and the height or length of the cylinder to cover the range $\{0.0, 11.0\}$ cm and plotted the numerical value of the output (representing the predicted grasp mode) over this

Table 6.2
Training examples for grasp mode selection neural network

Cylinder diameter (cm)	Cylinder height (cm)	Desired grasp modes
2	3	pp, lp
2	6	pg, pp, lp
2	9	pg, pp
4	3	pp
4	6	pg, pp
4	9	pg, pp
6	3	pg, pp
6	6	pg
6	9	pg

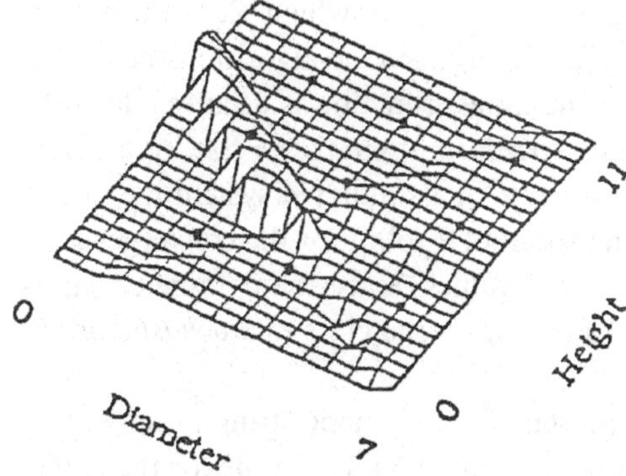

Figure 6.18
Result of simulation of grasp mode selector network showing generalization. The height of the graph indicates which sets of grasp modes were chosen. The nine training points are shown as black dots

entire range. The resulting three-dimensional plot is shown in figure 6.18. The predicted grasp modes agree with observations of human subjects. It should also be noted that attempts to train the network with fewer examples were not successful.

More complex neural networks were used to predict grasp modes for a variety of objects when the task to be performed was also taken into account (Bekey et al. 1993).

6.9 Evolutionary Algorithms

Another approach to robot learning is based on simulated evolution. We know that the evolutionary process in nature has been highly successful in developing organisms that are well suited to their environments. In contrast with other methods of learning or adapting, evolution works on a much longer time scale, requiring many generations. When used in computers, the succession of generations can be at an arbitrary time scale, in contrast with natural evolution, which is limited to real biological time. Like neural-network methods, evolutionary (or genetic) algorithms are basically parameter optimization methods. We have seen in neural networks that methods like back-propagation can be used to optimize the network parameters, the weights associated with each connection. Evolutionary algorithms (EAs) can be used to optimize any parameters, including neural-network weights or design parameters of engineering systems. Outstanding treatments of such algorithms and their application to robotics can be found in Nolfi and Floreano 2000; Hopgood 2000; Gomi 2000; Goldberg 2002; and Holland 1975.

As with biological evolution, computer-based genetic algorithms (GAs)[11] are based on a *fitness function*. The success of a given generation as compared to previous generations is measured by means of the fitness function, which plays the role of the performance index discussed in section 6.7.2. Hence, genetic algorithms belong to the class of hill-climbing methods. In nature, the fitness function is usually related to a survival skill of a species, such as its running speed, its strength, its size, or a coloration that enables it to blend with the environment. In a simulated system the fitness function could be the strength of a bridge, the speed of a robot, or the convergence of the output of a neural network to a set of desired values. Since the parameter variation includes some probabilistic aspects, GAs belong to the class of *stochastic gradient methods*.

In nature, the DNA of an individual is represented as a genetic string using a finite alphabet made of the elements adenine, guanine, thymine, and cytosine, or the letters AGTC. In computers we frequently use the binary alphabet $(0, 1)$ to represent all the adjustable parameters in the system being optimized. In other words, the system learns those parameter values that lead to an optimum value of the fitness function. We simulate N individuals, using a random selection of values for their parameters, and evaluate the fitness function for each. Thus, the selection of a set of values for the parameters of a system creates an individual; N individuals make up a generation. Now we need to look at ways that successive generations are created.

6.9.1 Selection and Reproduction

There are various strategies for selection and reproduction. Assume that we have evaluated the fitness of $2N$ individuals. We then select from these the N individuals

11. Some authors distinguish between *genetic algorithms* and *evolutionary algorithms* or *evolutionary programming*, depending on whether the parameters being optimized are binary strings (for genetic algorithms) or some more complicated representation. I make no such distinction in this book.

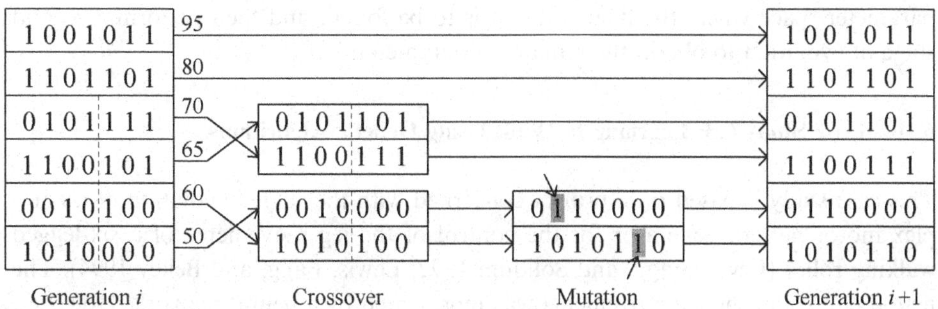

Figure 6.19
Reproduction in a genetic algorithm

with the highest fitness scores. (Alternatively, the fitness score could affect the probability of being selected, with the highest-scoring individual being most likely.) The highest-scoring individuals are immediately slated for the next generation. The remaining individuals are paired and passed to the next generation using a method known as *crossover*. In effect, we assume that each pair consists of a male and a female, and we allow their genes to cross over, as indicated in figure 6.19.

The leftmost column in the figure shows the parameter values of six individuals in the *i*th generation and the corresponding values of the fitness function for these individuals. In the version of the algorithm shown in the figure, the two highest-scoring individuals are simply transferred to the next generation. The remaining four are paired, and in each pair the bit string is divided into two parts, which then cross. In the crossover stage, each individual receives the first part of its bit string from one parent and the remainder from the other. Note that the crossover point could be located anywhere in the string. Then, the last two individuals are subject to *mutation*, that is, a random bit within the string changes from a zero to a one or vice versa. (Although it is 1 in 3 in the current figure, the probability of mutation is generally quite small.) The outcome of this process is a new generation.

Unfortunately, it is not at all clear that this process will converge, and if so, that it will do so rapidly. Hence, some of the same problems we saw with neural networks apply here as well. It has been said that "neural networks are the second best way of doing just about anything, and genetic algorithms are the third" (John Denker, quoted in Russell and Norvig 2002, 585). In the author's experience, both of these methods have been remarkably successful in enabling learning under difficult conditions, as is shown in the case study in the following section. In that study, an approach to increasing the rate of convergence of a GA through a method known as *staged evolution* is also presented. Basically, the idea is to first learn a basic set of parameters that enable us to get into the ballpark, that is, into a restricted area of the parameter space where the final solution is to be found, and then perform a second stage of evolution to obtain the remaining parameters.

6.10 Case Study 6.3: Learning to Walk Using Genetic Algorithms

This case study is based on a project concerned with the staged evolution of a complex motor pattern generator for the control of the leg movements of a six-legged walking robot (Lewis, Fagg, and Solidum 1992; Lewis, Fagg, and Bekey 1994). The neural control of the leg movements was represented by a neural network. However, the network parameters were adjusted using a genetic algorithm rather than backpropagation.

A method for generating and controlling the leg movements, such that they represent a specific gait pattern, is an important aspect of the design of legged robots (see chapter 9). Most designers of four-, six-, and eight-legged machines preprogram the sequence of leg movements. The project discussed in this case study was the first known attempt to apply GAs to the evolution of an actual robot, implemented in hardware (Nolfi and Floreano 2000).

A major problem with applying GAs to actual robots is that each parameter value for each individual needs to be tested in the real world. Such a process can be extremely time consuming. Hence, we have developed an approach we term "staged evolution" to speed up the process. We first provide a relatively easy set of challenges to the system, and the GA quickly converges to an appropriate set of parameters. Successive challenges reduce the size of the parameter space to "islands," as shown in figure 6.20. In the figure, the first challenge is simply to get a robot's leg to oscil-

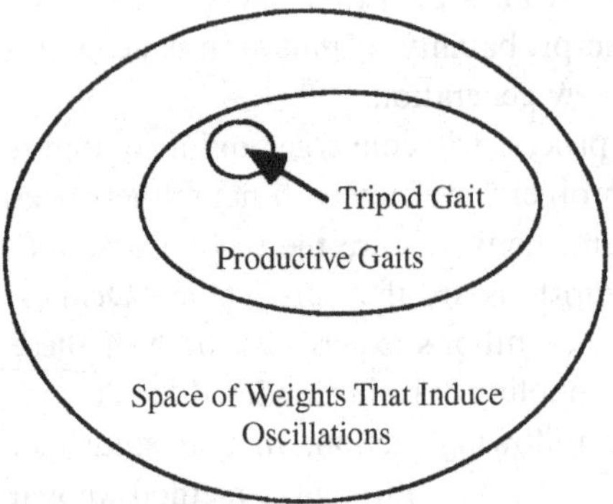

Figure 6.20
Islands of fitness as determined by a human designer. The GA guides the system from looser to more precise controllers as it evolves

late, based on the simple observation that if the legs cannot move rhythmically, the robot cannot walk. Once the robot's legs are able to move in an oscillatory manner, we move to the space of all possible gaits, and eventually to those parameters that represent the desired tripod gait.

6.10.1 Applicable Software

The specific robot used in these experiments, Rodney (figure 6.21), was a six-legged machine. Each leg had 2 degrees of freedom: elevation (rotation about a horizontal axis parallel to the body axis) and swing (rotation about a vertical axis), thus making it a 12-dof robot. Each of Rodney's legs was kinematically similar to those used in Genghis, discussed in chapter 9 (Brooks 1989). The body of the robot was approximately 35 cm long and about 12.5 cm wide. The legs were actuated by Futaba servo motors controlled by an on-board Motorola 68332 processor. Note that the antennae shown in figure 6.21(b) were strictly cosmetic and had no useful function. Rodney and other hexapods are discussed in more detail in chapter 9. The leg sequence that enabled Genghis to walk was originally hand-programmed. A later experiment used reinforcement learning to enable the robot to learn the desired sequence (Maes and Brooks 1990a). The only reinforcement was negative, based on the state of a switch that indicated when the robot's belly was touching the floor. Brooks and Matarić (1993) indicate that Genghis was successful in learning a tripod walking gait, but that the resulting gait was not as robust as that programmed by hand.

The GA simulator used in this study was known as GENESIS (General Neural Simulation System) (Grefenstette 1987); it used the genetic code to generate a neural-network description. This description was input into a neural-network simulator known as Neural Simulation Language (NSL), which simulated and then displayed

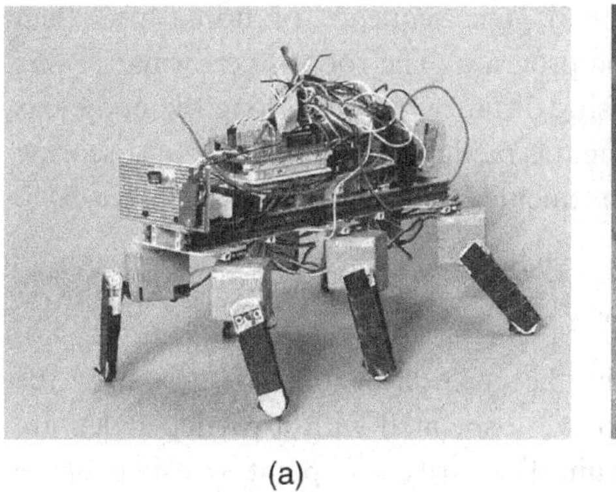

(a)

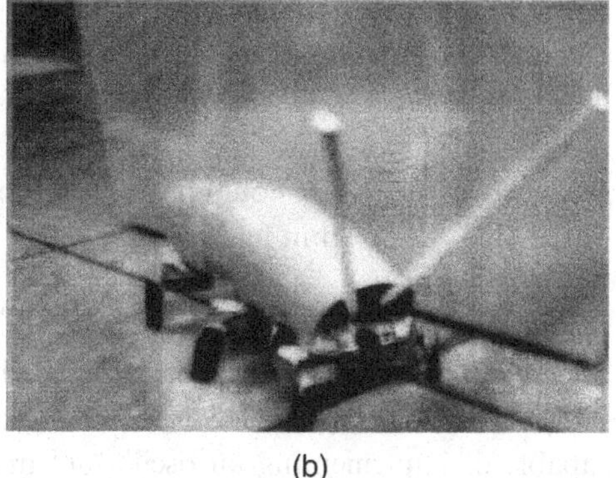

(b)

Figure 6.21
Rodney (a) without shell and (b) with shell, showing antennae (photographs from author's files)

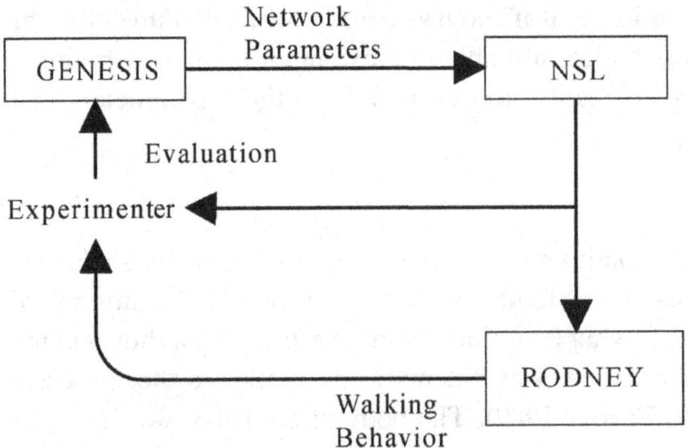

Figure 6.22
Experimental architecture for evolution of gait patterns

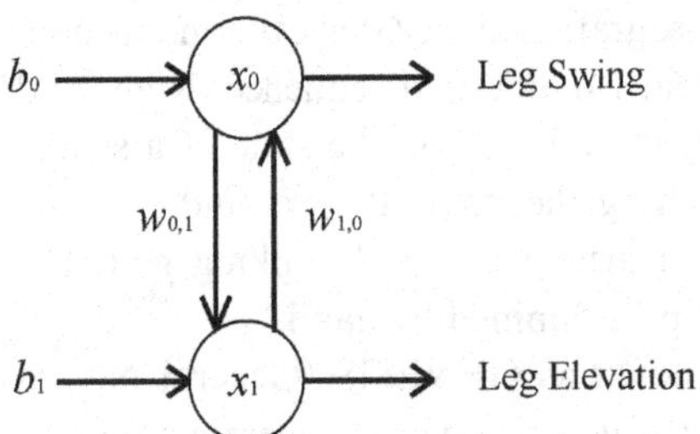

Figure 6.23
Two-neuron circuit for leg oscillation

the neural firing patterns (Weitzenfeld 1995). The sequence of firings was then downloaded to Rodney, which executed the program. The robot's performance was observed and scored by the experimenter using distance traveled along the body axis as the main fitness criterion. Clearly, automatic methods of measuring and scoring Rodney's performance could have been implemented. The fitness score was fed back to the GA, as shown in figure 6.22.

6.10.2 The Neural Model for Leg Movement

In the neural model employed in this study, the position of joint i is driven by the state of a neuron x_i. The two neurons that are associated with a particular leg are capable of implementing an oscillator circuit. To clarify this point, consider figure 6.23, which illustrates the circuit for one leg. Each neuron is represented by an integrator,

$$\frac{dy_i}{dt} = \sum_{j=0}^{1} w_{j,i}x_j + b_i, \quad i = 0, 1, \tag{6.24}$$

where the x_j represent the inputs to the neuron, y is the output of the neuron, $w_{j,i}$ is the weight from neuron i to neuron j, and b is the threshold value.

Now, it is clear from figure 6.23 that each neuron has only one input and one threshold value. Using the notation in that figure, we can replace (6.24), for the first neuron, with

$$\dot{x}_0 = w_{1,0}x_1 + b_0. \tag{6.25}$$

Similarly, the second neuron can be represented by

$$\dot{x}_1 = -w_{0,1}x_0 + b_1. \tag{6.26}$$

Differentiating (6.26) we obtain

$$\ddot{x}_1 = -w_{0,1}\dot{x}_0.$$

Substituting (6.25) into (6.26) yields

$$\ddot{x}_1 + w_{0,1}w_{1,0}x_1 = -w_{0,1}b_0.$$

If we let $b_0 = 0$ and $w_{0,1}w_{1,0} = \omega^2$, we obtain

$$\ddot{x}_1 + \omega^2 x_1 = 0. \tag{6.27}$$

A similar equation describes x_0. Both equations represent a harmonic oscillator. Given proper initial conditions, their solutions will be sine or cosine functions, and the outputs of the two neurons will be 90° out of phase. Hence, the circuit of figure 6.23 can lead to a sustained oscillatory movement the leg.

In the study, the neural states were initially set at random values. The first phase of genetic learning was devoted to finding a set of parameters that would result in oscillatory motion of the legs, when the outputs of the two neurons were connected to the leg joint drive motors. This was accomplished by using the neural output signal x_i to control a pulse-width-modulated signal that in turn controlled the position of a leg motor through the relation

$$\theta_i = m_i x_i + c_i, \quad i = 0, 1. \tag{6.28}$$

The constants m_i and c_i were adjusted so that a neural signal $x \in [0, 1]$ produced a leg joint movement $\theta \in [-60°, +60°]$. The resulting oscillations are shown in figure 6.24; it is evident that after a transient period, they are 90° out of phase. Note that if one unit is connected to the swing motor and the other to the elevation motor of the same leg, and they both oscillate out of phase, the leg will execute a stepping motion. If it

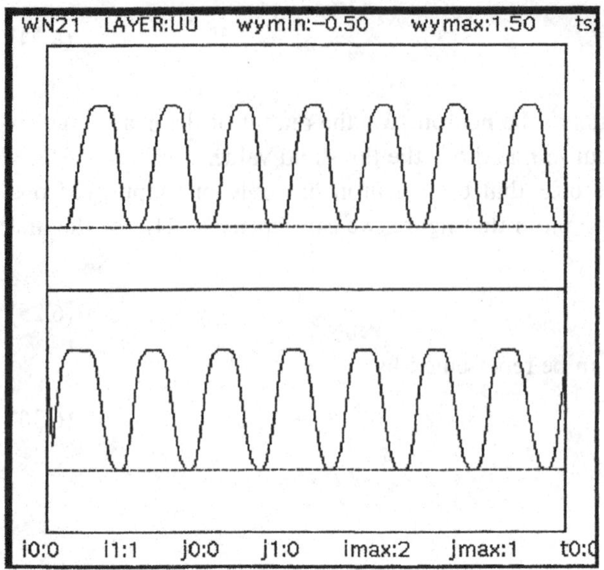

Figure 6.24
Outputs of leg movement neurons

is in touch with the ground, it will display both a swing and a stance phase (chapter 8). Hence, achieving oscillatory motion is the first step in the staged-evolution process described earlier in connection with figure 6.20.

The same parameter values were then replicated in Rodney's other legs. Note that although all the legs were now able to oscillate, the robot could not walk, since all the legs moved nearly simultaneously. Actually, it was rather humorous to watch Rodney on his belly, with all six legs flailing, but unable to move since its legs were not coordinated.

To approach the question of determining a desirable sequence of leg motions, we added connections among the various oscillators, as illustrated in figure 6.25. Note that all six leg oscillator neurons are shown. The connections between pairs of legs at the same level (i.e., front, middle, or rear) have weights denoted as A and B, and the connections between levels (front to middle and middle to rear) have weights denoted as C and D. The choice of equal weights in these locations was a reflection of the symmetry in the construction of the robot. Hence, in the second phase of learning, the robot was required to find values of A, B, C, and D so that it sustained a forward walk at the highest possible speed.

We now discuss the evolutionary processes leading first to the oscillator parameters and then to the leg sequence parameters, beginning with the fitness functions and then their use in a GA.

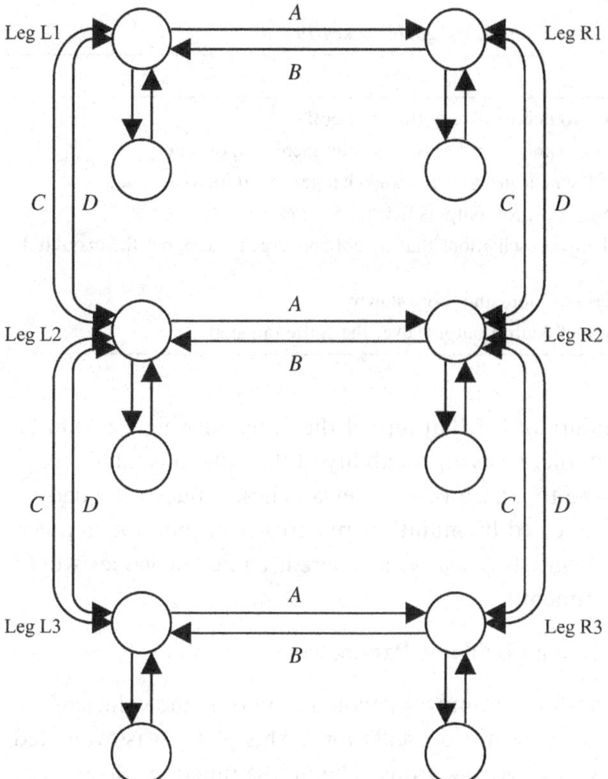

Figure 6.25
Complete neural network for hexapod walking

6.10.3 The Genetic Algorithm

The genetic string for this experiment had to include all eight of the parameter values to be learned through evolution, namely, $w_{0,1}, w_{1,0}, b_0$, and b_1 for the oscillator parameters, and A, B, C, and D for the interconnection weights between the legs. Each parameter was represented by eight bits using a Gray code.[12] These parameters required a total of 64 bits. We actually used a 65-bit string, where the 65th bit altered the mapping from the leg swing neuron to the joint actuator so that a value of one caused forward motion and a value of zero caused backward motion.

We used ten individuals in each generation. As in the algorithm depicted in figure 6.19, the two highest-scoring individuals were automatically promoted to the next

12. In a Gray code, a set of binary numbers is ordered such that only one bit changes from one entry to the next.

Table 6.3
Fitness function for oscillator evolution (from Lewis, Fagg, and Bekey 1994)

Score	Description
0	The outputs of the two neurons are both >0 or both <0
5	One neuron state is >0 and the other is <0 for an increment of time
10	For at least one of the neurons, $\dot{x}(t) > 0$ and changes to <0 (or vice versa)
25	Damped oscillations of neuron outputs that reduce to zero
40	Neurons display damped oscillations that do not converge to zero, but the magnitude of oscillation is <1
41–59	Oscillation that dies out more and more slowly
60	Sustained oscillation of neuron outputs over the entire range $[0, 1]$.

generation. There was a probability of 0.1 that any of the remaining eight would be paired for crossover. In addition, there was a probability of 0.04 that any single bit in a string would mutate from a one to a zero or vice versa. These values were chosen so that the weights were indeed affected by mutation and crossover, but not so much as to eliminate high-scoring individuals. Clearly, a different choice of values would change the outcome of the experiment.

6.10.4 Fitness Function for Evolving Oscillator Parameters

The first phase of the evolutionary algorithm was devoted to finding the values of the six weights in figure 6.23 leading to sustained oscillations. This phase was evaluated visually, by observing the behavior of two neurons. The fitness function was a score that enabled the experimenter to take the network through a series of stages leading to sustained and consistent oscillations. Since these evaluations were based strictly on observation, they did not require any understanding of the internal structure of the neurons. The stages and the corresponding fitness scores are shown in table 6.3. This table shows that even when the score equals 5, we are enforcing an out-of-phase relationship between the two neurons; progressively better approaches to sustained oscillations bring further increases in the fitness function.

In this first stage of evolution, only the first four parameters were adjusted, starting from random values in $[-8.0, +8.0]$. Given the GA parameters mentioned earlier, the first phase of evolution typically required seven to seventeen generations before at least half of the population began to oscillate over the entire $[0, 1]$ scale, thus ensuring that the legs would oscillate over their full range. In all cases, the parameters produced a solution in which the two neurons fired 90° or 270° out of phase. The two solutions occurred with equal probability. Figure 6.26 shows a plot of the best and average scores versus generation for the first phase of learning.

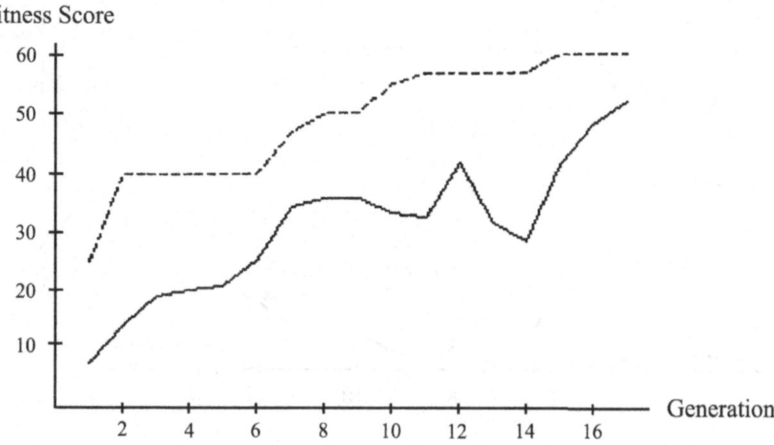

Figure 6.26
Fitness score versus generation for evolution of an oscillator. The solid line is the average performance of the population; the dotted line is the score of the best-performing individual in each generation. Note that the best individual required seventeen generations to reach a score of 60

Table 6.4
Fitness values for walking behavior (from Lewis, Fagg, and Bekey 1994)

Score	Behavior
0	No movement and no oscillations of legs
$10 + L - \frac{T}{10}$	Oscillations and movement backward
$10 + L - \delta\frac{T}{10}$	Oscillations and movement forward

6.10.5 Fitness Function for Walking

At the end of the first stage of evolution the robot has learned a set of parameter values that allow the legs to move in a sustained, full-amplitude oscillation. In the second stage, the values of the weights A, B, C, and D shown in figure 6.25 must be learned to obtain a proper walk, that is, consistent movement in the direction of the body axis. Thus, the genetic string for the second phase specifies not only the circuitry for the oscillator, but also the set of connections between the oscillators.

A single parameter specifies the value of several weights. For example, the connections that cross the midline from the left to the right side take on identical values, irrespective of the location along the spinal cord. Evaluation during this stage is considerably more simple than in the first. Scores are assigned as shown in table 6.4.

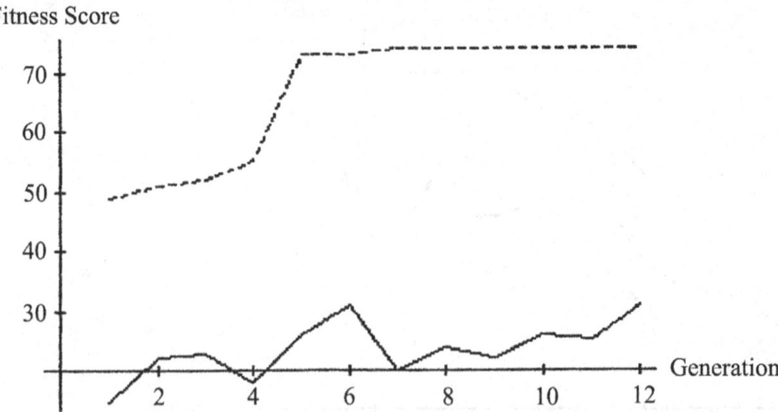

Figure 6.27
Fitness score versus generation for the walk controller. The solid line shows the average performance of the population; the dotted line shows the performance of the best individual in each generation

In this table, L is the number of centimeters walked in a straight line along the orientation of the initial axis of the body per unit time, T is the number of degrees turned during the walk, and δ is a number designed to distinguish forward from backward motion. Note that the T term is intended to favor solutions that keep Rodney oriented along the direction of travel.

Note that walking backward is rewarded, as is walking forward. The reasoning behind this stems from the fact that similar parameters are required for both. The only difference is the phase relationship of the swing and elevation neurons.

After stable, full-scale oscillations of the legs were reached, the second phase of evolution required ten to thirty-five additional generations before the fitness score reached a plateau. Figure 6.27 plots the best and average scores versus generation for the second phase of learning.

By the close of each experiment with the genetic algorithm, Rodney had learned to produce a *tripod gait*. In this gait, the left-front and left-rear legs move in phase with the right-middle leg. Thus, at any one time, at least three legs are on the ground. This gait is frequently seen in insects.

We indicated previously that Rodney had evolved a tripod gait by the end of each of the experiments. However, insects use other gait patterns as well (Beer 1990). One of the more common is the wave gait, in which corresponding pairs of legs move at the same time. With reference to figure 6.25, legs L1-R1 would move, followed by L2-R2, then L3-R3, and then the pattern would repeat. Thus, a wave gait is characterized by rear to forward waves of leg motions, with four legs on the ground at the same time. Since the tripod gait leaves only three legs on the ground at once, we would expect the wave gait to be somewhat slower. During several experiments, particular generations produced both individuals that performed the tripod gait, and those that generated the wave gait. In general, this splitting between two very different solutions lasted for several generations before the tripod gait came to dominate the population. That the tripod gate would eventually emerge as predominant would be expected, since with only three legs on the ground at the same time, the tripod gait tends to be more efficient than the wave gait. Once the tripod gait began to make fine adjustments to the weights, it very quickly outperformed the wave gait.

One of the surprising results of our experiments was that in all the experiments (in which $\delta = 2$ in the fitness function definition), Rodney preferred to walk backward. This preference appeared to be due to the angle at which the feet push against the floor in the particular mechanical structure of this robot. This result could not have been anticipated without a detailed analysis of the dynamics of the robot's interaction with the environment.

Further experiments, with $\delta = 4$, resulted in the domination of the forward-walking controllers. The results of these experiments indicated that around $\delta = 3$, the population would probably maintain an even distribution of both types of controller. Such results can give us insight into the redesign of the controller software, or even of the robotic hardware.

6.10.6 Conclusion

The case study described in this section has demonstrated a method for the practical application of GAs to real robots in which the controller is implemented using neural networks. It shows that it is indeed possible for a robot to learn to walk by means of proper leg coordination.

6.11 Case Study 6.4: Learning in the Tropism Architecture

The basic ideas of the tropism architecture, which is based on the "likes" and "dislikes" of the robot, were introduced in section 5.8. Learning in the tropism architecture takes place on two time scales: *ontogenetic* learning takes place within a given generation of robots, using a variation of reinforcement learning, whereas *phylogenetic* learning uses evolutionary algorithms to extend the learning over successive generations. The tropism architecture is shown in figure 5.12, which is repeated as figure 6.28 for convenience.

6.11.1 Ontogenetic Learning

The ontogenetic (individual) learning tropism architecture allows a robot to dynamically change its tropism elements, based on its experiences in the world. The robot

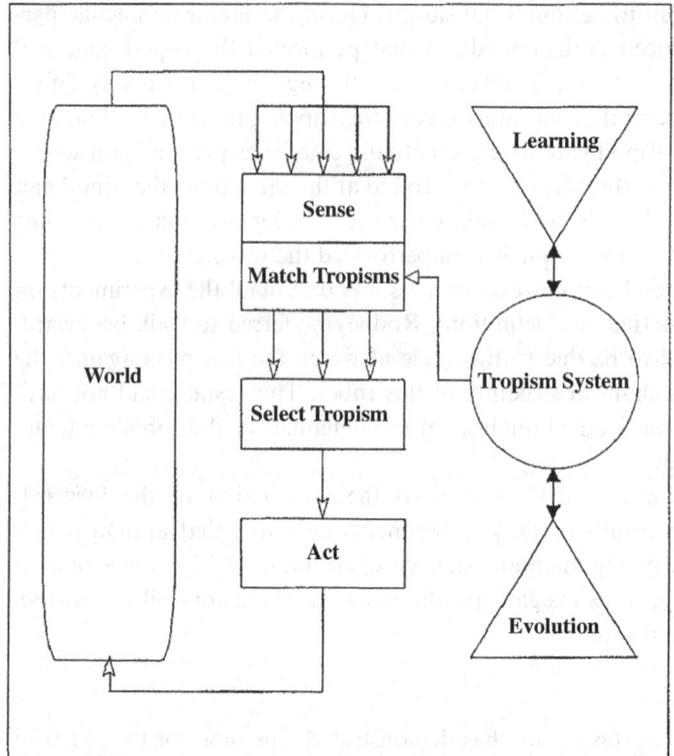

Figure 6.28
Tropism system architecture, showing two types of learning (phylogenetic and ontogenetic)

should be able to autonomously add tropism elements and modify the existing ones. Tropism elements are deleted through modification of the actions of the existing elements. The architecture makes it possible for the robot to learn in three types of situations: learning from perception, learning from success, and learning from failure.

6.11.1.1 Learning from Perception

Learning from perception occurs when a novel perceptual situation is encountered. A perceptual situation is novel for a robot the first time it senses a particular entity in a given state and finds no appropriate tropism for handling the situation. In order for the robot to deal with the novel situation, the learning tropism system must automatically develop a new tropism element that covers the novel situation. The system selects a random action for the novel situation and assigns an initial tropism value to it. The newly added tropism element may prove to be useful to the robot (if similar situations are encountered in the future), or it may not. (Both of these cases are handled by learning from failure and learning from success, as discussed in sections

6.11.1.2 and 6.11.1.3, respectively.) Denoting the sensed entity by ε and the entity's state by σ, the new tropism element will be associated with the random action α_{random} and have the preset initial tropism value $\tau_{initial}$:

$$\Sigma(t+1) = \Sigma(t) \cup \{\varepsilon, \sigma, \alpha_{random}, \tau_{initial}\}. \tag{6.29}$$

The predetermined initial tropism value can be replaced by a random setting of such value.

We assume that each robot can maintain only a finite number of tropism elements. This means that eventually the robot will reach this limit and then encounter a novel perceptual situation. In such a case an existing tropism element must be removed from the system to make space for the newly created tropism element. The element to be deleted is selected based on the chronological order of the element formation. Among all elements of the set of tropisms, the oldest one (i.e., the least recently created) will be deleted, and the new tropism element will then be added. Denoting the oldest element by $(\varepsilon^o, \sigma^o, \alpha^o, \tau^o)$, the new tropism system will be determined:

$$\Sigma(t+1) = \{\Sigma(t) - (\varepsilon^o, \sigma^o, \alpha^o, \tau^o)\} \cup \{(\varepsilon, \sigma, \alpha_{random}, \tau_{initial})\}. \tag{6.30}$$

The deletion of a tropism element in the tropism system architecture is analogous to "robot forgetting." The oldest learned concept is the first one forgotten. Two other methodologies could be utilized to select the candidate for deletion. In the first method, a count variable is associated with each tropism element, enumerating the number of times an element has been used. The deletion procedure would then select the least-used element. The second alternative would be to time-stamp each element as it is used. The deletion procedure would then select the most dormant element, that is, the element least recently used.

6.11.1.2 Learning from Success

Once a tropism element is selected by the robot and the associated action proves to be feasible and useful, the action will be called a success, and the robot can learn from the experience. Learning takes place by means of an increase in the value of the tropism element in question:

$$\Sigma(t+1) = \{\Sigma(t) - (\varepsilon, \sigma, \alpha, \tau)\} \cup \{\varepsilon, \sigma, \alpha, \tau + \tau_{increment}\} \tag{6.31}$$

The tropism increment value (i.e., the amount by which the value of a tropism element is increased each time it is successful) is predetermined, although it is possible to assign different increment values based on an action's outcome. Since there is an upper bound to the tropism value in an element, an increase of the tropism value beyond the maximum is not possible. Learning from success strengthens not only the elements that were initially part of the system, but also those that have been created as part of learning from perception.

6.11.1.3 Learning from Failure

Learning from failure takes place in cases in which an action selected by the robot proves infeasible. The randomness that is associated with action selection in learning from perception makes the robot's selection of infeasible actions possible. The robot does not know how to deal with a novel situation, and hence it makes a guess. Learning from failure enables the robot to recover from a wrong guess. In cases of failure, the action associated with the tropism element selected is replaced with a new random action. The tropism value is not changed, as it is most likely the initial value, since the element has not been used:

$$\Sigma(t+1) = \{\Sigma(t) - (\varepsilon, \sigma, \alpha, \tau)\} \cup \{\varepsilon, \sigma, \alpha_{\text{random}}, \tau\}. \tag{6.32}$$

The ability to learn from failure is also useful in cases in which the robot's world changes, so that actions once possible are no longer feasible. The tropism elements with which such actions are associated will eventually be modified by the system.

6.11.1.4 Tropism System Cognitive Architecture and Reinforcement Learning

The learning described in the foregoing can be considered a form of reinforcement learning, as described in section 6.4, since feedback from the environment is used to adjust parameters in the learning system. However, many reinforcement systems have a credit assignment problem (Minsky 1961): The feedback signal simply indicates that the performance is "good" or "bad" and does not indicate which individual action may require adjustment. In the Tropism System Cognitive Architecture, the feedback provided to the robot refers directly to the particular action in question, whose tropism value is then increased or decreased. It can also be noted that the feedback is obtained from sensors, so no teacher is required.

Tropism learning is a form of unsupervised learning. In principle, Q-learning could be applied to the tropism architecture as well.

6.11.2 Phylogenetic Learning

The learning described in section 6.11.1 concerns the changes in performance of the robots during a particular experiment. We are also interested in further changes in the tropism systems that can occur through evolution. Such changes, which are passed on to succeeding generations, are a form of phylogenetic learning and are achieved by means of the automatic updating of the tropism system architecture through the evolution of the system. Using genetic algorithms, as described in section 6.9, a colony of robots can be evolved for a number of generations, improving the performance of the colony. Techniques of fitness determination, selection, crossover, reproduction, and mutation are applied to the robot and the robot's chromosomal representation. The sequence of steps involved in evolution of robot colonies is shown in figure 6.29.

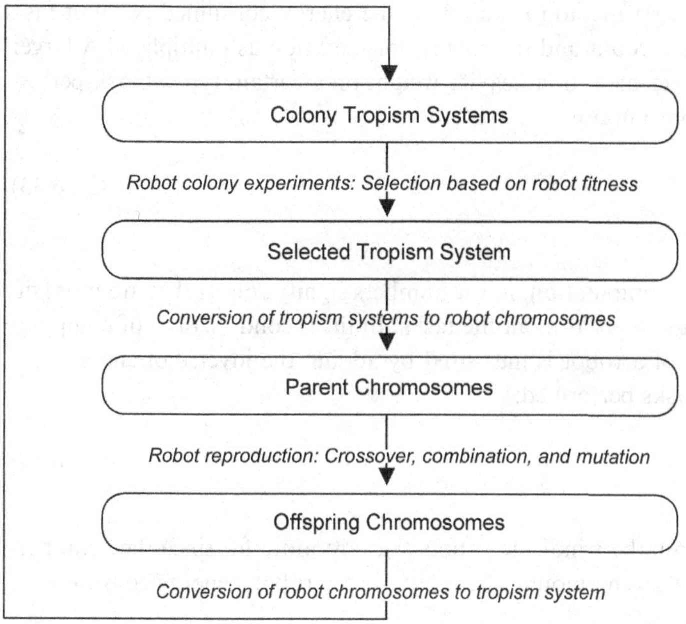

Figure 6.29
Evolution of robot colonies, resulting in phylogenetic learning (illustration courtesy of Arvin Agah)

Basically, evolution of a robot colony follows the same process as described in section 6.9. Here we highlight some of major features of the earlier discussion and the differences between that discussion and robot colony evolution.

6.11.2.1 Robot Selection Based on Fitness

Robots are selected for reproduction based on their fitness, which is measured by the number of objects they gather and the energy they consume during an experiment. Different types of objects may be gathered, which need to be counted separately. The selection process is based on a biased roulette wheel, as described in section 5.8. The portion of the wheel assigned to a particular robot is proportional to that robot's fitness. Selection by means of the roulette wheel gives preference to those robots having larger portions, as they are more likely to be chosen for reproduction. Assume that the robots perform n different tasks, $i = 1, 2, \ldots n$. For example, a task might be to gather small objects from a given area. For the i-th task, the number of actions (say number of small objects gathered) is denoted by p_i. The total energy consumed by the robots in the performance of all the tasks is denoted by e. Multipliers λ_i are used to assign different weights (strengths) to the different types of tasks the robot performs. An energy multiplier λ_e is used to assign a weight to the energy consumption of the robot. The robot's total fitness Φ can then be computed using two different methods. The first method is based on the energy consumed per unit task performed, using the task count and the energy consumption as multipliers. A larger multiplier implies the placement of a heavier weight on a certain type of task performance or the energy consumption:

$$\Phi_1 = \frac{\lambda_e e}{\sum_i \lambda_i p_i}. \tag{6.33}$$

In this method of fitness computation, lower numbers signify robots that are more fit, since they use less energy to perform all the tasks. In the second method of computation, the overall fitness of a robot is measured by adding the inverse of energy consumption to the total tasks performed:

$$\Phi_2 = \frac{1}{\lambda_e e} + \sum_i \lambda_i p_i. \tag{6.34}$$

In this method, larger numbers indicate robots that are more fit, since they can perform more tasks using a given amount of energy. As the robot's energy consumption increases, its fitness decreases.

The fitness function used in genetic algorithms is scaled during the determination of the most-fit parent robots. This fitness scaling is needed to prevent a few very fit individual robots from initially accounting for a significant proportion of the population, preventing formations and explorations of new characteristics. Another problem with the fitness function is associated with the later stages of the evolution, when the population average fitness and the population best fitness are close to one another. When this occurs, improvement in population performance may cease altogether. Linear scaling of the fitness values is a heuristic solution that can help alleviate both of these problems:

$$\Phi' = a\Phi + b, \tag{6.35}$$

where Φ' is a scaled value of the fitness function of (6.33) or (6.34). The coefficients of scaling a and b are computed at each generation, using the minimum, maximum, and average fitness values before the scaling.

6.11.2.2 Chromosome Changes during Evolution

Two methods are employed to alter the genetic characteristics of the individual robots from generation to generation: crossover and mutation, as described earlier. Both crossover and the mutation can result in robots' having chromosomal strings that are invalid once converted to the robots' cognitive architecture. The invalid tropism elements in the tropism system that result from these invalid strings will be unusable and remain as such until the next generation of the colony.

6.11.3 Results

In the case study explored in this section, both approaches to learning, ontogenic and phylogenic, were applied to large simulated colonies of simple robots (Agah and Bekey 1997c). A large number of experiments were performed, and the results were tabulated, plotted, and analyzed. Since robot actions were determined using the roulette wheel random selection mechanism, the robots could select incorrect actions. To average out some of the variability arising from this nondeterministic process, all experiments were conducted three times. In all of the plotted graphs shown in this section, each data point is the result of the averaging of these three separate, but similar, experiments.

The performance of both individual robots and the colony as a whole were measured, as indicated in the following. The total time of the experiments was set at a constant level for all experiments. The colony of simulated robots was embedded within an artificial world, containing obstacles, objects, and predators. Obstacles were either stationary or mobile and could not be moved by the robots. Robots had to navigate around the obstacles. Objects were entities that could be manipulated and processed by the robots. Small objects were gathered by the colony members, and large objects were decomposed into small objects that could then be gathered. Predators were hostile, mobile entities that could attack the robots or be attacked by the robots.

The tasks required of a robot consisted of gathering of small objects, gathering of large objects (requiring two robots), decomposing of dual objects, and attacking of predators. The task performance of the colony was measured in terms of the total number of gathered objects, the total number of decomposed objects, the total number of attacked predators, and the energy consumed. In addition, the changes in selected tropism values were used as an indication of the learning taking place in the robots.

6.11.3.1 Ontogenetic Learning Results

The experiments in ontogenetic learning were completed in simulation while the robots were performing gathering tasks. These experiments employed a colony of thirty robots, and the learning (i.e., adjustment of tropism values) took place over the course of thirty experimental trials. The tropism values acquired by a robot were retained at the end of each trial, the world resources were replenished, and the experiment was conducted for the next trial. The experiments included 1,000 small objects in the world to be gathered.

The results of typical experiments are shown in figure 6.30. In figure 6.30, the actual data are plotted using a heavy, solid line and the least-squares curve fitted to the data is shown as a dashed line. It should be noted that the vertical axes do not begin at zero, thus exaggerating the random variability between trials.

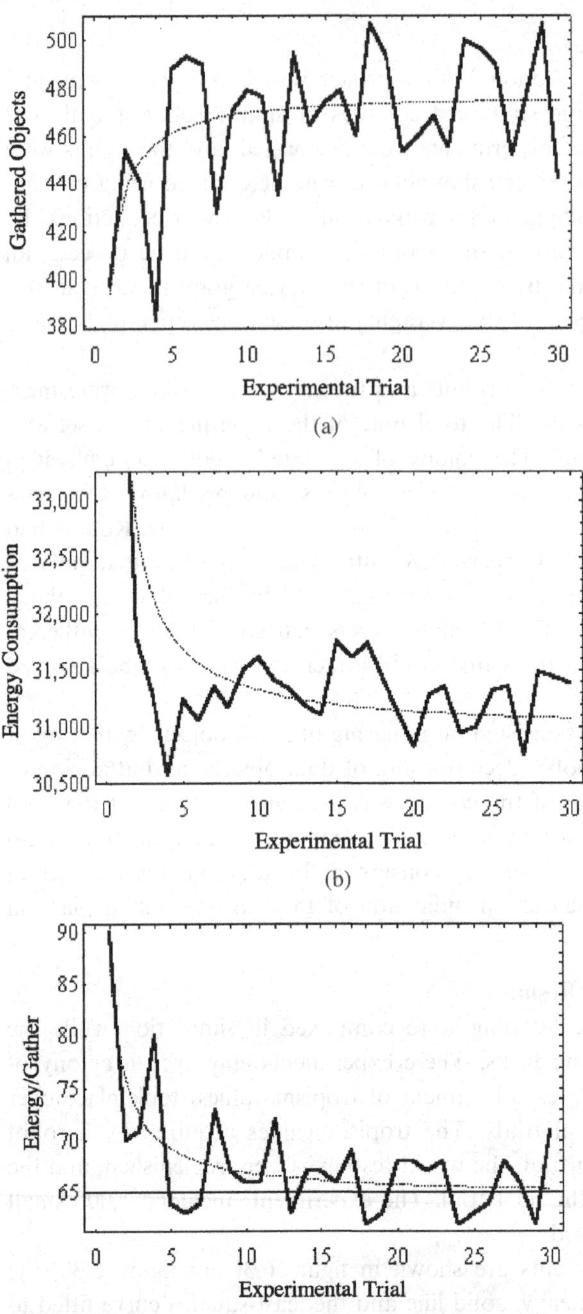

Figure 6.30
Ontogenetic learning: Improvement in colony performance with successive trials (illustration courtesy of Arvin Agah)

It can be seen in figure 6.30(a) that the task performance of the colony improved as a result of the learning but then leveled off, as the number of robots and the run time were limited so that the robots were not able to gather all the objects in the allowed time. The energy consumption of the colony trials decreased as a consequence of learning (figure 6.30(b)). Dividing the energy expenditure by the number of objects gathered provides a measure of the cost per unit of the task performed by the colony. As can be seen in figure 6.30(c), this measure decreased by about 25% over the thirty trials, as the robots became more efficient. The value of this measure seems to reach its minimum at about fifteen trials; that is, the robot colony learned as much as was possible during the first fifteen trials, with the given learning-system parameters.

It is evident from this figure that learning did indeed occur in the colony, since its ability to gather objects increased (on the average) with practice. The actual degree of improvement is a function of the parameters that specify the mapping from tropism to action, as well as the initial distribution of robots and objects in the world, the initial values and range of the tropism values, and other parameters, but the learning pattern they illustrate is typical of the results we obtained.

6.11.3.2 Phylogenetic Learning Results

A number of additional simulation experiments were performed using genetic algorithms to study the effect of simulated evolution on the ability of the robot colony to improve its task performance in the object-gathering task. In these experiments a colony with a population of eighty was evolved for sixty generations. The robots' world resources (objects to be gathered and energy available) were replenished at the end of each generation. One thousand small objects were uniformly distributed throughout the robots' world. The results of the experiments are shown in figure 6.31.

It can be seen in the figure that simulated evolution enabled the colony to increase the fraction of total objects it was able to gather from about 60% in the early generations to about 90% in later generations. The energy consumption of the generations decreased as a consequence of evolution, leveling off at about forty generations, as did the energy cost per object gathered.

6.11.4 Conclusion

Two classes of learning were examined in this study: phylogenetic learning of the colony through simulated evolution and ontogenetic learning of the individual robots through reinforcement learning, both in the context of a Tropism System Cognitive Architecture. This architecture, based on the likes and dislikes of the robot, enables the robot to transform its sensory information into action. The robot's dynamic tropism system learns from experience and changes tropism values, adds new tropisms, and extinguishes old and nonperforming ones.

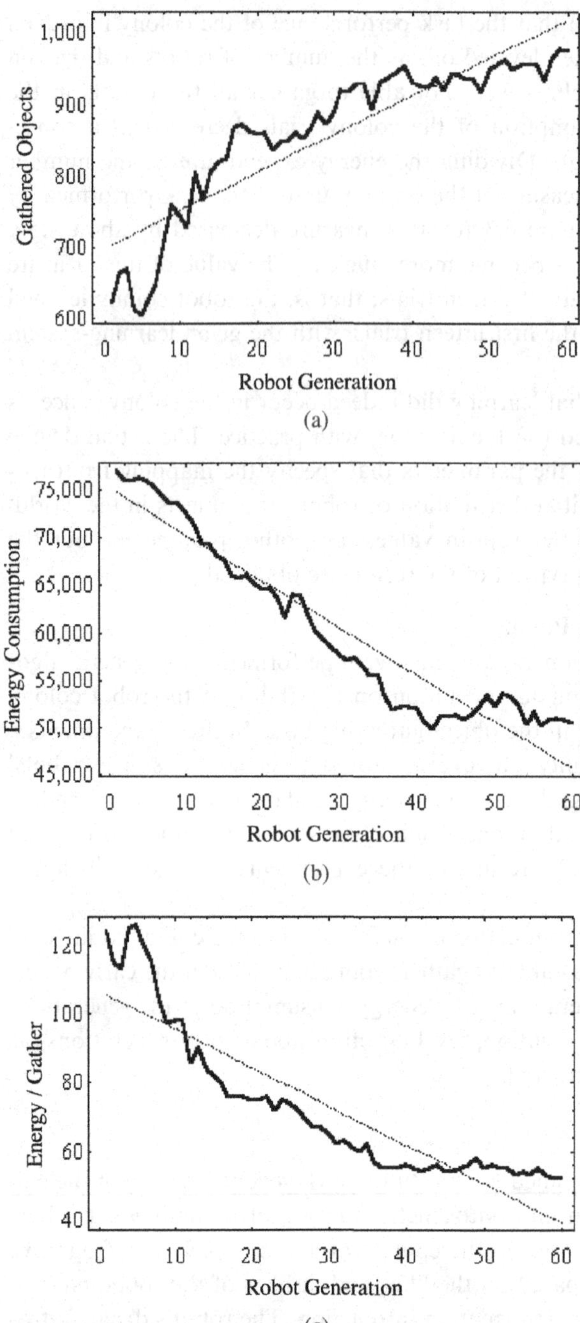

Figure 6.31
Phylogenetic learning: Improvement in colony performance with successive generations (illustration courtesy of Arvin Agah)

The simulated evolution of the robots, using genetic algorithms, generated better colonies (i.e., colonies that performed global tasks better, as defined by objective criteria). A major limitation of the work is that the results were obtained mainly through simulation, rather than through hardware experimentation. Hence, the hardware limitations of real robots were not considered.

6.12 Learning by Imitation

In recent years there has been increasing interest in robot learning by imitation, either of a human or of another robot. This approach to robot learning, also known as *learning from demonstration* or teaching by showing (introduced by an example in case study 5.2), is completely different from those reviewed in previous sections of this chapter. Its foundation is not in artificial intelligence, but rather in neurophysiology. We begin with the biological foundation. In humans, learning movements by imitating those of others is well accepted (see subsequent discussion). In robotics, imitation provides a way of teaching new skills to robots, particularly humanoid robots, without extensive programming.

6.12.1 Mirror Neurons and the Neurophysiology of Imitation

Imitation is a fundamental approach to the learning of movement in both humans and animals. Human babies imitate their mothers' movements and facial expressions; athletes and dancers attempt to imitate the movements of their teachers and role models; even workers on assembly lines imitate the motions of their mentors.

In recent years the neurophysiological basis of imitation has become both more clear and more mysterious (Arbib 2000). It is now known that there are neurons in area F5 of the premotor cortex, known as *mirror neurons*, that fire *both* when a particular movement is performed and when the same movement is seen in another individual (Rizzolatti et al. 1996). This is clearly remarkable, since it attests to a very close coupling of perception and action in the brain. Such neurons are known to exist in primates, including humans. The significance of these neurons is not known, but it is reasonable to assume that they have evolutionary importance, in enabling an individual to recognize particular motions that may have significance in avoiding danger or locating food sources. Mirror neurons provide the neurophysiological basis for imitation, since they indicate that imitation has a fundamental basis in the evolution of primates.

There is an even more remarkable story surrounding these neurons. In humans they are located in Broca's area in the brain, an area associated with the development of speech. Arbib (2000) has suggested that this location implies a very close coupling between imitation of movement and the development of speech. For example, the location of these neurons may be related to the role of gestures in communication in general and as adjunct to speech in particular. The existence of mirror neurons may also suggest that sign languages, such as American Sign Language, used by the deaf, may be a natural mode of communication, as natural as speech (Arbib and Rizzolatti 1996).

There are also psychophysical studies on imitation, many having their foundation in the pioneering work of the Swiss psychologist Jean Piaget (e.g., Piaget 1962). We can conclude from this very brief overview that imitation has a deep biological foundation. However, this does not mean that if we desire to build robots capable of imitating human movements, we must equip them with mirror neurons. On the contrary, the biological basis provides us with a justification for pursuing the subject, but we need to do it in a way that maximizes the use of resources and yields the best possible result.

6.12.2 Levels of Imitation

Given that imitation has a biological basis, we now must decide how to represent it in robots. One possible approach is to simulate behavior clear down to the neural level and include mirror neurons in the model. At the other extreme, one could design a robot system to perceive and imitate a behavior, such as stacking blocks, with no concern with fundamental mechanisms. In practice, most attempts to enable robots to learn by imitation fall somewhere between these two extremes. The work of Matarić and her students is fairly close to biology (Weber, Jenkins, and Matarić 2000; Amit and Matarić 2002; Matarić 2001). Based on the work of Mussa-Ivaldi, Matarić begins with the observation that vertebrates make use of stored motor programs called *motor primitives* (Mussa-Ivaldi, Giszter, and Bizzi 1994) and uses these primitives to represent the movements being imitated. She then attempts to break a complex movement into an assembly of primitive movements and to synthesize new movements from the primitives, as described in section 6.12.3.

Another biologically inspired approach to learning by imitation is seen in the work of Kuniyoshi et al. (2003), based on the imitative behavior of newborns (Meltzoff and Moore 1977). We describe this approach briefly in section 6.12.4.

The work of Atkeson and Schaal (1997b) falls at the other extreme of this continuum. Their goal is to reproduce a human motion, like juggling or balancing an inverse pendulum, without concern for its physiological foundations. Hence, they represent the motion by parametric or nonparametric mathematical models and proceed to identify the parameters of these models that yield the closest fit to human data.

We examine both types of approaches to learning by imitation briefly in the following paragraphs and then discuss some other results.

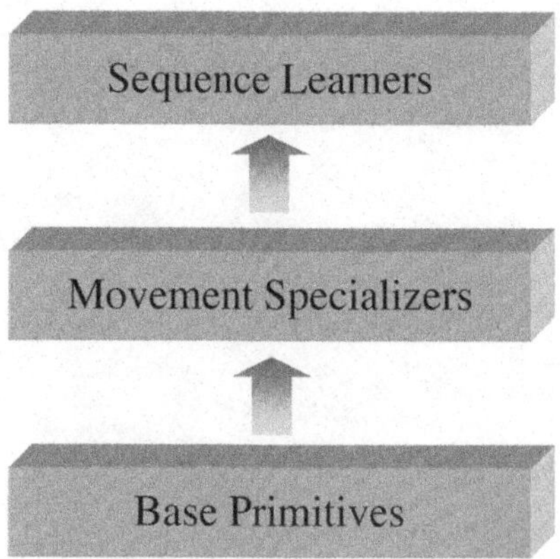

Figure 6.32
Architecture of Matarić approach to learning by imitation

6.12.3 Learning Movement Sequences from Demonstration

As indicated previously, there is extensive neurophysiological evidence for the existence of preprogrammed movements in vertebrates, sometimes known as *motor primitives*. The work of Matarić and her students is based on the assumption that imitation invokes motion primitives and that new movements can be learned by imitation if they can be synthesized from the primitives. This is an exciting approach to learning by imitation, which stays quite close to its roots in neurophysiology and yet provides a principled basis for the acquisition of new skills not initially in the imitator's (learner's) repertoire. The architecture of this approach is shown in figure 6.32.

Note that there are three layers in this architecture. The bottom layer, as shown in the figure, consists of *base primitives*. These elements encode and make possible the execution of simple movements, and they also recognize such movements when observed during a demonstration by a teacher. Thus, these elements are *visuo-motor primitives* that perform some of the same functions as the combination of mirror neurons and motor primitives in biological systems. At the next level are elements known as *movement specializers*, which become associated with specific movements occurring at the base primitives layer. At the highest level are *sequence learners* that use the activity of the movement specializers to to learn probabilistic models of movement sequences from multiple demonstrations (Amit and Matarić 2002; Weber, Jenkins, and Matarić 2000; Matarić 2001).

Of particular interest is the composition of complex movements from motion primitives and the learning of these movements by a robot (or simulated robot). Figure 6.33 shows the Matarić imitation model (Weber, Jenkins, and Matarić 2000). In the first stage of Matarić's model, movement primitives are extracted from a visual

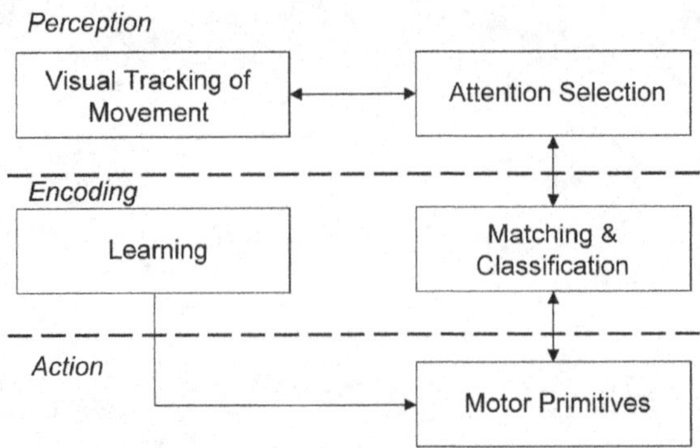

Figure 6.33
Imitation model used by Matarić et al. (illustration courtesy of Maja Matarić)

tracking system of human movements. Such extraction is not easy, since one must postulate coordinate systems for a tracking system, in such a way that the basic properties of a movement are preserved under a change of coordinates. The movement primitives can then be used to synthesize new movements through sequencing and/or superposition. The method was used by Matarić to enable an avatar to learn simple human movements from videotape recordings.

The first layer in figure 6.33 concerns perception. It serves to acquire and prepare motion information for processing into primitives. The second layer ("Encoding") takes the perceptual information and encodes it into appropriate primitives. It is this layer that determines how to represent motion segments so as to facilitate classification and learning. The third layer actually enables the robot to perform the imitation using the list of primitive segments provided by the second layer.

This model was used by Matarić and her students to enable a simulated robot to imitate the movement of a human subject observed through a video input. The model's motion-tracking system converted actual human movement into corresponding movements of a stick figure in 2-D. The "Attention Selection" component, as shown in figure 6.33, assumed in this case that only the end points of the arm were significant, since humans tend to focus on end-point motion, and the kinematics of the teacher and learner were in this case similar. Once the motion primitives were identified and classified, the information was sent to a simulated robot (an avatar named Adonis), which then imitated the motion. The avatar produced an approximate (but believable) imitation of the human motion. The implementation is shown in figure 6.34.

The approach described in this section is noteworthy because it attempts to stay close to biology, without attempting to replicate it down to the neural level. Thus, there are no mirror neurons in the approach of Matarić and her students, but the functional structure of the imitation system remains faithful to its biological roots.

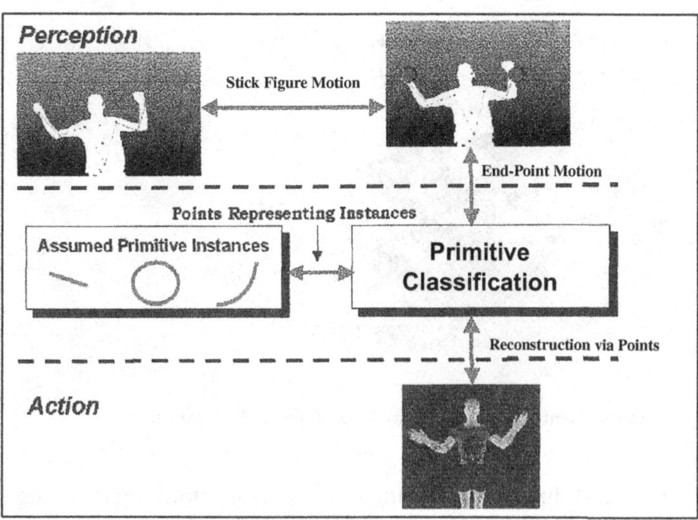

Figure 6.34
Implementation of Matarić's imitation model (illustration courtesy of Maja Matarić)

6.12.4 Developmental Approach to Imitation Learning

Kuniyoshi and his colleagues (2003) use the work of Meltzoff and Moore (1977) to suggest a quite-different approach to imitation learning of simple movements. Meltzoff and Moore indicated that fetuses engage in a large number of random movements. They then hypothesized that these self-exploratory movements can be reused by the newborn in imitating the movements of its mother or other humans. Kuniyoshi et al. use this idea as a basis for robot learning. They program a robot to perform a number of simple movements and allow the robot to watch them using its on-board camera. The visual input and associated arm movements are learned using a large spatiotemporal pattern memory neural network (with one thousand neurons). The robot is then placed in front of a human performing similar movements to those learned by the robot in self-observation. Even though the human movements are clearly different from the robot movements, the robot is able to identify the correct matching movement in at least 80% of the cases. The robot H3 is shown observing human movements in figure 6.35.

Although the details of this experiment are beyond the scope of this chapter, we can comment on a few features. The visual input to the neural network was selected to be in the form of optical flow, since optical flow plays a central role in the ability of primate brains to recognize movement. The optical flow array is represented by three overlapping circular receptive fields of twelve vectors each, in the major directions of the field, which yields a thirty-six-dimensional vector. The 4-degree-of-freedom

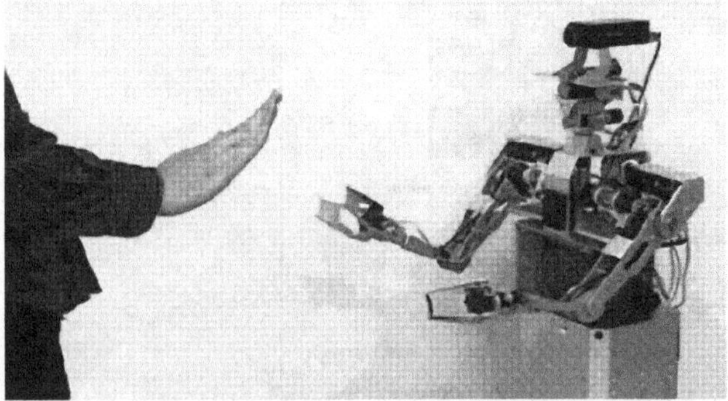

Figure 6.35
Robot H3 observing human hand movements (photograph courtesy of Yasuo Kuniyoshi)

arms were represented by a 150-dimensional joint angle vector. Both vectors were generated every 33 msec. The network made use of 850 recurrent neurons, for a total of 1,000 neurons, each connected to 499 other neurons.

Clearly, this was a major project. It showed that there are multiple ways of including biological data in a learning model. It is not clear from Kuniyoshi et al.'s paper whether the approach can be generalized to imitation learning of arbitrary movements.

6.12.5 High-Level Learning from Demonstration

In contrast to the approach described in the previous section, other investigators have enabled a robot to learn from demonstrations using mathematical models, with no attempt to replicate the biology. Rather, the goal is to enable the robot to learn a *control policy*, a mathematical representation of the task being performed by the teacher. The robot must learn from a small number of demonstrations and replicate the task, rather than attempting to follow faithfully the teacher's movement. This approach is grounded in the mathematics of system identification, the name given to a class of methods for constructing mathematical models of systems from input-output data, without a precise knowledge of the systems' internal structure. Note that this approach not only makes no attempt to emulate biology, it does not even attempt to mimic the movement of a teacher. Yet learning from demonstration has been highly successful in enabling a robot to learn such complex tasks as juggling and using two paddles to bounce a Ping-Pong ball back and forth between them (Atkeson and Schaal 1997a). Nevertheless, it is an attempt to learn by imitation, though in a different way than in the previous section. We summarize one experiment involving learning from demonstration in this section.

Consider the movement known as a *pendulum swingup task* (Spong 1995), illustrated in figure 6.36. This task is related to broom balancing and other inverted-

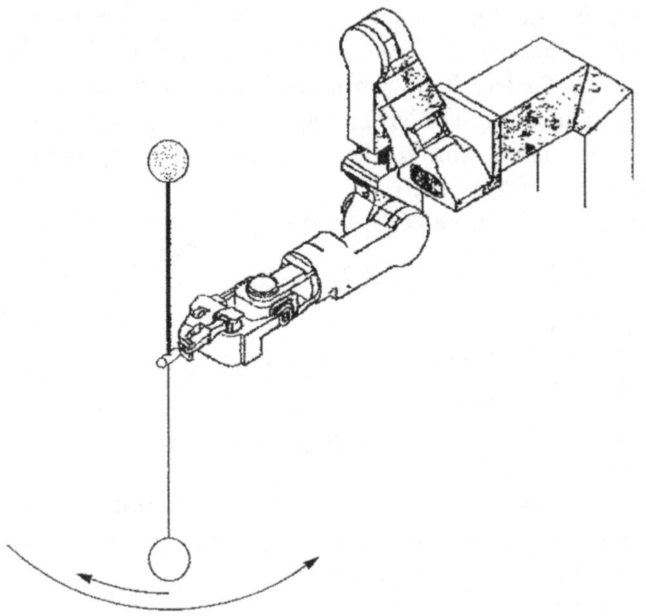

Figure 6.36
Sarcos robot arm performing a pendulum swingup task (illustration courtesy of Stefan Schaal)

pendulum situations. The hand (human or robot) begins with a pendulum hanging down in a stable position. The goal of the task is to move the hand such that the pendulum swings up and is then balanced in the inverted position. This is clearly a difficult task, both for humans and for robots. The goal is to enable a robot arm (in the case of the experiment discussed in this section, it was the Sarcos Dextrous Arm) to learn this task from human demonstrations.[13] In this section we summarize the approach and results on this task as obtained by Atkeson and Schaal (1997b).

The learning in Atkeson and Schaal's experiment was based on identifying the parameters of a mathematical model. Both parametric and nonparametric models were used. The parametric model was a discrete-time model of an idealized pendulum, with all its mass concentrated at the tip, attached to a hand that could move only along a horizontal line:

$$\dot{\theta}_{k+1} = (1 - \alpha_1)\dot{\theta}_k + \alpha_2 \frac{\sin \theta_k + \ddot{x}_k \cos \theta_k}{g}, \tag{6.36}$$

13. Atkeson and Schaal's implementation was complex and is beyond the scope of this chapter. The vision system used to observe the human demonstration and the robot control system had combined delays of some 0.12 seconds; a Kalman filter was used as a predictor to compensate for the delay. To enable the robot arm to follow desired hand motions, it was necessary to implement inverse kinematics and dynamics (see chapter 10).

where θ is the pendulum angle, $\dot{\theta}$ is the pendulum angular velocity, $\ddot{x}$ is the hand acceleration, α_1 is the viscous damping, and $\alpha_2 = \Delta g/l$, where Δ is the time step, g is the acceleration of gravity, and l is the length of the pendulum. Of course, this model was idealized, since the real pendulum did not have all its mass concentrated at the end point. The optimization program leading to the model parameters α_1 and α_2 used linear regression, implemented in MATLAB.

The so-called nonparametric models[14] were of the form

$$\dot{\theta}_{k+1} = f(\theta_k, \dot{\theta}_k, x_k, \dot{x}_k, \ddot{x}_k), \tag{6.37}$$

which, in the linear case, reduces to

$$\dot{\theta}_{k+1} = c_1\theta_k + c_2\dot{\theta}_k + c_3x_k + c_4\dot{x}_k + c_5\ddot{x}_k, \tag{6.38}$$

where the constants c_i are determined using Atkeson's locally weighted learning method (Atkeson, Moore, and Schaal 1997).

Given the model of either equation (6.36) or (6.37), it was still necessary for the robot to learn the swingup task. To accomplish this learning, Atkeson and Schaal used optimal-control techniques to minimize a cost function. A simplified version of this function (for one state variable) can be stated as

$$r(x_k, u_k, k) = (x_k - x_k^{\mathrm{d}})^2 + (y_k - y_k^{\mathrm{d}})^2 + \ddot{x}_k^2, \tag{6.39}$$

where the superscript d indicates desired values and the subscript k refers to the k-th state variable. The cost is made up of the squared deviation of the robot hand trajectory from its desired values in both x and y as well as the squared cost of control. This cost function clearly remains positive for deviations of the trajectory in either the positive or the negative direction and penalizes excessive control effort.

The results for the parametric case are shown in figure 6.37. Panel (a) of the figure shows the pendulum angle produced by the human teacher and in two successive robot trials. The first of these trials was an attempt to actually mimic the human trajectory by minimizing the cost function (6.39). This direct attempt to imitate the human motion failed to swing up the pendulum, but it provided data that the optimization program used to build a more accurate model of the robot, which then succeeded, as can be seen in the trajectory for the second trial. The traces in figure 6.37(b) are particularly revealing. Note that the hand position for the robot in the first trial (given by the dashed line) appears to be nearly identical to the hand position of the human teacher in the demonstration, but the motion failed. Clearly, simple imitation of the movement omits too many aspects of the task, such as the great difference in the way the human and the robot gripped the pendulum. The second trial, produced a hand

14. Clearly, this model has parameters to be identified. The label "nonparametric" arises from the fact that none of these parameters are related directly to the physics of the situation.

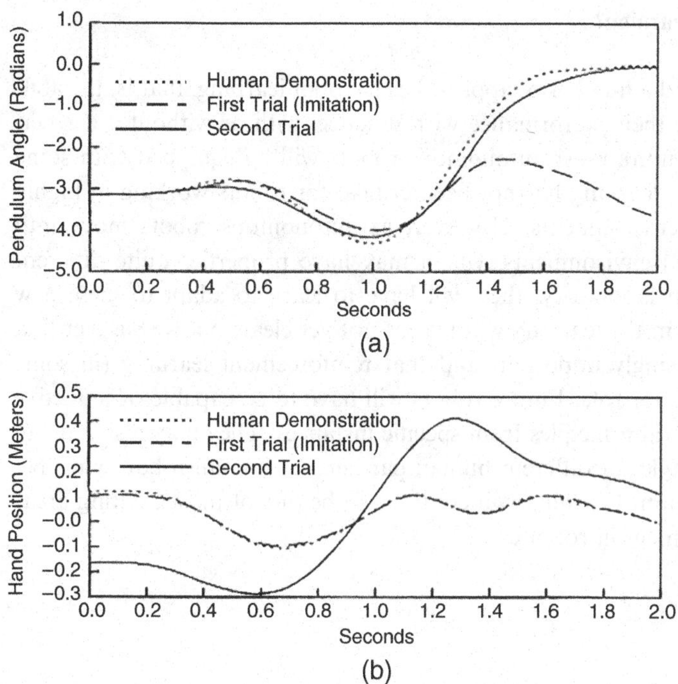

Figure 6.37
Results of learning swingup task using a parametric model: (a) pendulum angle produced and (b) hand position (illustrations courtesy of Stefan Schaal)

movement quite different from that of the teacher, yet succeeded in the swingup task.

This study shows that learning from demonstration can be highly successful, even for difficult tasks, without any reference to biology. On the other hand, the inspiration for such work arises from human experience with imitation.

6.12.6 Other Experiments in Learning by Imitation

Learning by imitation is a surprisingly old approach to teaching robots. Early work dealt with imitation of specific movements, such as block manipulation (Kuniyoshi, Inaba, and Inoue 1994).

In section 5.9, we reviewed an approach to learning the controller parameters needed for stable control of a helicopter using a method known as teaching by showing (Montgomery 1999). In this approach, a human operator provides the basic control signals, which are then used for reference. A fuzzy controller attempts to build a set of fuzzy rules to imitate the controls produced by the human. If the controller is faced with time-varying behavior, a neural network provides updates as needed. Clearly, this is another form of learning by imitation.

6.13 Whither Robot Learning?

This chapter has reviewed a number of approaches to robot learning, that is, the ability of robots to improve their performance with practice, with or without a teacher. We believe that in the future most autonomous robots will be equipped with some ability to learn. To date, learning has not been critical for robots working in highly structured, predictable environments. However, as autonomous robots move into increasingly unstructured environments, which may have properties quite different from those in the design laboratory, they will have to learn to adapt to these new circumstances. What form the learning will take is not yet clear, but we suspect that imitation will be increasingly important and that reinforcement learning (in some form) will also play a major role. Future robots will have to be capable of inductive learning, extracting general principles from specific instances. They may also have to adapt to the cognitive styles of different human partners, as human-robot teams become increasingly common. Learning will continue to be one of most exciting areas of research and development in robotics.

7 Robot Locomotion: An Overview

Summary

The purpose of this chapter is to illustrate the wide variety of mobile robots in existence and to analyze some of their respective advantages and disadvantages. We begin with wheeled robots, because they are most often used, and discuss some of their control problems. We also discuss tracked vehicles, the motion of snake robots, climbing robots, hopping robots, swimming robots, and flying robots. Although there are no biological prototypes for wheeled robots, biological models are discussed in connection with many of the other mobile systems in this chapter. The last section of the chapter considers self-reconfigurable robots that are capable of changing their modes of locomotion autonomously.

7.1 Animal Locomotion

From simple, single-celled organisms like amoebas to complex vertebrates, animals have the ability to move. They move in an astounding variety of ways: walking on land using legs, flying in the air with wings, swimming, and crawling. Since we look at animals as prototypes for the design of robots, we consider all these modes of locomotion in the book. Walking on legs is so important as a mode of locomotion in robotics that we devote whole chapters to biped, quadruped, and hexapod walking machines. Since we are concerned primarily with those aspects of animal locomotion that have been or could be emulated in robotics, we do not discuss movements like those produced by the pseudopods of single-celled organisms or the oscillations of cilia.

Animals do not use wheels for locomotion. Some animals are able to adjust their bodies into a nearly spherical shape and then move by rolling with the aid of gravity. Although there are no shafts and wheels as we know them in the animal kingdom, wheels are ubiquitous in human-designed systems. As might be expected, the vast majority of mobile robots move on wheels. Hence, we devote the major portion of this chapter to wheeled robots. This chapter also includes a brief discussion of swimming and flying robots. Our understanding of movement in some swimming animals has provided a great deal of background for the design of biologically inspired swimming robots. In fact, the work of Grillner has led to mathematical and computational representations of swimming in the lamprey eel, including its neural control (Grillner and Dubuc 1988). Since the study of swimming requires an understanding of both biology and hydrodynamics, it is not considered in this book. We include only some descriptions of simple swimming robots.

Similarly, flight in animals, whether birds or insects, is a complex process involving an interaction of biology with aeronautics, and we largely ignore it, in spite of its importance. We include a discussion of a robot helicopter and some fixed-wing aircraft in section 7.10. In the area of flying, Shimoyama and his colleagues have developed a flying microrobot that consists of a microchip with wings (Miki and Shimoyama 1999). Recent military projects have considered a number of miniature flying machines, about the size of a small bird, to aid in reconnaissance missions.

We also devote a section of this chapter to robots that emulate the crawling movement of snakes. Such robots may be important in such applications as inspection of pipelines or search and rescue in situations in which it is necessary to crawl through rubble.

7.2 Wheeled Vehicles

We have previously discussed the earliest mobile robots, beginning with Walter's (1950) tortoise. Nearly two decades elapsed between Walter's pioneering work and the Stanford work on Shakey in 1969 (Nilsson 1969), discussed in chapter 5 in connection with robot architectures (figure 5.2). Shakey was a three-wheeled robot, equipped with a video camera, a radio link to a remote PDP-10 computer, a range finder, and on-off touch sensors.

Note that this robot had two motorized drive wheels and a third unpowered caster wheel. Such a design is common in mobile robots today. The robot was designed to identify large colored blocks and move them through doorways and up ramps to desired locations. As indicated in chapter 5, its sense-think-act architecture was extremely inefficient. In addition, the robot suffered from numerous mechanical problems and was prone to failure. Nevertheless, it represented a major step forward in mobile robotics, since it attempted to couple mobility with intelligence.

Numerous wheeled robots were developed in the 1970s, including various rovers for planetary exploration at the NASA Jet Propulsion Laboratory (JPL). These developments culminated in the landing of the Sojourner robot on Mars in 2001. JPL robots generally have six wheels and several unique features discussed subsequently. Today the most common research robots are small, wheeled vehicles. We encountered some of them in chapter 1, including the Pioneer robots from ActivMedia Robotics (figure 1.1(b), figure 1.8) and the tabletop Khepera robot from K-Team in Switzerland, with two powered wheels and a skid located in the center of the rear portion (figure 1.1(d)). Thousands of these research robots have been sold, primarily to research laboratories worldwide. Here we describe some additional wheeled robots to highlight the variety of possible designs, even though some of these robots may no longer be commercially available.

7.2.1 ATRV Four-Wheeled Robots

ATRV (all-terrain robotic vehicle) robots are more rugged than the Pioneers just mentioned and are designed for outdoor activities in a variety of terrains. They are manufactured by iRobot Corporation in several sizes. The full-size ATRV illustrated in figure 7.1 is 65 cm high, 105 cm long, and 80 cm wide. The standard robot shown in the figure has the following features:

- *Sonars* It is equipped with eleven sonar transmitter-receivers, six forward facing, four side facing, and two rear facing.
- *Processor* The on-board processor is a Pentium-based ATX system.
- *Batteries* The vehicle is equipped with four lead-acid batteries, giving it 4–6 hours of operation.
- *Drive* Each wheel is operated by a high-torque 24 V DC motor. The drive wheels are differentially controlled, allowing for a zero turning radius by letting some of the wheels skid. This method of operation is known as *skid steering*.
- *Speed* The ATRV can translate at 2 m/sec and rotate in place at 70 deg/sec.
- *Payload* The robot weighs 118 kg and can carry a remarkable payload of 100 kg.

Figure 7.1
All-terrain robotic vehicle (photograph courtesy of iRobot Corporation)

Figure 7.2
Koala robot (photograph courtesy of K-Team S.A., Lausanne, Switzerland)

Many ATRVs are also equipped with laser scanners, vision systems, and wireless communication. As the figure shows, they have all-terrain tires.

7.2.2 Koala Six-Wheeled Robots

The Koala (see figure 7.2) is manufactured by K-Team, the same Swiss company that makes the tabletop Khepera robots. It is evident from the figure that this streamlined six-wheel vehicle is suitable primarily for indoor applications. Some of the features of this robot are

- *Size* The Koala is 32 cm long, 32 cm wide, and 20 cm high.
- *Weight* 4 kg with battery, capable of carrying a 3 kg payload.
- *Battery* The robot uses Nickel metal hydride (NiMH) batteries, giving it 4–6 hours of continuous operation, depending on the payload.
- *Drive* 2 DC motors with integrated incremental encoders.
- *Speed* Maximum speed ranges from 0.38 m/sec using the factory supplied PID controller (see chapter 4) to 0.6 m/sec directly.
- *Slope traversal* The vehicle is capable of traversing a 43° slope.
- *Sensors* 16 infrared (IR) proximity sensors, as well as sensors for battery and ambient temperature, motor torque, and global power consumption. Optional are six sonars and four longer-range IR triangulation sensors.
- *Processor* The standard processor is minimal, but PC 104 boards are available.
- *Software* One of the strengths of this robot is its software compatibility with the Khepera, for which simulation and other software packages are available.

7.2.3 Other Wheeled Robots

Of course, many other wheeled robots exist. Some university laboratories have fabricated their own mobile robots. Others, like the Yuta laboratory at Tsukuba University in Japan have fabricated numerous robots, some of which are commercially

Figure 7.3
Yamabico K-2 robot (photograph courtesy of Shin'ichi Yuta)

available under the name Yamabico. A typical Yamabico robot is shown in figure 7.3. It has two powered wheels as well as a third caster wheel (not visible in the picture). It also has a bumper and a number of sonar sensors, as well as a processor and a variety of other sensors and instruments.

iRobot Corporation manufactures several cylindrical robots, formerly associated with names like Denning. A typical robot of this class, known as Magellan, is shown in figure 3.13(b). Note that the robot is surrounded by both sonar and IR sensors.

The robot has a two-wheel differential drive, enabling it to turn in place. It is available with the usual complement of equipment described in previous sections in connection with the ATRV and Koala robot.

A number of larger wheeled robots have also been constructed:

- Several passenger automobiles have been developed using neural networks for navigation on highways (Thorpe, Jochem, and Pomerleau 1997; Dickmanns 1998).
- The Carnegie Mellon University Field Robotics Center has constructed several large robots for NASA, such as the Nomad, shown in figure 7.4. The Nomad was deployed in the Atacama Desert in Chile to test its performance and durability under extreme conditions.
- A number of unpiloted military vehicles have been produced, such as the U.S. HMMWV (High Mobility Multipurpose Wheeled Vehicle) (Chun and Jochem 1994). We expect that there will be many more unpiloted military ground vehicles in the future, some equipped with wheels and some with treads. These robots are

Figure 7.4
Carnegie Mellon University Nomad in the Atacama Desert (photograph courtesy of the William Whittaker)

frequently referred to as *autonomous ground vehicles* (AGVs) or *unpiloted ground vehicles* (UGVs).[15]

- In development and testing are various agricultural robot vehicles, road construction vehicles, and similar heavy-duty equipment.

7.2.4 JPL Rovers

The most dramatic wheeled robots of the past decade or so have been those designed for planetary exploration by the NASA Jet Propulsion Laboratory.

The rover robot Sojourner (figure 7.5) was the first autonomous Mars exploration vehicle. Its range was not very large, and it never left the surveillance of the lander ship, but it was certainly a pioneering robot in planetary exploration. The rover was constrained to a weight of 11.5 kg. Another 6 kg was allocated to lander-mounted rover telecommunications equipment, structural support of the rover, and its deployment mechanisms. The rover had a normal height of 280 mm, with ground clearance of 130 mm. Its stowage space in the lander allowed only 200 mm, forcing it to squat to a height of 180 mm when stowed. The rover was 630 mm long by 480 mm wide.

It can be seen from the figure that the rover had six wheels, two of which were powered. The front wheels were steerable.

15. The word "unmanned" is in common use to describe robot vehicles, on the ground (UGVs), in the air (UAVs), or underwater (UUVs). I have chosen to use the gender-neutral term "unpiloted," or simply to refer to such vehicles as "autonomous." Although it is not strictly correct to refer to the driver of a ground vehicle as a "pilot," it is better than other alternatives available at this time. Of course, I cannot change the name of a journal called *Unmanned Vehicles*, but I trust it will happen sooner or later.

Figure 7.5
Sojourner rover robot (photograph courtesy of NASA/JPL/Caltech)

The rover had a suspension system of a type known as a "rocker-bogey," a unique JPL design that allows the wheels to ride over a rock or similar obstacle that is about 1.5 times the vehicle's wheel diameter. It uses complex linkages and numerous actuators. The Sojourner mobility system used ten motors. The rocker-bogey suspension is clearly visible in another JPL vehicle, the rover robot Rocky 7, shown in figure 7.6.

Figure 7.7 shows Spirit, the NASA rover vehicle that landed on Mars in January 2004. The picture shows the six wheels, the rocker-bogey suspension, communication antennas, a large platform containing solar cells, and a mast in front containing a digging tool and various instruments.

As we have seen, the suspensions of vehicles traveling on rough terrain can present major problems in design. An analysis of the control of such vehicles with an actively articulated suspension, with particular attention to the Mars Sample Return Rover, can be found in Iagnemma et al. 2003.

7.2.5 Summary of Wheeled Robots

From the discussion and examples in this section it is evident that wheeled robots have certain aspects in common with one another:

• The motor drive to the wheels is provided by electric motors, usually servomotors because they are used in closed-loop control systems. They may be mounted directly

Figure 7.6
Rocky 7 vehicle showing rocker-bogey suspension (photograph courtesy of NASA/JPL/Caltech)

Figure 7.7
NASA Mars rover Spirit (photograph courtesy of NASA/JPL/Caltech)

on the drive shafts or connected to them by gears or chains. The control system for the wheel drivers is typically a PID controller (see chapter 4). Generally only one pair of wheels is powered.

• The robots carry an on-board processor, with software for such activities as navigation and obstacle avoidance.

• They have sensors, generally sonars for obstacle avoidance, IR sensors and/or laser scanners for longer-range detection of obstacles, and vision systems. One or more wheels are equipped with encoders to make odometry possible.

• They frequently have some form of communication with a larger computer, using wireless Ethernet or some form of RF radio communication.

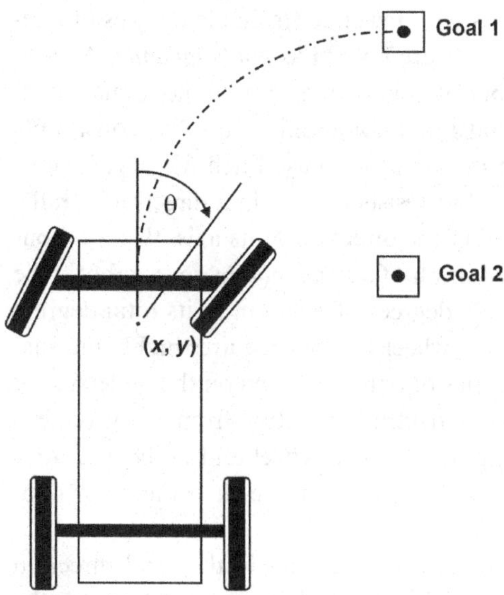

Figure 7.8
Typical nonholonomic vehicle

- Few of these robots are equipped with arms and grippers, mainly as a result of cost considerations, even though arms and grippers are highly desirable.

Navigation for autonomous ground vehicles is a continuing area of research. An excellent summary of work in this field can be found in Hebert, Thorpe, and Stentz 1997.

7.2.6 Control Issues

The control of wheeled autonomous robots involves a number of issues. Although it is not stated explicitly in the preceding sections, nearly all the robots discussed here use feedback control methods for their drive systems, including PID controllers. As discussed in chapter 4, these proportional-integral-derivative controllers have a number of advantages, from a control point of view, over simple proportional control.

One problem in the control of wheeled vehicles relates to the number of controllable degrees of freedom. Visualize a typical wheeled vehicle with front-wheel steering, as illustrated in figure 7.8. The usual goal of vehicle control is to reach a goal position (x, y) with a given orientation θ, say goal 1 in the figure. Note that this implies control over 3 degrees of freedom. However, we can only control 2 dof: vehicle motion and steering-wheel angle. Although this limitation is not a problem in reaching goal 1, it is clear that the vehicle cannot reach goal 2 (except by a series of forward and backward movements). Vehicles with fewer controllable degrees of freedom than the actual vehicle degrees of freedom are known as *nonholonomic*. Assuming the vehicle in question is a wheeled robot, it is known as a nonholonomic robot. Another way of looking at this is to say that a nonholonomic robot has constraints on some of its movements, in this case, on its lateral velocity. Each wheel can move freely in a direction perpendicular to its axis but resists motion in a direction parallel to the axis. Hence, its velocity is constrained in the direction of its axis; this is a nonholonomic constraint. As implied by the definition of *nonholonomic*, for a robot to be holonomic, it must have as many controllable degrees of freedom as its total degrees of freedom. To obtain such controllability in a wheeled robot requires highly unusual wheel and body structures, for example, a series of balls or rollers on the underside of the vehicle that make it possible to move in an arbitrary direction from any given initial position and orientation. Designs incorporating such wheel and body structures exist, but they are generally more complex and more expensive than those for nonholonomic systems.

It should be noted that a wheel can resist motion velocities in the axial direction only until the applied force exceeds the limit of tangential friction, after which the wheel will slip sideways. Such slippage is undesirable in passenger vehicles, because it causes tire wear. In an autonomous robot making long traverses in outdoor terrain, slippage will reduce the ability of the robot to use odometry to determine its position with respect to its starting location.

It is evident from the foregoing that four-wheeled vehicles with two-wheel drive and steering of two wheels that are parallel to the body axis will suffer from slippage of the front wheels when turning. A common solution to this problem is to use *Ackerman steering* (Dudek and Jenkin 2000). Assume that the vehicle's front wheels are steered. Then each of them rotates on a separate arm to allow the inner wheel to turn through a larger angle but travel a shorter distance than the outer wheel, as shown in figure 7.9. As the figure shows, this implies that there exists an *instantaneous center of curvature* (ICC), which must lie on a line through the rear axis of the vehicle, from which the turning radii of the two front wheels extend. Ackerman steering (also known as *kingpin steering*) is commonly used on automobiles and in larger robots.

As we have seen in some of the examples, a number of robots have two driven wheels and either a skid or a passive caster wheel to allow for a tripod of support. The control of such vehicles is somewhat simpler than for a four-wheel vehicle, since the third wheel can slip or move laterally. Such vehicles can turn in place but are still nonholonomic if they use standard wheels, since the same constraint on motion in the axial direction applies.

Many small three- and four-wheel commercial robots use *differential drives* on the driven wheels. A differential-drive robot has two driven wheels mounted on a com-

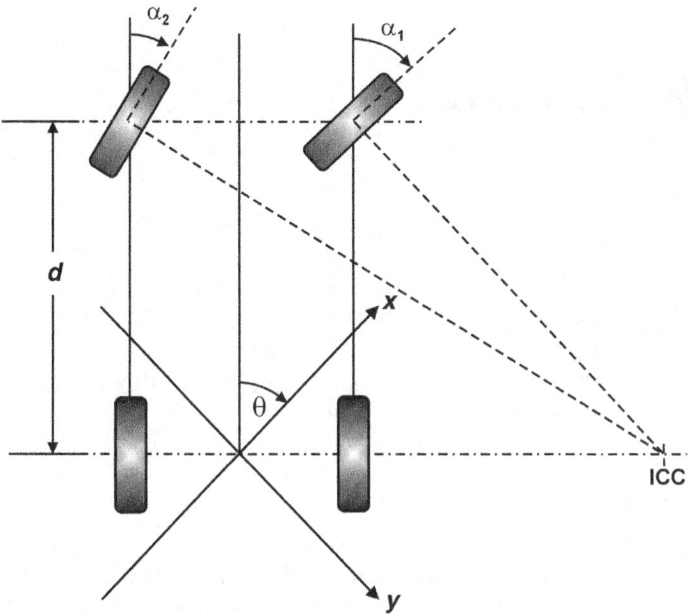

Figure 7.9
Ackerman steering

mon axis but driven by different motors. When the motors turn in the same direction at the same rate, the robot moves straight forward or backward. Turning requires the motors to turn at different rates. When the ground velocities of the two wheels are equal in magnitude but opposite in direction, the robot rotates in place. Differential drives are very useful for robots that have to maneuver in narrow corridors or similar environments.

Consider a three-wheeled vehicle (tricycle). Three options for power and steering are possible:

1. The rear wheels can be driven and provided with odometers, with steering provided by the front wheel. Differential drives are employed on the rear wheels.

2. It is also possible to provide both power and steering through the front wheel. In order for the vehicle to turn, a line through the front-wheel axis must intersect a line through the axis of the rear wheels, as shown in figure 7.10.

3. The structure can be reversed, so that the two co-axial wheels provide both steering and power, while the single wheel, now in the rear, is a passive caster. The driving wheels now must be differentially driven. This is the case with the Khepera robot, illustrated in figure 1.1(d).

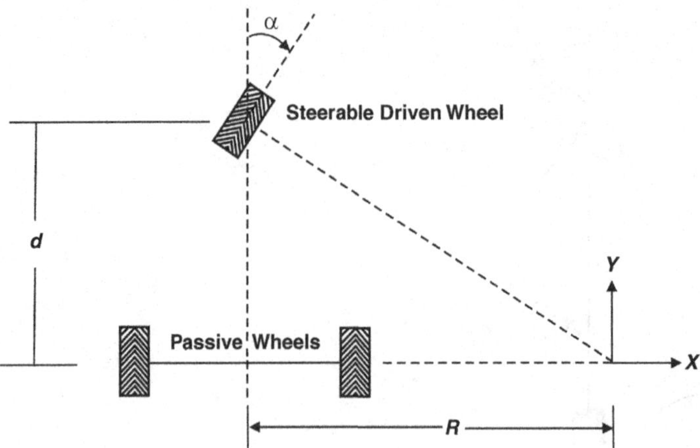

Figure 7.10
Geometry for three-wheeled robots

Figure 7.10 illustrates case 2 and shows the location of the ICC. If the steering (front) wheel is set at an angle α from the vehicle longitudinal axis and the forward velocity is v, the vehicle will rotate with angular velocity ω about the point R in the figure, where

$$R = d \tan\left(\frac{\pi}{2} - \alpha\right),$$
$$\omega = \frac{v}{(d^2 + R^2)^{1/2}}. \tag{7.1}$$

Equation (7.1) represents the forward kinematics of the vehicle. In general, solution of the inverse-kinematics equations (i.e., computation of the robot pose, given by its location (x, y) and its angle θ with the x-axis) is very difficult.

Some robots use so-called *synchronous drives*, in which all the wheels are steered and driven. In robots so equipped, all the wheels turn and drive in unison; they all point in the same direction and turn at the same rate. One can think of a synchronous-drive robot as an approximation to an ideal point robot, that is, a robot represented by a single point. Excellent discussions of the kinematics, dynamics, and control of wheeled vehicles can be found in Dudek and Jenkin 2000 and Borenstein, Everett, and Feng 1996.

Thus far we have emphasized wheeled robots ranging in size from perhaps 30 cm to the size of automobiles. Another area of great importance is the development of miniature robotic vehicles. Two examples of this very important area of research and development are cited in what follows.

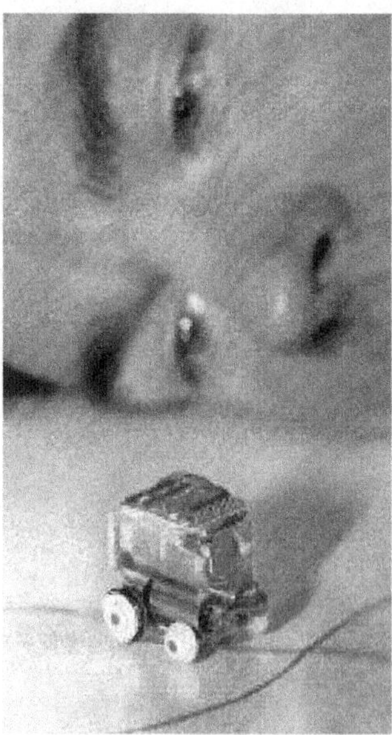

Figure 7.11
Miniature robot from Sandia National Laboratories (photograph courtesy of Sandia Laboratories)

Sandia National Laboratories has developed a miniature robot approximately 1 in^3 in size (about 2.5 cm on a side), as shown in figure 7.11. The University of Southern California Robotic Embedded Systems Laboratory has developed the Robomote, a tiny wheeled robot equipped with sensors and communication ability (see figure 7.12). The length of the vehicle is comparable to a U.S. 25-cent coin; its volume is approximately 47 cm^3; the wheels are about 2.5 cm in diameter. The “body” of the robot is basically a single printed circuit board containing an eight-bit microcontroller, an additional board containing a radio communications interface, an antenna, five infrared transmitters, four bump sensors, and a lithium-ion battery, which can keep it fully active for 3.5 hours. The wheels are equipped with encoders. Clearly, a great deal of engineering was required to achieve so much functionality in such a small size. We will encounter the Robomote again in connection with multiple robots in chapter 12.

7.3 Tracked Vehicles

In addition to wheeled vehicles, a number of track-driven robots have been constructed. Typically, these robots are designed for military purposes, for planetary exploration, or for hazardous environments. From a kinematic point of view, a tracked

Figure 7.12
Robomote, a tiny robot for experiments in sensor networks (photograph courtesy of Gaurav Sukhatme)

Figure 7.13
PackBot urban reconnaissance robot (photograph courtesy of Stefan Hrabar, reproduced with permission of iRobot Corporation)

vehicle can be considered a differentially driven vehicle. It can turn only by allowing slippage on the ground. Hence, treads on tracked vehicles may wear rapidly. The importance of tracked vehicles arises from the ability of tracks to climb over obstacles not passable with wheels.

An unusual tracked vehicle is the PackBot robot, designed and manufactured by iRobot Corporation, illustrated in figure 7.13. This robot has two sets of tracks: the larger or main tracks used for locomotion, and the smaller set located on two arms near the front of the vehicle. These separately articulated arms enable the robot to climb over obstacles and go up or down stairs. The robot is also extremely robust; I have seen it be thrown from a second-story window, fall on the ground, right itself

Figure 7.14
Talon robot used by U.S. Army (photograph courtesy of Foster-Miller, Inc.)

with the front tracks, and move away. (Of course, this version of the robot had no on-board sensors, which are not known for their resistance to impact forces.)

Owing to the additional cost of fabrication as well as the need to replace tread material, tracked vehicles are not used frequently in robotics. In general, tracked vehicles are slower than wheeled vehicles. They are, however, of great importance to the military, as exemplified by the U.S. Army's Talon robot, shown in figure 7.14. This robot has been used for several years to handle mines and clear dangerous ammunition. It can also be equipped with cameras (including night vision) and a variety of weapons. We expect to see increasing uses of robots in military operations, not only on land, but in the air and water as well. As with other military robots, the Talon operates partly in autonomous mode and mostly under control of a remote operator. As the figure shows, Talon is quite adept at climbing stairs. (Stair climbing requires either treads or legs.)

7.4 Legged Robots

Wheeled vehicles need to have all wheels in contact with the ground all the time.[16] By contrast, legged animals (and machines) keep only some legs in contact with the

16. An all-wheel-drive vehicle may be able to operate with one wheel in the air, but this is clearly not desirable. Also, an automobile may be temporarily airborne when going over a ramp at high speed, particularly in chase scenes in the movies. I do not recommend such activities for robots.

ground at any given time and may leave the ground completely for a time during certain gaits, such as galloping in horses. For certain types of rough terrain and large numbers of obstacles, legged vehicles have advantages over wheeled vehicles. Furthermore, we have biological models for bipeds, tetrapods, hexapods, and octapods. I have devoted separate chapters to biped locomotion (chapter 8) and to locomotion with four, six, and eight legs (chapter 9). Hence, I will considered these types of locomotion further in this chapter.

7.5 Hopping Robots

As an alternative to walking or running, some investigators have developed robots that move by jumping. Of course, there are biological models for jumping, like kangaroos, some insects, and even hopping birds. Here we present some hopping robots that do not have a direct biological prototype, as well as one that does.

Pioneering work on one-legged hopping was done by Raibert (Raibert, Brown, and Chepponis 1984; Raibert 1986; Nagakubo and Hirose 1994; Bahr, Li, and Najafi 1996). Since he also used his monopod results to build bipeds, we leave the discussion of his work to chapter 8. Consider first the hopping robots developed at the Jet Propulsion Laboratory and the California Institute of Technology (Fiorini and Burdick 2003). One of the designs for this type of robot is illustrated in figure 7.15.

There are a number of issues in the design of a hopping system, including

• *The propulsion mechanism* In the robot of figure 7.15, the propulsion mechanism is a combination of a compressed spring and a six-bar linkage mechanism. An electric motor compresses the spring-actuated "leg" using a power screw, which is driven

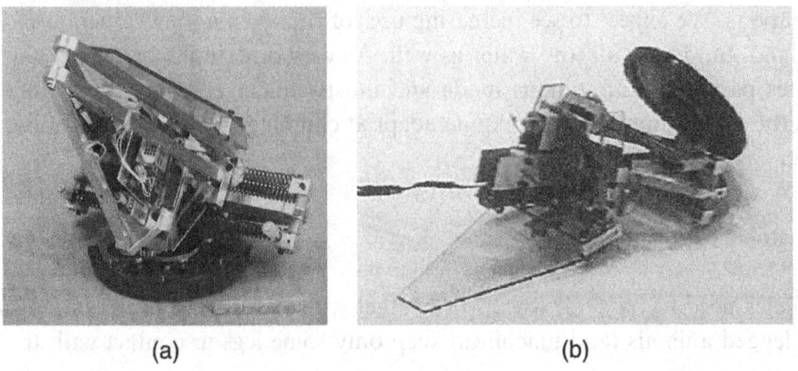

Figure 7.15
Caltech/JPL hopper: (a) compressed state, before takeoff, (b) with extended flap, in the process of righting after landing (from Fiorini and Burdick 2003, reproduced with permission of Kluwer Academic Publishers)

until a latch is engaged. A slight additional turn of the screw causes release of the spring. Since one end of the mechanism is on the ground, this stored energy produces kinetic energy in a hop. The device is shown in a compressed state (before jumping) in figure 7.15(a).

- *Steering* The structure of the robot in figure 7.15 is mounted at an angle of 50° with respect to a "foot." This orientation provides a takeoff angle for the robot and approximately maximizes the horizontal hopping distance. The orientation is adjustable by means of a steering mechanism.
- *Crash resistance* Since the robot in figure 7.15 flies and then lands at an unknown location, where there may be rocks or soft soil, it must be crash resistant. The sensitive portions of the mechanism are hidden behind a "crash cage," enabling the robot to jump to a height of about 1.2 m and a horizontal distance of 2–3 m without damage. The hopper carries a camera, mounted so as to resist damage during hopping.
- *Self-righting* The landing configuration is unknown a priori, so a hopping vehicle must be able to right itself. Fiorini and Burdick's robot rights itself to its original configuration by extending plastic flaps that initially turn it on its back and then turn it upright with an additional flap turns upright. The device is shown during the righting process in figure 7.15(b).
- *Power and drives* To minimize the amount of unnecessary weight, the robot uses batteries and a single motor with clutches, timers, and so on allow it to power different operations.

While different hopping robots employ a variety of mechanisms, the above issues apply to all of them.

An alternate approach to hopping robots, developed at Sandia National Laboratories, avoids the use of springs entirely, relying instead on a combustion-driven piston to achieve liftoff. Prototype hoppers have reached heights of about 30 ft (see figure 7.16). Another hopper jumps about 3 ft in the air some 6 ft from its starting point on each jump and can perform some 4,000 jumps on a single tank of gas (about 20 g of fuel). The hopper rights itself after each jump. It is designed to land on its piston, but not necessarily vertically. An internal microprocessor reads an internal compass and a gimbal mechanism rotates the offset-weighted internal workings so that the hopper rolls around until it points in the desired direction.

A biologically inspired approach to a hopping robot has been presented by Hyon, Kamijo, and Mita (2002). Basically, the robot developed by Hyon and colleagues is an electromechanical model of the hind leg of a dog. It has a planar design with three links (at the hip, knee, and ankle) and two actuators; the ankle is passive. A spring connects the thigh and heel, representing the ability of muscles and tendons to store energy. The robot is restrained and hops in a circle in a manner reminiscent of other robots in the MIT Leg Laboratory.

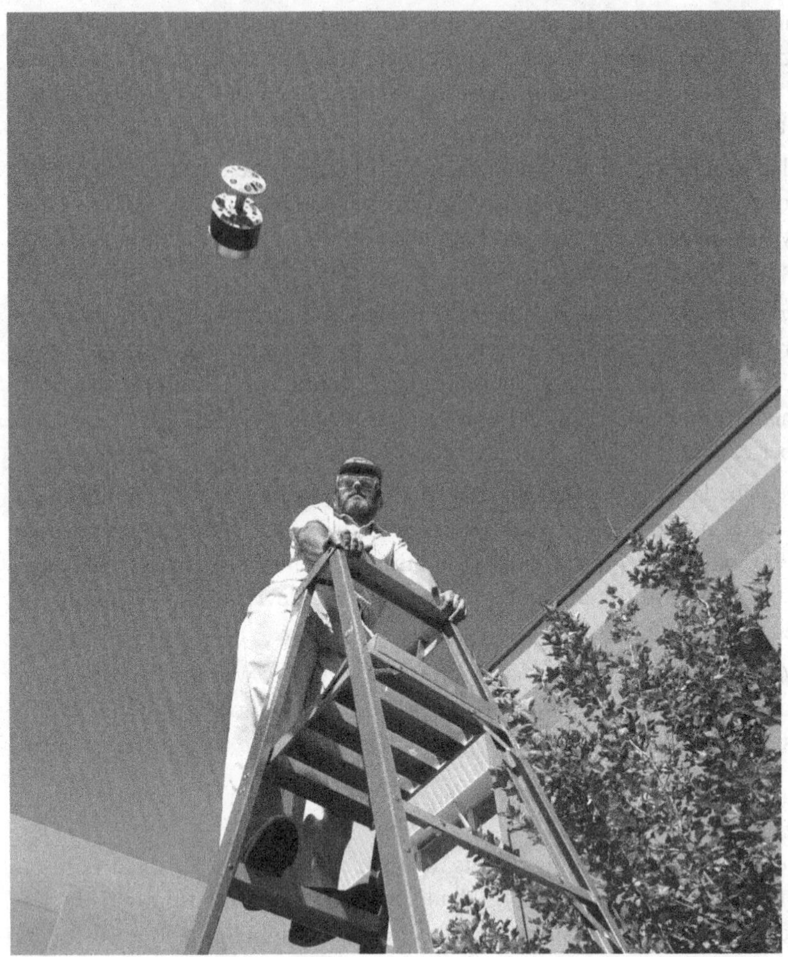

Figure 7.16
Combustion-powered hopping robot from Sandia National Laboratories executing a high leap (reproduced with permission of Sandia National Laboratories)

A number of other hopping robots have been developed, such as those described in Koditschek and Buehler 1991; Francois and Samson 1998; and Zeglin and Brown 1998. We anticipate that there will be interesting developments in the hopper area in the future, perhaps in the development of robots primarily for exploration of planets with lower gravitational forces than on earth.

7.6 Serpentine (Snake) Robots

Snakes are limbless vertebrates that have developed unique ways of locomotion. Living snakes have four major gaits:

1. lateral undulation (used by most snakes);
2. concertina gait (used for motion in confined areas such as narrow tunnels);
3. sidewinding (used by snakes that move on sand and other soft surfaces);
4. rectilinear motion (used by larger snakes, whose skin movements with respect to the skeleton enable them to ratchet along the ground) (Dowling 1997).

Designers of snakelike, nonlegged robots have adopted one or another of these methods to provide mobility for their machines (Ostrowski and Burdick 1996).

Recently there has been increasing interest in the design and fabrication of snakelike robots for a variety of applications. Larger snakes, it is hypothesized, could be used in search-and-rescue operations to crawl through rubble. Inspection of pipelines could be greatly facilitated if an autonomous snake robot, equipped with lights and cameras, could move through pipes and send back information on their condition. For medical applications, snakelike endoscopes could move through the intestines for both diagnostic and surgical purposes.

The key to developing robots with serpentine gaits is to mimic the multisegmented nature of the snake skeleton. Snakes may have 100–400 vertebrae in their spinal cords (Dowling 1997), and their articulation is very complex. The vertebrae form ball-and-socket joints with bone projections to prevent any torsional motion (which could damage the spinal cord). Lateral rotation is limited to some 10°–20°, and ventral rotation is limited to only a few degrees for most snakes. In spite of these rotational limitations, however, since snake skeletons have many segments, the combined motion of those segments can result in large excursions of the body.

7.6.1 Robot Snakes from Shigeo Hirose

The pioneer builder of snake robots is Shigeo Hirose at the Tokyo Institute of Technology, who has studied movements of living snakes and uses biological inspiration for his robots. Hirose photographed the movements of snakes and concluded that the

Figure 7.17
Hirose snake robot with adaptive cord mechanism ACM-III (photograph courtesy of Shigeo Hirose)

serpentine movements followed a curve (the *serpenoid curve*) in which the curvature along the snake's body varied sinusoidally. The serpenoid curve is given by

$$x(s) = sJ_0(\alpha) + \frac{4l}{\pi}\sum_{m=1}^{\infty}\frac{(-1)^m}{2m}J_{2m}(\alpha)\sin\left(m\pi\frac{s}{l}\right),$$

$$y(s) = \frac{4l}{m}\sum_{m-1}^{\infty}(-1)^{m-1}\frac{J_{2m-1}(\alpha)}{2m-1}\sin\left(\frac{2m-1}{2}\frac{\pi s}{l}\right), \tag{7.2}$$

where $x(s), y(s)$ represent the displacement in the x and y directions, respectively, measured along the curved body length s; l is the body length; and $J(.)$ are Bessel functions. Hirose found that the movement of natural snakes on constant-friction surfaces is very close to the serpenoid curves of equation (7.2). He then proposed that a multisegmented vehicle that followed the serpenoid curve could generate forward movement by applying internal torques to its segments, as a snake might do through muscle contractions. He described the resulting vehicles as containing *active cord mechanisms* (ACMs) (Hirose and Umetani 1976, 1981; Hirose 1993). One of his first snake robots based on the ACM principle is shown in figure 7.17. Note that this robot has wheels, but they are passive, being designed mainly to prevent lateral movement of the segments. The robot's motion is controlled entirely through internal torques, thus imitating the lateral undulation of many snakes.

Hirose has gone on to design and build a series of snakelike robots during the past 20 years, many of which are described in Hirose 1993. Some of his snakes have had

Figure 7.18
Robot Koryu-II from the Hirose-Yoneda Laboratory (photograph courtesy of Shigeo Hirose)

tactile sensors, allowing them to adjust to unknown environments. Most of Hirose's snakes were designed for movement on two-dimensional, nearly flat surfaces suitable for wheels. Later snakes were also capable of adjusting to terrain variations by means of relative vertical movement between the "vertebrae," thus allowing the snake to move in uneven terrain, as with the robot Koryu-II, shown in figure 7.18.

7.6.2 Other Snake Robots

Joel Burdick and his students at Caltech have studied the motion of snakelike mechanisms both theoretically and practically. Burdick coined the term "hyperredundant" for snake robots in view of their large number of degrees of freedom (Burdick, Radford, and Chirikjian 1993; Chirikjian and Burdick 1994, 1995). Although Burdick did build a remarkable robot named Snakey, his major contribution was in the theory of serpentine locomotion. We owe much of our knowledge of the dynamics of serpentine motion to Burdick and his students, particularly James Ostrowski (Ostrowski and Burdick 1996, 1998), Gregory Chirikjian (cited earlier in the paragraph) and Howie Choset (Choset and Burdick 2000; Wolf et al. 2003).

There has been considerable interest in the design of robot snakes with a mechanism strong enough to enable them to lift one or more of their segments. One such robot (figure 7.19) was developed by H. Ikeda and N. Takanishi (1987) at NEC in Japan. The six segments of this snake were connected with passive universal joints to prevent adjacent segments from twisting, while allowing bending and orientation (rotation about a lengthwise axis through the segment). This teleoperated robot (the Quake Snake) was intended for searching through rubble following an earthquake and was equipped with a small video camera in the "head."

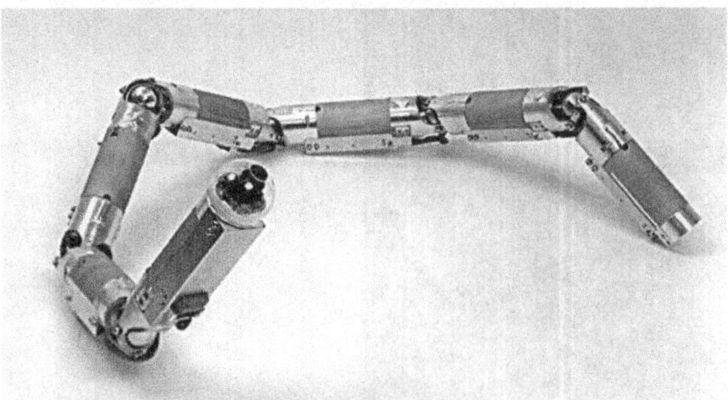

Figure 7.19
The NEC Quake Snake, with 12 dof (photograph courtesy of NEC Corporation)

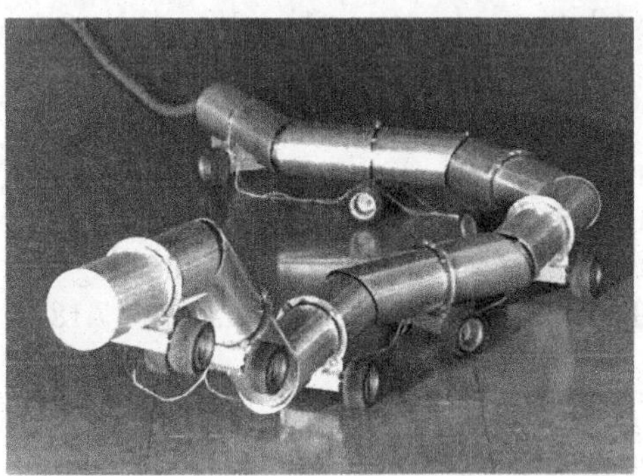

Figure 7.20
OBLIX snake robot from the Hirose-Yoneda Laboratory (photograph courtesy of Shigeo Hirose)

A somewhat different design was used in the snake robot OBLIX (figure 7.20), from the Hirose-Yoneda laboratory (Hirose and Umetani 1981; Hirose 1993). In order for OBLIX to move in three dimensions, a lightweight and high-output segment drive mechanism is required. OBLIX's oblique swivel-joint arm is a mechanism that alternately connects oblique swivel joints, which rotate around the axis that forms an angle from the central axis of the arm, and a coaxial swivel joint. The version of OBLIX (with drive wheels) depicted in the figure has a total of sixteen segments, an overall length of 1.6 m, and a total weight of 4.3 kg. The same principle employed in the design of OBLIX has also been used to design a highly flexible arm for machining operations.

The Institute for System Design Technology of the German research establishment GMD has developed several snakelike robots.The first snake (figure 7.21) used rub-

Figure 7.21
GMD-Snake robot (reproduced with permission of the Fraunhofer Institute for Autonomous Intelligent Systems)

ber joints to allow flexible bending (Paap, Dehlwisch, and Klaassen 1996). The segments of this snake were connected by means of cables to produce curvature along several segments at the same time. The goal was to develop a robot with high flexibility for such tasks as pipeline inspection.

The GMD-Snake consisted of equal sections. Up to 15 sections could be plugged together to form the complete snake. A special head element was plugged onto the front end of the snake. This element could carry different sensors and actuators. The prototype in figure 7.21 carried simple optical sensors used for orientation. The mechanical and the electronic design of the GMD-Snake was kept as simple as possible to achieve an optimized ratio of its weight and the force it could apply.

The specifications of this snake were as follows:

- *Length* 200 cm.
- *Weight* 3 kg.
- *Diameter* 6 cm.
- *Average power consumption* 15 W.
- *Creeping speed* 50 cm/min.

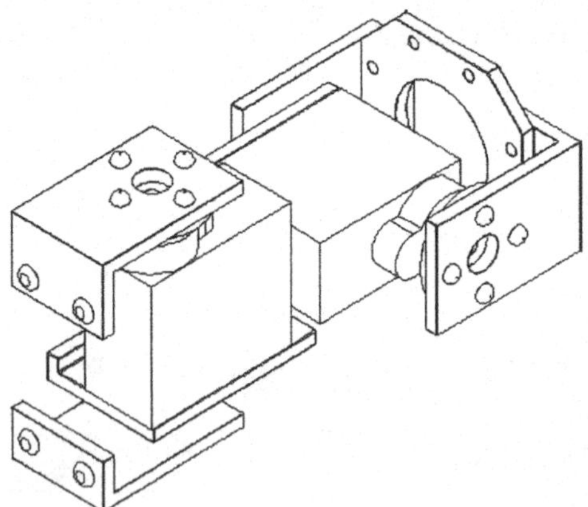

Figure 7.22
2-dof linkage used in Dowling snake design (photograph courtesy of Kevin Dowling)

Each section was driven by a separate processor (slave) that controlled the movement of the rubber joints and also polled their current-position sensors. All slaves were linked via a single serial bus to a centralized process controller that coordinated their movement. A more recent model made use of universal joints connecting five identical segments. In contrast with the first model (which could only be teleoperated), the current model can also operate autonomously, with the goal of inspecting damaged buildings (Paap, Christaller, and Kirchner 2000).

Kevin Dowling (1997) developed an outstanding snake robot (figure 1.1e) in the course of his doctoral research at Carnegie Mellon University. The robot consisted of ten segments, connected by 2-dof linkages made up of two servo motors mounted at right angles, each with approximately 170° of motion limited by the mechanics of the servos (see figure 7.22). The specifications of the robot were as follows:

- *Mass* The mechanism weighed 1.32 kg. The servos accounted for about two-thirds of the mass. The rest of the robot's mass was metal and hardware. Total mass, including wiring and controllers, was 1.48 kg.
- *Length* 102 cm (ten links, each 1.02 cm long).
- *Diameter* 6.5 cm.
- *Power* 24.2 W max total mechanical output. ~75 W max total electrical input.
- *Quiescence* 1.15 W (9.5mA@6VDC).

The 3-D snake link design utilized 2 orthogonal dof. Typical servo excursions were about 90°, but can be commanded to nearly 170°.

Dowling was able to obtain several gaits with his snake, including rectilinear, sidewinding, and lateral rolling. A wide variety of novel gaits and motions not existing in

Figure 7.23
S5, one of Gavin Miller's robot snakes (photograph courtesy of Gavin Miller)

nature were also produced, including lateral undulation, a variant of lateral rolling, a "ventral wave" gait, and a "butterfly" gait. Interested readers are referred to Dowling 1997 (his dissertation), which is available on the Carnegie Mellon University Robotics Institute Web site.

Another fascinating snakelike robot named PIRAIA has been under development by Martin Nilsson (1998) at the Swedish Institute for Computer Science. The snake consists of a number of identical segments and links. A failed segment can be removed, the remaining segments reconnected, and the snake can continue to function. The links are based on a roll-pitch-roll joint, which allows highly unusual gaits as a result of the rolling motion in the joints. For example, the robot can wrap itself around a tree and then "climb" the tree by rolling upward. Much of the emphasis in this project is on learning and self-awareness in robots.

Numerous other snake robot projects exist or have existed in Europe, Japan, and the United States. NASA laboratories, including the Jet Propulsion Laboratory and the Ames Research Center, have developed snakelike robots. We have cited work in Germany and Sweden, but snake robot projects also exist in France and England. An excellent and current catalog of snake robots is maintained by Rainer Worst in Germany at http://ais.gmd.de/~worst/snake-collection.html.

We end this brief survey of snake robots by citing the remarkable work of Gavin Miller, who has developed and fabricated a large number of snake robots as a personal research project, financed entirely by himself, since 1987. The research is described in the proceedings of two conferences (Miller 2000, 2002).[17] One of Miller's snakes, S5, is shown in figure 7.23. All of Miller's snakes carry their own power and control systems.

7.7 Underwater Robotic Vehicles

Thus far, this chapter has concentrated on the study of robots moving on the ground, whether on wheels or treads or crawling like snakes. Now we must look, albeit briefly, at robots that move and work under the water or in the air. We begin with underwater robots.

17. Pictures and movies of the snakes can be seen on Miller's Web site (http://www.snakerobots.com).

There are two major classes of underwater robots. The first involves attempts to make submarines and related vehicles autonomous. Clearly, this is of great interest to the military. We described the military interest in AGVs and UGVs, earlier in this chapter. Robot submarines are also known as autonomous underwater vehicles (AUVs). The second class of underwater vehicles is inspired by biology and attempts to create robots that move like fish, or crabs, or lobsters, or even water snakes. We consider biomimetic underwater robots later in this section.

Underwater vehicles differ from ground-based vehicles in a number of ways:

- The vehicle must move in all three coordinate directions, whereas UGVs operate in a plane. (They may encounter 3-D obstacles, but basically, ground-based robots move in two dimensions.)
- The density of the water in which they move changes as a function of depth, and with it, the control characteristics of the vehicle change as well.
- With increasing depth the water pressure on the vehicle increases. At a depth of only 33 ft, the pressure will be twice the normal atmospheric pressure. Hence, the vehicle structure must be designed to withstand these depths.
- The dynamics of underwater vehicles are highly nonlinear, thus making simple linear control nearly impossible. The situation is worse if the AUV carries one or more manipulators (say, to retrieve treasures from the ocean floor).
- The vehicles must be sufficiently waterproof to safeguard all the on-board instrumentation.

In spite of these difficulties, it is clear that the ocean is a vast reservoir of both living and inorganic resources and has a major effect on the earth's environment. Hence, a large number of autonomous underwater vehicles have already been built and we expect to see increasing activity in this field. Some thirty-four AUVs were developed during the 1990s in Europe, Asia, and the United States (Yuh 2000). Many of the issues concerning design and construction of AUVs are discussed in Yuh, Ura, and Bekey 1997. It should be noted that unpiloted underwater vehicles have been built for at least the past 30 years, but these vehicles have been controlled by means of attached cables or tethers. True autonomy, meaning no tethers but frequently human control of navigation and high-level behaviors using some form of communication (electromagnetic or acoustic), has been used only since about 1990. Current AUVs are able to dive to depths of 20,000 ft.

7.7.1 Dynamics

The dynamics of underwater robots are highly nonlinear and time varying. The interaction between the vehicle and the water, particularly in the presence of currents, may produce variations in hydrodynamic parameters. As noted previously, if the vehicle is equipped with manipulators, the dynamics are even more complex and depend on the degree of extension of the arms and any payload they may be carrying. The vehicle is controlled by means of a number of thrusters enabling motion in any desired direction.

The equations of motion of the vehicle are defined with respect to both a body-fixed and an inertial coordinate system, as shown in figure 7.24 (Yuh 2000), which indicates the names given to motion in the 6 degrees of freedom (surge, sway, heave, roll, pitch, and yaw). The equations can be written in vector-matrix form as

$$\dot{\mathbf{x}} = \mathbf{J}(\mathbf{x})\dot{\mathbf{q}},$$

$$\mathbf{M}\ddot{\mathbf{q}} + \mathbf{C}(\dot{\mathbf{q}})\dot{\mathbf{q}} + \mathbf{D}(\dot{\mathbf{q}})\dot{\mathbf{q}} + \mathbf{G}(\mathbf{x}) = \boldsymbol{\tau} + \mathbf{w}, \tag{7.3}$$

$$\boldsymbol{\tau} = \mathbf{B}\mathbf{u},$$

where $\mathbf{J}(\mathbf{x})$ is a 6×6 velocity transformation (Jacobian) matrix that converts the velocity from vehicle coordinates to earth-fixed (inertial) coordinates; $\mathbf{M}$ is an inertia matrix; $\mathbf{C}(\dot{\mathbf{q}})$ is a Coriolis and centripetal acceleration matrix; $\mathbf{D}(\dot{\mathbf{q}})$ is a damping

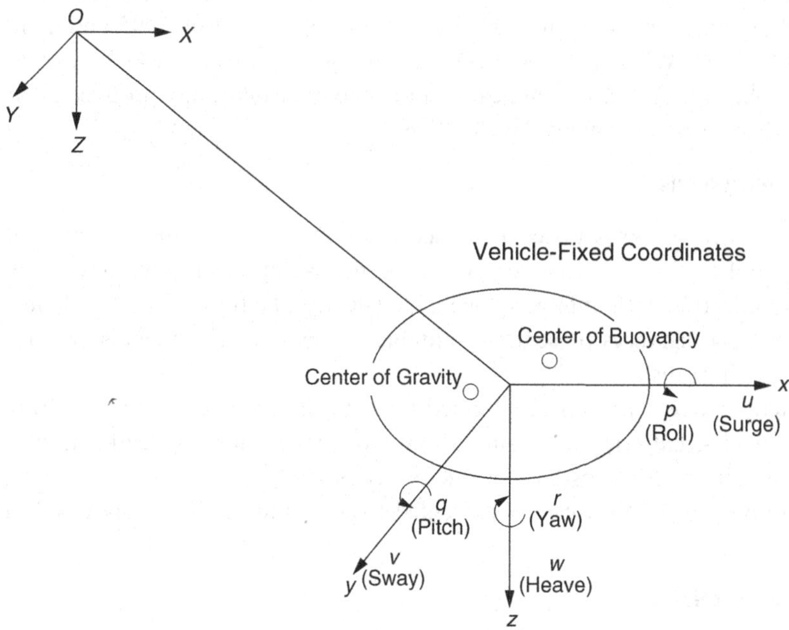

Figure 7.24
Coordinate systems for underwater vehicle (from Yuh 2000, reproduced with permission of Kluwer Academic Publishers)

matrix (including drag terms); **G(x)** is a restoring-force term, including the effects of buoyancy and gravity; τ is a 6×1 vector of control forces and moments; **w** is a vector of disturbances; **B** is a control matrix; and **u** is a vector of forces produced by thrusters.

The dynamics become even more complex as the vehicle moves under water and the effect of the mass of fluid displaced by the vehicle's movement needs to be considered. Clearly, simulation techniques are essential in the design process (Brutzman, Kanayama, and Zyda 1992; Choi and Yuh 1993; Yuh 1995).

7.7.2 Control

In view of the highly nonlinear nature of the dynamics of AUVs and the variability introduced by changes in buoyancy brought on the pumping of water into and out of reservoirs on the vehicle as well as manipulator motion, it should be clear that a simple linear PID controller will not produce satisfactory performance. It is essential to have some form of adaptive control to allow the system to track the changes in its environment and the vehicle's dynamics. Among the types of control that have been proposed and/or tried on AUVs are sliding-mode control (e.g., Healey and Lienard 1993), neural-network control (e.g., Ishii, Fujii, and Ura 1998), and fuzzy control (Kato 1995). Various control architectures have been proposed (e.g., Valavanis et al. 1997). Several excellent chapters on control of AUVs appear in Yuh 1995 and Yuh, Ura, and Bekey 1997. Whereas linear PID control may not yield satisfactory transient response, the addition of a nonlinear term to the controller may produce great improvement (Perrier and Canudas de Wit 1996).

7.7.3 Other Subsystems

Communication with AUVs was accomplished traditionally by means of umbilical cords with coaxial or fiber-optic cables. More recently acoustic modems have been used for communication with untethered vehicles. Clearly, the transmission of acoustic energy over long distances presents formidable challenges. RF modems are limited to relatively short distances.

Power for tethered vehicles can be supplied through the umbilical cord. For truly autonomous vehicles, batteries are required, the most common being lead-acid batteries or (for long life at high cost) silver-zinc batteries.

Several examples of AUVs are now presented to illustrate the issues discussed in the foregoing.

7.7.4 Example 1: ODIN

The Underwater Robotics Laboratory at the University of Hawaii has developed ODIN (Omni-Directional Intelligent Navigator), a spherical underwater robotic vehicle equipped with eight thrusters, four horizontal and four vertical, as shown in fig-

Figure 7.25
ODIN underwater robotic vehicle from the University of Hawaii (photograph courtesy of Junku Yuh)

ure 7.25. It should be evident from the location and direction of the thrusters that this is indeed an omnidirectional vehicle, capable of movement within all 6 degrees of freedom. ODIN can operate in either a tethered or an autonomous mode. It is controlled from a graphic workstation in the laboratory via a serial RS-232c line (Choi and Yuh 1993). The operating system is VxWorks. ODIN is equipped with an arm capable of only 90° of motion on a single axis. Communication takes place across an RF modem with a range of 480 m in an unobstructed area. Sensors include sonars for ranging, an inertial navigation system (providing angle and heading rates as well as three-axis linear acceleration) and pressure sensors.

7.7.5 Example 2: AUSS

The U.S. Navy Space and Naval Warfare (SPAWAR) Systems Center in San Diego, California, has developed the Advanced Unmanned Search System (AUSS), illustrated in figure 7.26. AUSS is designed to operate at depths up to 20,000 ft. It is 17 ft (about 5.2 m) long and 31 in (0.78 m) in diameter and weighs 2,800 lbs (1273 kg). While underwater, it communicates with the surface by means of an acoustic link, capable of sending compressed data at up to 4,800 bits/sec. On-board navigation makes use of Doppler sonar and a gyrocompass. Other sensors include side-looking and forward sonars and a CCD (charge-coupled device) camera. The vehicle runs on silver-zinc batteries with a running time of 10 hours; recharging takes 20 hours.

Figure 7.26
U.S. Navy Advanced Unmanned Search System (photograph courtesy of Space and Naval Warfare Systems Center, San Diego)

7.7.6 Example 3: ROMEO

Romeo (figure 7.27) is an AUV developed by the Naval Automation Institute (Istituto Automazione Navale) in Genoa, Italy, and used for bottom following (Caccia, Bruzzone, and Veruggio 2003). It is an open-frame vehicle equipped with four horizontal and four vertical thrusters (like ODIN) to guarantee full controllability of the vehicle. The intention of the designers was to use vehicle autonomy for control of the vehicle's heading, altitude, and speed, while employing a remote pilot to concentrate on the detection of interesting features from the on-board cameras and other instruments. During the bottom-following experiments the vehicle was equipped with two echo sounders (for different ranges), a compass, and a gyro. Using a previously obtained model of the vehicle's dynamics (Caccia, Indiveri, and Veruggio 2000), the designers were able to design velocity controllers that effectively reduced the heave and yaw dynamics to second-order systems.

7.7.7 Example 4: OTTER

The final example of an AUV discussed here is OTTER (Ocean Technology Testbed for Engineering Research) (figure 7.28) (Wang, Rock, and Lee 1996), developed jointly by the Aerospace Robotics Lab at Stanford University and the Monterey Bay Aquarium Research Institute (MBARI). This vehicle is an untethered submersible designed for ocean exploration. Its low-level control system is fully autonomous, but a remote operator provides high-level commands concerning planning or acqui-

Figure 7.27
Romeo AUV, developed by the Naval Automation Institute, Genoa, Italy (from Caccia, Bruzzone, and Veruggio 2003, reproduced with permission of Kluwer Academic Publishers)

sition of objects of interest under the water. The vehicle is 2.1 m long, 1 m wide, and 0.5 m high. It is designed to withstand depths of up to 1,000 m but has been used extensively only in a large test tank. The vehicle is equipped with eight thrusters as well as a motor for driving an arm and a pan-tilt mounting for an underwater camera. Sensors on the robot include a fluxgate compass, a two-axis inclinometer, an inertial navigation package with three gyros and three linear accelerometers, a depth sensor, a sonic ranging and positioning system, a manipulator torque sensor, leak detectors, and battery monitors. There are extensive on-board computer facilities. Communication was primarily via a tether as of 1996, when the paper cited was published, but acoustic modems were under investigation for autonomous dives.

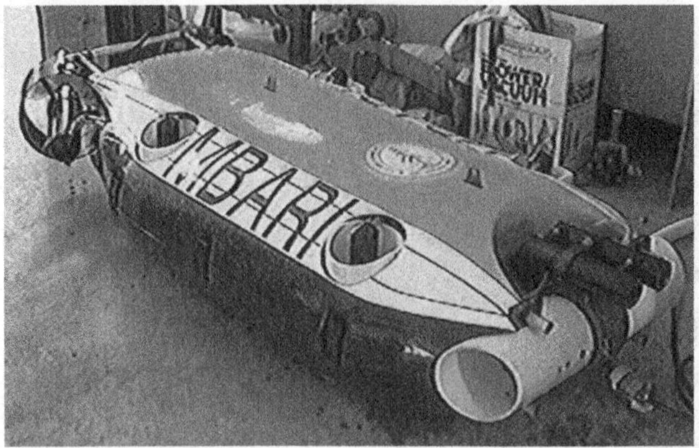

Figure 7.28
OTTER underwater robot (from Wang, Rock, and Lee 1996, reproduced with permission of Kluwer Academic Publishers)

7.7.8 Limited-Task AUVs

Some AUVs are designed for tasks such as temperature soundings or current measurements in the ocean, tasks for which such unpiloted vehicles are eminently suited. Ideally, such vehicles should operate autonomously for months at a time. However, the requirement to travel long distances and to dive to substantial depths puts a strain on their battery life. Researchers at Woods Hole Oceanographic Institution in Woods Hole, Massachusetts, the Monterey Bay Aquarium at Monterey, California, and other institutions have begun fielding small autonomous ocean-sampling robots that make use of clever principles to reduce battery drain (Fratantoni et al. 2000; Petit 2003). Pumps move oil from an external reservoir to an inner chamber in the robot, thus changing its buoyancy. The cylindrical robot glider Slocum, illustrated in figure 7.29, is equipped with small wings, so that as it dives or rises toward the surface, it also produces forward motion. Thus, the vehicle moves in a zigzag trajectory. Each time it rises to the surface, it sends information to central stations using satellite links. The on-board batteries are used for pumping oil, switching valves to control the direction of oil flow, and communication, not for propulsion, enabling vehicles like Slocum to stay at sea for months at a time and cover distances of hundreds of kilometers. Under development are robots that would use the temperature change from surface to depth to open and close the valves (say, by using the temperature to alternately melt and solidify waxlike compounds), thus further reducing the drain on the robot's batteries. Such thermal-glide vehicles could stay at sea for years at a time and cover thousands of kilometers and could dive to depths of 1500 m before returning to the surface (Graver et al. 2003).

Figure 7.29
Slocum glider being photographed by David Fratantoni (photograph courtesy of David Fratantoni)

7.7.9 Summary

As the foregoing review demonstrates, the field of autonomous underwater vehicles is very challenging and very active. We expect that as the needs for monitoring the ocean and extrating resources from the ocean floor increase, there will be a corresponding growth in the development and deployment of AUVs.

7.8 Biologically Inspired Underwater Robots

The second major class of underwater robots consists of machines inspired by fish and other water-based animals. Clearly, fish and mammals such as dolphins and whales are outstandingly suited for aquatic locomotion and tend to outperform any human-made vehicles in that area. For example, tuna can swim continuously at one to two body lengths per second but can burst to speeds in excess of ten lengths per second. They are also highly maneuverable. In hopes of finding ways of replicating the speed and maneuverability of fish in AUVs, a number of fish robot research projects are in process in Japan, Europe, and the United States.

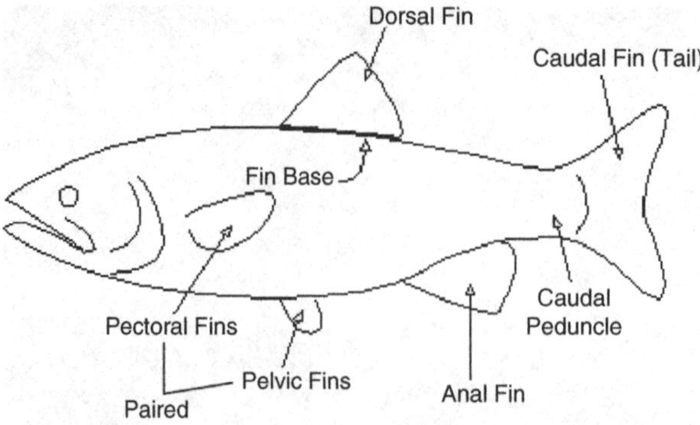

Figure 7.30
Body structure of a typical fish

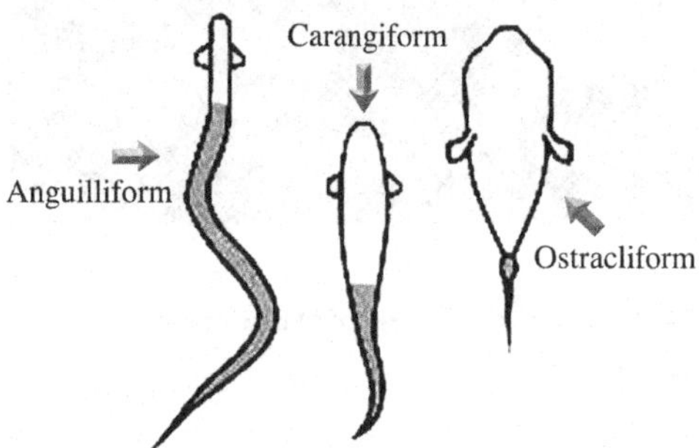

Figure 7.31
Major forms of fish structure (adapted from Breder 1926)

The general body structure of a fish is shown in figure 7.30. Locomotion is produced by the displacement of water by movement of the fins, primarily the caudal or tail fin. The other fins, while helping with propulsion in some fish, also assist in maintenance of body posture. In many fish both the peduncle (the body structure supporting the tail) and the tail move. Such fish are known as *carangiform,* and many common large and fast-swimming fish, like the tuna, salmon, trout, and swordfish, are of this type.

Anguiliform fish (such as eels) obtain propulsion by means of a continuous wave of muscle action, analogous to undulation in snakes. A third class, the *ostracliform* fish (such as the boxfish) oscillate their tail without moving their bodies. This classification (illustrated in figure 7.31) is due to Breder 1926, but other more recent classifications exist (e.g., Webb 1984). Most attention in classification of fish has been focused on periodic or sustained swimming, in which propulsive movements (primarily of the caudal fin) are cyclic and repetitive. Thus, carangiform fish produce repetitive bending of their bodies in backward moving propulsive movements which extend to the caudal fin.

In view of the importance of the caudal fin in producing locomotion in most fish, it is not surprising that the majority of robot fish have also concentrated on producing tail fin oscillation for propulsion. Combining biomimetic tail motion with a streamlined fish shape has produced very agile and fast underwater robots.

As noted previously, numerous laboratories in Japan, Europe, and the United States have developed robotic fish. We present the characteristics of a few of these robots in the following sections.

7.8.1 The MIT RoboTuna® and RoboPike

RoboTuna® and RoboPike were developed in the Ocean Engineering Laboratory at the Massachusetts Institute of Technology. By "reverse engineering" the hydrodynamic properties of living fish, investigators discovered that fish achieve much of their swimming efficiency by controlling the vortices around their moving bodies, and they included controls to this effect in their robotic fish (Gopalkrishnan et al. 1994). RoboTuna I® (figure 7.32(a)) derives all its propulsion from its caudal fin. The shape of the robot's body follows the morphology of a small bluefin tuna. Across the broad spectrum of fish form and movement, tuna are most desirable as a vehicle platform, as they are very streamlined, relatively rigid in the forebody, and propel themselves with low-amplitude movements in conjunction with a high-performance hydrofoil (caudal fin). Most tuna swim continuously at one to two body lengths per second but can burst to speeds in excess of ten lengths per second. Additionally, tuna are highly maneuverable compared to conventional underwater vehicle systems, largely because of the integrated nature of the main propulsor and maneuvering control surfaces.

The robot is not autonomous but instead is supported for swimming (in the MIT Towing Tank) by a vertical rod that carries signals and power. It is controlled by six powerful servomotors (rated at 2 horsepower each) and has force sensors placed at various locations along the path of the controlling tendons. The carriage is towed through the water, and swimming efficiency is calculated from the forces on the tendons. RoboTuna® follows the design of a carangiform fish, since both its peduncle and tail fin move.

Following the success of RoboTuna I®, the laboratory developed RoboTuna II®, with a complete set of fins and a different elastic body structure, shown in figure 7.32(b). Clearly, this is a very realistic-appearing robot. Like RoboTuna I®, it also swims supported in the MIT Towing Tank. RoboTuna II® was followed by RoboPike, with a similar elastic body structure made of three segments. The aim in

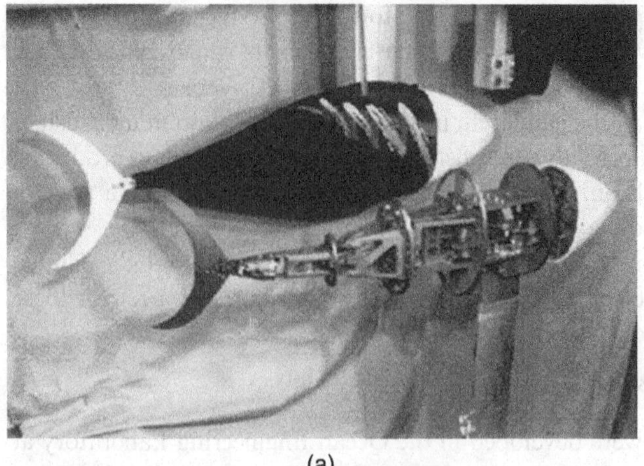

(a)

(b)

Figure 7.32
RoboTuna® from the MIT Ocean Engineering Department: (a) RoboTuna I® structure with and without "skin"; (b) RoboTuna II® (photograph courtesy of Michael Triantafyllou)

developing these robots is to learn more about the complex fluid mechanics that fish use to propel themselves. In the long run, propelling small autonomous vehicles via fishlike swimming could result in enormous energy savings for these vehicles and increase the length of time they can swim. In the meantime, by studying and building robots like RoboTuna® and RoboPike, the lab hopes to resolve Gray's paradox (i.e., the observation that fish don't seem to have enough muscle power to propel themselves at the speeds that they do).

The pike was selected for study and emulation since it has remarkable characteristics, including very quick turning and fast acceleration from a stop. In the wild, pike accelerate at rates from 8 to 12 Gs during a start from a standstill to 6 m/sec.

Figure 7.33
Draper Laboratory robotic tuna, VCUUVJ (copyright The Charles Stark Draper Laboratory, Inc., all rights reserved, reproduced with permission)

Although this may not be achievable in a robot, even half or a quarter of this acceleration would still demonstrate that flapping-foil propulsion is capable of producing higher accelerations than a propeller.

7.8.2 The Draper Laboratory Tuna Robot

In addition to the MIT Ocean Engineering Laboratory projects, the Charles Stark Draper Laboratory has also built a robotic tuna, in this case modeled after an 8 ft long, 300 lb yellowfin tuna, also to investigate fish propulsion. This type of propulsion produces vehicles several times more maneuverable than conventional unmanned devices which constitutes a significant advantage in tight quarters, and such vehicles could prove valuable for use in undersea exploration, recovering unexploded mines, laying cable, and many other practical applications. Named after the vorticity flow control mechanisms employed by fish to propel and maneuver, the Vorticity Control Unmanned Underwater Vehicle (VCUUV) (figure 7.33) mimics the form and kinematics of a yellowfin tuna.

7.8.3 The Mitsubishi Robotic Fishes

In contrast to the RoboTuna®, RoboPike, and VCUUV, the robotic fish designed by Mitsubishi Heavy Industries (MHI) are autonomous, relying only on a network of sensors in the walls of a large tank in which they swim. The ultimate goal of the Mitsubishi project is to improve the propulsion and steering of ships and submarines, in the belief that fitting them with oscillating fins rather than the usual propeller thrusters will improve speed as well as control. However, at this time the immediate goal is to build "aquariums" populated entirely by robot fish.

Figure 7.34
Mitsubishi Sea Bream robot (photograph courtesy of Mitsubishi Heavy Industries)

One of the first Mitsubishi robots, shown in figure 7.34, was modeled on a fish known as the sea bream. The robot weighs 2.5 kg (5.3 lb), is about a half meter in length, and can swim for a half hour before its batteries need recharging. Some observers have said that this robotic fish is so lifelike that only careful inspection of its eyes reveals that it is not living (CNN 1999). The robot's swimming motions are computer controlled.

Mitsubishi has also developed a robotic coelacanth, a nearly extinct fish sometimes referred to as a "living fossil." The artificial coelacanth is about 70 cm long and weighs 12 kg; its body is made partly of silicone plastic. The fish swims in a large water tank at the Aquatom, a science museum in Fukui prefecture in Japan. When the robot's batteries run low, it automatically swims to a recharging station.

Figure 7.35 shows the lead researcher of the Mitsubishi robot fish team, Yuuzi Terada, with one of his coelacanths (Yamamoto and Terada 1999; Terada 2000). The fish is propelled by elastic oscillating fins controlled by a computer. News reports have indicated that MHI plans to build a number of robotic fish in its Animatronic division, possibly to be sold or licensed along with an aquarium. This will be a purely entertainment application of the technology.

7.8.4 Other Biomimetic Underwater Robots

Early work in robotic fish was undertaken at the Free University of Brussels, Belgium, where a robotic fish was developed as early as 1992. The fish was characterized by a large, nearly rectangular caudal fin. Another tunalike robot was designed

Figure 7.35
Robotic coelacanth with developer Yuuzi Terada (photograph by Takano Akira, copyright NIPPONIA, No. 13, Heibonsha Ltd., 2000, courtesy of Yuuzi Terada)

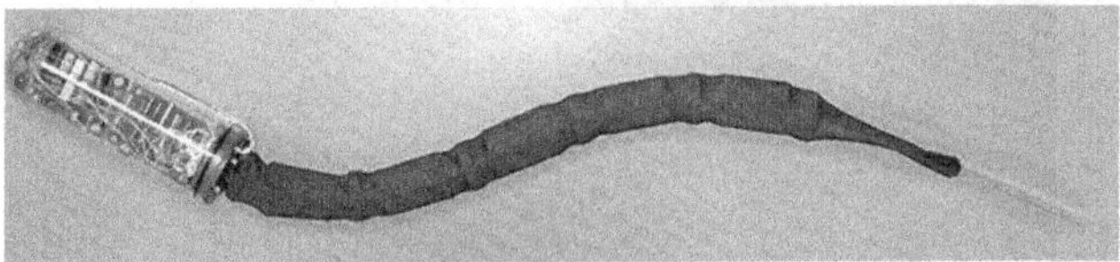

Figure 7.36
Robotic lamprey (photograph courtesy of Jan Witting, with permission of Joseph Ayers)

and built at the National Maritime Research Institute in Tokyo. Theoretical and experimental studies of propulsion in carangiform fish have been conducted at Caltech in Pasadena, California (Morgansen et al. 2001; Morgansen, Vela, and Burdick 2002).

In the Marine Biology Laboratory at Northeastern University in Boston, Joseph Ayers has developed robotic lobsters and lamprey eels. Ayers's robot lobster was introduced in section 3.8 in connection with a discussion of its shape memory alloy actuators, and illustrated in figure 3.8. The lamprey developed by Ayers is shown in figure 7.36. Like the lobster, the lamprey is controlled by SMA actuators. As shown in the figure, the electronics are located in a cylindrical module at the head of the robot.

A fascinating amphibian robot is Ariel (figure 7.37(a)), built by iRobot Corporation with major biological input from the PolyPEDAL Laboratory of Robert Full at the University of California at Berkeley (http://polypedal.berkeley.edu). Ariel is a

(a)

(b)

Figure 7.37
Unusual underwater robots: (a) Ariel walking in the surf zone (photograph courtesy of iRobot Corporation); (b) Swimming robotic snake HELIOS-I from Hirose-Yoneda Laboratory (photograph courtesy of Shigeo Hirose)

six-legged robot capable of walking either on land or underwater near the ocean shore (in the turbulent surf zone). It can climb over obstacles and resist the impact of waves (because of its steamlined body shape). The most interesting aspect of this robot is that its legs are invertible; that is, if it is flipped onto its "back" by the waves, its legs reorient so that the "back" becomes "front" (or "underside"), and the robot keeps walking. Ariel was developed to help locate unexploded mines in the surf zone.

As a final example of a biomimetic robot, we cite the water snake developed in the remarkable Hirose-Yoneda Laboratory in Tokyo (Takayama and Hirose 2001). This snake, known as HELIOS-I and shown in figure 7.37(b), has a hermetic 3-D active cord mechanism that can move both on the ground and in the water. However, unlike other robots in the traditional ACM series, its creation was based on the study of motion of a corkscrew-shaped microorganism called a spirochete rather than a snake. In spite of the great difference in size (spirochetes measure about 8 to 10 μm in length and HELIOS-I measures 1.7 m), the mechanical model successfully reproduces the unique spiral propulsion motion of these microorganisms.

7.9 Climbing and Other Unusual Locomotion Methods

We now consider two other methods of locomotion some animals employ on land: climbing and brachiation. Flies, lizards, and geckos are capable of climbing vertical walls with apparent ease. Monkeys use their long arms to swing from branch to branch as a way of moving rapidly through the forest.

7.9.1 Climbing Robots

Humans have always been fascinated by the ability of some animals to climb vertical walls and even to walk upside down on ceilings. Full and his colleagues at the University of California PolyPEDAL Laboratory have discovered recently that the feet of geckos are covered with huge numbers of tiny hairs that make use of van der Waals forces to enable them to cling to walls (Autumn et al. 2000). Attempts are already under way to build robots that, at least to some extent, emulate geckos and climb walls, and we expect to see many such efforts. The first such attempt is a robot named Mecho-Gecko, built by iRobot Corporation (figure 7.38), which climbs walls quite successfully. It does not have hairs as a gecko does, but its motion mimics the "peeling" method by which geckos move their feet. A series of peelable attachments provide the sticking power to allow the Mecho-Gecko to climb walls.

Without the ability to replicate the feet of geckos, designers of wall-climbing robots make use of suction cups or, if the vertical surface is metallic, electromagnets. This technology has applications in such robots as those for cleaning windows in

Figure 7.38
Mecho-Gecko wall climber (photograph courtesy of iRobot Corporation)

office buildings, those for inspecting exterior walls of large structures, and those for performing welding on the sides of ships. Let us look at three such systems.

The REST climbing robot (figure 7.39) was designed by researchers at IAI-CSIC (Instituto de Automática Industrial-Consejo Superior de Investigaciones Cientificas) in Spain to perform welding on the vertical surfaces of ships. Since the climbing surface on which the robot operates is ferromagnetic, the feet of the robot are provided with special electromagnetic grasping devices. This robot has six legs and 18 degrees of freedom. It carries it own processor and control system on board. The robot weighs 220 kg and is capable of carrying a payload of 100 kg on a vertical wall (or ceiling). The design of the legs makes it able to cross small obstacles (Armada et al. 1998).

The window-cleaning robot Sirius, developed by the Fraunhofer Institute in Germany and illustrated in figure 7.40, is intended for commercial use on large office

Figure 7.39
REST: A ferromagnetic surface-climbing robot (photograph courtesy of Manuel Armada)

Figure 7.40
Window-cleaning robot Sirius from Fraunhofer Institute for Factory Operation and Automation (IFF), Magdeburg, Germany (photograph courtesy Fraunhofer Institute IFF)

buildings. The robot uses suction cups to attach itself to the glass of the building's windows and a simple gait for locomotion, lifting and moving one leg at a time. To ensure that no water falls from the robot, it is equipped with a suction system that pulls the used water back into the machine to be filtered and reused. The robot is unique because of the combination of features as well as the wall-climbing gait. It is also suitable for cleaning and painting of ship hulls.

An unusual wall-climbing robot named NINJA has been developed at the Hirose-Yoneda Laboratory in Tokyo. NINJA-I (shown in figure 7.41) has four legs driven by three prismatic actuators each, whereas NINJA-II has articulated legs (Hirose, Nagakubo, and Toyama 1991). This robot has two unusual features. First, for attachment, instead of regular suction cups, it uses a valve-regulated multiple pad that consists of multiple suction pads and a set of valves for controlling the vacuum produced by each pad. One implementation has twenty suction pads, designed in such a way that there are no vacuum leaks. This system makes it possible for NINJA to

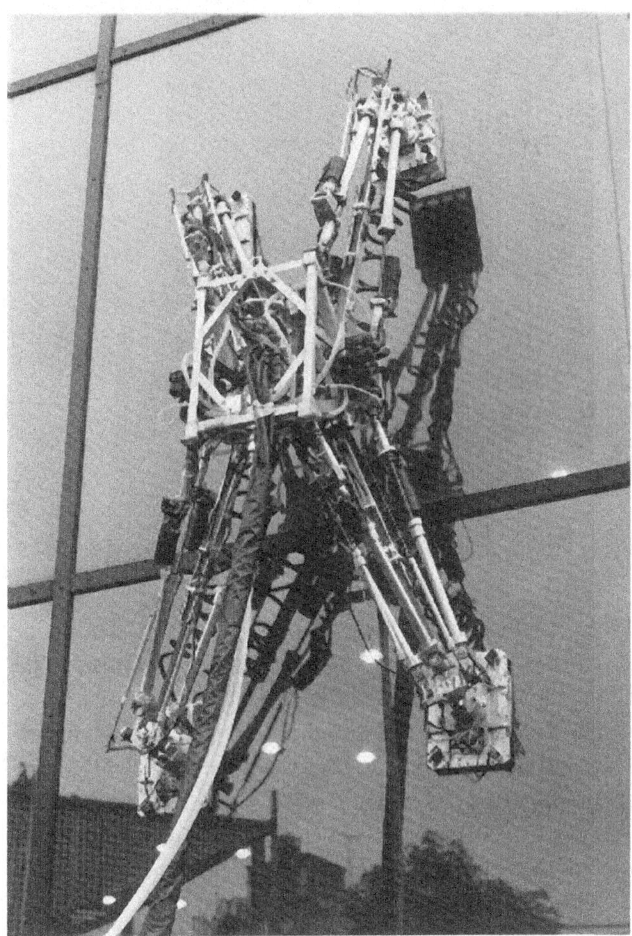

Figure 7.41
NINJA-I wall-climbing robot (photograph courtesy of Shigeo Hirose)

climb walls that have cracks or contours. Secondly, the robot's feet are compliant with the wall surface while always being oriented in the same direction as the body by a parallel movement mechanism. NINJA is quite large, being about 1.8 m long and 0.5 m wide, with a weight of 45 kg. It walks with an unusual gait involving the lifting of two legs on one side of the body at the same time, since the other two legs provide firm support (Nagakubo and Hirose 1994).

A unique climbing machine is the inchworm robot (figure 7.42) developed by Daniela Rus, originally at Dartmouth University and more recently at MIT. This robot follows Rus's desire to build robots on a minimalist principle, inspired by biology,

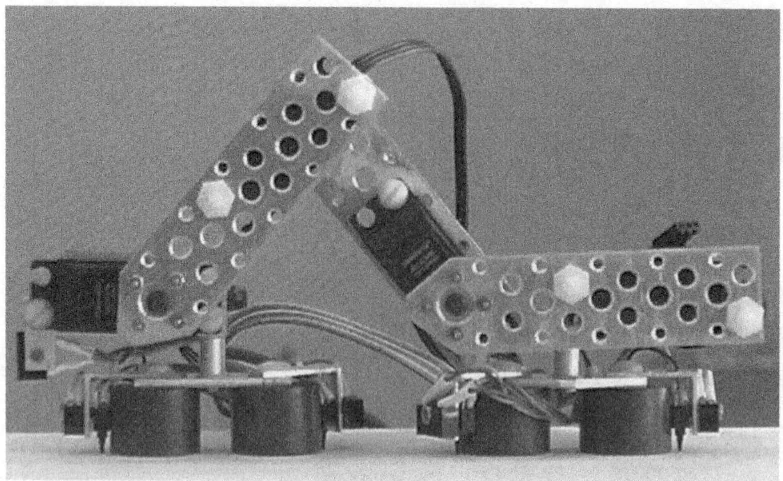

Figure 7.42
Inchworm robot (photograph courtesy of Daniela Rus)

but without a need to be faithful to the morphology of the inspiration. The inchworm robot has only four actuators, five IR sensors, and four touch sensors. The robot consists of four links, the first and last of which are two relatively large "feet." Each foot contains two electromagnets, connected by a trusslike mechanism. Each of the feet is capable of supporting the robot, so that with one foot firmly anchored, the truss structure raises the forward foot and moves it a small amount. With the forward foot in place, the rear foot can move forward a comparable distance. The robot is capable of climbing irregular metal structures like bridges or towers (Kotay and Rus 1996, 2000). Although the figure shows the robot on a horizontal surface, the flexibility of the robot's linkage is such that it can transition to a vertical position while moving, and even to an upside-down position while it walks on the ceiling.

Of particular interest is a robot capable of climbing between two vertical ladders. The robot has three legs, each with two links, two actuated joints, and two passive joints. The robot supports itself between the ladders on two of its legs. It then swings the third leg to a support position beneath one of the others, enabling it to move higher and raise the body. The third leg then swings under the other leg and the process is repeated. This example of a climbing robot was developed as part of a broader analysis of the control of multilimbed robotic systems (Dubowsky, Sunada, and Mavroidis 1999).

The design and development of climbing robots has been a challenge to robotics for some time. Hence, other climbing robots have been developed (e.g., Bahr, Li, and Najafi 1996), and we expect that others will be developed in the future.

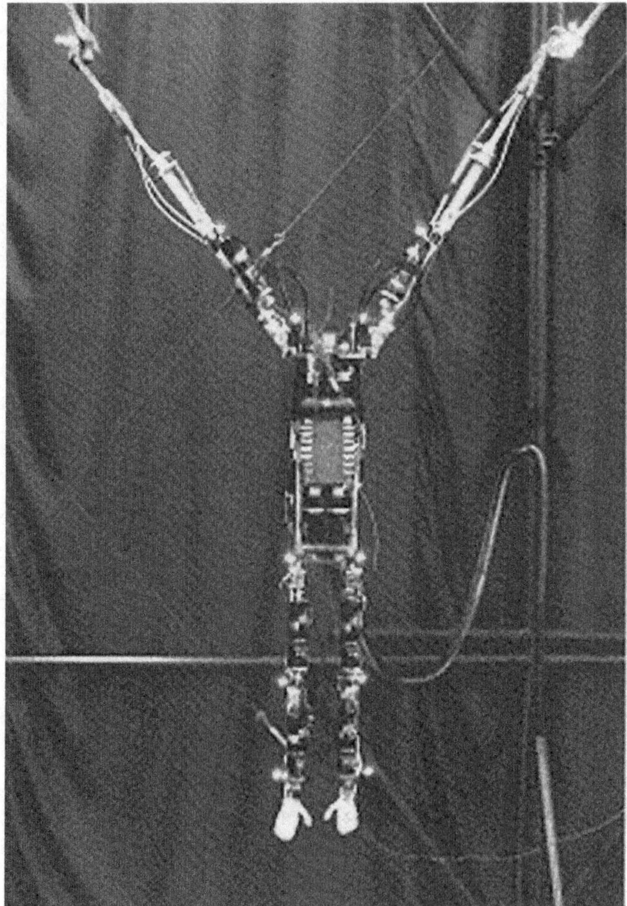

Figure 7.43
Brachiating robot (photograph courtesy of Toshio Fukuda)

7.9.2 Brachiating Robots

Brachiation refers to the way in which apes, such as gibbons, that have long arms move from limb to limb by hanging from one arm and swinging the body forward until the second hand can grasp a branch. Toshio Fukuda and his associates at Nagoya University in Japan built a series of brachiating robots, one of which is illustrated in figure 7.43. The robots hang by one arm from a support bar and initiate a swinging motion until the second arm is able to reach the support bar. At this point the first arm releases, and the momentum generated by the first swing generates a second swing; the process continues along a series of evenly spaced bars. The major difficulty is obtaining a sufficient elevation of the second arm in the first swing.

The brachiating robots are multilink structures. The first brachiator had only two links (Saito, Fukuda, and Arai 1994), whereas the model shown in the figure has thirteen links. Clearly, the control of such multilink systems is a complex control problem (Hasegawa, Ito, and Fukuda 2000).

7.10 Flying Robots

We have examined AGVs and AUVs in earlier sections of this chapter. We now focus on autonomous flying vehicles (AFVs). These robots move in three dimensions, like AUVs, and they have control difficulties similar to those of AUVs. Military AFVs may fly to altitudes at which the atmospheric density is sufficiently different from that at sea level to require a drastically different set of control system gains. Both fixed-wing and rotating-wing (helicopter) AFVs have been used in large numbers by the military services in recent years for such functions as mapping and reconnaissance. Using robot craft for these functions reduces the danger to human pilots, as well as dramatically reducing the cost of the vehicles.

Clearly, the control problems raised by a robotic air vehicle are quite different from those involved in land vehicles. First, to remain airborne, the vehicle must generate sufficient lift to overcome both drag and gravitational forces. Second, because of the difficulty of developing fully autonomous systems, most early models of unpiloted aerial vehicles (UAVs) (including drone aircraft) were teleoperated and required great concentration on the part of human operators. In recent years the trend has been toward greater autonomy, including such capabilities as the ability to adapt to environmental changes, cooperation with other vehicles, use of sensors and processors enabling some level of self-repair as well as planning alternative actions under uncertainty, and autonomous takeoff and landing in arbitrary weather conditions. Some of these requirements are currently (at least in part) beyond the state of the art.

7.10.1 Fixed-Wing UAVs

The military services in several countries have supported the development of a variety of autonomous and semiautonomous aircraft, primarily for reconnaissance purposes, four of which are shown in figure 7.44. We describe these medium-sized and small vehicles briefly.[18] These and other similar vehicles, described in Kirsner (2003), have been reviewed in such journals as *Unmanned Systems*.

The Eagle Eye (figure 7.44(a)) has a wingspan of about 15 ft and a length of about 18 ft. It is a tilt-rotor aircraft; that is, its rotors can pivot by 90°, enabling the vehicle

18. There is also a class of full-size robotic aircraft, either in existence or in development. Since such vehicles are intended purely for combat operations, they are not discussed here.

to perform vertical takeoffs and landings, as well as hover like a helicopter. With the rotors in a conventional, forward-pointing position, the vehicle flies like a conventional aircraft, at speeds up to 250 mph. It can fly at altitudes up to 20,000 ft (6,100 m) and stay aloft for 8 hours. Its navigation is fully autonomous, but it can tap into shipboard systems allowing landing on carrier decks. Since its major function is reconnaissance, it carries both day and night imagery sensors.

The Pegasus, also known as the X-47 (figure 7.44(c)) is both larger and faster than the Eagle Eye and is intended for both surveillance and combat missions. This UAV has a remarkable structure, with dimensions about the same in both axes (approximately 28 ft wingspan and length). It is probably intended for teleoperation.

The other two UAVs in the figure are much smaller. The Dragon Eye (figure 7.44(b)) is a twin-engine battery-powered vehicle with a 4 ft wingspan. It is made of lightweight composites and weighs only 5 lb. It can be disassembled and carried in a backpack. The Dragon Eye is launched by hand or by means of a bungee-cord slingshot. It is equipped with cameras and transmitters so that it can send back reconnaissance information to a soldier's laptop computer or virtual-reality goggles. It flies to preset GPS waypoints and connects with a ground station via wireless modem. The Dragon Eye is capable of fully autonomous flight. Its batteries give it a flight time of about 1 hour and a range of about 10 km.

The Aerosonde (figure 7.44(d)) is a joint development of Saab Systems and the Australian Defence Science and Technology Organization. In spite of its military sponsorship, the Aerosonde is also designed for a commercial market, primarily weather reconnaissance. The vehicle has a wingspan of about 2.9 m (10 ft). It is fully autonomous; high-level commands are provided from a remote ground station. The engine is in the rear of the fuselage, thus operating in a "push" mode. The airframe is a hollow, lightweight composite structure that makes it very light (about 15 kg) and gives it great range and endurance. The Aerosonde can stay aloft 20 to 30 hours and travel over a range of 2,000–3,000 km.

The four UAVs reviewed in this section give an indication of the wide range of autonomous and semiautonomous airborne vehicles either deployed or in development. Although the thought is clearly frightening, We see the possibility of future military engagements with large numbers of unpiloted vehicles bringing death and destruction to remote locations.[19]

7.10.2 Micro-UAVs

In recent years there has been an increasing interest in small and very small flying robots, ranging from aircraft the size of small birds to some the size of a housefly.

19. Unfortunately, as both Prometheus and Alfred Nobel discovered, technology can be used for both good and evil, and robotic technology is no exception.

(a)

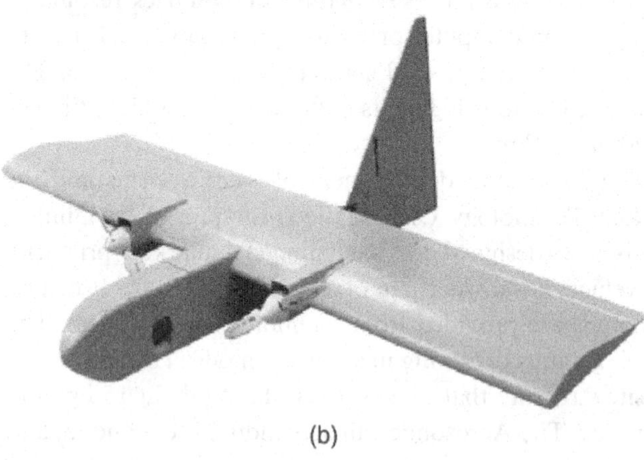

(b)

Figure 7.44
Autonomous flying vehicles: (a) Eagle Eye tilt-rotor UAV (photograph courtesy of Bell Helicopter Textron, Inc.); (b) Dragon Eye mini-UAV (photograph courtesy of AeroVironment, Inc.); (c) Pegasus delta-wing UAV (photograph courtesy of Northrop Grumman Corporation); (d) Aerosonde mini-UAV developed in Australia (photograph courtesy of Aerosonde, Pty Ltd.)

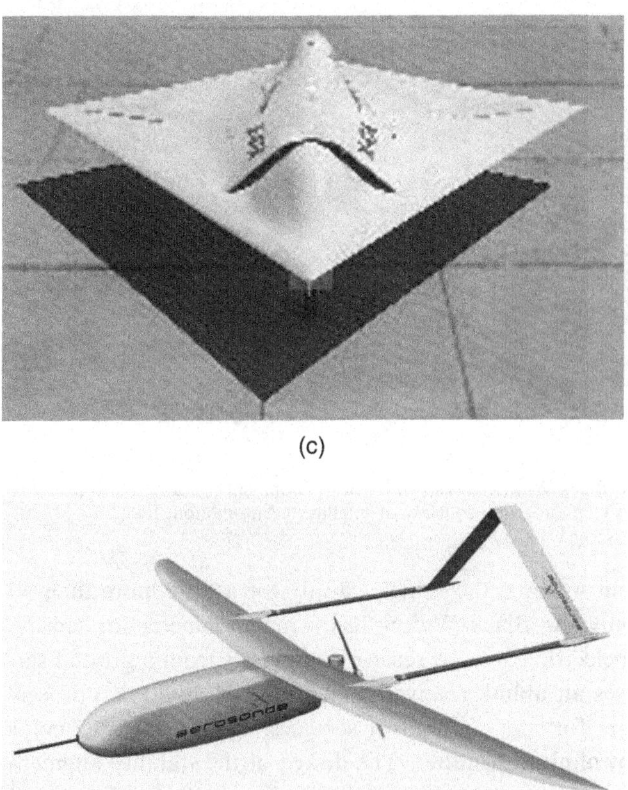

(c)

(d)

Figure 7.44
(continued)

The aerodynamics of such *micro-UAVs* (MUAVs or MAVs) make their design very challenging, since at low speeds it is difficult to generate sufficient lift with very small wings to stay aloft. Two examples of such vehicles are shown in figure 7.45, with human hands as a size reference. Both of these tiny airplanes have wingspans of 6 in (about 15 cm) and weigh under 100 gm. (To qualify as an MUAV, a vehicle's mass must be under 100 gm, and it must have a wingspan of not more than 6 in). Both can fly for 20–30 min.

The Black Widow MUAV in figure 7.45(a) is battery powered, flies at 20–40 mph, and has a range of about 2 km. The propeller is located toward the upper left of the figure. It is an unusually shaped vehicle whose total mass is about 80 g. Various optimization methods (including genetic algorithms) were used to optimize the selection of such vehicle parameters as propeller diameter and motor type. Using lithium-ion

(a)

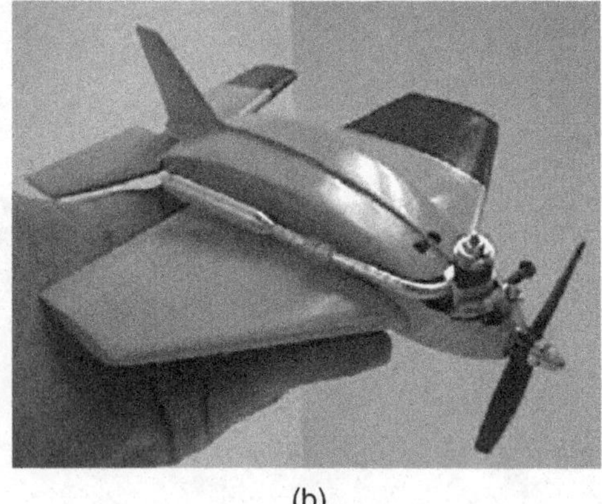

(b)

Figure 7.45
Micro Autonomous Air Vehicles: (a) Black Widow developed by AeroVironment (photograph courtesy of AeroVironment, Inc.); (b) IAI MUAV (photograph courtesy of Intelligent Automation, Inc.)

batteries with a draw of about 4 watts, the vehicle can fly for a little more than 30 minutes. To achieve autonomy the Black Widow has a magnetometer for sensing compass heading and a piezoelectric gyro. To receive commands from a ground station for vehicle control, it uses an uplink receiver (weighing 2 g), two microprocessors, and two 0.5 g actuators for moving control surfaces. It carries a 2 g color video camera and a 1.4 g downlink transmitter. The design of the stability augmentation system, for the camera and for the vehicle, were major challenges in view of the aerodynamics of small vehicles, which are highly susceptible to effects from wind gusts (see Grasmeyer and Keennon 2001).

By contrast, the MUAV from IAI, Inc. is gasoline operated. It weighs 90 g and has a 15 cm wingspan and a flight time of about 20 min. The developers believe that 1 hour endurance times are feasible. Its on-board autopilot includes servomotors, a video camera, and up- and downlink RF communications. The vehicle is expected to launch, cruise, loiter over an area, and return to base autonomously. The component weights are comparable to those of the Black Widow. Its physical appearance resembles that of a conventional airplane, which suggests that careful attention to design makes an MUAV possible in this small package.

It is evident from these two examples that current technology makes very small autonomous vehicles possible. In particular, the reduced size, power consumption, and increased power of microprocessors and the astounding decrease in size and weight of video cameras are two key ingredients in making these tiny vehicles feasible. Clearly, some aspects of their missions (such as image processing) are performed on the ground, but we expect increased autonomy from MUAVs in the near future.

Under development are even smaller vehicles, perhaps only 5 or 6 cm in length, and some researchers are suggesting that air vehicles the size of a housefly are possible. The possible uses of such "truly micro" vehicles are both exciting and frightening.

7.10.3 Rotary-Wing UAVs

We indicated previously that the aerodynamics of small vehicles, flying at low speeds, make their design difficult. This is true of both fixed- and rotary-wing (helicopter) aircraft. The dynamics of helicopters are quite complex, since they must include aerodynamics, blade bending (and possible oscillations), and the interaction among various control modes. Thus, an increase in forward velocity requires that the vehicle pitch forward. In the absence of control systems, such cross-coupling of effects might render the vehicle unstable. Hence, the resulting differential equations governing helicopter dynamics are highly nonlinear, and simple linear PID controllers are generally not satisfactory. It is well known that as a result of these problems, learning to fly radio-controlled helicopters is a difficult and time-consuming task. In recent years Yamaha has dramatically reduced the difficulty of operation by incorporating attitude control systems on board their unmanned helicopters.

A number of autonomous, semiautonomous, and radio-controlled (RC) helicopters have been built. Robot helicopters have numerous realized and potential applications, including agricultural spraying, search and rescue, inspection of pipelines and/or power lines, surveillance around sensitive facilities or along borders, and aerial mapping. Unpiloted helicopters with various degrees of autonomy have been built by companies in Austria, Canada, France, Germany, Israel, Italy, Japan, Korea, Sweden, the United Kingdom, and the United States. They range from large vehicles, comparable in size to small, piloted helicopters, to a variety of smaller vehicles with rotor diameters of only a few feet. In addition, numerous universities have research programs in autonomous helicopters, including the Technical University of Berlin, Simon Fraser and Waterloo Universities in Canada, and various universities in the United States, including MIT, Stanford, and USC. In most cases the platform used in these research programs has been a commercial RC helicopter that was then fitted with computers and additional sensors make autonomy possible. We illustrate these programs with four examples shown in figure 7.46. A fifth example, the USC AVATAR autonomous helicopter, was introduced in chapter 1 and discussed as a case study in chapter 5. We omit all vehicles with a purely military mission, but readers may find it of interest that in 2002 the U.S. Army established a research and development program designed to produce an unmanned combat armed rotorcraft (UCAR), with participation of major aerospace companies and helicopter manufacturers.

The RC helicopter in figure 7.46(a), the R-50, is manufactured by Yamaha Motor Company in Japan. It has been used extensively for distribution of agricultural

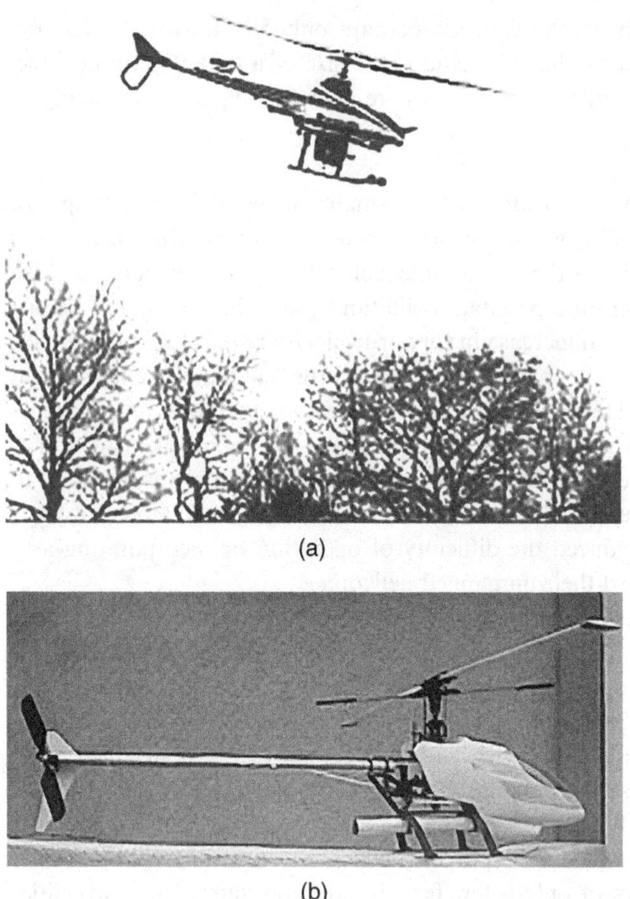

Figure 7.46
Four small RC helicopters used in autonomy experiments: (a) Yamaha R-50 helicopter from Carnegie Mellon University (photograph courtesy Omead Amidi); (b) Ikarus ECO-8 electric helicopter from the University of Southern California (photograph courtesy of Gaurav Sukhatme and David Naffin); (c) MARVIN autonomous helicopter from the Technical University of Berlin (photograph courtesy of Günter Hommel); (d) Yamaha "conventional" RMAX crop-dusting helicopter (photograph courtesy Yamaha Motor Co.)

(c)

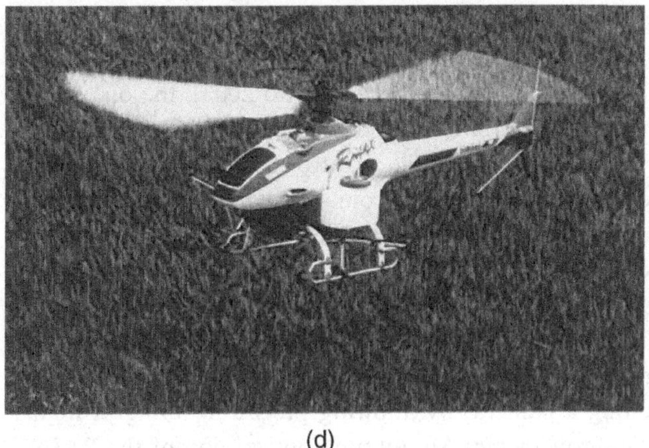

(d)

Figure 7.46
(continued)

chemicals. It was the first commercial unpiloted helicopter capable of carrying a payload of 20 kg. More than 1,000 units were sold between 1990 and 2002. The vehicle has also been used for autonomous robot experiments at several universities, including CMU, Georgia Institute of Technology (GIT), and the University of California at Berkeley (UCB). The CMU experiments emphasized vision-based autonomy and the development of "software-enabled control" (SEC). CMU was able to obtain nearly perfect altitude and hover control while following square and circular patterns (Amidi 1996; Miller et al. 1999). At UCB, several R-50s are used for studies of nonlinear control as well as experiments in formation control and multihelicopter

Table 7.1
Major characteristics of Yamaha R-50 and RMAX helicopters

Specification	R-50	RMAX
Empty weight	47 Kg	58 Kg
Payload	20 Kg	30 Kg
Flight time (with payload)	30 min	60 min
Main rotor diameter	3.07 m	3.115 m
Overall length	3.58 m	3.63 m
Body width	70 cm	72 cm
Overall height	1.08 m	1.08 m
Engine rpm	~9400 rpm	6350 rpm

cooperation (Kim, Shim, and Sastry 2002). All three institutions are devoting major efforts to the issues involved in control of RC-50s, using a variety of optimal control methods, fuzzy logic, and other approaches.

Figure 7.46(d) shows one of the latest versions of the Yamaha, the RMAX G0. Yamaha has introduced several versions of the RMAX (Sato 2003); the one shown in figure 7.46(d) is the "Conventional RMAX," and the data in table 7.1 refer to this vehicle. It is equipped with the Yamaha Attitude Control System (YACS), which includes three fiber-optic rate gyros and three accelerometers, which greatly simplifies operator training. In 2003 an agricultural RMAX was introduced. This vehicle was equipped with YACS augmented with GPS (YACS-G). This control system enables the vehicle to assume a hovering position when the operator releases the control stick. An "aerial RMAX" (introduced in late 2003) is designed for aerial photography. It improves on the agricultural model through inclusion of higher-frequency GPS and a method of mounting a camera, by allowing precise hovering at higher altitudes, and through other improvements. An autonomous version of the RMAX also exists, with the capability of flying over the horizon using programmed trajectories (which can be modified in flight). It has been used in Japan for observations of numerous volcanoes, including the Mt. Usu volcano in Hokkaido. In February 2002, following eruption of the Miyakejima volcano, when the entire island was evacuated due to high SO_2 concentrations, the autonomous RMAX was flown in with cameras and sensors for gas detection. The autonomous RMAX Type II G supplements the sensor data of other models with commands received from a ground station using wireless modems. It uses differential GPS (DGPS) to ensure the needed positional accuracy. Planned for the future is a new version of the autonomous RMAX which will be capable of higher speeds, a larger payload, and a longer flight time (2.5 hr) than the current model.

The conventional RMAX has a larger payload capability than the R-50 and significantly more endurance, as shown in table 7.1. Some of the research programs men-

Table 7.2
Major characteristics of the Ikarus ECO-8

Specification	R-50
Empty weight (without batteries)	1.27 Kg
Payload	1 Kg
Flight time (with payload)	10–12 min
Main rotor diameter	1.06 m
Overall length	91 cm
Body width	10 cm
Overall height	29 cm
Engine rpm	1800 rpm

tioned previously as employing R-50 aircraft are now switching to the RMAX. The RMAX is also used by NASA in the United States, at Linköpings University in Sweden, and in other organizations.

It can be seen from table 7.1 that the width and height of the two vehicles are about the same. The RMAX has a new engine that runs slower and enables double the endurance of the R-50. It is important to note that for the purpose of agricultural spraying, these vehicles are remotely controlled using a hand-held radio transmitter, at distances up to 150 m. Basically this is the visual range. The university programs that make use of the R-50 have developed various approaches to autonomy, both in control and in navigation.

Figure 7.46(b) shows the Ikarus ECO-8, one of a number of small, electrically powered helicopters being developed for research in formation flying at the University of Southern California (Alexander 1984; Fagg, Lotspeich, and Bekey 1994; Naffin and Sukhatme 2002); the major specifications of this helicopter are given in table 7.2. The figure shows it without computer or sensors, which have since been installed. A battery-powered helicopter like the ECO-8 is suitable for indoor operations. It should be noted that the ECO-8 is significantly smaller than the Yamaha vehicles; its weight without batteries is just over 1 kg, whereas the R-50 weighs almost 50 kg. Of course, the ECO-8's range and flying time are correspondingly limited. One of the possible applications of such helicopters is to fly into a building through open windows. As with the other vehicles discussed, the ECO-8 is radio controlled but is capable of autonomous operation with the on-board processor and sensor package installed. The vehicle's flight time capability is expected to increase significantly when lithium-polymer batteries are installed.

The final example of an autonomous rotorcraft is shown in figure 7.46(c), is known as MARVIN (Multipurpose Aerial Robot Vehicle with Intelligent Navigation). This helicopter was the 2000 winner in the annual International Aerial Vehicle Competition, sponsored by the Association for Unmanned Vehicle Systems International. It was the only vehicle that year to fly completely autonomously. It uses a commercial RC vehicle chassis with an empty weight of 6.5 kg and a rotor diameter of 1.8 m. This diameter is smaller than that of the Yamaha aircraft and larger than that of the Ikarus ECO-8. MARVIN (like most other autonomous helicopters) uses the following on-board systems:

• DGPS receivers, which allow the robot's position to be determined to within a few centimeters.

• A flux gate or similar compass enabling the orientation of the vehicle with respect to the earth's magnetic field to be determined.

• An IMU (inertial measurement unit) including three accelerometers (usually solid state) and three rotation sensors (effectively solid-state gyroscopes), providing rotation rates about the pitch, yaw, and roll axes.

• An rpm sensor on the main rotor.

• Ultrasonic proximity sensors to assist in the final phases of autonomous landing.

• Digital camera with automatic exposure control (all the vehicles discussed carry at least one digital camera).

• Communication modules, which may make use of wireless Ethernet or simple RF transmission, enabling the vehicle to communicate with a ground station, receive commands, and send camera and sensor data. The link with the ground is essential to enable manual control in the event of a failure of the autonomous system. (Such links are usually part of a vehicle's radio control system.)

• One or more processors, for software-based attitude control, sensor data processing, and navigation.

The most difficult issue in the design of autonomous helicopters is the flight control system, in view of helicopters' nonlinear dynamics and the degree of cross-coupling between control modes (Shim et al. 1998). In helicopters much of the control is obtained through adjustment of the pitch of the rotor blades, once every revolution (and hence referred to as *cyclic*). The cyclic change increases lift on one blade while decreasing it on the other, affecting both the vehicle's pitch and its roll. A control mode called the *collective* changes the pitch of both rotor blades by the same amount (collectively); this change in turn affects the thrust, thus increasing or decreasing the helicopter's lift. Tail rotor pitch affects the yaw. There is thus a great deal of cross-coupling between control modes. For example, changes in the thrust level (from the throttle or the collective) produce torques about the yaw axis, which need to be counteracted by the tail rotor to ensure that the vehicle's heading does not change. Consider figure 7.47, which indicates the various actuators needing control inputs in helicopter flight. As indicated previously, many of the control actions are coupled

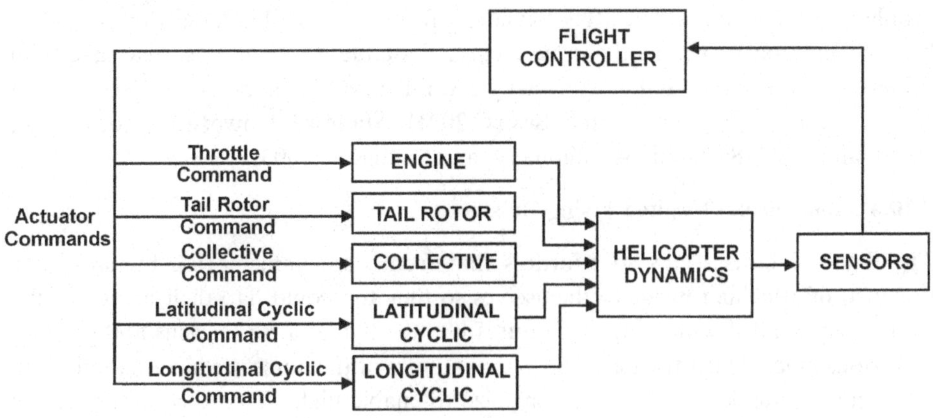

Figure 7.47
Helicopter flight control actuators

through the vehicle dynamics. The various sensors listed previously (in connection with MARVIN, specifically) provide inputs to the flight controller, which in turn provides the actuator commands. Hence, there are several approaches to helicopter control:

1. Detailed mathematical models have been developed for some helicopters and used as a basis for nonlinear controller design (e.g., Maharaj 1994). In some cases it has been possible to design a controller based on inverting the vehicle dynamics, but this is complex and difficult for helicopters. Detailed, model-based controllers have been more successful in simulation than in reality.

2. Large neural networks and combinations of neural and fuzzy controllers have been used to synthesize model-free controllers. In some cases these networks allow for rapid learning and parameter adjustment in flight (Montgomery and Bekey 1998; Buskey, Wyeth, and Roberts 2001). Such an approach was discussed in case study 5.2.

3. Some investigators have combined nonlinear control with human input, particularly in such tasks as landing control, with great success (e.g., Jones et al. 1998).

4. Heuristic approaches based on observation of behavior of human pilots while controlling RC helicopters have been used to adjust gains and other parameters in behavior-based controllers (Fagg, Lewis, and Montgomery 1993; Montgomery 1999).

Vision-based robot control has been an active research topic in the autonomous-helicopter area, particularly vision-based landing. Landing is a particularly difficult problem for helicopters, since the downdraft from the rotor blades creates an unstable air condition known as the *ground effect*. Notable results in this area have been achieved at Carnegie Mellon University (Amidi 1996), University of California at Berkeley (Sharp, Shakernia, and Sastry 2001), Stanford University (Rock et al. 1998), and USC (Saripalli, Montgomery, and Sukhatme 2003).

7.10.4 Biologically Inspired Flying Robots

The flight of birds seems so effortless that from time immemorial, humans have dreamed of attaching wings to themselves so they too could flap their arms and fly. Mythology is filled with winged creatures like the flying horse Pegasus and dragons of various kinds. Unfortunately, the aerodynamics of the horse are incompatible with the required muscle strength and wing size to enable flight. This is certainly true of humans as well. Men have attempted to strap large wings to their arms, but they have neither the required strength to fly nor the ability of birds to control the wing configuration as required to develop the necessary lift and thrust for flight.

The development of powered flight provided a basis for understanding the difficulty of flapping wings. In order to fly, a vehicle requires both lift and thrust. Birds provide both with their wings, but this has turned out to be very difficult in powered flight as well. Hence, the development of aviation has separated these two issues, using wings for lift and a separate engine or engines for thrust. Nevertheless, the dream of building aircraft with flapping wings continues to challenge and inspire aircraft developers.

Flying vehicles with flapping wings are known as *ornithopters*. In recent years there has been growing interest in ornithopters, particularly in small sizes in which the ratio of wing area to body mass can be sufficiently large to make flight of this kind possible. There has also been a revival of interest in building full-sized aircraft with flapping wings. Remotely controlled ornithopters were flown successfully in the early 1990s. As recently as 1998 the University of Toronto tested a full-scale piloted ornithopter that was able to leave the ground for short intervals, in a set of controlled hops (DeLaurier 1999). The wings of this aircraft had a span of 41.2 ft. They consisted of three hinged panels supported by center pylons and outboard vertical links. The center panel was moved up and down sinusoidally by the on-board 24 hp, three-cylinder engine. As it moved, it drove the outer segments on each side in a flapping motion. While such flight is clearly fascinating, it is definitely not autonomous and may never be practical because of the energy required to flap the wings.

On the other hand, small autonomous vehicles with flapping wings are possible and may be more efficient than propeller-driven fixed-wing vehicles of comparable size. Consider the birdlike craft shown in figure 7.48, developed by AeroVironment, Inc., Caltech, and the University of California at Los Angeles, and known as the Microbat. This hand-launched ornithopter has a 23 cm wing span and weighs 12 g.

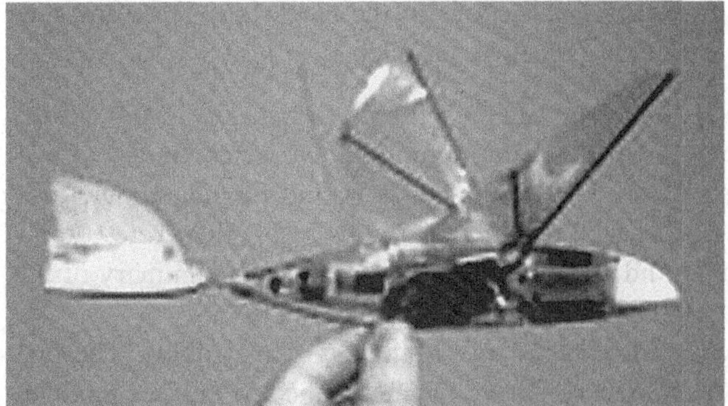

Figure 7.48
Microbat ornithopter developed by AeroVironment, Caltech, and UCLA (photograph courtesy of AeroVironment, Inc.)

It has flown for about 1 minute using a lithium-polymer battery at a flight speed of about 24 km/hr. The vehicle's wings flap at 20 Hz. The current version of the craft is radio controlled, but full autonomy is expected. For reconnaissance purposes, the vehicle will probably carry a tiny camera (Porsin-Sirirak et al. 2000).

Even smaller ornithopters are in development at the University of California at Berkeley, Georgia Institute of Technology, and elsewhere. One such fly-size vehicle is expected to have a wingspan of less than 2 cm and a total weight of 100 mg.

7.11 Self-Reconfigurable Robots

We conclude this chapter with a brief introduction to mobile robots capable of modifying their body structure. One of the dreams of robot designers has been the design of self-reconfiguring (morphing) robots, machines that can change their shape and function according to the environment and need. Such a robot should be able, for example, to change from a walking machine into a snake, or it should be able to change the function of legs to arms. Such adaptability is characteristic of living cells that can be organized into a vast array of creatures, from insects to elephants. In addition, such a robot should be more fault tolerant than a fixed-structure robot, since it should be able to move working modules to replace faulty ones. From the biological model it is evident that such reconfigurability requires

- a large number of building blocks ("cells");
- a method for connecting individual cells;
- "blueprints" for organization;

• a message-passing system to provide instructions to the blocks and to receive any necessary feedback from the central organizing controller.

A recent article refers to such robots as "shape shifters" (Mackenzie 2003). When such a system can change its structure autonomously, we refer to it as *self-reconfigurable*. The first such systems were proposed by Toshio Fukuda about 1990. His "cellular robotic systems" consisted of individual units (cells) that were coordinated for a given task (Fukuda and Kawauchi 1990). An excellent summary of the state of the art up to 2001 can be found in Rus and Chirikjian 2001.

At present there are basically three types of self-reconfigurable robots. The first kind, known as *lattice robots,* consist of stacks of interconnected cubical modules that can attach to one another at specific discrete locations. In the second kind, known as *chain robots,* individual modules connect to one another in sequence, at one or more points. The third kind are capable of specialized function changing of their parts, such as transformation of a hand into a foot. Let us look at some examples of these three types of reconfigurable robots.

The Telecube lattice robots (figure 7.49) developed by Mark Yim at the Xerox Palo Alto Research Center are capable not only of connecting to one another, but of extending their sides to effectively double their length. Yim has shown that the telecube can recofigure from one shape into any other (Zhang, Roufas, and Yim 2001; Yim, Zhang, and Duff 2002; Yim et al. 2003).

Another form of lattice robots was developed in the laboratory of Daniela Rus. Her lattice robots, which she has termed Crystals, can attach themselves to similar

Figure 7.49
Telecube reconfigurable robot (photograph courtesy of Palo Alto Research Center Incorporated)

units. Each Crystal consists of *crystalline atoms* that are actuated by expansion and contraction of their faces. This process allows an individual atom to move in general ways throughout a 3-D structure; that is, it is not restricted to moving on the surface and can travel throughout the volume of a Crystal. Note that lattice robot modules need communicate only with their immediate neighbor modules (Rus and Vona 2001). The requirement to execute shape changes has led to research in distributed algorithms for planning (Pamecha, Ebert-Uphoff, and Chirikjian 1997; Rus and Vona 2001).

Chain robots are made of modules that attach to or detach from other modules in sequence, with at least one module always attached at a minimum of one point to the rest of the system. The chains may also branch to form other sequences. It has been suggested that a way of visualizing such a robot is to consider a person who clasped his hands together, then disconnected one arm at the shoulder, thus producing a new arm twice as long as the original. Clearly, such reconfiguration is useful only if the individual modules can perform useful tasks or can be programmed to change from one task to another. Some of Rus's robots are made of units called *molecules*, each of which consists of two atoms. The atoms have connection points enabling them to connect to other molecules and a rotational degree of freedom with respect to one another. Thus, a robot can be assembled from multiple molecules. The molecules can move by tumbling (Rus et al. 2002).

Other noteworthy chain robots have been constructed at the Xerox Palo Alto Research Center by Yim and at USC by Shen and Will. Both of these systems feature units that can swivel with respect to their neighbors and provide communication and power at the connections. Both systems can morph into a variety of robots. Yim's PolyBots can form snakes, moving wheels, and four-legged walking machines, as illustrated in figure 7.50. Other configurations like lizards or machines with up to twelve legs are possible.

At present Yim has assembled some systems consisting of a few PolyBots and others with up to 100. Each PolyBot module is about 5 cm on a side and contains a microprocessor, motors, mating connectors for connection to other modules, proximity sensors, and angular-position sensors. There are also latches for securing the modules to one another. Since the modules must find one another autonomously and make secure connections, the proximity and angular-position sensors on the modules are essential. Ultimately, one can visualize such systems with hundreds or thousands of modules (Yim, Zhang, and Duff 2002; Yim et al. 2003). In order to change gait, each module consults a *gait control table*.

The approach taken by Shen and Will is similar to Yim's, but with some significant differences. Shen and Will's robot is named CONRO (for Configurable Robot) (Castano, Shen, and Will 2000; Castano, Behar, and Will 2002; Shen, Salemi, and Will 2002). CONRO's modules communicate with one another during docking using

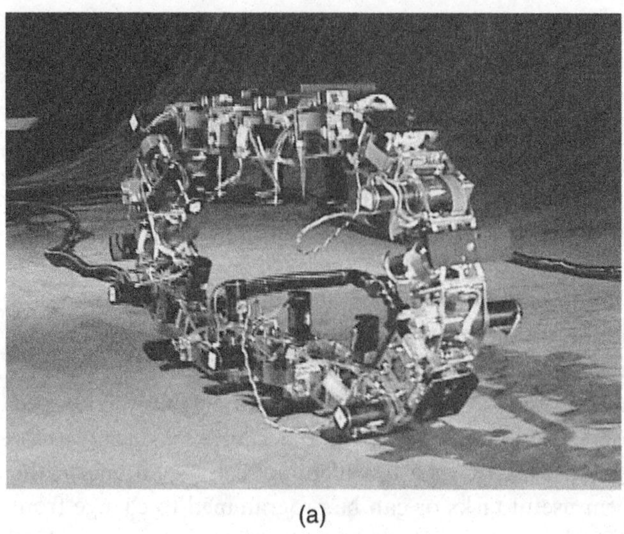

(a)

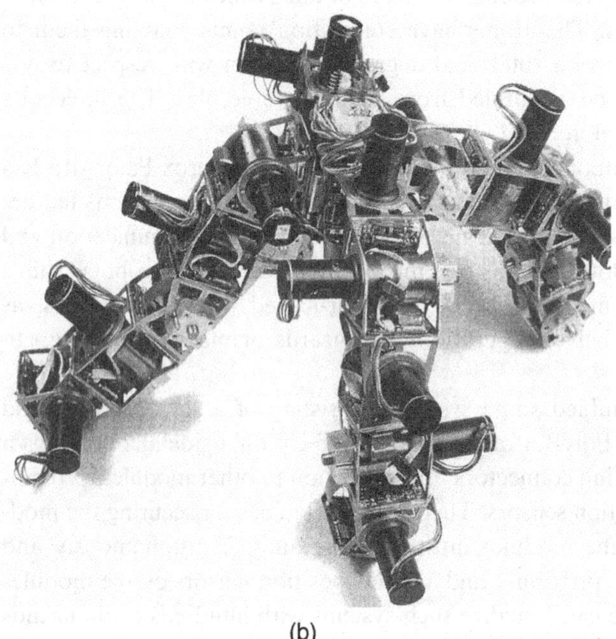

(b)

Figure 7.50
Reconfigurable PolyBot robots: (a) moving wheel; (b) four-legged walker (photographs courtesy of Palo Alto Research Center Incorporated)

(a)

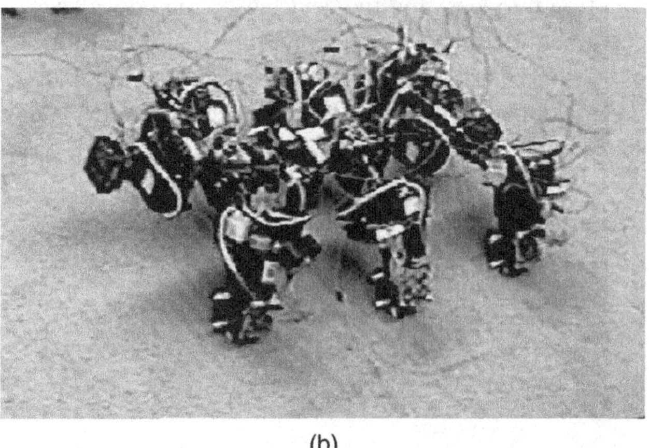

(b)

Figure 7.51
CONRO self-reconfigurable robot: (a) a CONRO module showing the latching pins; (b) hexapod configuration (photographs reproduced with the permission of the USC Information Sciences Institute)

infrared beams. Each CONRO module is self-contained; it carries its own microprocessor, motors, batteries, infrared transmitter-receivers, and docking connectors. Each module is 10 cm long, weighs about 100 g, and has 2 dof, pitch and yaw, allowing it to assume one of 255 positions for connecting with another module. Figure 7.51 shows a CONRO module and a complete robot, configured as a six-legged walker.

The CONRO system is highly adaptable, as a result of its distributed architecture and decentralized control system. Rather than relying on a central controller to send messages to each module, CONRO relies on signals analogous to biological hormones. A module's reaction to a "digital-hormone" signal depends on its current state. Each module constantly monitors its neighbors to determine what its current role is (e.g., a leg or a head or a spine). If its position changes, so will its behavior. Thus all CONRO modules get the same message, but their current state and those of their neighbors determine how they will act on it. As a result, the system can respond to such stimuli as a snake robot's breaking apart; it will start crawling as two snakes. Stick a snake's tail in its "mouth" and it will figure out what happened and start rolling like a wheel or tank tread (without the tank). Attachment of snakes to the side of a snake leads to their transformation into legs, and the device gets up and walks.

As a final example, consider a robot capable of reconfiguring its legs into arms (or vice versa), that is, to use the same hardware for locomotion and manipulation. The duality between these two functions has fascinated researchers for many years. Finger manipulation of a sphere, for example, can be viewed as legs walking on a

Figure 7.52
TermnatorBot emerging from a concrete block (photograph courtesy of Richard Voyles)

sphere. Some animals use their front paws as if they were hands. Similarly, Richard Voyles at the University of Minnesota has developed a small, centimeter-scale robot (named TerminatorBot) with two 3-dof arms that can also be used for locomotion, as shown in figure 7.52. In this figure the the robot is seen emerging from a cement block where it may have been hiding or through which it could be entering an area covertly (Voyles 2000). The arms/legs can fold back into a stowed configuration, allowing for ballistic deployment without damage to the robot. For grasping, the arms are equipped with force/torque sensors.

7.12 Concluding Remarks

This lengthy chapter was designed to provide an overview of the many aspects of mobility in robotics, on land, under water, and in the air. Clearly, this is a very active area of research and development in the field, and we expect continued progress. We believe that developments in the near future will be primarily in three areas: (1) continued miniaturization, with fully functional mobile robots perhaps only 1 cm^3 in size; (2) dramatic increases in robot intelligence, allowing robots to adapt to new and unstructured environments; and (3) increased reliability and robustness, particularly in larger robots, with such features as self-diagnosis and repair. It should also be noted that an examination of legged locomotion in robots was not included in this chapter, since it is the subject of the next two chapters.

8 Biped Locomotion

Summary

The goal of this chapter is to explore the control of locomotion in two-legged robots. We begin with a quick look at animals that stand and/or walk on two legs. We then turn to human locomotion, since humans are the model par excellence for biped locomotion. The subsequent sections of the chapter present a number of examples of biped robots and analyze the difficulties associated with their design and implementation. The final section of the chapter deals with mechanical aids to human walking, since a number of such systems are based on robotic models.

8.1 Standing and Walking on Two Legs

Control of locomotion in legged animals and robots requires consideration of two problems: (1) stability control, to ensure maintenance of a vertical posture, and (2) motion control, to allow forward motion at various speeds. Animals with six or more legs have guaranteed *static stability* (i.e., the ability to stand in a stable posture). The stability of four-legged animals, on the other hand, is conditional on the location of their center of gravity as soon as they lift one leg off the ground. Clearly, static stability is even more difficult to ensure in a two-legged creature. Nevertheless, requiring only two legs for locomotion on the ground provides a number of evolutionary advantages, such as allowing use of the upper extremities for grasping tools and assisting in support, as well the ability to move through more restricted terrain than is possible for animals with four or more legs. It is interesting to note, however, that the largest land animals are quadrupeds, since it is easier to carry a large body weight with four legs rather than two.

Humans are not the only creatures capable of locomotion on two legs. Kangaroos and related marsupials hop on two legs, but they use their massive tails as static stability aids. Dinosaurs also used their huge tails to assist in balance during bipedal locomotion. Birds have only two legs, and they move on the ground either by hopping or by walking. Their stability is aided by the forward-backward distribution of body weight, since there is some compensation for the forward position of the head by the position of the tail. Further, birds have active control systems that enable them to maintain static stability even when perched on relatively small support surfaces, such as tree branches (or telephone wires). The forward-backward movement of a bird's head and tail are very evident when it walks. Many quadrupeds are capable of moving for short distances on their hind legs, but this is not their normal mode of locomotion. Even insects (such as cockroaches) are capable of running on two legs for short distances. Nevertheless, the only creatures apparently evolved for primary locomotion on two legs are some of the great apes and human beings. As with four- and six-legged animals, legs make it possible for bipeds to move over uneven and difficult terrain, to navigate steps, and even to hop to cover changes in ground elevation.

In view of the previous comment on static stability, it should be evident that (1) to ensure lateral stability, humans must place their center-of-gravity vector between their feet; and (2) some form of active control is required to maintain forward-backward stability. These issues are discussed in section 8.2. It should also be evident that the design and construction of two-legged robots is considerably more difficult than that of machines with four or more legs. The static-stability problem for biped robots is usually solved in one of two ways. In toy walking machines, it is common to have very large feet to ensure static stability. In other biped robots, active control is used; that is, sensors are employed to measure the deviation of the robot's axis from the local vertical and appropriate compensating signals are sent to actuators. We discuss these issues in sections 8.5 and 8.6. During motion, the support forces of the legs, momentum, and inertial forces are summed to produce dynamic stability.

8.2 The Nature of Human Walking

Human walking is a surprisingly complex process, requiring the coordinated activity of some twenty-eight muscles in the lower extremities, as well as additional muscles in the trunk. Stable human locomotion requires a periodic motion of the legs, which produce a series of strides resulting in a normal free-gait velocity of approximately 80 m/min for adults on a smooth, level surface. Men can walk approximately 5% faster than women. Normal persons can increase and decrease their normal walking speed by about 50%. Increases of speed of more than 50% are either uncomfortable or lead to running, whereas decreases below 50% of normal result in disruption of the walking rhythm (Perry 1992). During locomotion each leg can be in one of two states: *stance* and *swing*. During stance the foot is in contact with the ground and the

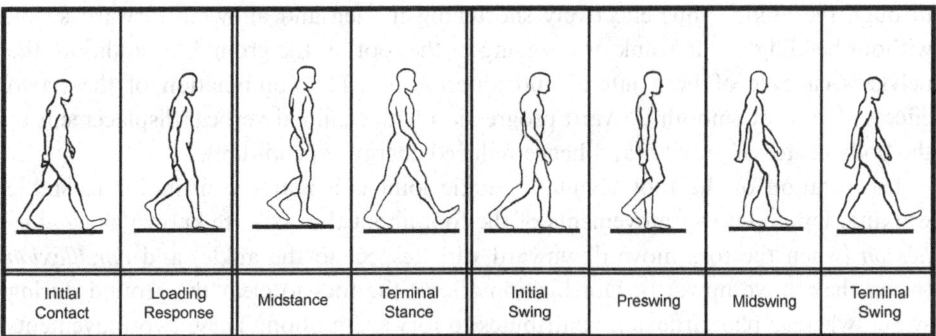

Figure 8.1
Stages of human walking (adapted from Perry and Gronley 1989)

body center of mass moves forward over it, while the opposite (*contralateral*) leg swings forward, as illustrated in figure 8.1 (Perry and Gronley 1989). Stance begins when the heel touches the ground at the end of swing and ends when the toes leave the ground. Note that stance consists of two phases: *single-limb stance* (when one foot is on the ground) and *double-limb stance* (when both feet are on the ground). Double-limb stance occurs both at the beginning of a gait cycle and at its midpoint, when the foot that has been in swing contacts the floor.

It is evident from figure 8.1 that normal walking involves coordinated motion at the hip, knee, and ankle. None of these joints are simple pin or hinge joints. Furthermore, muscles attach at various points along the rigid skeletal supporting structure and often perform more than one function (e.g., a muscle may be responsible for deflection of the toes as well as ankle rotation). Some muscles cross more than one joint. This complex architecture makes mathematical modeling and analysis extremely difficult but also provides the redundancy that makes walking possible (though perhaps difficult) following disease or injury.

It is interesting that during normal human gait, the time spent in stance is approximately 60% of the gait cycle, and that the 40-60 ratio of swing to stance is nearly independent of the specific limb dimensions of the walkers at the customary 80 m/min rate of walking.

Energy consumption in normal walking increases linearly with speed. Moreover, a large fraction of the energy expended during walking is required for the vertical motion of the body center of mass, and a relatively smaller fraction for forward propulsion at normal walking speeds (Perry 1992). Thus, the ability of humans to walk long distances is related to the remarkably small vertical motion of their center of gravity, which, in turn, is due to appropriate motions of the pelvis and the legs. Upon the beginning of swing, the foot dorsiflexes (rotates about an approximately horizontal axis through the ankle), thus effectively shortening the leg and allowing forward swing without bending of the trunk or dragging of the foot on the ground. In addition, the pelvis is capable of both lateral and anterior tilt. The combination of these two effects allows for smooth forward progression with minimal vertical displacement of the body center of gravity (and hence reduced energy expenditure).

The rotation of the foot about the ankle joint is a major contributor to stable walking. Up-and-down movements of the foot about this axis are known as *plantarflexion* (when the toes move downward with respect to the ankle) and *dorsiflexion* (when they move upward). Dorsiflexion allows the toes to clear the ground during swing, whereas plantarflexion contributes to forward motion. These two movements effectively shorten and lengthen the leg during walking, thus reducing the up-and-down movement of the body's center of gravity. The importance of this mechanism becomes clear when we note that failure of the neural control of these movements requires a person so handicapped to move the swinging leg laterally in order for the toes to clear the ground. Such a handicap can frequently be corrected through electrical stimulation of the dorsiflexor muscles in the lower leg at the beginning of swing.

8.3 Musculoskeletal Dynamics

The leg and torso movements associated with walking are produced by a large number of muscles. The major muscle groups that come into play in walking are illustrated schematically in figure 8.2.

Several points can be noted from this drawing. First, the lever arm for different muscles may be quite different, depending on the point along the bone where the muscle is attached. Thus, production of a given torque on a leg segment depends not only on the force produced by the muscles attached to that segment, but also on the attachment point. Second, some muscles (such as the hamstrings) cross two joints between their attachment points, which greatly complicates the analysis of their effects. Note also that only eight of the twenty-eight muscles of the lower extremity are shown for simplicity and that some of the names represent multiple muscles with different attachment points on the bones. For example, the word *vasti* refers to four muscles responsible for extension of the lower leg (*vastus lateralis*, *vastus medialis*, *vastus intermedius*, and *rectus femoris*). The muscle denoted as *soleus* refers to the combined action of the soleus and gastrocnemius muscles that extend (plantarflex) the foot, as when one is standing on one's toes. In view of the complexity of the anatomy, most investigators include only a few of the muscles in their models, as in the figure (Jalics, Hemami, and Clymer 1997).

Human muscle dynamics are highly nonlinear. The force exerted by a skeletal muscle is a function of both the muscle length and its velocity of contraction. Hence,

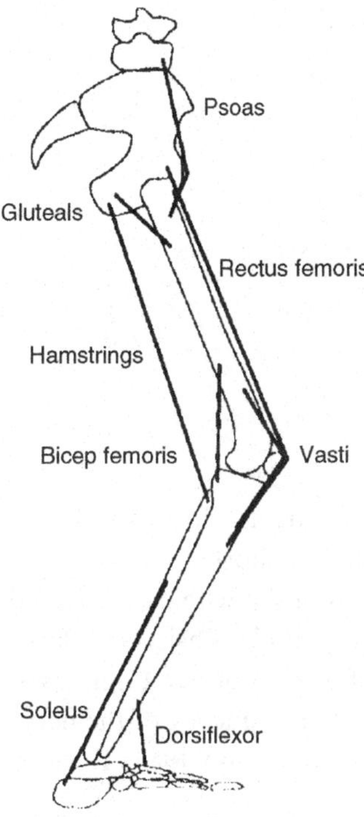

Figure 8.2
Major muscle groups responsible for walking (adapted from Jalics, Hemami, and Clymer 1997)

the properties of muscle are frequently studied under *isometric* (constant-length) or *isotonic* (constant-tension) conditions. The first physiologist to model muscle behavior was A. V. Hill (1938). Hill characterized the muscle as consisting of an active, force-producing element (known as the *contractile element* or CE), a spring in series with CE (the *series elastic element*, or SE), and a spring in parallel with these two components, known as the *parallel elastic element* or PE), as illustrated in figure 8.3.

By applying constant-force (isotonic) inputs to active muscle Hill was able to measure the initial velocity of contraction, which is approximately constant. He then plotted the values of this force-velocity relation and fitted the data to a hyperbola:

$$V_{\mathrm{CE}} = \frac{b(P_0 - F)}{F + a}, \tag{8.1}$$

where

F is the applied isotonic force;

V_{CE} is the resulting initial rate of shortening of the muscle;

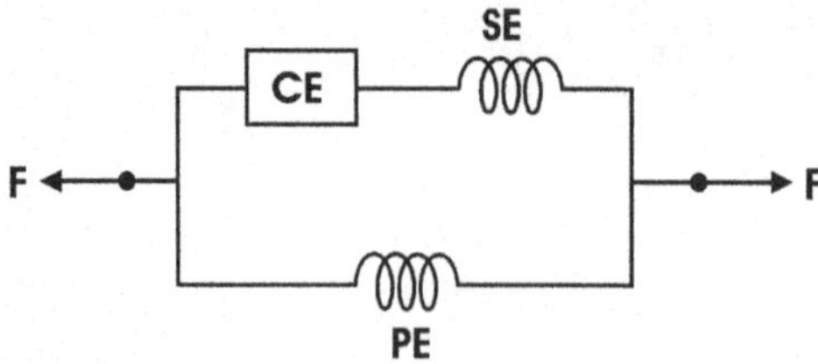

Figure 8.3
Hill muscle model

P_0 is the maximum isometric force produced by the muscle;

a, b are empirical constants.

Equation (8.1) is a mathematical model of the contractile element, which is still used to describe the force-velocity relation of skeletal muscle during shortening.

Note that the Hill equation describes the gross properties of the whole muscle and not the activation of the muscle by its neural input. Skeletal muscle consists of a large number of muscle fibers, and the neural input consists of a train of nerve impulses. Hence, the force produced by the muscle depends both on the frequency of the nerve impulses and on the recruitment of muscle fibers as well (McMahon 1984; Krylow, Sandercock, and Rymer 1995).

8.4 Control of Human Locomotion

In view of the multiple-linkage system of the legs and the multipurpose nature of many of the muscles in the lower extremities, the control of human walking and running is a complex process. The control system for these activities is hierarchical in nature, including local reflexes, loops that close through the spinal cord, and inputs from higher centers (including the motor cortex). Figure 8.4 diagrams the walking-control system. Although the diagram omits many of the details of the process, it illustrates some of its complexity.

Consider some aspects of figure 8.4. Central to this figure is a central pattern generator (CPG). This neural structure is responsible for producing the periodic patterns of excitation that make periodic movements (such as constant-speed walking) possible. In animals the CPG has been known and studied for a number of years, whereas in humans even its existence was controversial until recently. Clearly, a CPG, or a structure with the functions of a CPG, must exist in the spinal cord to make possible such phenomena as walking while asleep. As illustrated in the figure, the CPG is subject to various influences from higher centers. Local CPGs may also be present

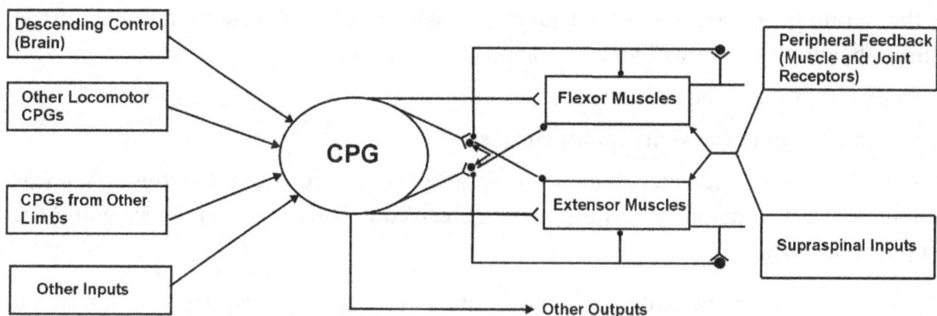

Figure 8.4
Major influences on the control of human walking (adapted from Grillner 1981)

within the legs of animals. Note that the CPG is shown as producing output signals to the leg muscles by way of the so-called alpha motoneurons. The contracting muscles provide feedback by means of a number of specialized sensors, including muscle spindles and Golgi tendon organs, as illustrated in figure 2.6.

The muscle spindles are specialized stretch receptors located in parallel with the muscle fibers. As the muscles shorten (when activated) or lengthen (say, as a result of contractions of antagonist muscles), the firing frequency from the output fibers of the spindles changes. The output nerve fibers from the spindles (termed *gamma fibers*) carry this information to the spinal cord and the brain. When a muscle is stretched (say, when a physician taps the associated tendon during a muscle test), the signal from the spindle initiates a reflex arc through the spinal cord that causes the muscle to shorten, resulting in a twitch. This sequence of events does not involve the motor cortex at all. On the other hand, voluntary movements are produced by descending (efferent) signals to the muscles along alpha motoneurons. It has been postulated that one of the functions of the muscle spindles is to effectively control the compliance of the muscle (Houk, Rymer, and Crago 1991).

A second set of sensors are shown in figure 2.6: the Golgi tendon organs. As the name implies, these structures are located within the tendons that attach the muscles to the skeleton. It has been suggested that these structures sense the force exerted by the muscle and act as safety "fuses," capable of disconnecting further inputs to the muscle when injury is possible (Houk, Crago, and Rymer 1980).

It is evident from the above discussion that mathematical representations of the control of walking are difficult to produce, since they would have to include

- the contractions of all the relevant muscles in the legs;
- the local reflexes associated with these contractions;

• the nature of the neuromuscular junctions (where nerve fibers stimulate the muscle through complex electrical and biochemical processes);

• feedback through the spinal cord;

• voluntary inputs via alpha motoneurons;

• the influence of the cerebellum in adaptation to environmental changes (such as changes in the slope of the terrain or the perceived friction between the shoe and the supporting surface).

Hence, most mathematical models concentrate on some limited aspects of walking control.

The preceding paragraphs have emphasized the neural control of locomotion in steady state, that is, at a constant speed, while moving in a straight line and on level terrain. Changing terrain or a desired change in speed requires major adaptations in the pattern of walking or running. Moving in a curved path requires the leg(s) on the outside of the curve to turn during swing and to traverse longer distances than the inside leg(s), so the stride length of the two (sets of) legs will no longer be equal. Walking uphill or downhill requires a change in body posture to ensure static stability. Walking along a constant-elevation contour on the side of a hill requires dramatic adjustments of leg and shoulder or hip position to maintain stability. Stability in walking humans is taken for granted, but stability during running has been described as a pattern of "controlled falling." Changes in speed also require complex neural mechanisms. Bipeds have only three modes of locomotion: walking, running, and jumping. A further function of neural control is to allow locomotion in the presence of obstacles. Both animals and humans use vision to step around obstacles.

Pathological gait in humans following disease or injury is the subject of much contemporary research (e.g., Bekey et al. 1992; Perry 1992). The variety of potential disturbances to the neural control of locomotion is enormous, ranging form damage to motor neurons controlling a single muscle to the global effects of cerebral palsy. Pathology gives rise to remarkable compensatory mechanisms to allow for locomotion, often with unusual body postures or the use of muscles for unaccustomed functions.

As indicated previously, control of locomotion involves issues of stability and forward progression. Standing stability in human beings is accomplished by means of a control system involving spinal reflexes and sensors in the leg muscles, as illustrated in figure 8.5. The arrows point to symbolic groups of muscles at the knee and hip, respectively.

Note in reference to this figure that a forward or backward rotation of the rigid body will result in stretching of specific muscles (e.g., the gastrocnemius and soleus muscles during forward lean, or the anterior tibialis muscle during backward lean).

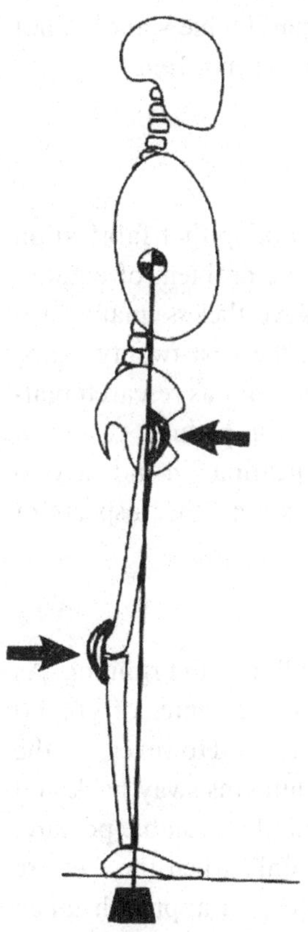

Figure 8.5
Reflex control of standing posture (from Perry 1992, reproduced with permission of SLACK Incorporated)

This stretch will be sensed by the muscle spindles and result in a reflex arc that sends a stimulus to the muscles, causing them to contract, thus returning the body to an upright position. Since activation of this control loop requires deviation from the local vertical, a standing human being cannot be completely rigid and vertical but must *sway* forward and backward. This sway is of the order of 1–2 cm at the body's center of pressure for young adults and is normally not noticeable (unless exaggerated by alcohol!).

Humans are capable of only two gaits: walking and running. (Jumping is a mode of locomotion for short distances, but cannot be considered a gait.) As indicated earlier in this chapter, human gait during walking consists of two phases: stance and swing. The stance period, in turn, is divided between single-limb stance and double-limb stance. During double-limb stance, when both feet are on the ground, the body can shift its weight from support of one leg to the other. When double stance is not present in locomotion, a person is running rather than walking (Perry 1992).

8.5 Robotic Models of Biped Locomotion

It should be clear from the preceding discussion of human walking that fabrication of two-legged robots involves a number of difficulties, since the problem of equilibrium must be solved as well as that of forward propulsion. Nevertheless, many such systems have been developed in research laboratories during the past twenty years. We review some of them here. Most biped machines have been built as research platforms for experiments in robotics or for entertainment in motion pictures or television; it appears that none have been used for practical applications in industry to date. In this chapter we concentrate on issues of biped walking; other aspects of humanoid robots are discussed in chapter 13.

8.5.1 The Biped Robot Stability Problem

The stability problem in bipeds is different during standing, walking, and running. As indicated previously, during stable standing, the projection of the center of gravity of the body lies between the feet, thus providing lateral balance. However, in the forward-backward plane, the body is statically unstable, and humans sway back and forth. True static stability, enabling an animal or a robot to stand in a stable posture, requires at least three legs. It can be obtained with two legs only when the feet are very large in comparison to those of humans and other bipeds, an approach taken by some toy manufacturers. Walking and running are statically unstable. To prevent the body from falling forward while moving, it is necessary to balance inertial, frictional, and gravitational forces, thus obtaining *dynamic stability*. To maintain a stable vertical posture, many robots have been provided with sensors to measure joint angles in the legs and reaction forces at the feet. Appropriate feedback controllers can then be used to return the system to a vertical position if it starts to lean in one direction or another. Clearly, this is an attempt to imitate the postural control system in humans.

Dynamic stability during walking or running requires the actual measurement of the trunk angular acceleration, velocity, and position with respect to the ground. In humans this measurement is obtained using the semicircular canals and the otolith organ in the head. To obtain such information in biped robots, it is necessary to equip them with gyroscopes to obtain the angular velocities in each axis, or in some cases, with complete inertial platforms that provide accelerometers as well as gyroscopes in all three rotational axes. The latter are expensive but have been used in recent biped robots, such as the Honda P-3 robot (Inoue et al. 2000) and the MIT M-2 (Hu, Pratt, and Pratt 1998).

8.5.2 Control of Leg Movements during Walking

As we have seen, anthropomorphic legs are complex nonlinear systems with many degrees of freedom. To produce an appropriate sequence of joint angles to allow for walking, several approaches have been taken:

• *Simple oscillator control* If the legs are sufficiently simple and have few degrees of freedom, it is possible to control their movements by treating them as oscillators to produce a stable periodic walking sequence.

• *Direct kinematics* If the legs are sufficiently similar to human legs, it is possible to record the hip, knee, and ankle joint movements in human walkers (under various circumstances) and use them to control the corresponding joints in the robot. This approach was taken in the design of the early Honda P-2 and P-3 robots.

• *Inverse kinematics* One can record only the desired sequence of ground contacts by the feet and the motion of the body, then use inverse kinematics to compute the corresponding joint angles.

Clearly, all three of these approaches have advantages and disadvantages, as we show in the discussions of various robots in the next section.

8.6 Some Biped Robots

Biped robots are attempts to imitate some aspect of human locomotion. However, not all of them are "humanoids" in the sense of resembling humans. As we show later, some biped locomotion machines are basically only powered legs, with no attempt to make them resemble the complete structure of the human body. On the other hand, in recent years there has been an explosion of interest in humanoid robots that, in turn, has stimulated significant work on biped locomotion, as illustrated by the P-series and the ASIMO walking machines developed by the Honda Research Institutes (Inoue et al. 2000), the WABIAN robots developed at Waseda University (Hashimoto et al. 2000), the M-2 biped from the MIT Leg Laboratory (Hu, Pratt, and Pratt 1998; Paluska et al. 2000), and robots from Fujitsu, Sony, and other laboratories. In this chapter we discuss the control of locomotion of these and other robots. Further details on humanoid robots can be found in chapter 13.

8.6.1 Early Biped Robots

Early biped robots relied on large feet and very slow walking for static balance, much in the way some toy robots do now. One of the first dynamic walking bipeds was built by Miura and Shimoyama (1984). They represented the body during the

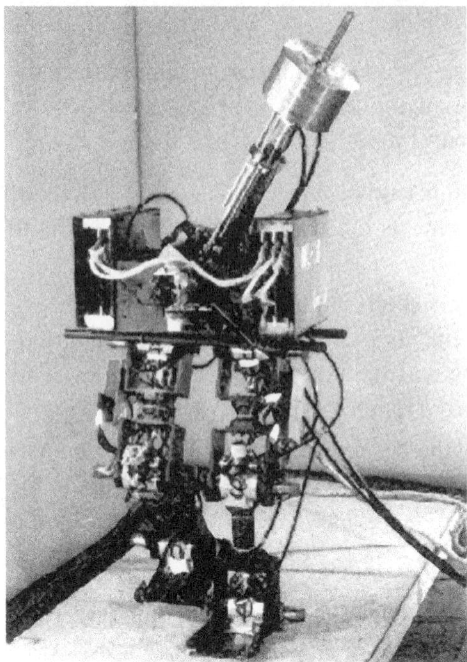

Figure 8.6
Waseda biped robot, 1986 (photograph courtesy of Atsuo Takanishi)

single-limb support phase (or single-limb stance) by means of an inverted pendulum. Another very interesting early implementation is due to McGeer (1990), who introduced the idea of *passive dynamic walking*: A totally passive-legged structure developed by McGeer was able to walk stably down an inclined plane.

The late Ichiro Kato was a pioneer in mobile robotics. He and his associates from Waseda University demonstrated a biped walker known as WHL-11 during the 1985 Exposition in Tsukuba, where it walked several kilometers (figure 8.6) (Kato et al. 1983; Takanishi et al. 1985). The machine weighed 40 kg and had 10 hydraulically powered degrees of freedom and relatively large feet (to aid in static stability). The robot's stance phase was quite long, but once during each step, as the foot swung forward, the machine tipped forward. Then, as the moving foot contacted the ground, the machine would regain equilibrium by quickly transferring the weight to that foot. The authors termed this robot's gait *quasi-dynamic*, since the robot regained its stability after each step passively, without any active control mechanisms. It was aided in this process by its large feet. During the design phase, the machine was modeled as an inverted pendulum.

8.6.2 Raibert's Monopod and Bipeds

Marc Raibert designed and fabricated a number of legged robots during the 1980s. One of his major contributions to legged robotics was the separation of stability from propulsion, as demonstrated first in a one-legged hopper (figure 8.7) and later in bipeds and tetrapods.

The machine shown in figure 8.7(a) had two major parts: a body and a springy leg. The body carried the actuators and instruments required for the machine's operation. The leg was able to telescope to change its length. The machine was not self-contained; it required a cable to connect it to a power source and a computer. Figure 8.7(b) is a diagram of the machine, showing the major components: a two-axis gyroscope needed for posture control; a gimbal connecting the leg and the body, effectively the device's "hip"; hydraulic actuators to control the angle between the leg and the body; a pneumatic actuator for controlling the leg length; and a small "foot," about 1 cm^2. Friction between the foot and the ground prevented slippage during the stance phase. Since a one-legged machine cannot be statically stable, Raibert's first hopping machine was attached to a radial arm that provided the necessary stability and restricted motion to a circular path. Later experiments were concerned with an unrestrained robot hopping in 3-D.

In normal operation each leg of this robot alternated between periods of support and periods of flight, corresponding to the stance and swing periods of normal bipedal gait. Since with a single leg there is nothing corresponding to heel strike and toe off in humans, Raibert referred to the transitions as *touchdown* and *liftoff* respectively.

One of Raibert's most important contributions was the decomposition of the control problem into three parts (postural-stability control, hopping-height control, and forward-speed control):

1. *Postural-stability control* Balance control of a one-legged robot (or, equivalently, a human on a pogo stick) is basically the control of an inverted pendulum. The control system can manipulate the attitude of the body only when the foot is in contact with the ground (i.e., during stance). The hip actuators are used to obtain a desired angle between the leg and the body:

$$\begin{aligned} f_1(t) &= K_P(\theta_P - \theta_{P,d}) + K_V(\dot{\theta}_P), \\ f_2(t) &= K_P(\theta_R - \theta_{R,d}) + K_V(\dot{\theta}_R), \end{aligned} \tag{8.2}$$

where

θ_P, θ_R are the pitch and roll angles of the body;

$\theta_{P,d}, \theta_{R,d}$ are the desired values for the pitch and roll angles with respect to the leg;

(a)

Figure 8.7
(a) Raibert's one-legged hopping machine and (b) a diagrammatic representation of it (photograph and illustration courtesy of Marc Raibert, reproduced from Raibert 1986 with permission of the MIT Press)

f_1, f_2 are hip actuator forces needed to obtain the desired pitch and roll angles;

K_P, K_V are the position and velocity control system feedback gains (see chapter 4).

2. *Hopping-height control* The only way a one-legged machine can move is by hopping. The hopping-height control part of the control system regulated the flow of compressed air into the lower part of the leg actuator. Air trapped in the cylinder was compressible, thus making the leg "springy." Raibert found experimentally that losses in the air cylinder (resulting from sliding friction and acceleration and deceleration of the remaining mass of the leg) resulted in about a one-third decrease in the height of each succeeding hop. The leg actuator was designed to just compensate for these losses. Providing a fixed thrust on each hop could be used to adjust the hopping height.

3. *Forward-speed control* Raibert found that the position of the foot when it touched the ground was critical in determining the acceleration the system would ex-

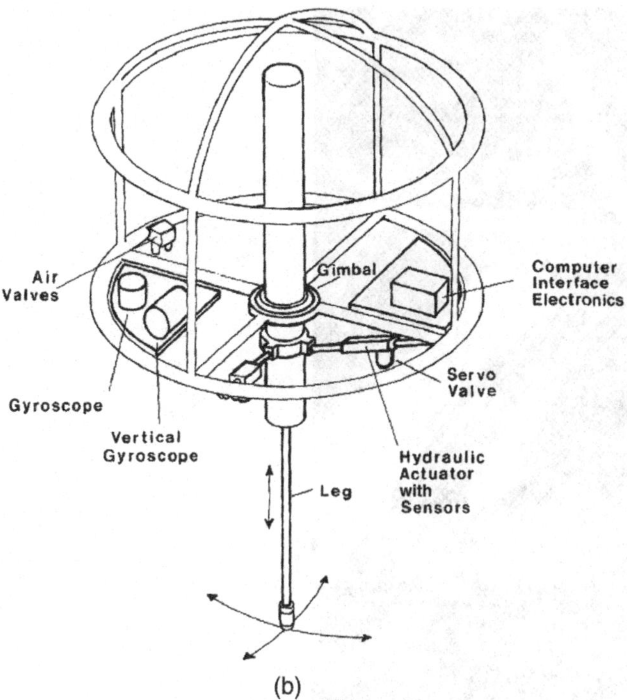

Figure 8.7
(continued)

perience during the support phase of the gait cycle. The error in the desired forward velocity was used to calculate a change in the position of the foot from the nominal position. The details of the algorithm can be found in Raibert, Brown, and Chepponis (1984).

Raibert obtained biped mobility by joining two of the hoppers in a long body, as shown in figure 8.8.

However, as pointed out by Pratt (2000), Raibert concentrated on running (rather than walking) machines, so that his biped performed a "bouncing run" as it transferred its weight from one leg to the other. Walking in a stable manner would have required a more complex leg structure. Raibert used hydraulic actuators at the hips to obtain high power and good velocity control, and he used air pressure to provide the compressibility leading to the bounce of the robot's springy legs. The hydraulic and pneumatic hoses for these actuators required that the system be tethered and connected to appropriate pumps. Such a restriction was not imposed on Kato's original bipeds or in later work in this field.

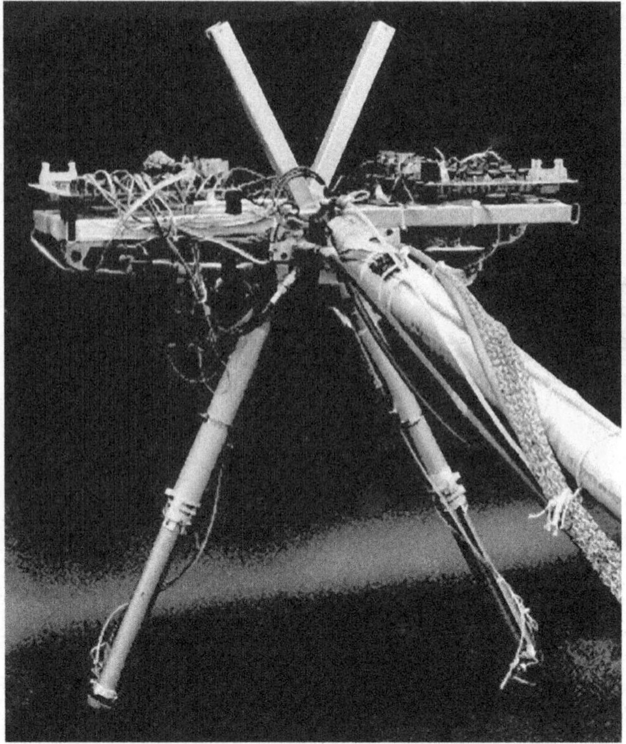

Figure 8.8
Raibert's bouncing, running biped (photograph courtesy of Marc Raibert)

In order to travel on rough terrain, a legged robot must adjust its step length so that the feet land on suitable places. An analysis of this problem for the Raibert biped appeared in Hodgins and Raibert (1991).

8.6.3 Walking Legs

Several biped robots constructed during the 1980s and 1990s are basically pairs of walking legs, with only a single link for the torso and no head or arms. The emphasis in these projects was the study of gait control. The robots were frequently supported in some way to reduce the stability problem. Typical of these robots were those constructed in the laboratories of Hemami (Jalics, Hemami, and Clymer 1997), Zheng (Golden and Zheng 1990), and Gruver (Shih and Gruver 1992). The SD-2 biped robot developed by Y. F. Zheng and his students at Ohio State University is shown in figure 8.9. This robot consists mainly of two legs with joints at the hips and ankles. The robot has no knees, but the hip and ankle joints have 2 degrees of freedom each, allowing for motion in both the frontal and sagittal plane. The feet have a flat bot-

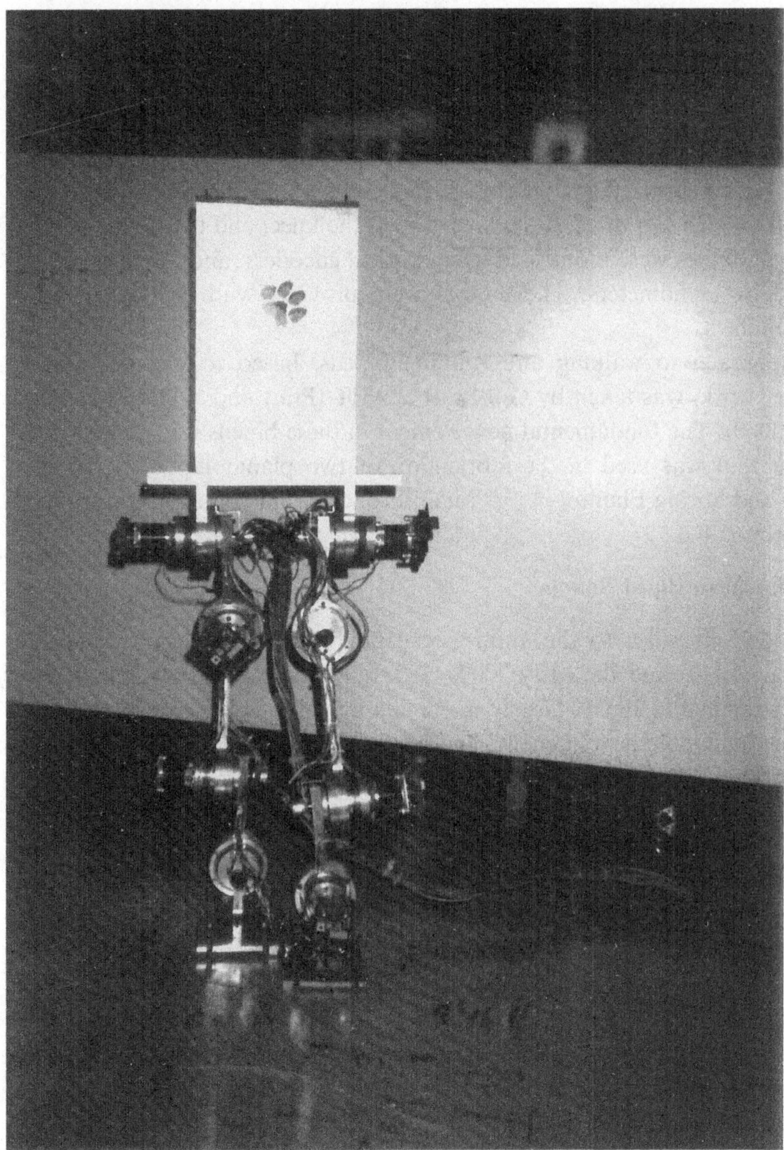

Figure 8.9
SD-2 biped robot (photograph courtesy of Yuan Zheng)

tom equipped with force sensors, enabling the system to compensate for an unbalanced position and retain static stability. The torso is rigid. An umbilical provides power to the on-board computer and motors. This 8-dof robot was able to adjust its own parameters, so it could walk on inclined surfaces. The kinematics and dynamics of machines of this type are discussed in section 8.7.

The biped robot constructed by William Gruver and his students (Shih and Gruver 1992) had 6 dof in each leg: three at the hips, one at the knee, and two at the ankles. The relative joint angles were monitored using optical encoders, and the joint speeds were measured by tachometers. The motors were provided with velocity control using servo amplifiers.

A unique approach to walking and running bipeds, based to some extent on Raibert's earlier work, was taken by Gill Pratt at MIT (Pratt and Williamson 1995; Paluska et al. 2000). The fundamental new element in these bipeds was a new series-elastic actuator that was used in the fabrication of two planar bipeds (known as Spring Turkey and Spring Flamingo) and an anthropomorphic biped known as M-2 (described in section 8.6.5).

8.6.4 Other Japanese Biped Robots

We now turn our attention to the anthropomorphic biped robots constructed in Japan in the late 1990s and the early 2000s. Several biped walking machines were designed and fabricated in Japan, largely under support of a major government program to support humanoid development. The first of these remarkable machines was the Honda P-series of humanoids. Initially these biped robots were larger and heavier than a man. They were not autonomous but required a sophisticated teleoperated control system under human supervision. Nevertheless, as biped machines, these robots were a great advance in the state of the art. They had an outstanding ability to maintain their balance and were able to walk up and down stairs. More recently Honda introduced a smaller humanoid known as ASIMO, Sony introduced the Sony Dream Robot (SDR), Waseda University developed a series of humanoids under the generic name WABIAN, and the ERATO (Exploratory Research for Advanced Technology) project funded by the Japanese government developed a humanoid named PINO. All these machines are bipeds, and they demonstrate that the original problems of balance and control have been largely solved. The physical structure and major properties of these robots and other humanoids are presented in chapter 13.

8.6.5 Pratt's M-2 Headless Robot with Series-Elastic Actuators

As with the walking-legs robots described previously, the M-2 robot developed by Pratt at MIT has no degrees of freedom above the waist, but humanlike degrees of freedom below the waist, as shown in figure 8.10. It is approximately human-sized and weighs 28 kg (Paluska et al. 2000).

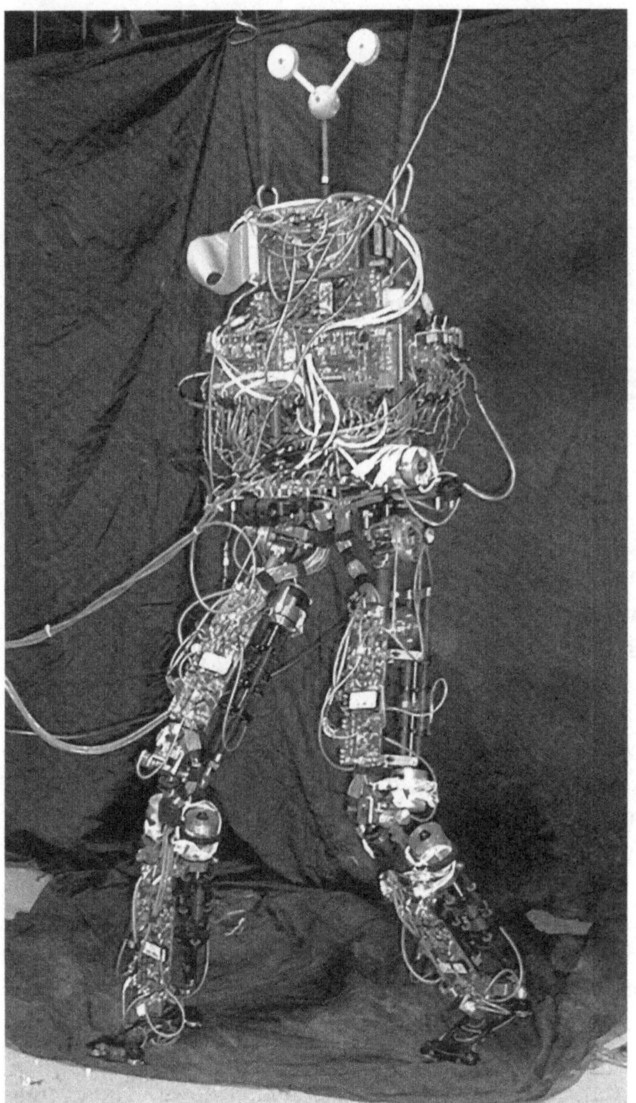

Figure 8.10
M-2 walking machine developed by Pratt (photograph courtesy of Gill Pratt)

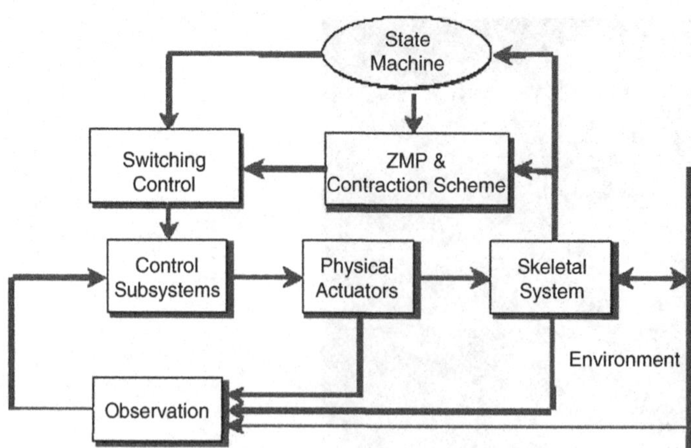

Figure 8.11
Overall structure of M-2 control system (illustration courtesy of Gill Pratt)

All the actuators on this robot are Pratt's series-elastic actuators (Pratt and Williamson 1995). Other unique features include universal joints in the ankles and a variety of sensors, including vestibular sensors and load cells in the relatively large feet for balance. The sensors provide inputs to a nonlinear controller.

The stability of biped machines is clearly a difficult problem. Pratt and his students performed careful control system designs for a two-legged robot named Spring Flamingo, a predecessor to the M-2 humanoid (Hu, Pratt, and Pratt 1998; Hu and Pratt 2000). The goal of the control system they developed was to ensure both gait stability (i.e., the gait should converge to a stable periodic pattern) and postural stability (i.e., the robot's pitch, roll, and height should be bounded within a predefined range of values). To ensure postural stability in the M-2, the authors made use of the zero-moment point (ZMP) concept,[20] introduced by Vukobratović (Vukobratović and Juricić 1968; Vukobratović et al. 1989). The *zero-moment point* is a point on the floor that is determined by the gravitational force and the inertial force of a walking robot. If the zero-moment point remains within the area touched by the foot of the robot's rear leg, a bipedal walking robot stays upright without falling down. The overall structure of the control system for the M-2 robot is shown in figure 8.11.

There are five main components in the system of figure 8.12: a state machine (discussed later), a ZMP computation module, a state observer, control subsystems, and

20. The zero-moment point concept was originally introduced by Vukobratović in 1968 and then elaborated in one of the volumes of his series of books on robotics (see Vukobratović 1989). It has reappeared in the robotics literature and is used in many biped robot designs to ensure stability, including the Honda and Waseda University robots discussed in chapter 13.

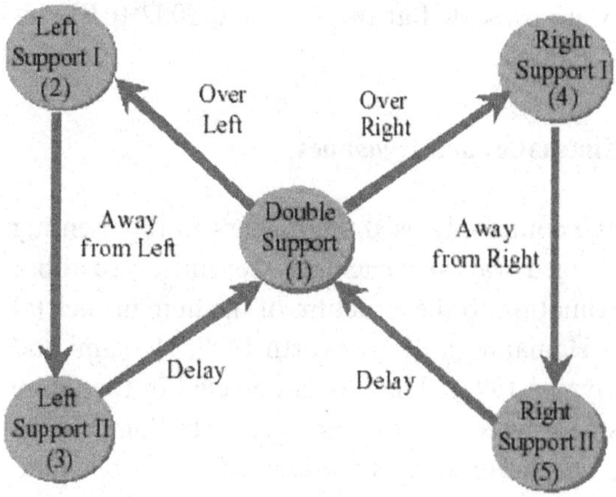

Figure 8.12
State diagram used with control system of figure 8.12 (illustration courtesy of Gill Pratt)

a switching-control module. As I indicated in connection with human walking, the dynamics of a walking robot change as it moves from single-limb to double-limb support. For this reason it is necessary to switch controllers as the system state changes. The various states are shown in the state diagram of figure 8.12. (The single-limb support and double-limb support phases of human walking were discussed in section 8.2.) Note that the output of the state machine (which indicates which state the robot is in) is then used to switch among the appropriate controllers. Computation of the ZMP shows it to be in the support plane of the two feet and adjacent to the toe of the rear foot (Hu and Pratt 2000), so the biped is stable in the double-limb support phase. It should be noted that since the control system switches between different controllers, in totality, the system is nonlinear.

8.6.6 Other Walking Biped Projects

In the preceding sections we have presented a few selected analyses of walking in biped robots. These specific projects were selected because they illustrate important principles, but they clearly do not cover the field. Numerous papers on biped walking have been published during the past twenty years. Since the structure of a biped robot involves many degrees of freedom and the dynamics of walking are described by nonlinear differential equations, many of these papers are highly mathematical in nature. One of the earliest treats biped dynamics using minimum energy criteria and their relation to postural and gait stability (Frank 1970). System theory has been used as a basis for the derivation of the equations of motion of a biped (Hemami and Wyman 1979). Inverse kinematics and dynamics of biped robots have been discussed in a number of papers (e.g., Shih, Gruver, and Lee 1993). More recently, several important papers in this field were presented at the 2000 and 2002 IEEE International Conferences on Humanoid Robots.

8.7 Mathematical Models of Biped Kinematics and Dynamics

In this section we introduce some of the commonly used approaches to representing the kinematics and dynamics of two-legged walking machines. Several biped robot models are based on a five-link approximation to the structure of the human skeletal system, such as the one in figure 8.13 (Hemami and Farnsworth 1977; Hemami and Wyman 1979; Jalics, Hemami, and Clymer 1997). The model depicted in the figure consists of a rigid trunk and two legs composed of two segments each. Some of the problems with this model are evident: The entire torso is represented by a single link, so head movement cannot be represented explicitly; there are no feet, so the contribution of plantarflexor and dorsiflexor muscles to both stability and propulsion cannot be included; and there are no arms, so the effect of arm swing on walking cannot be included. Nevertheless, the model has been very useful in providing an analytical basis for the design of walking machines. The discussion that follows is based primarily on Jalics, Hemami, and Clymer 1997.

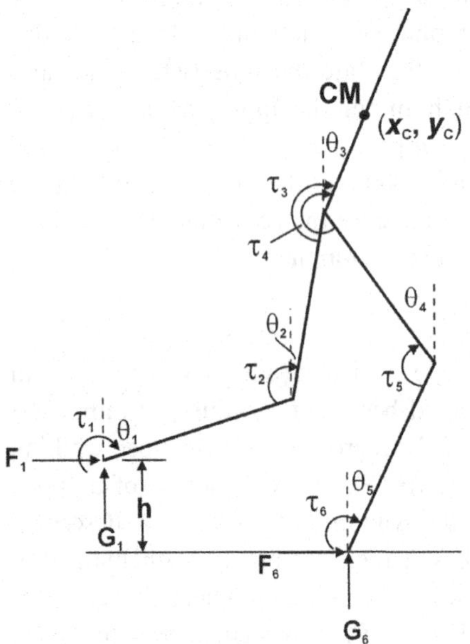

Figure 8.13
Five-link approximation of a biped robot

It can be noted that during each stance phase, this model is basically an inverted pendulum. The inverted-pendulum model is common in the study of biped mobility.

The position of each link in figure 8.14 is determined by the angle θ_i of the link with respect to the local vertical, so the kinematic equations can be written directly. The dynamical equations require consideration of the mass and moment of inertia (about the center of mass) of each link, the friction in the joints, and the actuator responsible for producing movement. Since the robot is not a model of human physiology, there is no need to model human muscle properties precisely, but they are included here. (It is common to assume that all the actuators are located in the joints and contain their own power sources.)

The model of figure 8.14 has 7 dof: the 5 rotational dof (specified by the joint angles) and 2 translational dof (specified by the coordinates x_c, y_c of the center of mass of the torso:

$$\mathbf{W} = [\theta_1 \ \theta_2 \ \theta_3 \ \theta_4 \ \theta_5 \ x_c \ y_c]^T. \tag{8.3}$$

Equation (8.3) gives the state vector of the system. The dynamics of the system can be written in terms of the state **W** as

$$\mathbf{I(W)\ddot{W}} + \mathbf{B(W,\dot{W})\dot{W}} + \mathbf{G(W)} = \frac{\partial \mathbf{C(W)}}{\partial \mathbf{W}}^T \mathbf{\Gamma} + \mathbf{E}\tau, \tag{8.4}$$

where

I(W) is the inertia matrix;

$\mathbf{B(W,\dot{W})}$ is the Coriolis acceleration matrix;

G(W) is the gravity matrix;

τ is the torque matrix, multiplied by coefficients **E**;

C(W) is a set of ground constraint equations that relate the state variables to the positions of the feet with respect to the center of mass of the torso;

$\mathbf{\Gamma}$ is a matrix of ground reaction forces, defined as

$$\mathbf{\Gamma} = [F_1 \ G_1 \ F_6 \ G_6]^T. \tag{8.5}$$

The F and G terms in equation (8.5) are defined in figure 8.14 and represent frictional and gravitational forces respectively.

Clearly, this is a nontrivial set of equations, but it is not yet a complete set. Equations (8.4) and (8.5) describe the dynamics of the biped but do not include the muscle actuators to produce motion. As we have seen, the force produced by a muscle depends on its neural input, the muscle length, and the rate of change of the muscle length (Winter 1984, 1987). This relation can be modeled as (Jalics, Hemami, and Clymer 1997)

$$f = k(r + a\ell^2) + \beta\dot{\ell}u(-\dot{\ell}), \tag{8.6}$$

where

f is the force generated by the muscle;

k and a are constants;

r is the neural input along the alpha motoneuron (see chapter 2);

ℓ is the change in muscle length from its resting value;

$\dot{\ell} = d\ell/dt$ is the rate of shortening of the muscle;

β is a constant representing the viscosity of the joint;

$u(\ )$ is the step function.

This equation can be used to compute the muscle forces in the eight leg muscles shown in figure 8.2. Now, to compute the torque produced by each of these muscles at a particular joint, it is necessary to include the moment arm for each muscle and the direction in which the force is applied. The moment arm depends on the point of attachment of the muscle tendon to the appropriate bone. Both the attachments and the approximate directions are shown in figure 8.2 where the muscles are represented simply by dark lines.

Finally, to complete the model it is necessary to add a controller to the musculoskeletal dynamics. The inputs to the controller are sets of desired states $\mathbf{W}_d(\mathbf{t})$ and the actual states $\mathbf{W}(\mathbf{t})$. For example, for constant-speed walking, the desired states are periodic.

We have summarized the major ingredients of a mathematical representation of biped walking to highlight the complexity of such a representation and the large number of parameters to be estimated before solution of the equations is possible. Nevertheless, a number of investigators have developed and solved similar equations. One can conclude that the human nervous system is indeed a remarkable controller, since it clearly does not solve these equations explicitly and yet is able to maintain the body's stability and forward progression during walking.

8.8 Modeling Compensatory Trunk Movements While Walking

When humans and animals walk, their trunks also move to compensate for changes in the location of the body's center of gravity. Such movements are particularly pronounced in birds. Anyone who has watched a pigeon walking will have noticed the dramatic back and forth movement of the bird's head. The Humanoid Research Laboratory at Waseda University in Japan has done a careful study of trunk compensatory movements and included them in a mathematical model of its WABIAN RII robot. A description of the WABIAN family of robots can be found in chapter

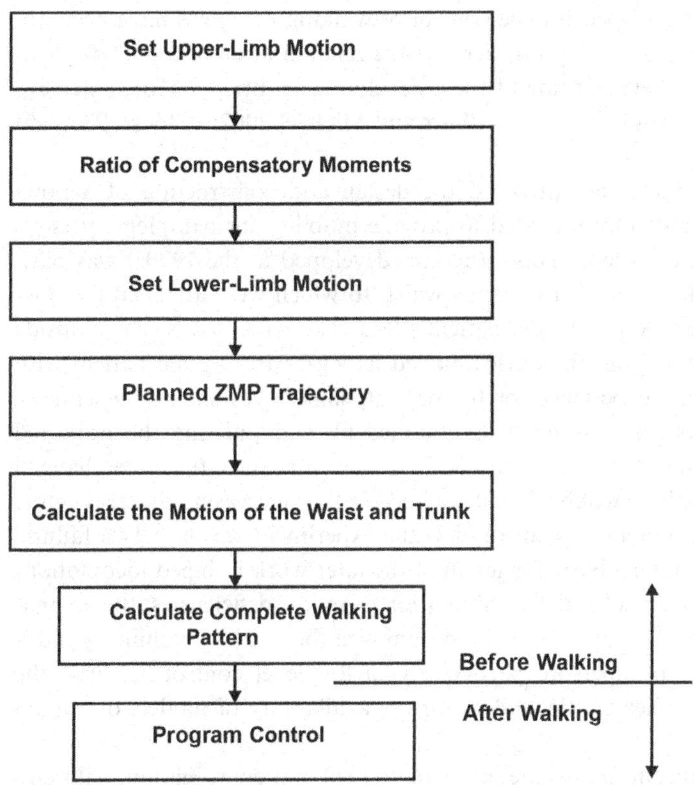

Figure 8.14
Control structure of WABIAN RII robot (reproduced with permission of Waseda University)

12. A block diagram of the control structure of the WABIAN-RII robot is shown in figure 8.14.

Several comments need to be made about figure 8.14. Note that in addition to pre-computing the motion of the upper and lower limbs (thus providing for natural-appearing leg movements and corresponding arm swings), the program includes the calculation of a planned ZMP trajectory. (The ZMP was discussed in section 8.6.5.) The trajectory of the ZMP can be controlled by the angular trajectory of each of the joints.

8.9 Mechanical Aids to Human Walking

We have indicated previously that legged vehicles have been developed mainly for research purposes, with limited commercial applicability to date. On the other hand, some of the principles developed for the control of walking machines have been applied successfully in the design of controllers for the rehabilitation of motor deficiencies in humans. We now review some of these developments briefly. More extensive treatments are given in such references as Shurr and Michael 2002; Billock 1996; and Wilson 1998.

A significant early experiment involved the design and construction of a powered exoskeletal biped structure intended to provide mobility for paraplegic persons (Vukobratović and Juricić 1968). This structure (developed in the 1970s) was basically a metal frame fitted around a patient's waist, to which were attached two mechanical legs that were strapped to the patient's legs. The patient was placed inside the structure. When turned on, the device moved its legs, carrying the patient with it. This was a frightening experience for the patient, since he or she had essentially no control over the system. Stability was obtained by manipulating the projected center of gravity and the support area provided by the feet. Since this area changed with every step, the machine wobbled as it walked, further increasing the discomfort for the patient. From a practical point of view the experiment was largely a failure, but it provided the theoretical basis for much of the later work in biped locomotion.

In human-machine systems for the rehabilitation of motor deficiencies, the emphasis is on coordination and control in order to minimize the need for voluntary intervention of the handicapped person, particularly in low-level control. Clearly, the design of such devices is heavily dependent on the availability of models of human gait.

In this section we concentrate on the issues of control and coordination, with emphasis on the problem of assisting the walking of partially paralyzed persons.

8.9.1 Above-Knee Prostheses

It is well known that adequately designed passive below-knee prostheses can improve the gait quality of an amputee to such a degree that the effects of amputation are hardly observable (Perry 1992). However, this is not the case with above-knee amputation. The loss of the knee entails instability, since the rigid knee assures the upright position of the body. Passive knee-locking devices (the most primitive of which rely on the amputee to pull a cord to unlock them) cannot fully replace the active control of rigidity, since they operate not on sensory feedback, but on measurement of terminal extension angles. Such a solution reduces the reliability of the knee as a weight-carrying mechanism and increases the metabolic effort required to maintain ambulation. In addition, for those equipped with such devices, special skill and considerable effort are required to master such obstacles as rough ground, ramps, and stairs. For these reasons, it makes sense to develop active prostheses for above-knee amputations.

Several approaches to the design of active above-knee prostheses have been tried (Shurr and Michael 2002). They range from simple solutions involving relay-controlled knee-locking mechanisms to controllers affecting joint transition trajectories, step length adjustments, and gait mode selections and microprocessor-controlled knees. The *friction brake knee* is activated by weight bearing during stance that compresses a spring and causes the knee to press against a brake bushing, thus providing the needed stability. When the knee is unweighted, the spring releases the brake and the knee can be flexed. The human knee is not a simple hinge but has a variable center of rotation. In an attempt to capture this feature, there are *polycentric* knees that apparently offer some advantages over single-axis designs (Shurr and Michael 2002).

Technically, the control of a one-joint dynamical system is trivial, but a human-machine system with an active knee is quite different from passive engineering systems. Not only must such a system perform satisfactorily from an engineering point of view, but it must be acceptable to the amputee. In order to be accepted by those who use it, an assistive system must satisfy following conditions:

- *Sensor-driven control* The locomotion process requires monitoring of relevant environmental features as well as measurement of knee and ankle behavior and automatic adjustment as needed without intervention by the user.
- *Control matching* Steady-state locomotion relies on reflex actions, which are triggered by such events as heel contact, full knee extension, and shift of the body axis. If control of the prosthesis differs drastically from natural control, the mismatch between human and machine prevents the amputee from properly integrating the assistive device into his or her reflex and automatic system.
- *Adaptivity* The system should adapt automatically to changes in slope, to climbing stairs, or to walking on a level surface.
- *Security* The wearer must feel completely secure and not be concerned about a possible failure of the prosthesis.
- *Constraints* The assistive system must be compact, its weight cannot be burdensome, and it must be highly reliable and sufficiently autonomous so as not to put an additional cognitive load on the wearer. The issue of reliability cannot be overemphasized; if assistive devices are unreliable, persons with disabilities will not use them.

These requirements make the design of an above-knee prosthesis a difficult and complex problem.

8.9.2 Intelligent, Sensor-Based Above-Knee Prostheses

One of the most advanced contemporary solutions is the C-Leg (Computerized Leg) system (figure 8.15), developed by Otto Bock HealthCare in Germany (Agah and

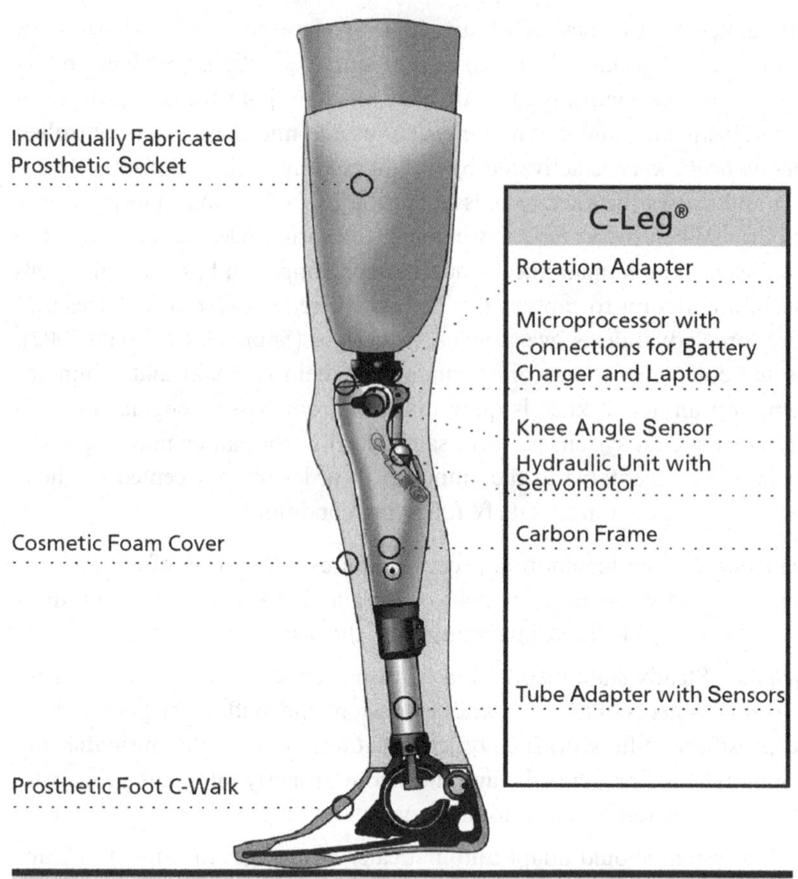

Figure 8.15
Otto Bock C-Leg prosthesis (photograph courtesy of Otto Bock HealthCare)

Bekey 1997d; Näder and Näder 2002). To overcome some of the problems mentioned in the preceding section, this leg features a hydraulic knee controlled by a microprocessor in both swing and stance.

The C-Leg is a highly active prosthesis. It can be referred to as a "robotic leg," since it uses sensors, a processor, and actuators. As shown in figure 8.15, it is equipped with two types of sensors: strain gauges in the shin tube for measuring the anterior and posterior moment of flexion at the ankle and angle sensors to measure the instantaneous knee angle and its rate of change. The signals from these sensors are input into the microcontroller at an update rate of 50 Hz; the microcontroller then calculates the proper resistance to movement and opens and closes hydraulic

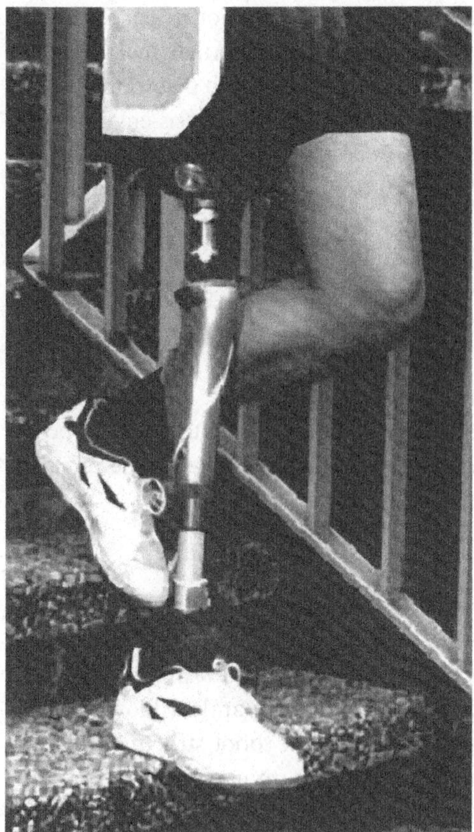

Figure 8.16
Amputee with Otto Bock C-Leg descending stairs (photograph courtesy of Otto Bock HealthCare)

valves as necessary to obtain the desired flexion and extension of the knee. The hydraulic resistance during stance stabilizes the knee during heel strike, without the need for an actual mechanical lock. Then, at the end of the stance phase, the resistance is removed, and the knee flexes effortlessly during swing. Clearly, the valves require energy to operate, so the C-Leg is provided with a rechargeable lithium-ion battery. When fully charged the battery provides enough energy for a full day of operation. The shell of the leg is made of a carbon fiber composite. The C-Leg is adjusted and calibrated with the aid of a PC that plugs directly into the knee socket and a proprietary software package. According to the manufacturer, the C-Leg allows for walking on irregular terrain as well as participation in such sports as bicycling and cross-country skiing. Figure 8.16 shows an amputee equipped with the C-Leg descending stairs.

An alternative approach to using sensory inputs for control of a leg prosthesis relies on pressure sensors distributed over the insole of the foot. Angle measurement sensors are also provided, as well as a pendulum-type transducer to determine the local vertical. In this way, data on the sensed limb attitude and prosthesis-environment interaction are provided to the prosthesis's microprocessor. Control of the prosthesis is performed using "artificial reflexes," that is, motor responses to sensory inputs designed to mimic the normal response of the leg. (Note that "artificial reflexes" are essentially the same as behavior-based responses used by robots: They require sensors for actuation, and they involve a very close coupling from sensation to action.) The reflexes are implemented using simple if-then production rules. For example, toe liftoff is interpreted to indicate the start of the swing phase for the leg, thus causing the actuator to begin extension of the knee.

Clinical evaluation tests of a system of this type were quite encouraging (Popović and Schwirtlich 1993). It took the amputee only a few trials to integrate the prosthesis system into his way of controlling ambulation.

8.9.3 Reflex-Controlled Orthoses

Even more complicated problems are encountered when the anatomy of a limb is preserved but there is a motor deficiency due to an impaired neuromuscular tract. Devices fitted over an existing but nonfunctional limb are called *orthotic devices* or simply *orthoses*. A straightforward approach for patients with paralyzed lower limbs, would be to build an active skeleton, like some of the biped robot structures we saw earlier, attach the impaired limbs to it, and let the structure carry the person. Although technically feasible, this approach is rejected by persons with disabilities, since it does not use internal sensory, motor, and control resources and they feel helpless. Hence, patients tend to select either some form of strap-on orthosis, if one can be fitted to them, or simply to use a wheelchair. Some manufacturers provide modular orthoses, but in many cases they are custom made and fitted to the individual. Various methods of bracing, ranging from below the knee to whole leg and even the torso, have been developed; the classic work in this field is Perry and Hislop 1967, reprinted numerous times since its original publication.

An alternative approach is to attempt to activate fully the remaining neuromuscular resources of the disabled person and use them for his or her rehabilitation. By means of functional electrical stimulation (FES) of the neuromuscular system at different levels with multichannel stimulators, functional reflex responses may be evoked and the impaired motor control improved, even reorganized and reconstructed (Popović, Tomović, and Tepavac 1989). However, FES-evoked functional motions should not be compared to smooth, dynamic, efficient motor acts in unimpaired persons. Externally activated motor responses are usually jerky and badly coordinated, with inadequate dynamics. From the metabolic point of view, they are inefficient, so fatigue appears much sooner than normally. Besides, stability problems arise as a result of these motions, and mechanical walking aids are needed to enable the patient to maintain an upright body position.

In some cases it is possible to use FES up to a point and then augment it with external sensors and actuators. Such a combined rehabilitation method has been given the name *hybrid assistive system* (HAS) (Popović, Tomović, and Tepavac 1989). Implementation of such a system requires active braces, but with only a partial power supply and much simpler control, since it only supplements, but does not replace, the biology. As a result of its supplemental nature, the active orthosis can be built as a modular, light, easily fitted mechanical structure (Popović and Schwirtlich 1993). The principles of reflex control, using both natural and artificial reflexes, in orthotics and prosthetics were presented by Bekey and Tomović (1986).

It should be clear from the discussion in this section that the control of prosthetic and orthotic devices has a great deal of overlap with robotics.

8.10 Concluding Remarks

In this chapter we have presented various aspects of locomotion with two legs, both in humans and in robots. It should be clear that the anatomy and control aspects of human locomotion are much too complex to be emulated in biped robots. Nevertheless, current robots are capable of a normal-appearing gait, not only on level surfaces, but on ramps and stairs as well. Some of the lessons learned from the construction of biped robots can be applied to the design of assistive devices for persons with disabilities, since these systems must be simple, reliable, and easy to control.

9 Locomotion in Animals and Robots with Four, Six, and Eight Legs

Summary

Beginning with the automatons built in the nineteenth century, people have been fascinated by legged machines, especially when they display some autonomy. In this chapter we summarize some basic principles of legged locomotion in animals and then discuss the application of these principles to the design and fabrication of legged robots. We begin with insect locomotion and six-legged machines and present a number of examples of such robots. We then turn to four-legged animals and robots. Again, we illustrate the nature of the robots by discussing various examples, with particular attention to issues of static and dynamic stability. Finally, we present several eight-legged robots.

9.1 Introduction to Legged Locomotion in Animals

From simple, single-celled organisms like amoebas to complex vertebrates, living things have the ability to move by means of internally generated forces. In previous chapters we have discussed swimming and flying robots as well as biped locomotion. In this chapter we focus attention on the locomotion of legged animals and robots with four or more legs.

Walking is a method of locomotion employed by legged land animals and humans, as well as legged robots. It is a complex process. In animals and humans it requires the coordination of numerous muscles to maintain a stable posture while providing forward progression. The remarkable series of coordinated actions that results normally requires no central control from higher centers, depending instead on spinal feedback and large numbers of local control and feedback systems. Stable walking requires the generation of systematic periodic sequences of leg movements at various speeds of progression. At slow velocities, stable walking is characterized by static stability, in which the center of mass of the body remains within the polygon of support formed by the legs in contact with the ground. Animals with six or more legs have guaranteed static stability if they leave at least four feet in contact with the ground. The stability of four-legged animals, on the other hand, is conditional on the location of their center of gravity as soon as they lift one leg off the ground. Quadrupeds have active control systems that shift their body positions appropriately to ensure that the vertical projection of their center of gravity onto the support surface falls within the triangle of support from the legs. As we saw in the previous chapter, this situation is more complex in humans, since true static stability in a biped occurs only during standing and then only in the lateral plane (unless the biped has very large feet). Birds walk or hop on two legs, but they use head and tail movements to assist in stability; similar compensation occurs in kangaroos. During motion, the support forces of the legs, momentum and inertial forces are summed to produce dynamic stability. In both humans and animals, each leg alternates between stance, the time its end point (foot, paw, or hoof) is in contact with the ground, and swing, the time it spends in forward motion. The elastic properties of the legs are also important. The design of walking machines must consider these same factors, as discussed in the following. The fact that locomotion is possible without central control is evident in quadrupeds with a broken spinal cord. A *spinal animal* can run by shifting its weight appropriately, thus triggering reflexes that cause the leg muscles to contract. The movement is awkward, but remarkable.

9.2 Neural Control of Locomotion

Nervous system locations and mechanisms of control of rhythmic limb movements during locomotion in higher vertebrates have been the subject of intensive study. These mechanisms and neural connections are very complex and widely distributed in the CNS. Hence, this section we provide only an overview of the major neural control issues. Two CNS mechanisms appear to be involved in locomotion control. There are centers within the nervous system capable of producing the periodic discharges of nerve impulses associated with walking or various running gaits; they are usually referred to as central pattern generators (Grillner 1981). CPGs are located primarily in the spinal cord but also appear elsewhere in the nervous system. In turn, the level of activity of CPGs is controlled by higher centers in the nervous system and influenced by sensory feedback from peripheral receptors in the limbs. The cerebellum is responsible for fine control of locomotion; the loss of cerebellar function does not prevent locomotion, but movements become less coordinated. The decision to initiate locomotion is made somewhere else in the brain, probably in one or more brain stem locations. Basically, the function of the CPGs is to activate alpha motoneurons in reciprocal patterns, producing alternating sequences of flexion and extension in the various limb muscles. The role of CPGs in animal locomotion continues to be a subject for research (e.g., Cohen 1992; Collins and Richmond 1994; Cohen and Boothe 1999).

The issue of coordination between limbs in locomotion is also complex, and the phenomenon can be explained in a number of ways. It probably involves some form of reciprocal inhibition between pairs of CPGs, phase delays in coupling between CPGs, and a variety of connections between CPGs. These mechanisms have been modeled in invertebrates. In addition, the control of locomotion is strongly influenced by feedback from peripheral receptors. Such factors as speed, load, and limb position provide afferent inputs, which affect the CPGs. The interaction of these input and feedback signals in the control of locomotion was illustrated in figure 8.4.

Control of locomotion on non-level terrain requires a variety of adjustments to maintain stability. For quadrupeds, walking uphill requires that the rear legs extend further back than during level walking and conversely for down hill walking. Walking along a constant-elevation contour on the side of a hill requires dramatic adjustments of leg and shoulder or hip position to maintain stability. This is evident in the humorous comment that cows in Switzerland have shorter legs on one side of the body.

To change speed, animals with four and six legs exhibit a variety of gaits. It now appears that the switch from one gait pattern to another occurs, at least in part, by means of mechanisms that attempt to keep the muscular forces and joint loads in the legs within particular limits.

Finally, consider the control required to walk safely in a field with many obstacles. Animals and humans can certainly use vision to step around obstacles. However, for quadrupeds this applies only to their front legs; some exquisite form of coordination must exist between front and back legs to ensure that the latter also avoid the obstacles that the front legs avoid as a result of vision. Clearly, a very complex problem in inverse kinematics is being solved by the nervous system in this process.

9.3 Walking Multilegged Robots

Both the anatomy and the control of locomotion in multilegged animals are so complex that we cannot emulate them in hardware. Hence, the construction of robot walking machines is an attempt to isolate from the complex of locomotion properties cited in the foregoing a set sufficient to produce stable forward motion with a minimum of hardware and software. Clearly, such an abstraction of properties from those developed by evolution over millennia is not an easy process, and many of the walking machines that have been built, whether biped, quadruped, or hexapod, have not been highly successful. Most walking machines are only statically stable, whereas a few others display both static and dynamic stability.

The designers and builders of these machines have usually taken one of two approaches. Some have concentrated on some aspect of kinematics and dynamics, including the design and fabrication of the mechanical substitutes for skeletal, joint, and muscle function (e.g., Waldron et al. 1984). Other robot builders have devoted major attention to the control of the leg movement sequence, attempting to imitate the variety of gait patterns exhibited by animals during locomotion (e.g., Chiel et al. 1992). These patterns have fascinated robotics researchers for many years.

Numerous experimental walking machines with four, six, or eight legs have been developed since the 1950s, in the hope that they would be used in a broad range of applications. Such machines have been proposed and applied primarily as vehicles for operations on rough terrain or in hostile environments. In general, legged locomotion machines have not yet found permanent applications, either in industry or as vehicles for hostile environments. In the vast majority of practical applications, wheeled or tracked vehicles have advantages over legged vehicles in terms of power requirements and complexity and provide lower-cost solutions to the need for locomotion, even in difficult environments (as on the surface of the moon, for example). Animals and humans, on the other hand, employ legged locomotion because of the incredible adaptability and versatility this method of locomotion provides. Legged machines do not require a continuous, unbroken support path. Legs make it possible to move on smooth or rough terrain, to climb stairs, to avoid or step over obstacles, and to move at various speeds. A comparison of wheeled and legged locomotion in simulated Mars environments has confirmed the observation that legged vehicles have advantages over wheeled ones only in the presence of large numbers of obstacles that can be surmounted by the legs (Sukhatme and Bekey 1995). However, recent developments in legged robots, often based on biological inspiration, have led to significant improvement in their speed and stability, so there are reasons to believe that significant applications of legged robots will follow in the near future.

In the remainder of this chapter we review the major characteristics of some of the many walking machines constructed throughout the world. We cannot cover them all, but those selected for discussion display the major features utilized by most other legged robots. We concentrate on physical robots and do not discuss simulated robots (except in a few rare instances) because we believe that ultimately robotics requires embedding in hardware. While offering an overview of the field, we also provide some examples from the simulation and fabrication of walking machines in the USC robotics laboratory. In all these systems, we are concerned with the issues of architecture of legs and complete systems, control of motion of a single leg, and coordination of leg movements.

9.4 Six-Legged Walking Machines

Large, six-legged walking machines have been built in the United States, in Japan, and in the former Soviet Union since the 1970s, if not before. As indicated in the previous section, the goal of these machines has been locomotion over rough, irregular terrain.

Some of the earliest experiments in walking machines were carried out in the former Soviet Union (Okhotsimski and Platonov 1973), where a six-legged vehicle was built in the early 1980s (Devjanin et al. 1983). In the United States, the Odetics six-legged walking machine (known as The Functionoid) (figure 9.1), also from the 1980s, was designed for inspection of power plants in Europe (Russel 1983). This machine was able, under remote control, to climb stairs and walk up steep ramps.

A remarkable series of large-scale hexapod walking machines was built at Ohio State University in the 1970s and 1980s. Robert McGhee developed a large six-legged walker in the late 1970s (McGhee and Iswandhi 1979; McGhee et al. 1984). This walker (depicted in figure 9.2(a)) was followed by the development of a much larger hexapod, capable of carrying a person over rough terrain (McGhee et al. 1984; Waldron et al. 1984; Waldron and McGhee 1986; Song and Waldron 1989). This machine, known as the Adaptive Suspension Machine (ASM) (figure 9.2(b))

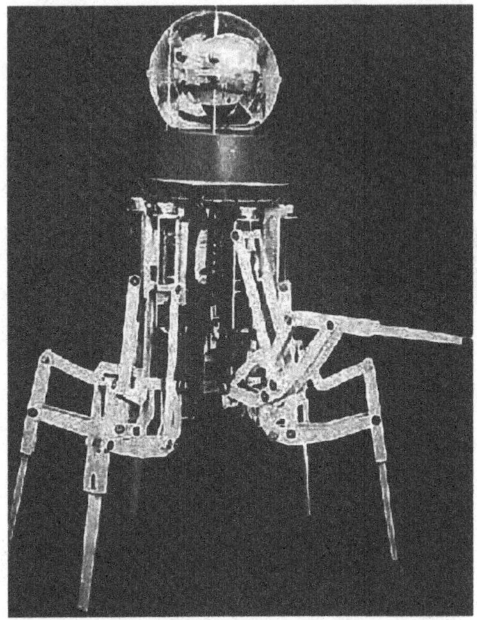

Figure 9.1
Odetics walking machine The Functionoid (reproduced with permission of Iteris Holdings, Inc.)

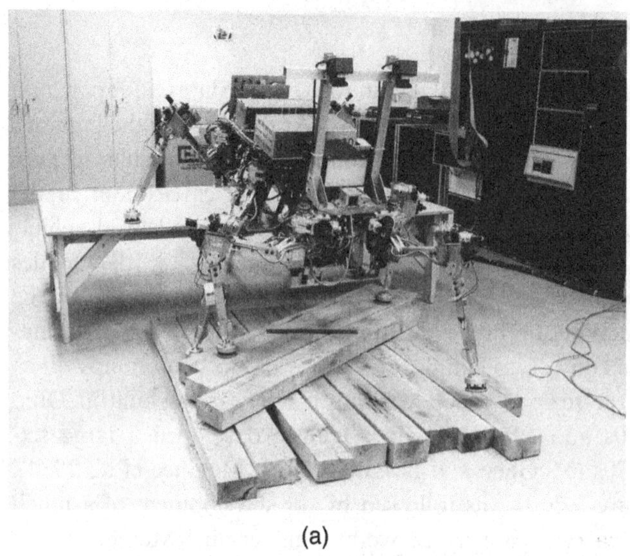

(a)

(b)

Figure 9.2
The Ohio State University hexapods: (a) the early hexapod, walking over a table; (b) the Adaptive Suspension Machine (photographs courtesy of Robert McGhee and Kenneth Waldron, respectively)

was about 5 m in length and weighed some 2,700 kg. Each leg of the ASM was basically a pantograph mechanism. The legs were driven hydraulically, as required to obtain the power levels required to move such a massive machine.

9.4.1 Insect Locomotion

Insects have six legs, a fact that gives them clear stability advantages over four-legged animals. For this reason they have been studied extensively and used as models for the design of walking machines. Much of the basic work on insect locomotion was done by Holk Cruse in Germany (Cruse 1976, 1979, 1990; Cruse et al. 1995) and Hillel Chiel, Randall Beer, and their colleagues in the United States (Beer 1990; Chiel et al. 1992). Mechanical insects were developed on the basis of this work by Roger Quinn and Roy Ritzmann (1998). It is interesting to note that insects can adapt to the loss of two of their six legs without much apparent loss of performance (Delcomyn 1989). Much of the work on the neural control of insect locomotion is summarized in Beer 1990 and Beer, Ritzmann, and McKenna 1993. Slowly walking insects are known to use a wave gait (also known as a *metachronal wave gait*), in which pairs of legs move in sequence from the front to the back, with the front legs moving as soon as the rearmost finish their swing. When they walk faster, insects typically switch to a tripod gait, in which the front and back legs on one side of the body move synchronously with the center leg on the other side. Since three legs move, the other three are on the ground and provide a stable tripod platform. We encountered insect wave gaits and tripod gaits previously in case study 6.3, which concerned learning to walk using genetic algorithms.

Insects can walk on irregular surfaces; many insects can walk on vertical surfaces and even upside down (such as flies on a ceiling). The challenge to designers of six-legged walking machines is emulating the walking patterns of insects and some of their neural-control mechanisms without implementing the actual structure of the insect's anatomy. Hence, a number of the early robotic "insects" have little resemblance to the creatures that inspired them.

9.4.2 The Case Western Cockroaches and Their Control

The two insects that provided most of the inspiration for early projects in robotics were the American cockroach (*Periplaneta americana*) and the walking stick (*Carausius morosus*). The cockroach was studied extensively by Pearson (1973), and the locomotion of the walking stick was studied by Cruse (1979). Based on Pearson's work, Beer, Chiel, and their colleagues in the Biologically Inspired Robotics Laboratory at Case Western Reserve University (CWRU) developed a neural-network model of cockroach walking (Beer, Chiel, and Sterling 1989), illustrated in figure 9.3. Note that at the center of each of the six legs is a pacemaker neuron, whose output oscillates at a steady rhythm, determined by the excitation and inhibition it

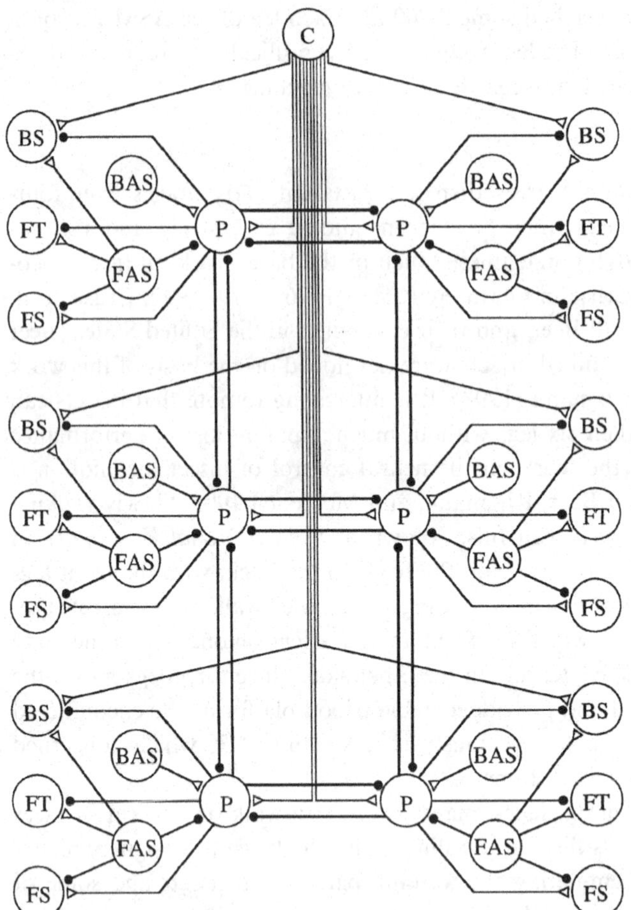

Figure 9.3
Neural-network model of cockroach walking. Note the command neuron (C) and six leg controllers with pacemakers (P), foot motor (FT) neurons, backward-swing (BS) and forward-swing (FS) neurons, and backward-angle (BAS) and forward-angle (FAS) sensors. Excitatory connections are denoted by open triangles, inhibitory connections by filled circles (from Beer and Chiel 1995, reproduced with permission of MIT Press)

Figure 9.4
Hexapod robot controlled by a neural network from cockroach simulations (photograph courtesy of Roger Quinn)

receives from the command neuron, from other pacemakers (to prevent adjacent legs from swinging at the same time), and from the forward- and backward-angle sensors. The forward-angle sensor inhibits the pacemaker when the leg has reached its maximum forward position, and the backward-angle sensor encourages the pacemaker when the leg is at its extreme backward position. Using simulation, this model was shown capable of generating a range of statically stable cockroach gaits, similar to those observed in living insects.

To determine whether the result obtained with the simulation would also hold for a physical robot, several walking machines were constructed using the controller model of figure 9.3. The first of these hexapods is shown in figure 9.4. It is approximately 50 cm long and 30 cm wide and weighs about 1 kg. Note that each of the hexapod's six legs had 2 degrees of freedom: swing about an axis perpendicular to the body and a radial degree of freedom along the leg axis. Potentiometers mounted in parallel with the drive motors provided sensory feedback. The resulting distributed architecture was quite robust and able to repeat the various gait properties obtained in simulation (Quinn and Espenschied 1993). The computer implementing the controller and the twelve servos for driving the leg motors were located remotely from the robot; an umbilical cable can be seen at the rear of the machine in the figure.

A more recent version of a cockroach-inspired robot, also from the Biologically Inspired Robotics Laboratory at CWRU, is Robot III, shown in figure 9.5. This robot is based on the kinematics of the cockroach *Blaberus discoidalis* (Ritzmann et al. 2000). It has a total of 24 degrees of freedom with 5 for each front leg, 4 for each of the middle legs, and 3 for each of the rear legs. The robot is pneumatically actuated using off-the-shelf cylinders and blocks of three-way pneumatic valves. Pulse width

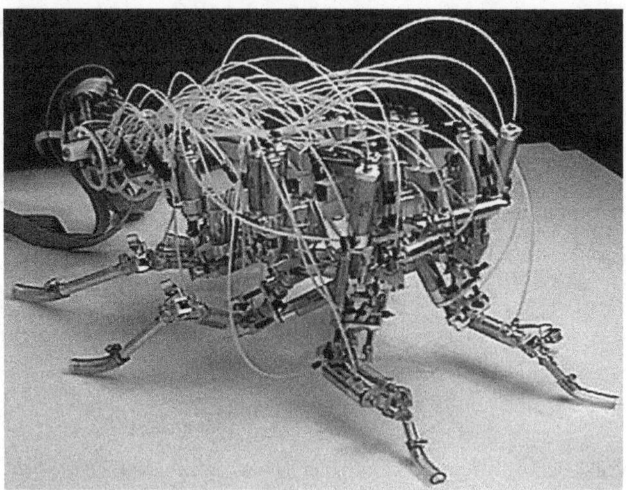

Figure 9.5
Robot III, modeled on the cockroach *Blaberus discoidalis* (photograph courtesy of Roger Quinn)

modulation of the valves is implemented for variable-position control of the cylinders. The structure of the robot is constructed from high-grade aluminum alloys. The robot is capable of jumping as well as walking.

9.4.3 Genghis

One of the earliest autonomous six-legged robots was Genghis (figure 9.6), a walking machine constructed at MIT (Brooks 1989). It is interesting that although this machine has a distinctly insectlike appearance, its creator, Rodney Brooks, clearly states, in the paper just cited, that this was not deliberate. Rather, since Raibert (1986) was already building robots with one, two, and four legs, a six-legged machine appeared to be a reasonable challenge, especially since the previous hexapods (like those built at Ohio State University, as mentioned earlier) were complex and difficult to control.

Brooks demonstrated as early as 1986 (Brooks 1986) that complex control systems for mobile robots can be built using finite-state machines (FSMs). A network of such machines can be viewed as an idealized neural network that translates sensory inputs into motor outputs. Brooks later added more complex input and output circuitry, as well as internal clocks, to the FSMs, producing augmented FSMs (AFSMs), which are interconnected to provide for a variety of behaviors in a multilayer structure known as the subsumption architecture (see chapter 5). The AFSMs in the Brooks architecture can be triggered by sensory events, by a clock, or a by computed func-

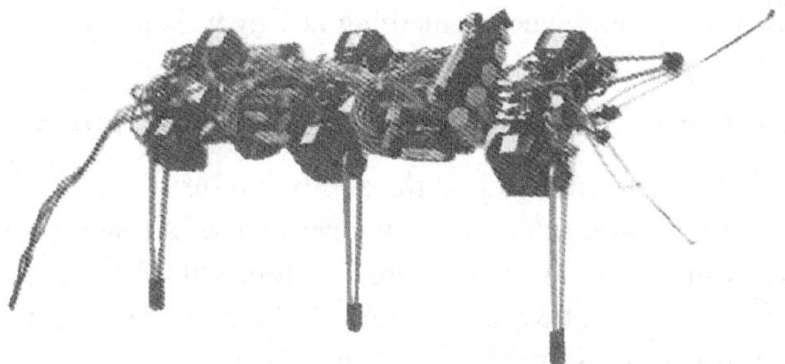

Figure 9.6
Genghis (reproduced with permission of Rodney Brooks)

tion from other AFSMs, thus leading to a rich repertoire of behaviors in robots built from them. The subsumption architecture makes it possible to augment these networks of AFSMs by adding new AFSMs in an incremental way, thus increasing the number of tasks a robot is able to perform. It should be noted that the subsumption architecture makes it possible to add AFSMs without changing any of the existing structure. The AFSMs are connected with one another using message-passing wires. The final configuration of Genghis used fifty-seven AFSMs, concerned with such functions as handling sensor inputs, moving the legs, walking, and steering. Higher-level behaviors subsumed lower-level ones, as is typical of subsumption architectures.

Genghis was about 35 cm long and weighs about 1 kg. Each rigid leg could move about its shoulder mount in 2 degrees of freedom. The sensors included motor servos providing approximate force measurements (by examining motor currents), two front whiskers, two four-bit inclinometers to provide pitch and roll information, and six forward-looking passive IR proximity sensors. On board were four linked 8-bit microprocessors. The robot was completely autonomous; power was provided by on-board batteries carried between the legs.

Genghis demonstrated the following behaviors:

- standing up
- walking
- force balancing on the legs to allow for walking on rough terrain
- leg lifting (when encountering an obstacle)
- monitoring whiskers to anticipate obstacles
- pitch stabilization, to improve its stability beyond what simple force balancing on the legs made possible

• prowling (in which the robot walked only when something nearby was moving, as sensed by the proximity sensors)

• steered prowling (which made it possible for the robot to follow a moving person)

Each behavior added to Genghis's repertoire required the addition of new AFSMs.

It is interesting that the control system resulting from the network of AFSMs was highly distributed and that behaviors such as walking were produced with little or no central coordination, in contrast with the cockroach models in the previous section, which had a strong central controller. These results emphasize the two-way interaction of biology and robotics: We can use biology to inspire our robot designs, and we can also propose new hypotheses concerning living-system behavior from observations of the behavior or our robots.

9.4.4 Rodney

As noted in chapter 6, the University of Southern California Robotics Research Laboratory fabricated a six-legged robot in the early 1990s. Each of the robot's legs had 2 degrees of freedom: elevation (rotation about a horizontal axis parallel to the body axis) and swing (rotation about a vertical axis), thus making it a 12-dof system. Each was kinematically similar to those used in the Brooks robot Genghis described in the previous section; the USC robot was named Rodney.[21] Rodney (figure 6.21) was approximately 35 cm long and about 12.5 cm wide. The legs were actuated by Futaba servo motors controlled by an on-board Motorola 68332 processor. As pointed out in chapter 6, the antennae shown in this figure were strictly cosmetic and had no useful function.

Genetic algorithms (described in chapter 6) were used to enable Rodney to learn to walk (Lewis, Fagg, and Bekey 1994). The sequence of leg movements was initially random, making walking impossible. Following the first phase of evolution, the robot developed a wave gait. As noted earlier in the chapter, in this type of gait, corresponding pairs of legs on opposite sides of the body move at the same time. This gait, commonly observed in insects, enables them to keep four legs on the ground at one time. Further evolution resulted in a tripod gait. As mentioned earlier, tripod gaits (in which three legs are on the ground at the same time) are used by insects when moving rapidly.

9.4.5 Walking-Stick Insect Models

As noted previously, in addition to the cockroach, the walking-stick insect *Carausius morosus* has been studied extensively as a model for hexapod robots. The biological

21. Naturally.

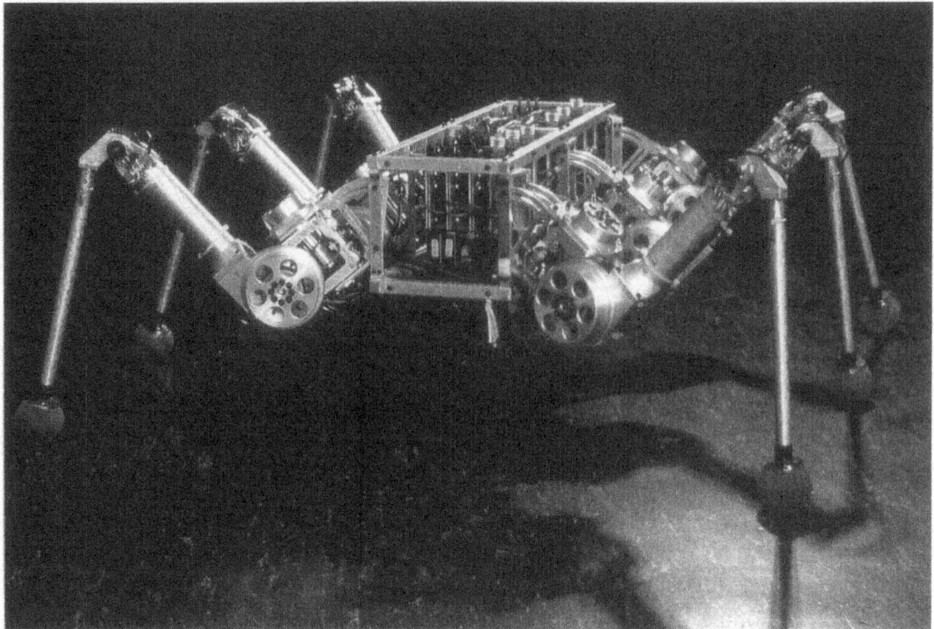

Figure 9.7
Six-legged robot inspired by walking-stick insect (reproduced with permission of Friedrich Pfeiffer)

background comes largely from the work of Bassler (1977, 1983), Cruse (1976, 1979, Cruse et al. 1995) and Dean (Dean and Wendler 1983; Dean et al. 1999); a robot based on this background was designed and constructed at the Technical University of Munich (TUM) (figure 9.7) (Pfeiffer et al. 1990; Pfeiffer, Eltze, and Weidemann 1995). The design of the legs of this robot was strongly influenced by the kinematics of the insect. The living walking-stick insect has an astounding load-to-weight ratio of the order of forty to one; that is, it can carry a weight forty times its own body weight. (If humans had such abilities, a 70 kg man could carry a weight of 2,800 kg!) Other insects (such as ants) also have remarkable load-to-weight ratios. By contrast, an industrial robot can carry less than one-twentieth of its weight. TUM was able to achieve a load-to-weight ratio of six to one in its stick insect robot through a combination of leg design, the latest in motor design (rare-earth Neodym magnets), and harmonic drives.

Figure 9.8 shows the kinematics of the robot's legs. It can be seen from the figure that the leg segments (the robot's femur and tibia) move in a plane that rotates about an axis oblique to the body. The respective angles of orientation and rotation are shown in the figure, along with the ground reaction force F_z as well as the lateral forces F_x and F_y acting on the legs.

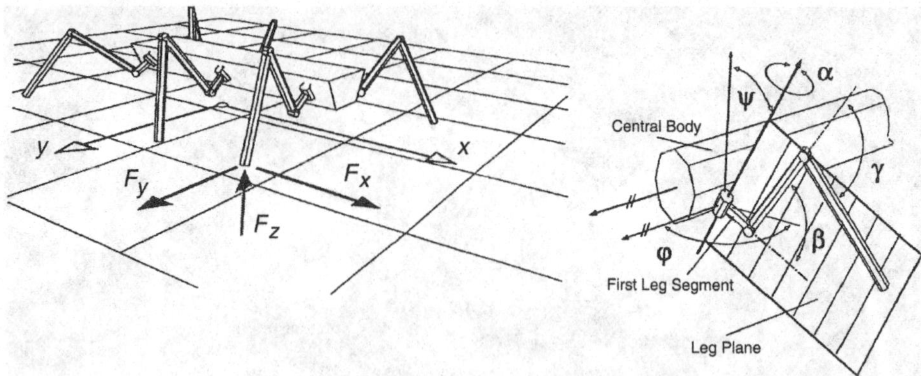

Figure 9.8
Leg design for the TUM walking machine (illustration courtesy of Friedrich Pfeiffer)

The details of the design of the legs are given in Pfeiffer, Weidemann, and Eltze 1993 and Pfeiffer, Eltze, and Weidemann 1995, including such items as gearing and motor selection. The control system is also biologically inspired and includes a leg coordination module that embodies single-leg controllers. The latter include swing-stance control as well as control of retracting and reswinging in the presence of obstacles. The implementation of these concepts required six on-board computers. The machine can walk in the both of the gaits observed in the walking stick in nature: a wave gait and a tripod gait.

The TUM hexapod is one of the most carefully designed robots in existence at this time. It is based on a skillful combination of biological inspiration and first-class engineering design.

9.4.6 Ambler, a Very Large Hexapod

In the late 1980s roboticists at Carnegie Mellon University, directed by W. L. Whittaker, built a very large, autonomous six-legged walking machine, Ambler (figure 9.9), as a research prototype for a planetary exploration robot (Bares and Whittaker 1993). Note that each of Ambler's legs had a rotary degree of freedom at the "hip," followed by 2 prismatic dof. Although it is not clear from the figure, the three legs on each side rotated about a single "stack" axis. An arched space within the robot body allowed all the legs to turn full revolutions. After six leg recoveries, each leg had completed a full revolution about its respective shaft; hence this design, pioneered by Shigeo Hirose (1994), is termed a *circulating walker.* The legs were orthogonal, in the sense that the support actuators were gravity loaded, whereas the planar actuators were orthogonal to the gravity vector and thus needed only to overcome friction and inertia in order to move.

Figure 9.9
Ambler (photograph courtesy of William Whittaker)

One of the unique features of Ambler was its ability to use long strides. Using a *circulating stride*, the robot placed the recovering legs ahead of the supporting legs; this allowed for long strides and a large foothold selection area, which was required to maintain the desired stability margins. Turning was accomplished by adjusting the step length on one side of the body.

Ambler, which had a variable height because of its prismatic legs, was 4.1–6 m high, including a range finder, about 4.5 m wide, and about 3 m long. With its maximum payload, its mass was 3,180 kg. In tests, it was able to step over a 1 m boulder. Although Ambler did not travel to Mars (despite the origin of the research project that produced it), it pioneered a number of design features for legged robots.

9.4.7 RHex, a Fast, Biologically Inspired Hexapod

In contrast to the robots presented in the foregoing sections, RHex is a hexapod that can run as well as walk. Its design is inspired by the ability of yet another cockroach, the so-called death-head cockroach *Blaberus discoidalis*, to move rapidly on rough terrain. Much of our understanding of cockroach and other insect structures comes from the work of Robert Full at the University of California at Berkeley. Cockroach legs are arranged in a sprawled posture, spreading out in order to confer static stability at rest and during low-speed locomotion (Full et al. 1998). The biological literature also shows that the insects' legs exhibit significant compliance.

RHex was designed with inspiration from biology, but with major emphasis on its engineering aspects. One of the major goals in designing the robot was simplicity. Hence, the robot was built with sprawled legs with only a single degree of freedom at the hip, but carefully tuned compliance. Thus, although its body shape and legs have no resemblance to those of the cockroach, it has stability properties similar to those of a cockroach. The legs are forced to track a feedforward reference signal; in a tripod gait, this means three legs at a time. The controller uses a proportional-derivative (PD) control law (see chapter 4). In addition, the leg-body mechanism moves like a spring-loaded inverted pendulum (SLIP). In this sense, RHex has some similarity to Raibert's robots, discussed in the previous chapter (Altendorfer et al. 2001; Saranli, Buehler, and Koditschek 2001). There is evidence in the biomechanics literature that agile locomotion, as in a bouncing gait, is accomplished by causing the body to behave as if it were mounted on a SLIP (Blickhan and Full 1993).

The robot resulting from the design features in the previous paragraph (i.e., sprawled legs, tuned compliance, feedforward signal tracking, PD control and SLIP features) runs at speeds several times its body length per second over highly irregular terrain. It has been constructed with several types of legs, including a simple "compass leg," a four-bar linkage leg as described in (Altendorfer et al. 2001), and most recently a "half-circle" leg. Figure 9.10(a) shows RHex with half-circle legs. It is able to run using either a metachronal wave gait or a tripod gait.

Recent versions of RHex are about 51 cm in length and 12 cm in height, with a body mass of 7.5 kg. The legs are about 12.5 cm long when unloaded and have a spring constant of 1900 N/m. This "springiness" is essential to obtaining the compliance characteristics required for stability during motion.

RHex is also able to move up and down stairs autonomously (Moore et al. 2002; Campbell and Buehler 2003). A waterproof version of RHex has walked successfully under water. With appropriate balancing mechanisms, RHex is able to run on its two rear legs, as apparently some cockroaches can.

9.4.8 The Sprawl Family of Hand-Sized, Fast Hexapods

The influence of Robert Full is also seen in the development of a family of remarkable, very fast, miniature hexapods developed in the laboratory of Mark Cutkosky at Stanford University. The robots are basically the size of a human hand or smaller and bear such names as Sprawl, Sprawlita (shown in figure 9.10(b)), Mini-Sprawl, and Sprawley-Davidson.[22] The names arise from the fact that these robots, likc others we have seen, achieve much of their stability through the sprawled position of their legs (in contrast with earlier robots, many of which had legs directly under the body).

22. For those roboticists who might prefer motorcycles over cockroaches.

(a)

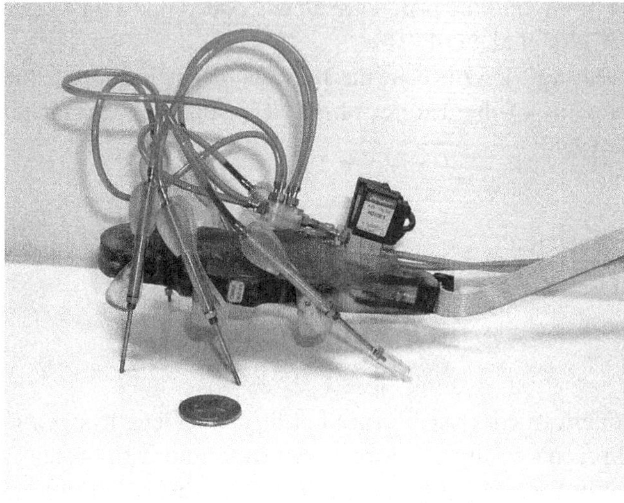

(b)

Figure 9.10
Two very fast cockroach-inspired hexapods (a) RHex fitted with half-circle legs (photograph courtesy of Martin Buehler); (b) Sprawlita (photograph courtesy of Mark Cutkosky)

In common with RHex, these robots feature compliant legs and very little active control. They have one active degree of freedom per leg, a prismatic joint (visible in the figure). A passive hip joint provides compliance about the rotary axis of the servo. The prismatic joints are pneumatically powered, using two-way valves that pressurize three pistons at once (one tripod of support). The servos are commanded and held at a certain angle. The two tripods are alternatively activated and deactivated. Whole-body motion results from this sequential activation and deactivation as well as from the compliance in the hips and the legs. Here again we see the crucial importance of "springiness" to achieving a largely passive control system inherent in the robot's structural design.

A unique feature of Sprawlita is that its body is shape deposition manufactured (SDM). This means that the entire body is one piece. The servos and wiring are embedded within the plastic of the body. Each of the legs is also one piece: the servo attachment, the compliant hip joint, and the piston are all embedded in a single piece of plastic (Bailey et al. 2000; Clark et al. 2001).

As a result of the design features described in the previous two paragraphs, the latest Sprawl design achieves many of the features observed by Full in cockroaches (Full and Tu 1991; Full et al. 1998):

- self-stabilizing posture
- thrusting and stabilizing leg function
- passive visco-elastic structure
- timed, open-loop/feedforward control
- integrated construction

Yet to be achieved are the American cockroach's speed of fifty body lengths per second and the death-head cockroach's ability to traverse uneven terrain with obstacles up to three times the height of its center of gravity without appreciable slowing. Nevertheless, it appears that recent hexapods have demonstrated many of the features of their living prototypes, suggesting that high-performance larger robots may not be far behind.

9.4.9 Final Comments on Hexapods

We started this section with a discussion of insect locomotion. However, it is evident from the examples given in this chapter that robot hexapod locomotion is not necessarily inspired by insects. In some cases, the basic inspiration and implementation are based on biology, as in the work described in section 9.4.2. In other cases the basic designs did not start out with a biological motivation, but the final product has insectlike appearance and behaviors, as with Genghis and Rodney. In the case of RHex and the Sprawl robots, biological inspiration provides the motivation and some basic features, but, as the developers of RHex say: "Pending formal understanding, empiricism and intuition about biological function remain subservient to engineering practice in our work" (Altendorfer et al. 2001, 209). Finally, some hexapod robots, like the Adaptive Suspension Machine or the robots built by McGhee have no biological foundation, except for the obvious increase in static stability and robustness arising from building machines with more than four legs.

It is also important to note that the few examples given above do not cover the entire field of hexapod robots. For example:

• Researchers at the University of Karlsruhe have developed a series of hexapods named LAURON (Legged Autonomous Robot Neural Controlled). As the name implies, these robots are controlled by means of neural networks (Berns 1994). The latest of these is LAURON III (Gaßmann, Scholl, and Berns 2001). An earlier version (LAURON II) learned leg coordination by means of an "adaptive heuristic critic" method, a form of reinforcement learning, which is discussed in chapter 6 (Ilg and Berns 1995). LAURON II was used as a platform for the study of a variety of robot learning concepts, including both neural networks and reinforcement learning (Ilg et al. 1997).

• Hamlet is a hexapod walker developed in New Zealand (Fielding, Damaren, and Dunlop 2001) that uses combined position and force control to achieve robust walking over rough terrain.

• An amphibious hexapod named Ariel (discussed briefly in section 7.8.4) was built by iRobot Corporation in Boston (with inspiration from Full's PolyPEDAL Laboratory) to search for unexploded land mines in the shallow-water surf zone. Ariel's legs have been designed to minimize drag. The robot is able to reverse its leg positions and operate when inverted. It walks like a crab. Ariel can carry a payload of 6 kg at depths of up to 25 ft (Voth 2002).

• Katharina is a six-legged robot built at the Fraunhofer Institute for Factory Operations in Magdeburg, Germany. This robot is shaped like a hexagon, with evenly spaced legs around the body. The legs are equipped with force as well as position sensors. The robot has been used for drilling operations (Schneider and Schmucker 1999).

There are many more, but those described in the preceding list and text illustrate the major principles used in hexapod design and implementation.

9.5 Locomotion in Four-Legged Animals

The locomotion of quadrupeds differs both from that of insects and from that of humans in a number of significant ways. As compared to that in humans and other bipeds, static stability in quadrupeds is enhanced by the increased number of support points that result from having four legs in contact with the ground rather than two and by the horizontal posture of the quadruped spine, which decreases the potential moment of the upper portion of the body about its center of gravity. In addition, quadruped locomotion is characterized by a number of different periodic sequences of leg movements (termed gaits), such as crawl, walk, trot and canter, which differ in the sequence in which the legs contact the ground. The transition from one gait pattern to another is related to speed and efficiency (energy consumption per unit distance traveled). In addition, the ground contact area of the paws or hooves of many quadrupeds is small in relation to their body weight, when compared with human feet. On the other hand, quadrupeds (also known as tetrapods) are less stable than hexapods. As noted previously, raising one leg from the support surface during stance requires that a quadruped animal position its center of gravity over the triangle of support formed by the remaining three legs to ensure static stability. Clearly, insects can raise two legs and retain four-point support.

The study of quadruped gait patterns began in the late nineteenth century with a remarkable series of stop-motion photographs taken by Eadward Muybridge (Muybridge 1899/1957), using a large number of still cameras triggered in sequence. Muybridge undertook the task of obtaining these photographs at the request of the governor of California to determine whether, during trot, all four of a horse's legs could be off the ground at the same time (reported by Raibert [1986]). (As Muybridge's photographs showed, they could.) Muybridge subsequently photographed the walking and running of many mammals, including humans, elephants, and cats. Muybridge defined a "footfall formula" to indicate which of an animal's feet are on the ground at each stage of the gait cycle. The fundamental work on the gaits of horses comes from Hildebrand (1965).

Quadruped locomotion is highly variable and highly dependent on the species in question. Energy consumption is nearly linear with velocity in both quadrupeds and humans.

As with hexapods, at low velocities, only one of a quadruped's legs is off the ground (in swing) at any given time, to ensure static stability. At higher velocities, the support forces of the legs are summed with inertial and gravitational forces to produce dynamic stability, which is possible with more than one leg in the air. Long-legged quadrupeds are capable of keeping all four legs in the air during certain gaits.

9.6 Four-Legged Walking Machines

A human-operated mechanical horse was patented by Lewis Rygg in 1893, as reported by Raibert (1986). This machine depicted in the patent had stirrups that doubled as pedals, so that the rider could power the machine's walking, clearly an inefficient (and probably impossible) procedure. It is possible that no actual ma-

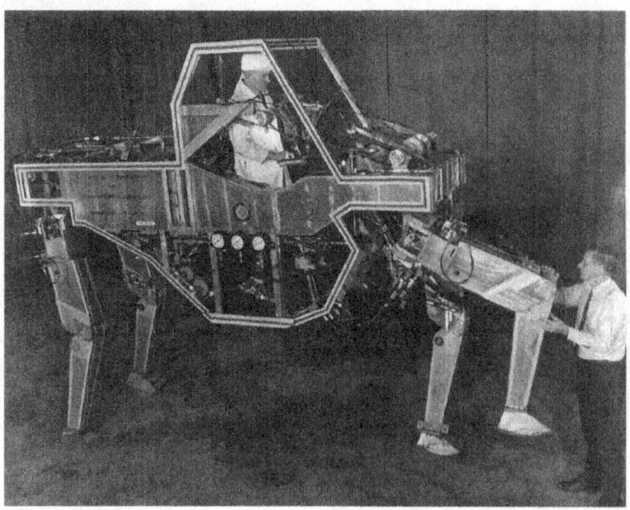

Figure 9.11
General Electric human-controlled quadruped (from Liston and Mosher 1968, reproduced with permission of General Electric Company)

chine based on the patent design was ever built. However, human-controlled, if not human-powered, machines were constructed as early as the 1950s, when General Electric Company built a large, hydraulically powered quadruped machine controlled through a master-slave arrangement by a human operator. The front legs of the machine followed movements of the arms of the operator; its rear legs followed the movements of the operator's legs. A later version of such a "quadruped truck" (shown in figure 9.11) was built at General Electric about 1968. This machine was about 11 ft tall and weighed 3,000 lb.

9.6.1 The Phony Pony

The first autonomous quadruped robot in the United States was constructed in the 1960s at the University of Southern California (Frank 1968); it was dubbed the Phony Pony (figure 9.12). The four legs of this robot had two joints each (hip and knee). They were identical to one another, and the front and back pair were mounted and controlled in the same way.[23] This was a highly nonbiological structure, since the architecture of the front and back legs of living quadrupeds and their support structure differ markedly from those of the robot. Nevertheless, the robot was capable of emulating a number of quadruped gait patterns, including crawl, walk, and

23. This fact made our laboratory the subject of numerous jokes, mostly emphasizing that this robot proved that engineers did not know the difference between the front and the back of a horse. Of course, students used somewhat less polite language.

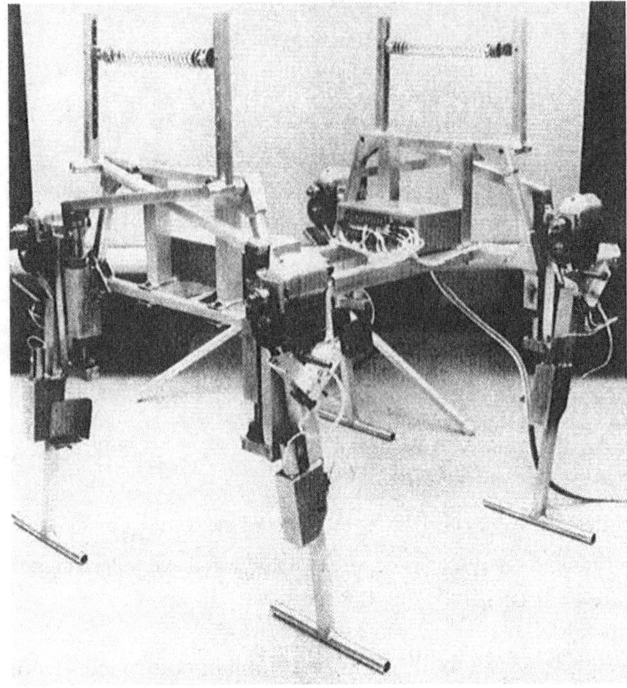

Figure 9.12
The Phony Pony (from Bekey and Karplus 1968, reproduced with permission of John Wiley & Sons, Inc.)

trot, but at a very slow speed. It was able to maintain static vertical stability in these gaits by means a unique spring-restrained "pelvic" structure and by virtue of its wide feet. The machine was not capable of high-speed gaits (such as canter or gallop) in which all four legs may be off the ground for short time intervals.

This robot was constructed before the existence of microprocessors, so it was controlled by means of a remote minicomputer (located on the second floor of an adjacent building!). The cable connecting the robot with the computer is visible in the right edge of the figure. The robot was controlled by a finite-state machine using sensory feedback on the state of its joints, without any internal model of its kinematics or dynamics.

Although the Phony Pony itself was constructed by Andrew Frank (1968) as part of his Ph.D. dissertation,[24] much of the theoretical work was done by Bob McGhee

24. I was chairman of Andrew Frank's doctoral committee, but his work was directed largely by Bob McGhee and Rajko Tomović. McGhee and I obtained a visiting-scholar grant from the National Science Foundation that enabled Tomović to spend a year at USC; Frank then spent a postdoctoral year with Tomović in Yugoslavia.

and Rajko Tomović, who proposed a finite-state machine as a model for human locomotion (Tomović and McGhee 1966). This work was then elaborated as a control method for quadruped locomotion (McGhee 1967a, 1967b) and used by Frank for control of the Phony Pony. The theory was further elaborated in McGhee and Frank 1968 with respect to quadruped "creeping gaits," that is, gaits that can be executed while keeping at least three feet on the ground at all times (Tomović 1961). (Although in theory there are many possible quadruped gaits, only six can be considered creeping gaits.) Frank and McGhee (1969) then established under what conditions a trot can be made stable using a control system whose only inputs are leg joint angles.

The control of the Phony Pony probably represents the first application of finite-state automata to robot walking, a number of years before the development of the hexapod Genghis at MIT.

9.6.2 Raibert's Quadrupeds

We encountered Raibert's unique inverted-pendulum approach to locomotion in one-legged and two-legged robots in chapter 8. Raibert also connected two of his bipeds to obtain a four-legged running machine, illustrated in figure 9.13 (Raibert 1986).

This machine was used only for gaits in which pairs of legs work together, as in the trot. If both legs of a pair strike the ground at the same time and leave for the swing portion of the cycle at the same time, they can be considered a single *virtual leg*, the control of which is equivalent to controlling a biped. By incorporating this principle in his design, Raibert was able to apply the theory of his one-legged hoppers to the quadruped.

The machine shown in figure 9.13 had an aluminum body on which all the components shown in the figure were mounted: hip actuators, computer interface electronics, and gyroscopes. Note that each leg has 2 dof: one moves the leg forward and backward and the second one (within the telescoping leg) changes its length. There is also an air spring within the leg, to control compliance in the axial direction. Sensors are used to measure leg lengths, the pitch and roll of the body, and various actuator parameters. The overall length of the vehicle was 1.05 m; its overall width was 0.35 m. The machine ran with a distinct bounce resulting from the springiness in the legs. Of course, it was connected to a remote computer using an umbilical.

9.6.3 The TITAN Robots

Shigeo Hirose of the Hirose-Yoneda Laboratory at the Tokyo Institute of Technology is a pioneer in the design and fabrication of quadruped robots dating back to the 1970s. The earliest machine, built in 1976, had long, thin legs like some spiders;

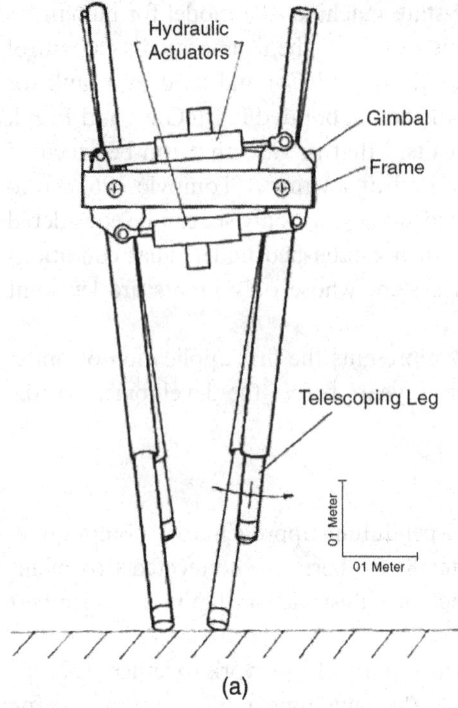

Figure 9.13
Raibert's quadruped: (a) rear (or front) view; (b) lateral view (illustrations reproduced from Raibert 1986, courtesy of Marc Raibert, reproduced with permission of MIT Press)

it was named KUMO-1. Its total weight was 14 kg and the legs were 1.5 m long. A successor to this robot, PV-II, had legs were based on pantograph mechanisms. PV-II was the first sensor-based robot to climb stairs successfully (Hirose and Umetani 1978). Since the 1980s the Hirose-Yoneda Laboratory has developed a number of quadruped walking machines, under the name TITAN; these robots were discussed briefly in section 1.8.3. The latest robots in the TITAN series are shown in figure 9.14. The major features of these quadruped robots are summarized in the following sections.

9.6.3.1 TITAN VI

TITAN VI was capable of running on flat ground (using a trot gait) as well as walking up steep stairs, as shown in figure 9.14(a). Some of its major features are the following (Hirose et al. 1991):

- The robot's legs have 3 dof. Two degrees of freedom are at the hip (swing away from the body and rotation about a vertical axis through the hip); the third degree

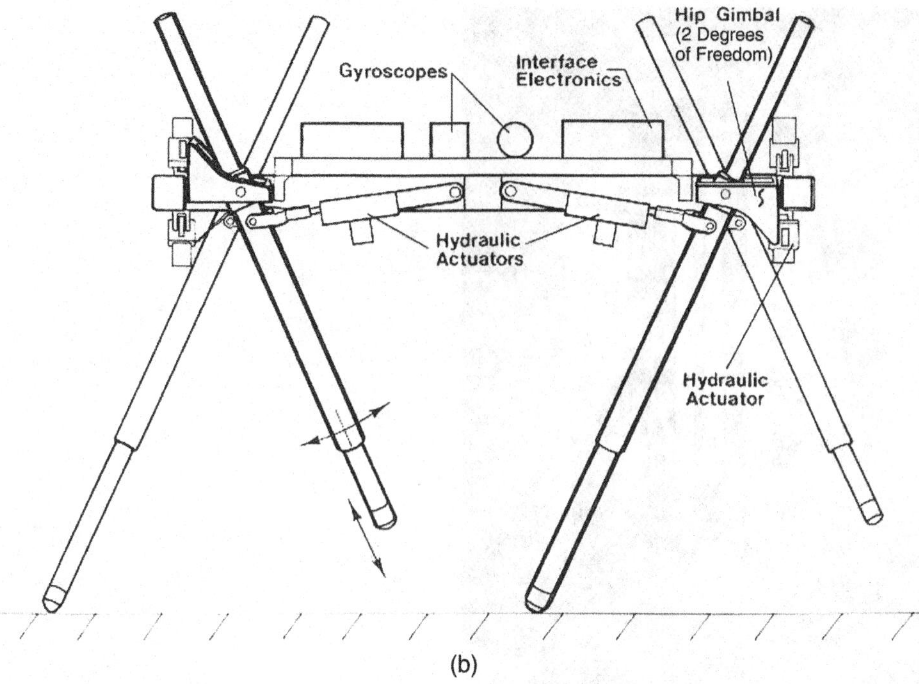

Figure 9.13
(continued)

of freedom is prismatic to facilitate stair climbing. Movement is obtained with twelve 120 W DC motors.

- The height of the machine is 1.5 m, its left-right width is 1 m, and its length is 1.5 m.
- The robot weighs 190 kg.
- Sensors for force on the robot's hips are located on the backs of its legs to detect ground contact and measure the support weight.

9.6.3.2 TITAN VII

TITAN VII was designed to assist in transportation tasks on very steep slopes. As of the writing of this book, the robot could move up slopes as steep as 70° suspended on wires. Its major features are the following (Hirose, Yoneda, and Tsukagoshi 1997):

- Prismatic actuators are used to provide the high torques required for climbing. One is clearly visible in figure 9.14(b).
- The legs themselves are extensible to assist in providing stability on slopes.

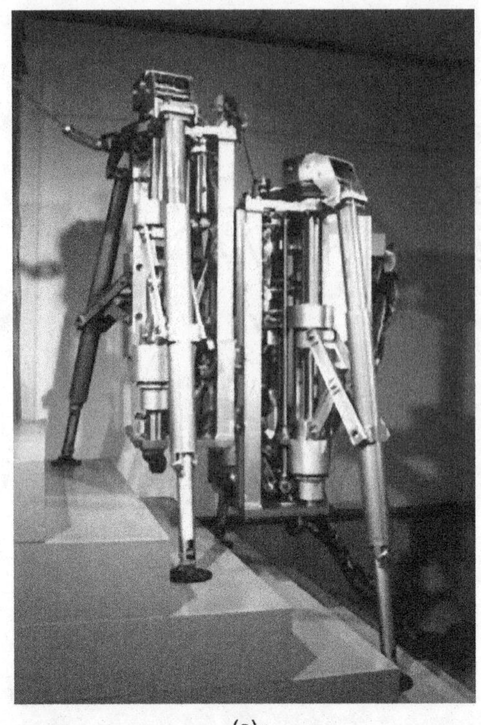

(a)

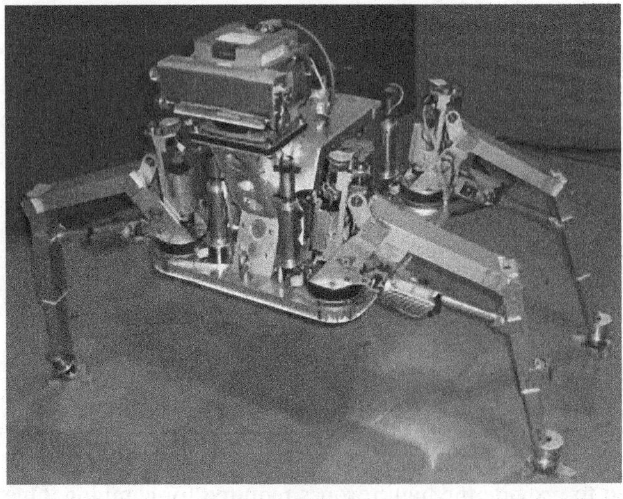

(b)

Figure 9.14
The TITAN robots: (a) TITAN VI, (b) TITAN VII, (c) TITAN VIII, (d) TITAN IX (photographs courtesy of Shigeo Hirose)

(c)

(d)

Figure 9.14
(continued)

• To allow for adaptation to irregular and sloping terrain, rocker-bogey suspensions (see section 7.2.4) are used on the feet.

• The complex ankle mechanism is equipped with a variety of sensors, to detect ground contact and applied force.

• The robot moves using a variety of the crawl gait, which has enabled it to climb a 15° slope autonomously.

• The robot weighs 60 kg.

9.6.3.3 TITAN VIII

TITAN VIII, a behavior-based robot with very tight coupling of perception to action, was designed for fabrication as simply as possible and hence at the lowest possible cost. Its major features are the following (Arikawa and Hirose 1996):

• Each leg has 3 degrees of freedom. Two of them (knee flexion-extension and hip flexion-extension are controlled by wire-pulley systems, as can be seen in figure 9.14(c).

• The third degree of freedom rotates the planar wire-pulley system.

• The robot walks with its legs jutting out on the sides, as in the figure, with the lower leg approximately vertical to the ground. This posture separates movements that are influenced by gravity from those that are not.

9.6.3.4 TITAN IX

TITAN IX was designed to assist in the identification and removal of unexploded land mines. Its major features are the following (Kato and Hirose 2001):

• At least one foot is equipped with a tool change mechanism to allow the robot to use appropriate tools, say, for cutting or manipulation.

• To provide the needed dexterity and range of motion, the knees of the robot are double jointed, as can be seen in figure 9.14(d).

• The robot is equipped with a provision for master-slave (teleoperation) control of the tool-carrying arm. Under such control, the sensed force is be reflected back to the remote master arm.

It should be evident that the Hirose-Yoneda Laboratory has extensive experience in the design and fabrication of quadruped robots, probably the most experience (and success) of any robotics laboratory in the world.

9.6.4 The Scout II, a Simple Quadruped Designed for Bounding

The Scout II (figure 9.15) was developed in the Ambulatory Robotics Laboratory at McGill University in Canada. It is approximately 40 cm high, 50 cm wide, and 85 cm long. This robot is remarkable in a number of ways.

Figure 9.15
Scout II quadruped robot (photograph courtesy of Martin Buehler)

First, the design of its legs is very simple. In contrast to some of the complex, multi-dof legs we have encountered in many of the legged robots discussed previously, the four legs of Scout II have 2 dof per leg, but only a single actuator (Poulakakis, Smith, and Buehler 2003), which controls the *active* degree of freedom: leg rotation at the hip in the sagittal plane. There is also a *passive* degree of freedom, namely, the compliance of the leg. Thus, each leg is a SLIP (which we encountered previously in connection with the hexapod RHex). This is not surprising, since R. H. Full influenced the design of both robots. Note that the SLIP character of the legs shows the influence of Raibert's work.

Second, the control of the robot's motion is largely due to intrinsic mechanical feedback, and there is no active state feedback (like velocity, for example) to ensure stability. (The issue of intrinsic mechanical feedback is raised again in section 9.6.7.) Consider the basic bounding gait, also used by Raibert. The various phases of the gait cycle during bounding of Scout II are illustrated in figure 9.16. Note that bounding is symmetrical on the two sides of the robot, so only one side (sagittal) view is needed.

The sequence of events and actions is as follows. The controller consists of two independent virtual controllers, one each for the front and back legs, respectively. These controllers detect only two states for the legs: stance and flight (rather than swing), which is analogous to Raibert's view of his monopod. Let us begin the cycle at phase C in the figure, where both legs are in flight. (Since the front leg has just lifted off the ground, it is not evident that it is in flight). During flight, the controller servos the flying leg to a desired (and fixed) touchdown angle (as, for example, with the front leg during the transition from phase C to phase D). During stance the

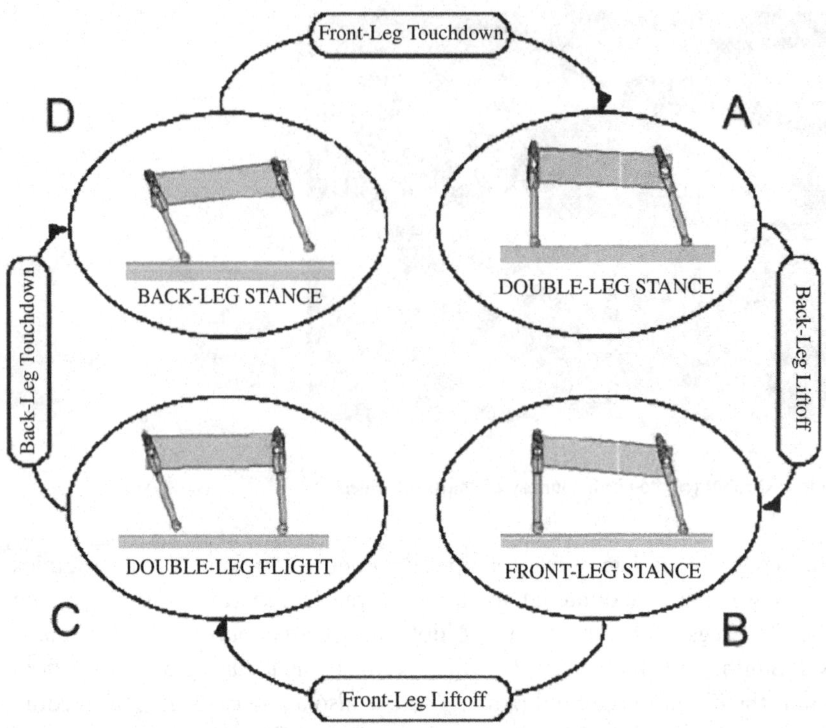

Figure 9.16
Sequence of events during bounding (illustration courtesy of Martin Buehler)

hip actuator swings the leg back with a constant torque until it reaches a preset limit. This action is evident in the change of the rear-leg orientation from phase D to phase A.

The most important part of this sequence of actions is that it results in robust bounding, with speeds in excess of 1 m/s, without any task-level feedback (like forward velocity). The bound is stable, but the stability is inherent in its dynamical structure. A mathematical analysis of this process is given in Poulakakis, Smith, and Buehler 2003, which also describes asymetrical gaits like the half-bound and the rotary gallop. Scout II is probably the only physical robot capable of galloping at the present time, though such gaits have been simulated (e.g., in Koenig and Bekey 1993).

Sensors on the robot include four linear potentiometers (one on each leg), encoders on each of the four Maxon DC motors used for hip rotation, and two range-finding lasers for pitch measurement. Current experiments are concerned with enabling Scout II to climb stairs dynamically, rather than statically (one step at a time) as other robots have done.

9.6.5 Other Research Quadrupeds

The robots described above are only a small sample of a large number of quadruped robots under development in research projects around the world. Here is a sampling of some of the others (see figure 9.17):

1. Puppy, also known as the Running Dog Project, was developed in the Artificial Intelligence Laboratory at the University of Zurich (Iida and Pfeifer 2004) to explore biomimetic principles in legged-robot design. Hence, the design was based on the anatomy of the dog, including joints, muscles, and dimensions. The body structure has a strong resemblance to that of a dog, as seen in figure 9.17(a). The skeleton consists of twenty-eight passive joints, each of which provides 1 rotational degree of freedom. Artificial muscles connect the leg and body segments. An additional six muscles control the neck. The head includes a binocular active vision system, with four servomotors and two miniature cameras. All the motors are controlled by an external computer through a communication interface with microcontrollers. The robot runs with a bounding gait.

2. Tekken I and II, as well as another quadruped named Patrush, were developed at the University of Electro-Communications in Tokyo (Kimura, Akiyama, and Sakurama 1999; Fukuoka, Kimura, and Cohen 2003), in collaboration with Avis Cohen at the University of Maryland. Cohen is well known for her work on central pattern generators, which play an important role in these robots (Cohen and Boothe 1999). The implementation of the robot includes a nervous system model with a neural oscillator that provides the robot with some ability to adapt to irregular terrain, both in walking and in running. Tekken II is shown in figure 9.17(b).

3. Warp I (figure 9.17(c)) was developed at the Royal Institute of Technology in Stockholm (Ingvast et al. 1993). The robot is about 1 m tall; it is clearly biologically inspired, with a major emphasis on mechatronic issues. It was designed to walk on rough terrain.

4. Geo II, developed at the University of Illinois Urbana-Champaign, in collaboration with Iguana Robotics, Inc., is unique among quadruped robots in that it has a flexible spine, as can be seen from the top view in figure 9.17(d). The spine has a universal joint for flexibility as well as a twist degree of freedom. Its predecessor (Geo I) was built at the University of Southern California to study evolution of gait. Geo I also had a flexible spine and could walk with outspread legs like a salamander, then tuck the legs under the body and walk with a mammalian gait. Geo II has been used to study hypotheses about the role of CPGs in the postural reflex and various gait adaptation mechanisms (Lewis and Bekey 2002).

Of course, there are many more quadruped robots, such as the SILO4 robot designed and fabricated in Spain. SILO4 will be used as a platform for experiments

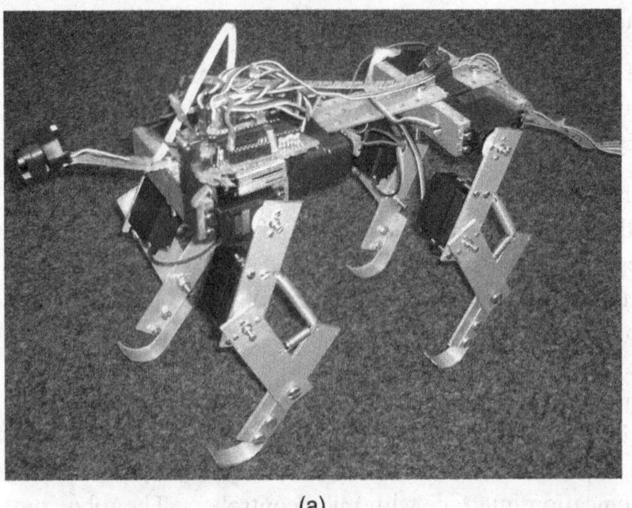

(a)

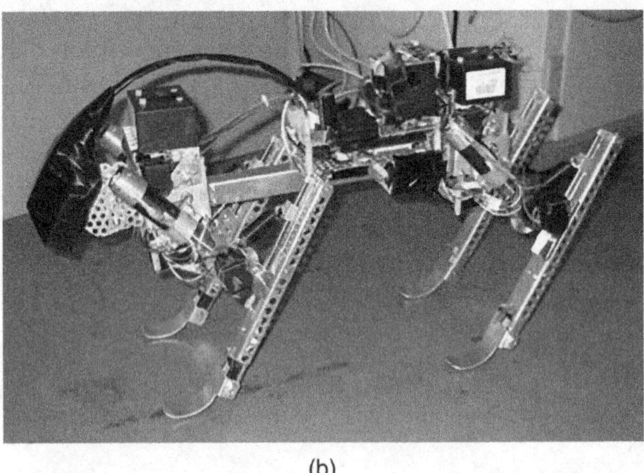

(b)

Figure 9.17
Various quadruped robots: (a) Puppy from Switzerland (photograph courtesy of Fumiya Iida and Rolf Pfeifer); (b) Tekken II from Japan (photograph courtesy of Hiroshi Kimura); (c) Warp I from Sweden (photograph courtesy of Johan Ingvast, copyright Carl Tillberg); (d) Geo II from the United States (photograph courtesy of Anthony Lewis)

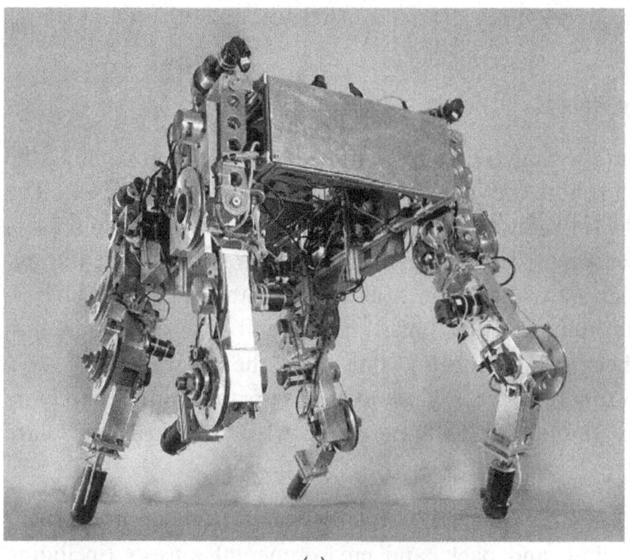

(c)

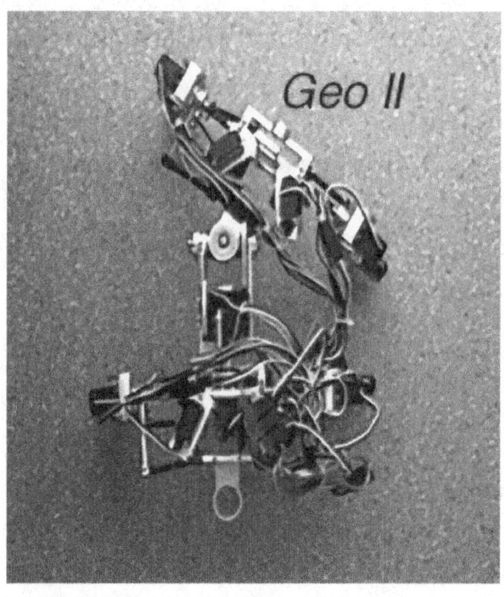

(d)

Figure 9.17
(continued)

in terrain adaptation, sensor integration, artificial intelligence, and other areas (Gonzalez de Santos et al. 2003).

9.6.6 AIBO and Other Toy Robots

The AIBO "pet robots" from the Sony Digital Creatures Laboratory are among the most interesting and creative robots ever constructed (Fujita and Kitano 1998). The third-generation AIBO ERS-7 (introduced in chapter 1) has a rich collection of sensors and behaviors. The major sensors and other hardware features are illustrated in figures 9.18(a) and 9.18(b), which show the robot from the front and back, respectively.

It is evident from the figure that the ERS-7 has a large number of sensors. Actuators provide leg, head, and body movements and thus allow the robot to engage in a variety of behaviors, from sitting on its rear legs and waving with a front leg to lying down and standing up again, autonomously. Some of the AIBO's major features are as follows:

- Inputs to the robot can be provided via speech (it has a speech recognition module), touch sensors on the head, chin, and back, and environmental sensors (including proximity, ground contact, and vision).
- The speech recognition module enables it to understand some one hundred words or phrases.
- It can be trained to recognize its owner's face and voice.
- Its behaviors can be set so that it "grows up" from puppy to adult over a period of six weeks.
- It has various facial expressions to express emotions.
- It can play with a ball or bone autonomously.

In addition to being used as pets, AIBOs are used in robot soccer competitions (Asada and Kitano 1999a). They are built with an open architecture known as OPEN-R, which provides standard interfaces for a number of input and output devices (see section 5.10).

Following Sony's introduction of AIBO, numerous other robot dogs have been produced, with names like Poo-Chi, i-Cybie, Tekno, and RoboK9. It is evident that robotic technology has a major market in entertainment, particularly as robots become more autonomous and acquire more ability to learn.

9.6.7 Stability and Control

We have previously indicated that quadrupeds are conditionally statically stable. When they lift one leg off the supporting surface, the downward projection of their center of gravity must lie within the triangle of support provided by the remaining

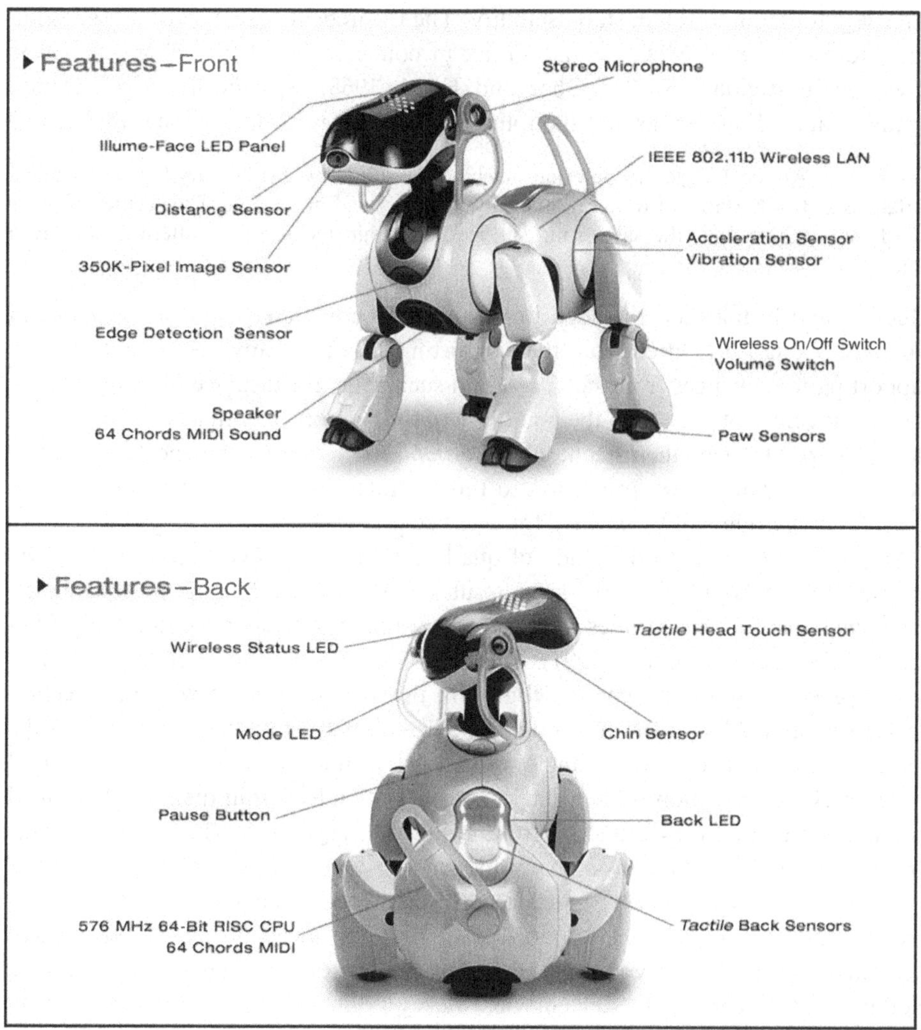

Figure 9.18
Major hardware features of AIBO ERS-7, seen from (a) the front and (b) the rear (illustrations courtesy of Sony Corporation)

three legs for them to retain static stability. The theoretical basis for this observation dates to the work of McGhee and Frank in connection with the Phony Pony, as described in section 9.6.1 (McGhee and Frank 1968). Specifically, McGhee and Frank's paper (1968, 334) states the following theorem regarding static stability:

Theorem 1 An ideal legged locomotion machine supported by a stationary horizontal plane surface is statically stable at time t if and only if the vertical projection of the center of gravity of the machine onto the supporting surface lies within the support pattern at the given time.

McGhee and Frank then extended this theorem to cover *creeping or crawl gaits,* in which only one leg is lifted from the supporting surface at any one time. Since the support pattern changes from step to step in such gaits, the distance from the projection of the center of gravity to the boundary of the support pattern also changes from step to step. One can then define a *static-stability margin* as the shortest distance from the center-of-gravity projection to this boundary. Clearly, if the gait is stable, the stability margin will be positive for every step.

Major contributions to the study of quadruped stability have also been made by Hirose (1984), who added to the basic results of McGhee and Frank by considering the permissible foot trajectories during stable walking. Consider the diagram of figure 9.19 (Hirose and Kunieda 1991). This figure shows a succession of steps taken by a quadruped using a crab-type crawl gait. In this type of gait the walking direction deviates by an angle α from the straight line determined by the front-rear axis of the vehicle. Note the sequence of triangles of support at times t_1, t_2, t_4, and t_5. At each of these times the projection of the center of the gravity falls within the triangle formed by the three legs on the ground. At time t_3, all four legs are on the ground. Hence, the gait depicted is a statically stable walk.

Hirose and Kuneida (1991) go on to investigate the reachable area of each foot and develop *generalized standard foot trajectories.* The issues of foot placement and step length are particularly important during walking on irregular terrain (see, e.g., Hodgins and Raibert 1991; McHenry and Bekey 1995).

In recent years there has been increasing emphasis on higher-speed locomotion of quadrupeds, using such gaits as trot or bound and even gallop (Schmiedeler and Waldron 1999). We have seen in some of the previously discussed examples (such as Scout II in section 9.6.4) that dynamic stability in certain gaits, like the bound, can arise from mechanical feedback inherent in the construction of the system, without any additional feedback control system as such. It now appears that some of the robots discussed in this section are capable of high-speed gaits such as the bound or gallop. It is interesting that as recently as 1999, however, it was still possible for researchers in the field to say that "given that we have not yet been able to operate any artificial vehicle in a gallop, there is a long way to go for practical applications of

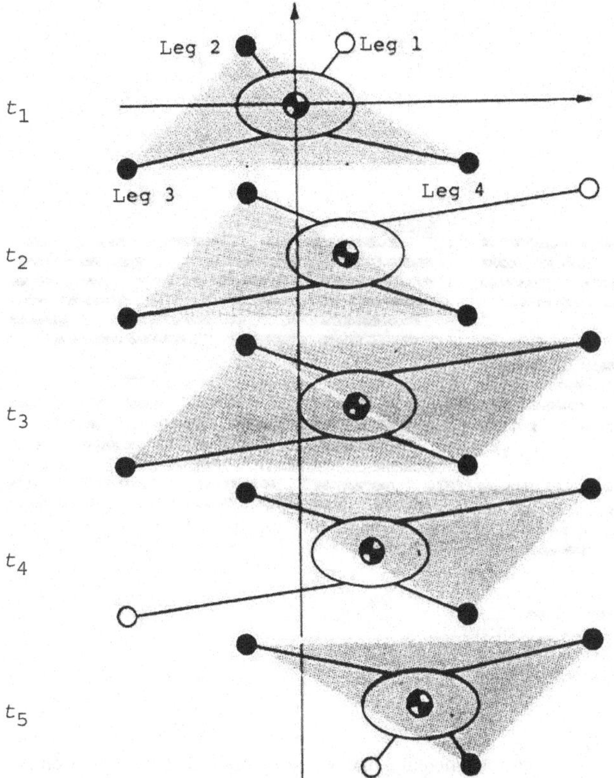

Figure 9.19
Stability support planes for a crawl gait. Filled circles represent feet in contact with the ground; open circles represent feet in swing. Triangles of support at each time point are shaded (illustration courtesy of Shigeo Hirose)

this technology" (Schmiedeler and Waldron 1999, 1233). The technology of legged machines, influenced by biology, is clearly evolving rapidly.

9.7 Finite-State Models of Legged Locomotion

Beginning with the work of Tomović and McGhee (1966), the complexity of control of legged-robot locomotion has been reduced by representing the robot's legs as finite-state machines. The point of view behind this representation regards locomotion as a sequence of events, rather than a continuous dynamical process. The finite-state machine point of view was presented earlier in the chapter, in connection with the hexapod Genghis and the quadruped Phony Pony. The advantage of this point of view is that it makes possible the description of any gait pattern as a sequence of

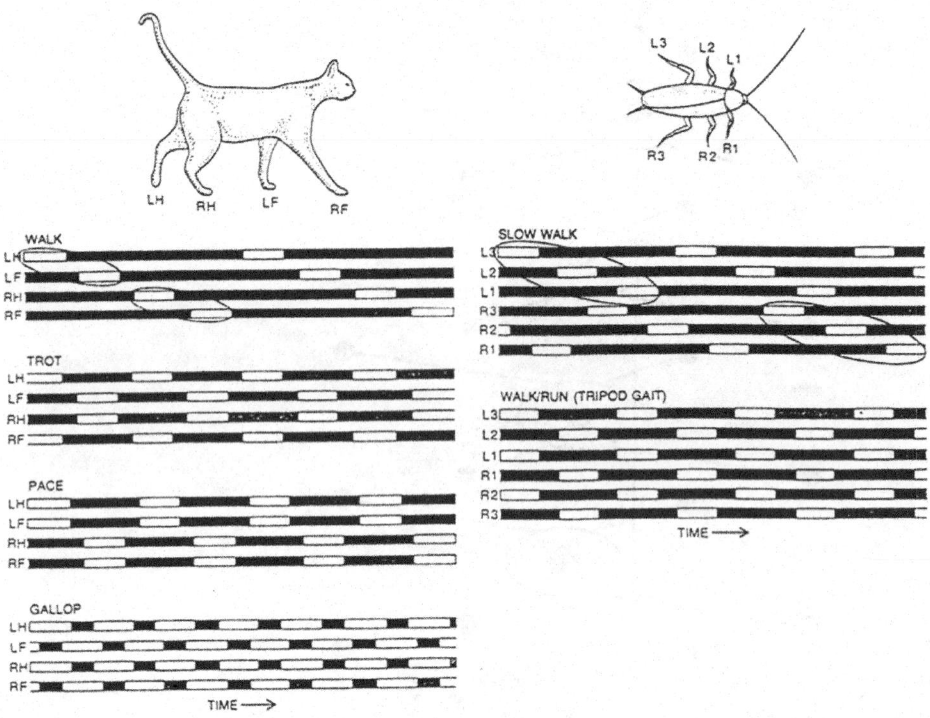

Figure 9.20
Swing and stance patterns of cat and cockroach for different gaits. Solid bars indicate that a leg is on the ground (stance); unfilled bars indicate that the leg is in the air (swing). (From Pearson 1976, reproduced by permission of the artist, Alan Iselin.)

states, and the control of the gait then requires only a finite algorithm for description. At the simplest level, the stance and swing portions of a leg's movement can be regarded as two states and used to describe the timing of various gait patterns. This approach was used in the cockroach models described in sections 9.4.1 and 9.4.2. Figure 9.20 illustrates these states for a quadruped (cat) and an insect (cockroach) for various gaits. A colleague and the author have also used the finite-state machine approach for the study of human walking and the development of human rehabilitation devices (Bekey and Tomović 1986). In a study of equine locomotion, the author and a graduate student used an augmented FSM to represent the walk, trot, and canter of a simulated horse (Koenig and Bekey 1993).

Although such a two-state description (involving swing and stance) is sufficient for classification of gaits, it is insufficient for the development of a control algorithm that enables representation of hip, knee, and other joint movements. To make such a model possible, Tomović and McGhee (1966) proposed the representation of joint

Table 9.1
Definition of cybernetic actuator

Input	Actuator state	Output
00	0	Free
01	1	Decreasing
10	2	Increasing
11	3	Locked

Figure 9.21
Meno, a quadruped robot, in a simulated Mars environment (photograph from the author's file)

motions by so-called *cybernetic actuators*, devices that transform discrete inputs into continuously varying outputs, as shown in table 9.1. Although the name given to these actuators is somewhat unfortunate, the important contribution of Tomović and McGhee's early investigations lies in the simplification that they produce, by transforming a problem of nonlinear dynamics into one of sequential decisions, triggered by events recorded by sensors. In the case of Frank's Phony Pony, these events were hip and knee angle limits, which triggered the transitions to new states.

9.8 Case Study 9.1: Control and Stability in the Quadruped Meno

Several quadrupeds were fabricated in the Robotics Research Laboratory at USC during the 1990s. Here we describe some of the major sensing and control aspects of a small four-legged machine, Meno (Sukhatme 1997), designed to walk in a simulated Mars terrain, as illustrated in figure 9.21. One of the goals of the research was to build an inexpensive, lightweight machine capable of walking in uneven terrain.

Meno was a 12-dof autonomous robot.[25] Each leg was a rotary-rotary-prismatic (RRP) 3-dof system in which the first two links were in the horizontal plane. This orthogonal leg design was modeled after Ambler (see section 9.4.6). Its body was built of aluminum tubing. Its total mass was approximately 5 kg, with a length of 0.18 m and a width of 0.15 m. Since the legs were prismatic, the height was adjustable from 0.14 m to 0.29 m. The limbs were actuated by off-the-shelf servomotors.

Although a small robot, Meno was equipped with a large number of sensors:

- *Foot switches* to provide an indication of contact with the ground
- *Foot obstacle sensors* switches to indicate when the advancing foot contacted an obstacle during its swing phase
- *Foot retraction microswitch* to provide an indication that the foot was fully retracted
- *Potentiometers* on each of the rotary joints, to measure joint angle
- *Compass* to measure yaw
- *Two-axis inclinometer* to measure roll and pitch of the body with respect to the local vertical
- *Sonar* to measure the distance to obstacles

Onboard computing was done on a custom PC board, built around a commercial Motorola microprocessor. Although computing was on board, power was provided by means of a tether for long traverses.

The robot walked in a crawl gait, with careful attention to static stability and stability margins. As with other quadrupeds, the walk consisted of two phases in which three legs were on the ground, while the fourth leg recovered and moved to a forward position. These two phases were followed by a phase with all four legs on the ground while the robot's center of mass moved forward with respect to the legs. This cycle was then repeated. When turning was desired, during the four-leg support phase, the body was turned by 5° clockwise or counterclockwise. Figure 9.22 illustrates the above sequence of states during stable locomotion. The states are labeled S1 to S6. Assume that S1 and S2 are the two phases just mentioned, in which the robot has three legs on the ground while the fourth leg moves forward, thus reaching state S3, in which all four legs are on the ground. The body then adjusts to reach state S4, and the process continues. Note that upon reaching state S3, it also possible for the robot to initiate a right or left turn by entering into a sequence of states denoted by rectangular boxes TR or TL, respectively. Each of these boxes summarizes additional state

25. I use the past tense because Meno, like so many other laboratory prototype robots, met an untimely death as a result of disuse and cannibalization of some its parts.

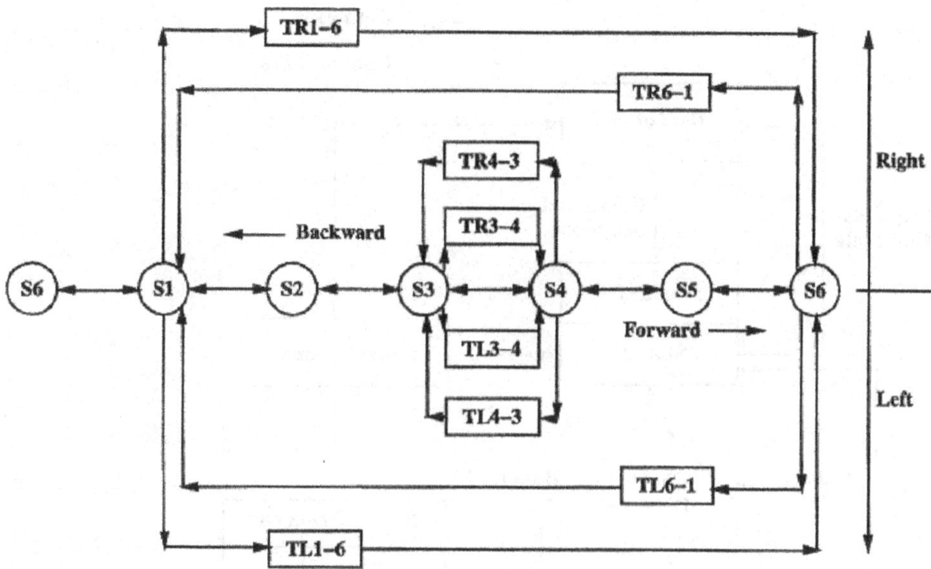

Figure 9.22
State graph representation of Meno's gait (from Sukhatme 1997, reproduced with permission of Kluwer Academic Publishers and Gaurav Sukhatme)

changes required to complete a turn of 10° in either direction. It can also be seen that turns can be initiated whenever four legs are on the ground (i.e., in states S3 or S6). Finally, it can be noted that the diagram is symmetrical, since the robot itself is symmetrical, so that it can walk forward or backward; this is shown by the arrows in the figure.

9.8.1 Stability and Control

Experiments with Meno showed that the stability margins (i.e., the distance from the projection of the center of gravity to the edge of the triangle of support) were quite small. Since there were small unmodeled errors, attempts to generate walking without feedback control almost always resulted in instability while a leg was recovering (moving forward). To solve this problem it was necessary to install a feedback control system that used the inclinometer outputs to indicate the amount of pitch and roll when a leg was lifted from the ground. If this amount was greater than a given threshold, small corrections were made at every step so as to tilt the center of mass away from the leg being recovered. This certainly slowed down the walk, primarily because of the low frequency of response of the inclinometers, but it nearly always produced a stable walk. Since this was a heuristic method, stability could not be guaranteed.

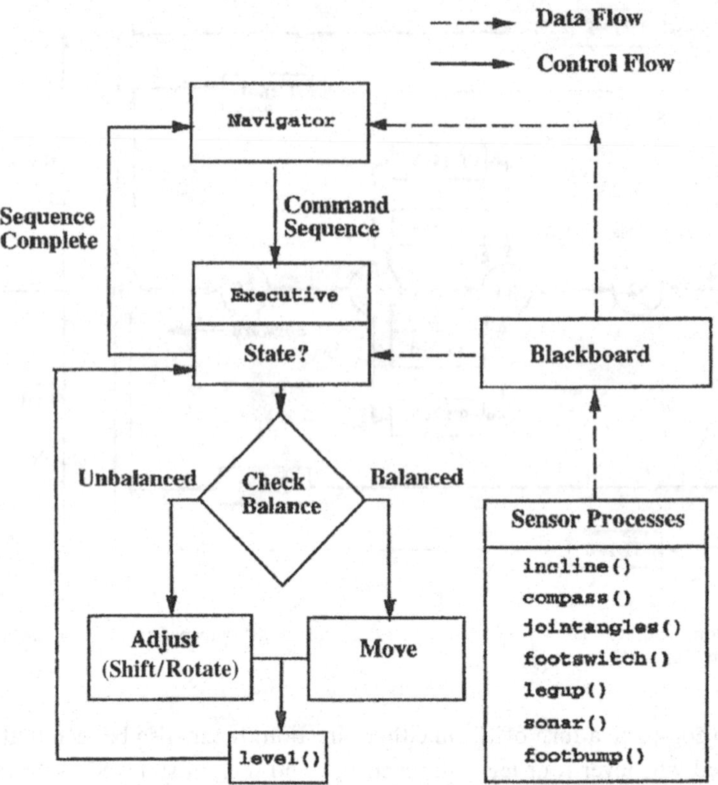

Figure 9.23
Control system architecture of Meno (from Sukhatme 1997, reproduced with permission of Kluwer Academic Publishers)

The complete control system for Meno is shown as a set of interacting processes in figure 9.23. The fast-responding sensor processes are shown in the lower right, with one process for each sensor. The executive shown in the center of the picture receives target information from the navigator (see section 9.8.2) as well as information from a global data structure (the blackboard) that is updated by the sensor processes. Given the appropriate state, the executive commands a guarded move of a leg; the leg is lifted off the ground a small distance and the inclinometers are sampled. If the pitch and roll angles are within a given limit, the leg is raised further. If at any point the angles exceed their threshold values, the leg is lowered, and the body is shifted away from the moving leg. This procedure is shown in the diagram by "check balance," with the consequence of the procedure being movement of the leg or adjustment of the body while keeping it level.

9.8.2 Navigation

Meno was designed to walk autonomously toward a goal while avoiding obstacles. There are several parts to the procedure for accomplishing this:

1. The robot is given a goal in world coordinates as well as its initial position. The kinematics of the robot are used to measure distance traveled (i.e., to use dead reckoning, as described in chapter 14).

2. As it walks toward the goal, its foot obstacle sensors may be activated by an obstacle. When this occurs, the first reaction of the robot is to raise the leg higher and attempt to step over the obstacle. If this is successful, then the traverse continues.

3. If the leg is fully retracted and the obstacle is still in the way, the robot backs up and turns, keeping track of its position and orientation, so that it can reorient itself toward the goal when clear of the obstacle.

4. The sonar on the vehicle is oriented so as to detect large obstacles that cannot be surmounted. As soon as such an obstacle is detected, the robot turns. If no obstacles are detected (either by the sonar or the foot obstacle sensors), the robot will continue forward, provided the goal is within 30° of the current forward direction, as determined by the on-board compass. If this is not the case, the robot turns in place until it is aligned with the goal direction. This strategy has been used by other reactive systems (e.g., Gat 1995).

9.8.3 Results

The strategy described above was quite successful in enabling Meno to walk stably toward a goal in the presence of multiple obstacles. The absolute value of the body pitch and yaw did not exceed 4°. While navigating toward a goal in a rock-strewn sandbox some 3 m on a side, the robot showed dead-reckoning errors between 10% and 15% of the length of the traverse.

Meno was not the most advanced quadruped built, even in the mid-1990s. It has been presented here because its design and implementation had a number of features in common with those of other quadruped walking machines. We are certain that readers can think of ways in which its performance could be improved. (We certainly hope so.)

9.9 Eight-Legged Walking Machines

Biological prototypes for eight-legged animals include spiders and other arachnids as well as lobsters and similar underwater walkers. From a biological point of view, the large number of legs of such creatures may be related to the need for greater stability than is possible with fewer legs. It may also be that eight legs are an evolutionary solution to the need to adhere successfully to vertical surfaces (for example, while weaving webs), in contrast, say, to the wall-climbing ability of geckos. In any case, there are at least thirty thousand varieties of spiders in the world, so eight legs are clearly useful for animals of a certain size.

Most investigators studying legged locomotion have worked with machines having two, four, or six legs. Hence, there are relatively few eight-legged robots. Two successful ones are the eight-legged Carnegie Mellon University robot Dante, designed for exploration of Mt. Erebus (Krotkov, Simmons, and Whittaker 1995) and the robot lobster we encountered in chapter 3 in connection with the use of shape memory alloys as actuators for its legs.

9.9.1 Dante

Dante (figure 9.24) was an eight-legged robot designed and constructed at Carnegie Mellon University under support of NASA. It was capable of climbing very steep slopes while tethered. It was deployed for the exploration of Mt. Erebus, a volcano in the Antarctic. The interior of the volcano had patches of snow, as well as large and small rocks and slopes of 50–90°. The exploration was carried out during the summer, with temperatures averaging −20°C and winds of 10–20 km/hr. This was probably the most challenging environment ever faced by a robot. The robot was approximately 3 m in length and 2 m in width and had a mass of 400 kg. The legs were pantographic in structure (the four-bar linkage is visible in the figure). The legs were arranged in two groups of four on an inner and an outer frame. This design greatly simplified the walking, since the robot basically had only one gait. To walk, all four legs of one frame would lift simultaneously and reach forward, while the remaining four legs supported and propelled the body. Thus, static stability was ensured by the four legs on the ground, assisted by a rappelling cable. To turn, the robot could rotate one of the frames with respect to the other.

The tether was used to support the robot against gravity, as well as providing power and communications. Its sensors included three cameras to allow for depth triangulation, as well as laser range finders. This was a remarkable machine, all the more so since it was designed and built over a period only ten months.

In December 1992, Dante was flown to the McMurdo station in the Antarctic, and eventually it was taken to the volcano by helicopter. On January 1, 1993, after a number of technical problems were solved, the robot was launched from the rim of the crater. It rappelled to about 10 ft below the rim when a fiber-optic line (used for communication with the crew) tore. It was impossible to repair it under the conditions at the volcano, so the mission was aborted, and the robot was dragged back to the rim and disassembled for shipment back to the United States. Although the complete mission was not accomplished, in many ways the robot was a success, since its

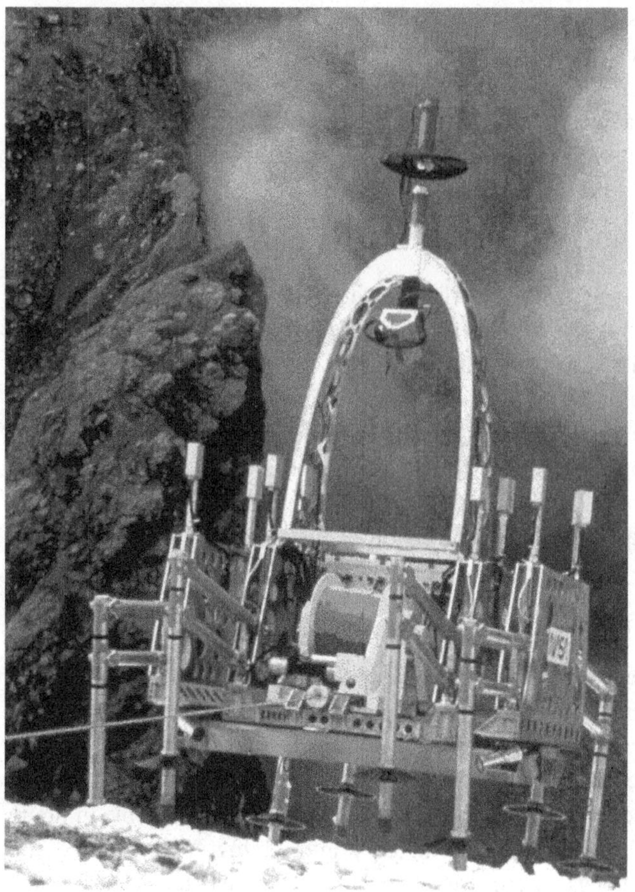

Figure 9.24
CMU volcano-exploring robot (photograph courtesy of William Whitaker)

eight-legged gait was successful, and the software and sensors worked perfectly (Wettergreen, Thorpe, and Whittaker 1993).

Following the failure of Dante to complete its mission, a new version, Dante II, was used in an attempt to explore Mt. Spurr, a volcano in Alaska. Dante II was structurally similar to Dante I, but rather than having four legs on each side, it had four legs each in front and back. This configuration provided more stability. The legs were also significantly stronger than those on Dante I. Dante II was equipped with eight cameras. While descending into the crater on July 31, 1994, one of its legs was struck by a falling boulder. Nevetheless, thanks to its remaining seven legs, the robot was able to reach the crater floor, collect gas and water samples for analysis, and send back video. During its climb out of the volcano, the robot lost its footing and fell to the crater floor. Attempts to rescue it with a helicopter also failed when the tether broke; the robot fell to the crater floor and was badly damaged. Nevertheless, the mission was considered a success because of the amount of data collected and the experience gained on the use of multilegged robots in truly harsh environments (Apostolopoulos and Bares 1995; Bares and Wettergreen 1999).

9.9.2 The Robot Lobster

Another remarkable eight-legged robot is the lobster designed and built at Northeastern University by Joseph Ayers and his associates (Ayers 2001; Safak and Adams 2002). This machine was introduced in chapter 3 as an example of the use of shape memory alloys for robot actuators. The robot is illustrated in figure 3.8. Each of its eight legs has 3 dof: the protraction-retraction (PR) joint, the elevation-depression (ED) joint and the extension-flexion (EF) joint, for a total of 24 dof. The ED joint carries the weight of the robot while a leg is in contact with the ground. Lobsters can walk either forward/backward (in the robot version, using the PR joints) or laterally (in the robot version, using the EF joints).

9.9.3 Other Eight-Legged Robots

As indicated previously, there are relatively few eight-legged robots besides the two described above. In England an eight-legged teleoperated robot named Robug IV (figure 9.25(a)) was used to study walking over unknown and irregular terrain by employing feedforward neural networks to represent terrain surface contours (Erwin-Wright, Sanders, and Chen 2003). Four of the legs are clearly visible in figure 9.25(a). Each leg has two links and four actuated joints: abductor, hip, knee, and ankle. Since there are eight legs, there are thirty-two embedded controllers, each with its own microprocessor. Pneumatic actuators are used to provide a high torque-to-weight ratio.

Scorpion was an eight-legged robot developed jointly by researchers in Germany and the United States, including Joseph Ayers (mentioned in the previous section in connection with the robot lobster) and Frank Kirchner at the University of Bremen. It is illustrated in figure 9.25(b). Its structure was modeled after that of living scorpions; it was designed to survive and travel in the Mojave Desert of California for periods of weeks, using a solar panel to recharge its batteries.

The robot was 450 cm long, 200 cm wide and 300 cm high, being about the size of a dog. It weighed 3.5–5 kg depending on its sensor and other equipment complement, including its battery, and had a maximum speed of 20 cm/s. The robot's walking was controlled based on central pattern generators and behavioral models of basic motion patterns. Actions were selected using a finite-state machine. Scorpion was equipped with a variety of sensors, including load-pressure sensors in each foot

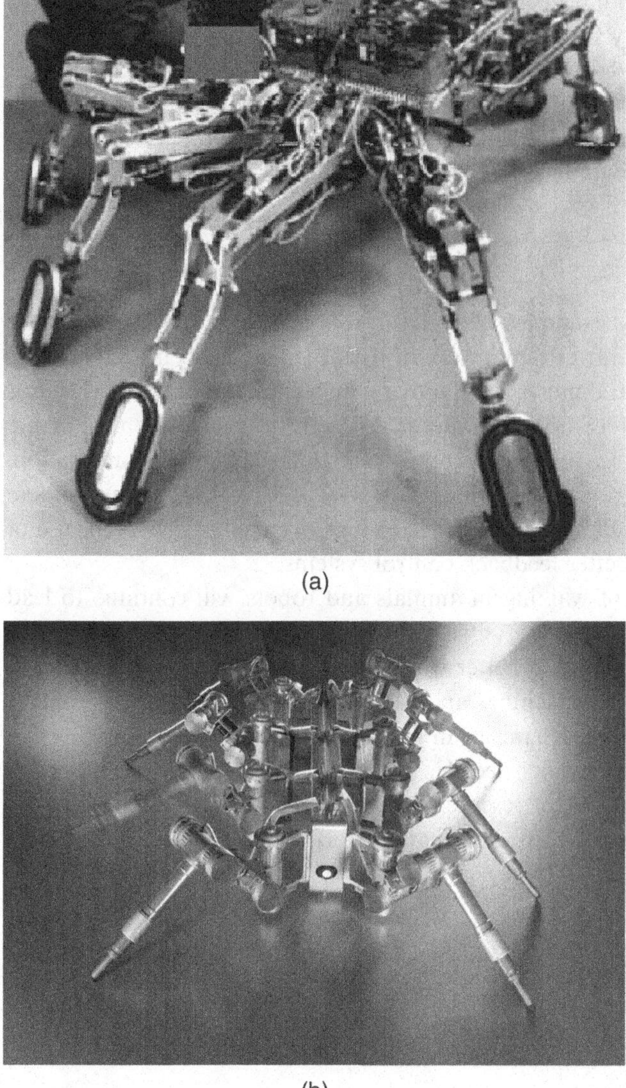

(a)

(b)

Figure 9.25
Other eight-legged robots: (a) Robug IV (photograph courtesy of Stephen Erwin-Wright and David Sanders), (b) Scorpion (photograph courtesy of Frank Kirchner, Bremen University, reproduced with permission of Dieter Klein)

and three-dimensional inclinometers, in addition to the more standard joint angle, motor current, and proximity sensors.

To the best of our knowledge, the above are the major eight-legged machines built in research laboratories up to 2003. Other small eight-legged robots have been built for teaching or demonstration, such as the OCT-1 developed by Applied AI Systems in Ontario, Canada, or a robot made from LEGO MINDSTORMS kits.

9.10 Concluding Remarks

It is evident from the many examples of walking robots discussed in this chapter that the field has moved forward dramatically in recent years. We believe that these advances, which have produced much faster moving machines, are due largely to our increased understanding of locomotion in animals. The results have been dramatic. Current generations of legged robots not only walk but run, bound, and gallop. They can move at speeds of several body lengths per second. Further, many of these advances have come as a result of structural design improvements (such as the use of compliant legs) rather than better feedback control systems.

It appears that the study of walking in animals and robots will continue to lead to fruitful insights in both directions. As we learn more about the neural control of locomotion, we will be able to build better walking machines. Conversely, the construction and study of walking machines and their control will provide hypotheses for the study of neural control mechanisms in living beings.

We expect that during the coming decade, practical and useful legged machines will begin to appear, moving out of research laboratories into the marketplace. As we have shown in this chapter, it appears that the technology now exists to make this possible.

10 Arm Motion and Manipulation

Summary

Mechanical arms designed to move and position objects are also known as robot manipulators. In this chapter we review the kinematic structures of both human and robot arms and describe the ways in which both are controlled. The current status of manipulation robots is illustrated with a number of examples. We then present a variety of coordinate systems and structures for robot arms. Mathematical descriptions of their forward and inverse kinematics are not discussed in detail; only the basic principles are presented. We then consider arm dynamics and control, including the use of neural networks for arm control. The final section of the chapter provides an introduction to arm prosthetics and orthotics.

10.1 Human Arms and Robot Arms

Robot arms, also known as robot manipulators, are mechanical structures designed to carry loads from one point to another. They are commonly used in industry, in which the major applications are welding, spray painting, palletizing, and assembly. Thus the "load" carried by the manipulator may be a welding torch, a paint spray gun, an automobile engine, or a microchip, since robot manipulators are used in the assembly of large structures like automobiles as well as of smaller structures such as electronic circuit boards. Unimation, Inc. of Danbury, Connecticut, made the first successful installation of an industrial robot arm in the early 1960s. Its hydraulically driven arm had 5 degrees of freedom and carried a payload of 1–2 kg along a predetermined path. This description of the first successful robot arm points to some of the essential elements of a robot manipulator: the number of degrees of freedom, the nature of the actuators, the payload capacity, and the control system. The preceding specification does not indicate the *structure* of the manipulator, which we will consider briefly later.

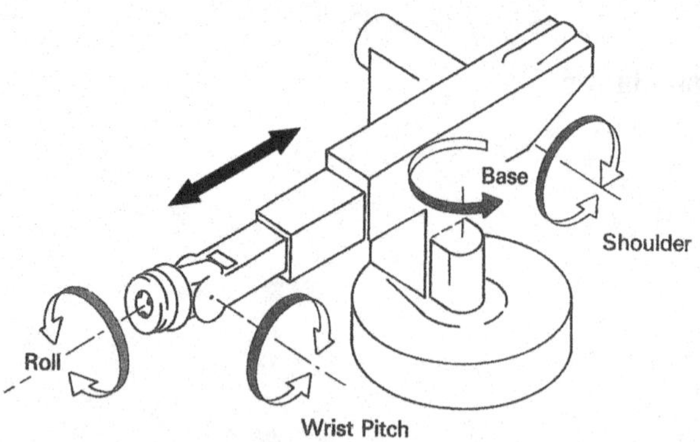

Figure 10.1
A typical robot manipulator

The structure of robot arms is frequently (but not exclusively) modeled after human arms, with rotary joints referred to as "shoulders," "elbows," and "wrists," even though the mechanical joints do not precisely mimic those in the primate upper extremity. Of course, other types of joints are possible, the most common of which is the telescoping or *prismatic* joint. Figure 10.1 shows a typical structure of a robot manipulator. Note that this manipulator has a rotational degree of freedom at the base (which turns the entire structure about a vertical axis), a rotational degree of freedom at the shoulder (which moves the entire arm about a horizontal axis), a prismatic degree of freedom that lengthens the arm, and two rotational degrees of freedom at the wrist.

To place an object in a particular position and orientation within its reachable space, a manipulator must have at least 6 dof to describe the three translational and three rotational positions. The manipulator in figure 10.1 has only 5 dof, so there are positions and orientations it cannot reach. As we show in the next section, the human arm has more than 6 dof. This is evident since (for example) it is possible to move the elbow without disturbing either the position or the orientation of the hand. Manipulators with more than 6 dof are said to possess *redundant degrees of freedom*; an elephant's trunk has many redundant dof.

The major focus of this book is on mobile, autonomous robots, rather than industrial manipulators. We include this brief chapter because many mobile robots have arms to enable them to both move and manipulate. Furthermore, there is a relationship between walking and manipulating. Anyone who has held a child's feet and let her "walk" using her arms has seen this connection. Thus, much of our understanding of degrees of freedom and joint structures in legs applies to arms as well. We believe that in the future many more mobile robots will include at least rudimentary manipulators.

In the following sections of this chapter, we first review the anatomy and control of the human arm, in order to provide a basis for evaluation of robot arms. We then consider various aspects of robot arm design and control. The subsequent section presents the basic principles of the mathematical analysis of robot arms, including both kinematics and dynamics. The final section of the chapter discusses artificial arms for people with missing or nonfunctional upper extremities, since some of these devices are based on robotic principles.

10.2 Control of Arm Motion in Humans

10.2.1 Structure and Movement

Most primates share a similar arm structure, consisting of two nearly rigid links (the upper arm and the forearm), three joints (the shoulder, elbow, and wrist), and a remarkable end effector, the hand. We consider both human and robot hands in the next chapter. In this chapter, we concentrate on the arm as an instrument for reaching and manipulating objects at some distance from the body.

Consider first the possible movements of the arm about the shoulder joint with the elbow and wrist joints being fixed, as shown in figures 10.2(a) and 10.2(b). These figures illustrate the motion of the arm with respect to the shoulder in the frontal and lateral planes. Arm *flexion* and *extension* are shown in figure 10.2(b), and figure 10.2(a) illustrates *abduction* and *adduction*. Rotation of the arm, as shown in figure 10.2(c), involves motion about both the shoulder and elbow joints. In addition, the entire shoulder can move up and down in movements known as *shrugging* or *depression*. The shoulder can also be moved forward or backward (*protrusion* or *retraction*).

Movement of the links with respect to the joints (such as the movements illustrated in figure 10.2) is produced by the contraction of muscles attached to the bones. Since muscles exert forces only by contracting, bidirectional motion of a link requires at least two muscles, with attachments on opposite sides of a joint. Some muscles span a single joint, and others span two joints, so that the relation between muscle contraction and link movement is quite complex. Furthermore, the joints are not simple hinges. Rather, they are complex structures of bone, cartilage, fibrous tissue, and ligaments, capable of supporting a variety of movements. The shoulder joint, for example, supports the upper-arm bone (the *humerus*) and enables it to move in the ways illustrated in figure 10.2.

The forearm consists of a pair of bones, known as the *radius* and the *ulna*. These two bones are connected by fibrous tissue that limits their relative movement but still allows flexion and extension (figure 10.3(a)) and rotation (figure 10.3(b)).

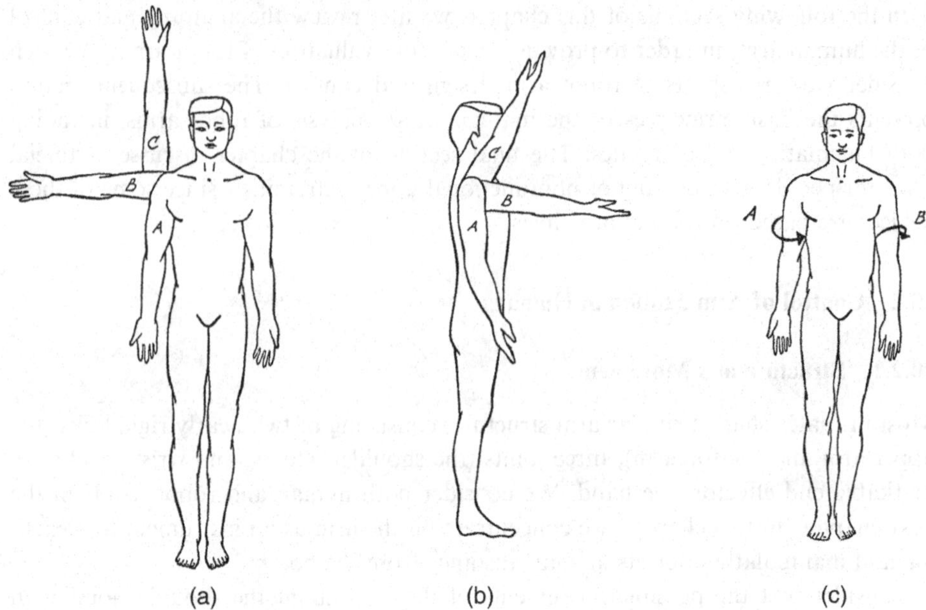

Figure 10.2
Major movements of the human arm: (a) abduction (A to C) and adduction (C to A); (b) flexion (A to C) and extension (C to A); (c) medial (A) and lateral (B) rotation (from Murray 1969, copyright 1969 by Doubleday, a division of Random House, Inc.; reproduced with permission of Doubleday)

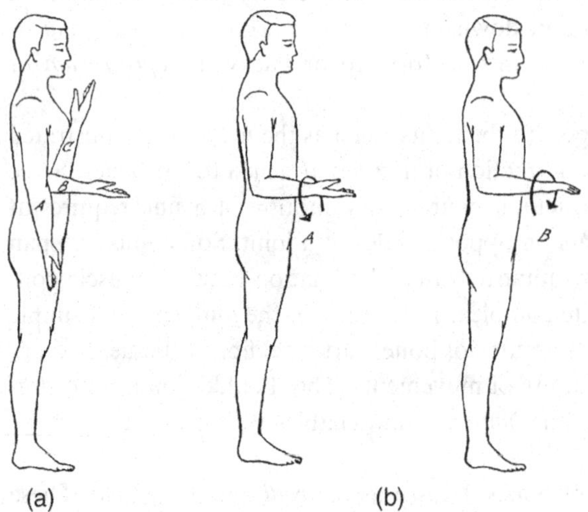

Figure 10.3
Forearm movements: (a) flexion and extension; (b) rotation (from Murray 1969, copyright 1969 by Doubleday, a division of Random House, Inc.; reproduced with permission of Doubleday)

The preceding discussion highlights the complexity of the possible movements of the human arm, made possible by the presence of multiple muscles with varying attachment points to the bones and the complex structure of the joints. There is clearly some redundancy in the system, so even with damage to a particular muscle, the system may still allow for all or nearly all of the normal movements, although sometimes with less available range or force. As with other redundant systems, there is more than one combination of movements, for example, that will enable the hand to reach a desired position and orientation.

Clearly, it is neither necessary nor desirable to duplicate the exact structure of the human arm in the design of a robot manipulator. Rather, the goal of manipulator design is simply to enable the manipulator's end effector to reach a target within a given space with the desired orientation. As we show in section 10.3, the design of these devices can therefore be considerably simpler than that of the human arm.

10.2.2 Control

One can study the control of human arm movement (also known as *motor control*) at many levels: for example, neural firings, the anatomy of muscles, the biomechanics of the musculostekeletal systems, learning of movements, and the relation of perception to action. Regardless of how we look at the system, it is evident that human arm movement is amazingly complex. Here we first examine some of the psychophysical and biomechanical aspects of movement; these results are drawn mainly from Schaal 2002.

One of the first observations that can be made (alluded to in the previous section) is that there is a great deal of apparent redundancy in the arm control system. There appear to be many more muscles than needed to produce movement with the relatively limited number of degrees of freedom present in the human arm. There are many ways in which a given target can be reached, but there seem to be certain preferred paths for the hand to follow. The path is *nearly* straight, but the velocity profile has a characteristic bell shape. In reaching for a target, the movement time MT has been found to depend both on the target width W and on the distance D to the target, according to the following empirical relationship:

$$MT = a + b\log_2\left(\frac{2D}{W}\right),$$

where a and b are constants. This expression is known as *Fitts' law*; it has proven surprisingly robust. As we show in the next chapter, reaching and grasping are correlated under certain conditions, which then removes some of the redundancy from the system.

Primates have a great deal of ability to devise new arm movement strategies when faced with new tasks, which suggests that learning needs to be considered in computational models of movement control systems. As we show subsequently, every reaching movement requires the solution of a problem in inverse kinematics, since the relation between starting and target hand coordinates needs to be transformed into joint coordinates and ultimately muscle commands. Some investigators have suggested that such a transformation requires the formulation of internal models (see section 10.9). Coordinated movement of the two arms has also been postulated to result from some form of internal dynamical model. Finally, there is the issue of hand-arm coordination (or, more generally, perceptual-motor coupling), which is another complex problem, one that is evident in such acts as juggling or interacting with a moving object (Sternad et al. 2000).

At the anatomical level, control of movement in human extremities is obtained through contraction of *skeletal* muscles (so named because these muscles are attached to the bone structure of the skeleton, in contrast with heart muscle, for example). But what controls the timing and force of the contraction to produce the desired movement? We have discussed the neural control of muscle contraction at some length in chapter 2 and in chapter 8 (the latter in connection with leg movements), so only a brief review of basic principles is undertaken here.

The muscle itself is composed of a large number of fibers, each of which is innervated with a nerve fiber, a branch of an alpha motoneuron that brings control signals from the central nervous system. A single alpha motoneuron fiber and all the muscle fibers it innervates are known as a *motor unit.* Since the degree of precise control required for a particular muscle depends on the muscles, the number of muscle fibers in a single motor unit varies. As with other neural signals, information in alpha motoneuron fibers is frequency modulated; that is, the strength of the stimulating signal depends on the frequency of nerve impulses arriving at the motor junction. It is significant that muscle is controlled in a feedback loop, and hence it depends on signals from muscle spindles and Golgi tendon organs (see chapter 2). Muscle dynamics are highly nonlinear and exhibit relations among length, velocity, and force.

Human motor control is a complex and multifaceted problem and at the current state of knowledge, it is not possible to "reverse engineer" the neuromuscular system. Even if it were possible, the resulting system would be incredibly complex. As noted previously, however, the control of robot arms need not emulate the control of the human neuromuscular system to provide those arms with the ability to conduct useful manipulation.

10.3 Robot Manipulators

As indicated previously, the first successful industrial robots were made by Unimation, Inc. The Unimation 2000 was an arm capable of moving in 2 or 3 degrees of freedom and opening and closing its gripper. It had no sensors and was used primarily for "pick-and-place" operations. Contemporary manipulators made by such companies as KUKA, Brown-Boveri, Hitachi, Panasonic, and Fujitsu are equipped with sensors (vision and touch), can move in 5 or 6 dof with great precision, may use a variety of end effectors (rather than simple grippers), and are programmed externally using computers. For a comprehensive analysis of robot manipulators and their applications, the reader is referred to Craig 1989; Klafter, Chmielewski, and Negin 1989; and Niku 2001. In this section we provide a brief overview of robot structures, sensors, actuators, and control.

10.3.1 Manipulator Structures

We have encountered robot manipulator structures previously in chapter 3 in connection with a discussion of actuators. The structure of robot manipulators consists of links and joints. The joints may be revolute (R) or prismatic (P), so that robot manipulator structures may be described by such codes as RRP or RRR indicating the joint sequence form base to wrist. Examples of two typical structures are shown in chapter 3. The manipulator in figure 3.2 displays four rotary joints, so it is denoted by RRRR. The manipulator in figure 3.3 is denoted by RPPR.

The structure of a manipulator can also be described in terms of the axes along which movement is possible. Thus, we can have rectangular robots, cylindrical robots, spherical robots, and so on. Regardless of the specific structure, it is evident that none of the robots illustrated in figure 10.4 can reach every point in its surrounding space. The design of the joints puts constraints on the ability of a robot arm to reach certain points in the space around the robot. Hence, any work for the robot must be located within its accessible space. Note also that none of the three-coordinate systems depicted in this figure resemble those governing the human arm.

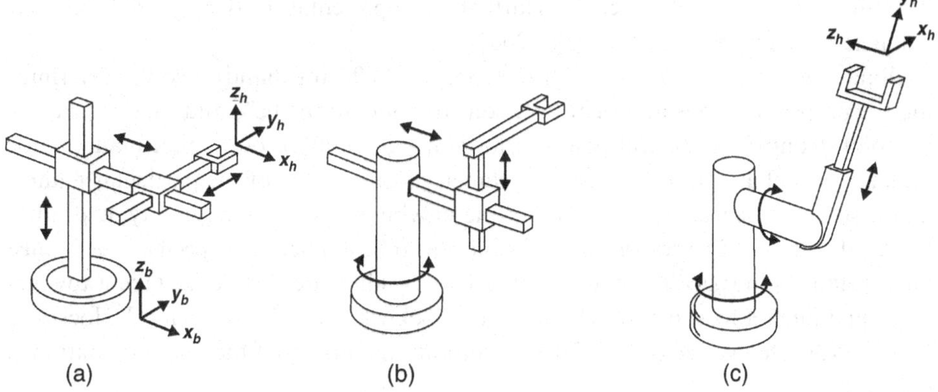

Figure 10.4
Robot coordinate structures: (a) rectangular, (b) cylindrical, and (c) spherical

10.3.2 Sensors

Contemporary robot manipulators are equipped with vision to enable them to adapt to changes in the environment in which they operate. Prior to the use of vision systems (and integrating them into the control of manipulators), parts had to be placed very precisely on conveyor belts, or the robot arm would not pick them up. Vision processing enables the arm to find the parts and move toward them accordingly. Tactile sensors on the grippers or end effectors enable the system to grasp parts just firmly enough to avoid slippage, but not so tightly as to fracture the object being held. (See chapter 11 for a further discussion of grippers and other end effectors for robot manipulators.)

10.3.3 Power and Control

Most robot manipulators are driven (actuated) by electric motors located at the joints. Very large industrial robots or cranes may employ hydraulic drives to obtain the torques needed to handle large and heavy objects. Small manipulators, such as those found attached to mobile robots, are generally driven by DC electric motors.

10.3.4 Coordinate Frames and Transformations

It can be seen in figure 10.4 that the manipulator hand or end effector moves with respect to a coordinate system fixed to itself. An inertial coordinate system is shown at the base of figure 10.4a, and is assumed to apply to the other two structures as well. To determine hand movements with respect to the inertial coordinate system, it is necessary to perform a series of coordinate transformations, moving backward from the hand frame, joint by joint, to the base frame. As the manipulator structures become more complex, these transformations also become increasingly messy. To facilitate these transformations, it is common to use a standard set of notations and definitions, known as the Denavit-Hartenberg representation (Craig 1989; Klafter, Chmielewski, and Negin 1989; Niku 2001).

Obtaining a desired movement of the robot end effector (hand) involves determining the proper rotations and/or translations to apply to the base and each intermediate joint. Control of the end-point position and orientation of a robot manipulator requires careful attention to issues of geometry (kinematics) of the particular manipulator structure as well as the inertia of the moving mass, friction, gravity, and other forces (dynamics). The transformations must also be applied in a specific order, since finite rotations in space are noncommutative. To illustrate this point, take a toy airplane and pitch it forward by 90°, then yaw clockwise by 90°, and then roll clockwise by 90°. Now change the order of these rotations and note that the final orientation is different.

Before we examine issues of manipulator kinematics and dynamics in more detail, let us look at some typical machines and their features.

10.4 Some Typical Robot Arms

We encountered three industrial manipulators in chapter 1, shown in figures 1.1(a), 1.7(a), and 1.7(b). We begin our examination here with another look at one type of industrial manipulator and then review several other types.

10.4.1 The KUKA KR 16 Multipurpose Industrial Robot

The KR 16, illustrated in figure 10.5, is an industrial robot manufactured by KUKA Roboter Gmbh in Germany. It is a 6-rotary dof manipulator, with rotation of the base about a vertical axis, rotation of the "shoulder" and "elbow" joints, and 3 dof in the wrist. It is capable of carrying a payload of 16 kg with a maximum reach of 1.6 m. The maximum angular speed depends on the axis, and ranges from about 150°/sec for the base to over 300°/sec at the wrist. The robot is multipurpose, being suitable for material handling, machining, palletizing, assembly, welding, and other applications, depending on the choice of end effector and availability of software. Similar robots are manufactured by other companies, such as Panasonic, ABB (Asea Brown Boveri), and Fanuc.

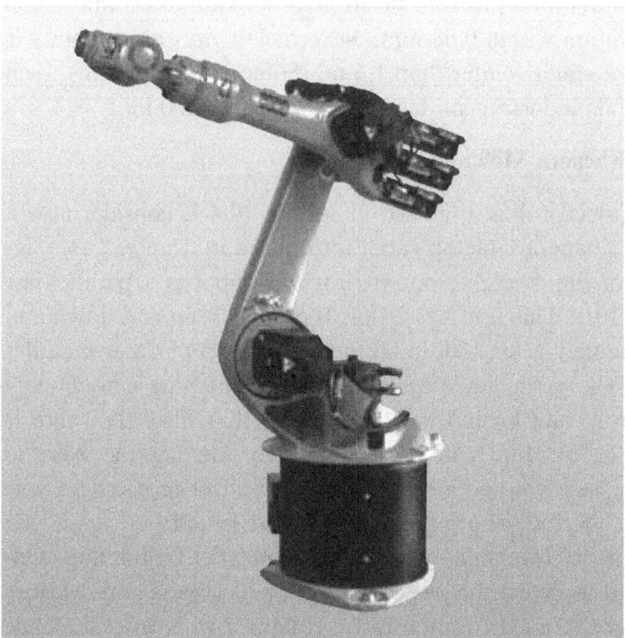

Figure 10.5
KUKA KR16 industrial robot (photograph courtesy of KUKA Roboter GmbH)

Figure 10.6
Gripper for Khepera robot (photograph courtesy of K-Team, S.A., Lausanne, Switzerland)

Position accuracy for manipulators is generally defined in terms of their repeatability. The KR 16 arm has a position repeatability of ± 0.08 mm. This means that the arm can *return* to within 0.08 mm of a previous position; it does *not* mean that it can reach a new commanded position within 0.08 mm. Nevertheless, this repeatability is remarkable for a robot with a reach greater than 1.6 m. (Some industrial arms, such as those used in the automobile industry, can handle payloads of 1000 kg.)

10.4.2 The Gripper on the Khepera Mobile Robot

In contrast to the large industrial robot discussed in section 10.4.1, consider now a small research machine, the Khepera tabletop robot introduced in chapter 1 (see figures 1.1(d) and 1.10). One of the available options for this robot is a parallel-jaw gripper, illustrated in figure 10.6. This arm has 2 dof: It can pitch up and down and close or open the jaws. It mounts as an additional module on top of other modules with the same form factor. The pitch range exceeds 180°, so the jaws can actually touch the ground both in front and behind the robot. Note that since the robot is mobile, additional degrees of freedom can be obtained by moving the base to approach an object to be grasped from an arbitrary angle, in effect providing a yaw degree of freedom. However, the robot's gripper cannot rotate in roll.

The gripper turret contains its own microprocessor for position regulation. Onboard sensors allow the robot to detect the presence of gripped objects and measure their electrical resistivity. The robot's drive motors are equipped with absolute angle encoders, which enable them to measure the size of a gripped object. The payload capacity is 50 g.

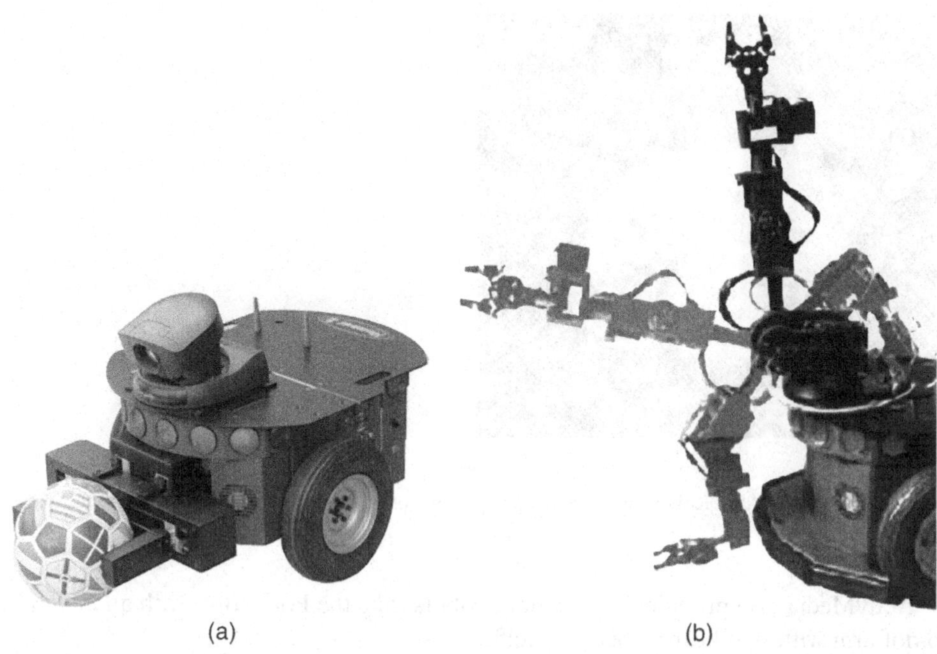

(a) (b)

Figure 10.7
Arm and gripper for Pioneer robots: (a) Pioneer 2 floor-level parallel-jaw gripper; (b) Pioneer 3 5-dof research arm (photographs courtesy of ActivMedia Robotics)

10.4.3 Arms and Grippers on ActivMedia Research Robots

ActivMedia robots (see figures 1.1(b) and 1.8) are in wide use in robotics research laboratories in the United States. ActivMedia Robotics makes two types of manipulators available as options on these robots, as illustrated in figure 10.7.

The Pioneer 2 parallel-jaw gripper is similar to the Khepera gripper, but instead of a rotary (pitch) degree of freedom, it has a vertical (z-axis) degree of freedom, enabling it to rise 9 cm with a payload of 2.5 kg. Front and rear break beams between the jaws indicate when an object is in grip position. The jaws are 3 cm tall and 9.5 cm deep. The grasping pressure is under software control and varies between 0.5 lb (170 g) and 5 lb (1.7 kg). The jaws are actuated by DC motors to move to fully open or fully closed position in 3 seconds.

The Pioneer 3 arm has 5 dof plus grip, all driven by DC motors. The arm can reach up to 50 cm from the center of its rotating base to the tip of its closed fingers, allowing the robot to pick up objects from the floor. Its joints are rotating base, pivoting shoulder, pivoting elbow, rotating wrist, pivoting gripper mount, and gripper fingers. The payload is 150 g. The arm can move from full extension to full flexion in 1 second. It has a positional repeatability of ± 1 cm.

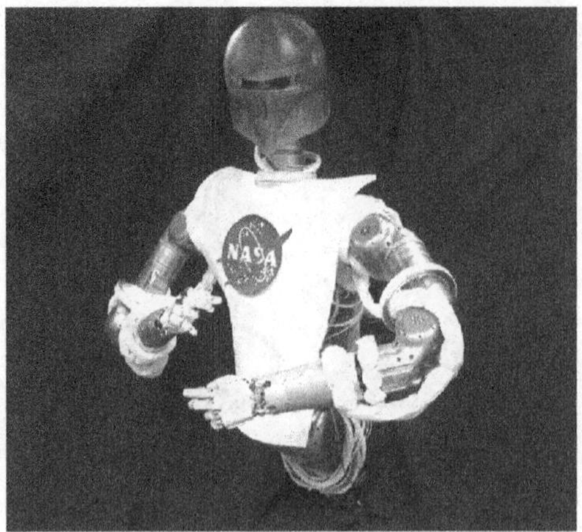

Figure 10.8
Arm structure of NASA Robonaut (photograph courtesy of Robert Ambrose, NASA Johnson Space Center)

ActivMedia also provides some mobile robots (like the PowerBot) with an integral 6-dof arm with significantly better specifications.

10.4.4 Arms on Humanoid Robots

The arms and grippers on the robots described in the preceding sections are strictly functional. Arms on humanoid robots are designed to mimic both the function and the gross structure of the human arm. As we indicated earlier in this chapter, the detailed anatomy and control of the human arm are very complex, and it is not necessary to duplicate them to obtain comparable appearance and degrees of freedom in robotic arms.

Consider first the NASA Robonaut, illustrated in figure 10.8. (This robot is discussed in greater detail in chapter 13. Here we concern ourselves only with the structure of its arms.) The anthropomorphic structure of the Robonaut's arms is apparent from the figure. Each arm has 5 dof plus 2 additional dof at the wrist. It is human in scale, with a one-to-one strength-to-weight ratio. The kinematics of the arm are Cartesian for ease in computation of inverses and control. Each joint has sixteen sensors, including dual six-axis load cells to enable the robot to monitor forces interacting between its arms and the environment it contacts.

As is shown in chapter 13, major advances in humanoid robotics have come from Waseda University in Japan. One of the issues addressed in the design of the Waseda mechanical impedance adjustment (MIA) manipulator (figure 10.9) is minimization of possible dangers to humans in the environment. Ideally, humanoid arms are lightweight manipulators that can provide strength while exhibiting compliant motion.

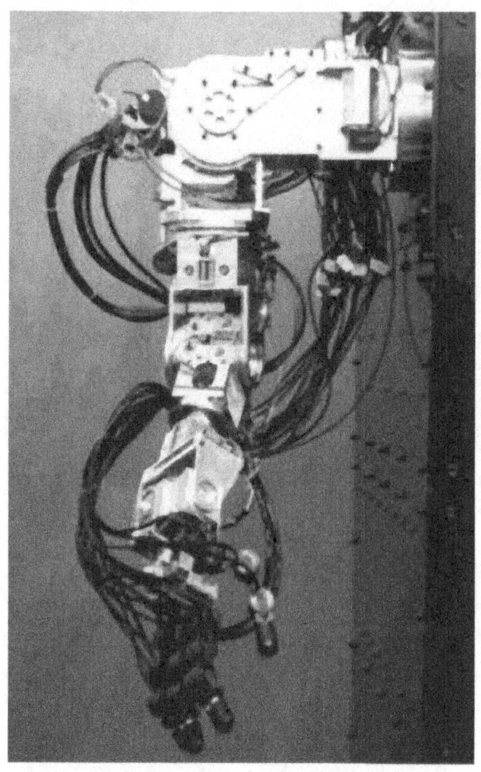

Figure 10.9
MIA arm developed in the Sugano Laboratory at Waseda University (photograph courtesy of Shigeki Sugano)

Seeking a compliant, yet strong robot arm, Waseda developed a 7-dof anthropomorphic manipulator that includes a shoulder, elbow, and wrist, all with MIA. The arm in the figure was built for a humanoid robot named WENDY (Waseda Engineering Designed Symbiont) designed for human-robot symbiosis.

Instead of using an active (motor-driven) approach to compliance, in which performance is limited by the response of servo motors, Waseda uses MIA, a passive compliance control method in which a linear spring and brake dynamically adjust the compliance in each arm. The most impressive feature of the MIA is that an extremely simple servo system is required to adjust the compliance or damping coefficient, because each component takes charge of the function independently. This control method can change the spring constant and damping coefficient in the joints to those appropriate for interactive tasks. The result is a force-controlled robot that can safely cooperate with humans while carrying out advanced dextrous manipulation tasks (Morita, Iwata, and Sugano 1999, 2000).

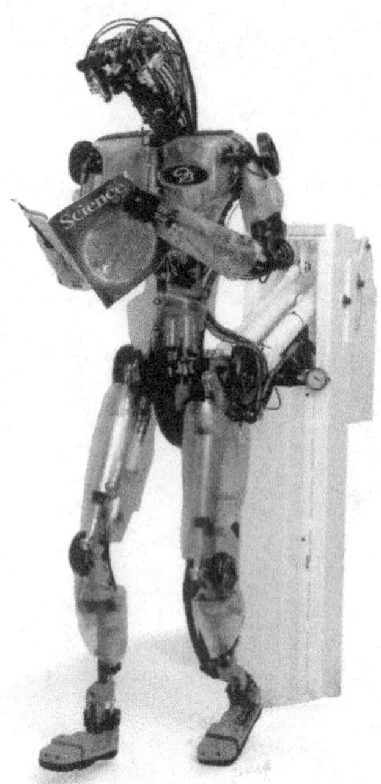

Figure 10.10
DB robot, manufactured by Sarcos Inc. (photograph copyright ATR Computational Neuroscience Laboratories, courtesy of Mitsuo Kawato)

Consider now the DB (Dynamic Brain) robot, developed jointly by the ERATO program in Japan and Sarcos, Inc. in Salt Lake City, Utah. The robot, built by Sarcos, is shown in figure 10.10 (see also chapter 13). This robot has arms with shoulders, elbows, wrists, and simple "hands" without fingers. The arms have 7 dof:

- arm flexion and extension
- arm adduction and abduction
- medial and lateral rotation
- elbow flexion and extension
- wrist flexion and extension
- wrist abduction and adduction
- wrist rotation

The first four of these degrees of freedom can be compared with the human degrees of freedom shown in figure 10.2. As indicated in chapter 13, this robot is hydraulically driven. Each degree of freedom has a position sensor and a load sensor (Atkeson et al. 2000).

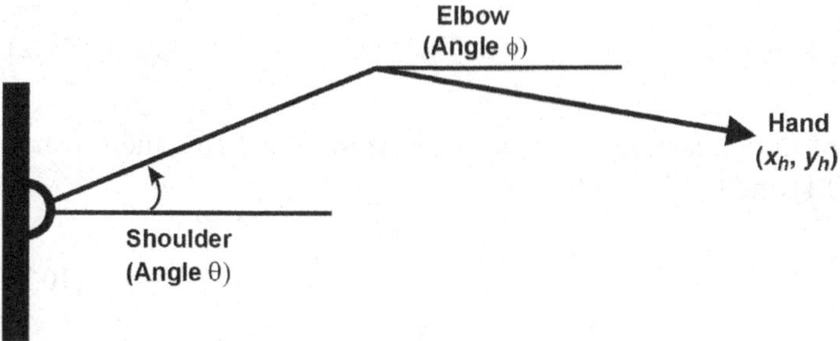

Figure 10.11
Two-link planar manipulator

10.4.5 Other Manipulators

In the foregoing sections, we have illustrated some of the existing manipulators. Particularly in the area of industrial robots, the number of available robot manipulators is quite large. As with other examples in this book, we have attempted to illustrate principles, without an exhaustive coverage of a large field.

10.5 Forward Kinematics of Manipulators

Consider the 2-dof manipulator shown in figure 10.11. If we rotate the shoulder and elbow joints the end point of the hand location (x, y) will change. In fact, the only way to change the hand position is by applying torques to the joints, since those are the only possible actuation inputs. The transformation from these angles (θ, φ) to the resulting hand location (x, y) is termed *forward kinematics*. The term refers to the geometric relationship between coordinate axes, independent of dynamics (i.e., inertia, friction, or applied torque). (The latter are treated in connection with dynamic transformations, as discussed in section 10.7.)

For the simple manipulator in figure 10.12, the computation is straightforward. We begin with the rotation of the shoulder joint, θ. Let us place the origin of the coordinate system at the point where the manipulator attaches to the wall. Then it is clear from the figure that the elbow position is given by

$$x_e = l_1 \cos\theta,$$
$$y_e = l_1 \sin\theta, \tag{10.1}$$

where l_1 is the length of the shoulder-to-elbow link. Similarly, the hand position is given by

$$x_h = x_e + l_2 \cos \varphi,$$
$$y_h = y_e + l_2 \sin \varphi, \tag{10.2}$$

where φ is the angle of the forearm with the local horizontal and l_2 is the forearm length. Combining (10.1) and (10.2) yields

$$x_h = l_1 \cos \theta + l_2 \cos \varphi,$$
$$y_h = l_1 \sin \theta + l_2 \sin \varphi. \tag{10.3}$$

Thus, given the joint angles and the segment lengths, it is possible to compute the resulting coordinates of the end effector. In general, the motion will take place in three dimensions, so the hand position will be given by (x, y, z) and its orientation in space by the three angles. To be able to control these 6 dof, an RRR manipulator will require six independent rotations. These may be obtained through a number of alternative designs. For example, the shoulder may have 2 dof, the elbow 1, and the wrist 3.

The transformation presented in equations (10.1)–(10.3) can be generalized to an arbitrary number of joints. Assume that there are three joints and that we need to define the hand location in x, y, and z. Let us denote the joint angles by θ_i, $i = 1, 2, 3$, and the Cartesian coordinates by $x = x_1$, $y = x_2$, $z = x_3$. Then we can define the vectors

$$\boldsymbol{\theta} = \begin{pmatrix} \theta_1 \\ \theta_2 \\ \theta_3 \end{pmatrix}, \quad \mathbf{x} = \begin{pmatrix} x_1 \\ x_2 \\ x_3 \end{pmatrix},$$

and replace (10.3) with the general relationship

$$\mathbf{x} = \mathbf{F}(\theta), \tag{10.4}$$

where **F** indicates a vector-valued function.

A typical industrial manipulator may have 3 rotational degrees of freedom at the wrist. There are several representations for these rotations; the most common are pitch-yaw-roll (PYR) and Euler angles. Detailed analyses of manipulator kinematics are found in such references as Craig 1989; Klafter, Chmielewski, and Negin 1989; and Niku 2001.

10.6 Inverse Kinematics

If we desire to place the end point of the manipulator of figure 10.12 in a desired goal location (x_g, y_g), we need to invert the relationships in the previous section to calculate the joint rotations that will accomplish this task. The transformation we seek is

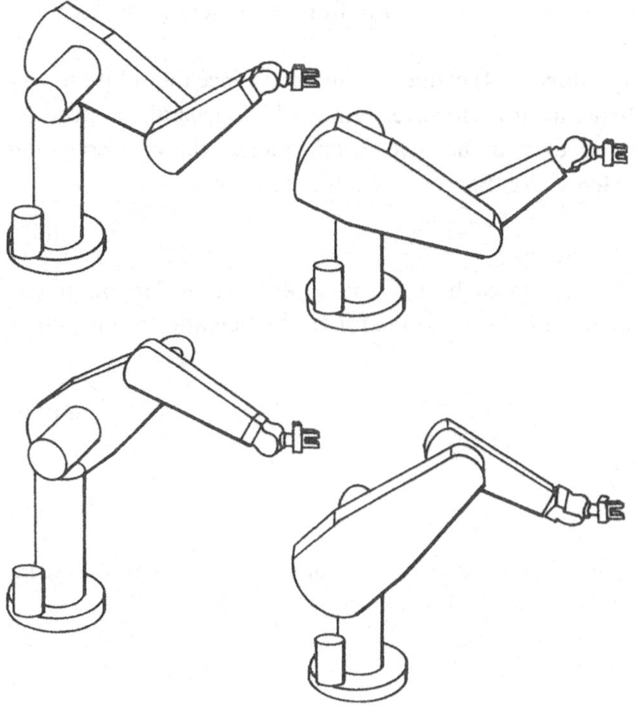

Figure 10.12
Alternative manipulator configurations yielding the same end effector position and orientation (adapted from Craig 1989)

from the goal position to the joint angles. Clearly, this is the usual goal of manipulation: We desire to reach some point in space, and we need the joint angles that will enable us to achieve this goal. Computation of this transformation is referred to as *inverse kinematics.*

We have noted previously that the solution of the inverse-kinematics problem may not yield a unique solution. For example, all the four joint arrangements of the RRR robot shown in figure 10.12 place the hand in the same position and orientation.

Furthermore, even for the simple planar example of the previous section, it is very difficult to invert the forward kinematics of equation (10.3) analytically, because the desired variables are expressed in terms of sines and cosines. Such transcendental, nonlinear equations may not have closed-form solutions. Fortunately for the robotics field, closed-form solutions for the inverse of the forward-kinematics equations do exist, but they are complex, and we do not present them here; They can be found in the textbooks referred to at the close of the previous section. Furthermore, in practice, inverse-kinematics expressions can be solved on computers using a variety of available software packages.

As indicated previously, equation (10.4) cannot be inverted, since the solution may not be unique (or may not exist at all). However, although in general the position equation (10.4) cannot be used to find the inverse kinematics, the corresponding rate expression is invertible. Hence, we differentiate (10.4) to obtain

$$\dot{\mathbf{x}} = \mathbf{F}'(\theta)\dot{\theta}. \tag{10.5}$$

The terms in the matrix $\mathbf{F}'(\theta)$ are given by $\partial F_j/\partial x_k,\ j,k = 1,2,3$. This matrix is known as the *Jacobian* and denoted by $\mathbf{J}(\theta)$, so the forward kinematics can be represented as

$$\dot{\mathbf{x}} = \mathbf{J}(\theta)\dot{\theta} \tag{10.6}$$

and the inverse kinematics by

$$\dot{\theta} = \mathbf{J}^{-1}(\theta)\dot{\mathbf{x}}. \tag{10.7}$$

The solution of (10.7) is an important issue in the control of complex industrial manipulators or multijointed legs in walking machines.

10.7 Dynamics

Thus far we have been concerned with purely geometric or kinematic relationships. The expressions for these relationships are sufficient for control of a manipulator if it moves very slowly, so that inertial effects can be neglected. In contemporary systems, whether free standing or mounted on a moving robot, we cannot ignore inertial effects, since we want our systems to move as rapidly as possible. The rotational dynamics for a manipulator can be summarized in a state space (vector-matrix) equation:

$$\tau = \mathbf{I}(\boldsymbol{\theta})\ddot{\boldsymbol{\theta}} + \mathbf{V}(\boldsymbol{\theta},\dot{\boldsymbol{\theta}}) + \mathbf{B}(\boldsymbol{\theta},\dot{\boldsymbol{\theta}}) + \mathbf{G}(\boldsymbol{\theta}), \tag{10.8}$$

where

τ is the torque vector, whose components are the n individual joint torques;

$\boldsymbol{\theta}$ is the vector of n joint angular positions;

$\mathbf{I}(\boldsymbol{\theta})$ is the $n \times n$ *inertia matrix* of the manipulator. The components of this matrix depend on the mass distribution along the manipulator links and on the payload being carried;

$\mathbf{V}(\boldsymbol{\theta},\dot{\boldsymbol{\theta}})$ is an $n \times 1$ vector of centrifugal and Coriolis acceleration terms;

$\mathbf{B}(\boldsymbol{\theta},\dot{\boldsymbol{\theta}})$ is an $n \times 1$ vector of friction terms;

$\mathbf{G}(\boldsymbol{\theta})$ is an $n \times 1$ vector of gravity terms, which depend on the orientation of the links with respect to the gravity vector.

Depending on the structure of the manipulator, there may be an analogous set of equations relating forces to Cartesian displacements.

Equation (10.8) may be used to estimate the total torque (and hence the power) required to drive the manipulator. It is important to note that this equation depends on the values of the inertia, centrifugal, and gravity terms, which, in turn, depend on time, since the robot's orientation in space, as well as its velocity and acceleration, are changing. The equation is very complex, so that in practice various simplifications are made to obtain approximate solutions for special cases. We have introduced the equation here in general form to give the reader some feeling for the nature of the manipulator control problem.

10.8 Manipulator Control

Given a desired trajectory of the end effector in space, the goal of the control system is to compute the torques and/or forces at the joints needed to move the manipulator through this trajectory. Assume for the moment that we have a position control system. If there are n joints, we need to compute n desired joint positions for each "instant" of time, where the increment between these instants depends on the sampling rate of the system. Clearly, the sampling rate must be fast enough compared to the highest frequencies present to avoid aliasing. In practice, it is desirable to use sampling rates five to ten times the highest frequencies present in the system. The architecture of a joint position control system is shown in figure 10.13. Note the sequence of operations indicated in this figure:

- The initial state of the system is assumed to be known, and a desired end-point trajectory in Cartesian coordinates, $\mathbf{x}(kT)$, is computed and sampled at the desired update frequency $(1/T)$. The index k is a counter for the sampling times.
- The inverse-kinematics block computes the corresponding desired joint angles $\boldsymbol{\theta}(kT)$ (with the necessary additional computations that may be required in view of the nonuniqueness of the solutions, as described in section 10.6).
- The desired joint angles are used as inputs to the each of the n actuator servos that compute the actual joint positions at each interval.

The control method pictured in figure 10.14 is known as *resolved motion position control* because it specifies the desired path in Cartesian coordinates, then uses inverse kinematics to obtain the corresponding angular displacement of the joint servo drives. An alternative method that avoids the problems of inverting position kinematics, is known as *resolved motion rate control.* The architecture is very similar to

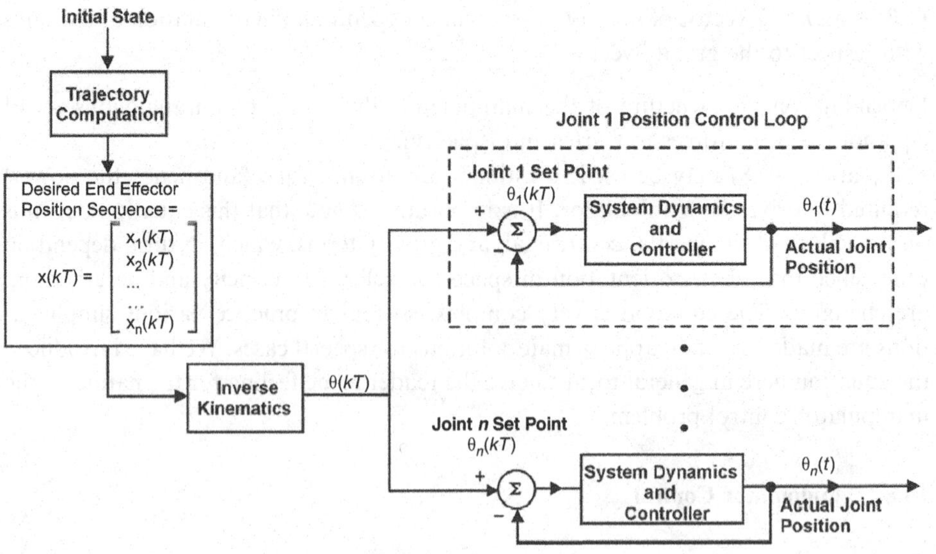

Figure 10.13
Resolved motion position control

that in figure 10.13, except that Cartesian *rate* commands are multiplied by the inverse Jacobian to obtain the desired joint angle rates. The servos then produce a joint velocity that must be integrated to obtain the actual joint angle.

The bottom line here is that manipulator control is not easy, particularly when one is dealing with large manipulators moving at high rates. When small arms are mounted on mobile robots, or when mobile robot arms move slowly, most designers return to the standard PID controllers.

10.9 Alternative Approaches to Manipulator Control

As the preceding discussion has indicated, control of multilink manipulators is a difficult problem. The mathematical models are highly nonlinear, and the problems are made particularly difficult by the need to perform transformations from end effector coordinates to joint coordinates. There are several approaches to handling this complexity.

For certain industrial manipulators, such as those used for spray painting, a learning approach avoids the need to program and solve the equations of motion. An experienced painter moves the end effector as if she were moving the sprayer manually. The system records the joint angles resulting from this movement as the sprayer is moved back and forth over an object to be painted; these time histories can then be

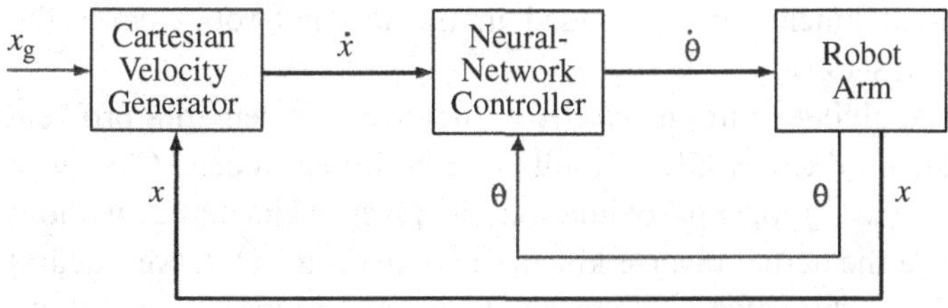

Figure 10.14
Inverse kinematics using neural network

used to drive the robot. Commercial manipulators are generally equipped with so-called teach pendants, basically small joysticks used to move the end effector in the coordinate directions. An operator uses the teach pendant to drive the end effector to points along a desired trajectory. As each point is reached, the operator pushes a button, and the corresponding joint angles are recorded for future control. Most manufacturers of manipulators also have available software packages that solve the inverse-kinematics problem and make the programming of trajectories much simpler.

From the research point of view, there are other approaches to the control problem. We consider two of these briefly, in connection with problems of inverse kinematics and inverse dynamics, respectively.

10.9.1 Neural Networks for Inverse Kinematics

In chapter 6 we saw that artificial neural networks can be used to model the input-output behavior of very general classes of systems, without detailed mathematical models. In other words, neural networks are *universal approximators* to the behavior of systems. This suggests that they can be used to approximate inverse kinematics without actually performing the matrix inversions associated with inverse kinematics. Figure 10.14 illustrates such an approach, which is particularly useful for complex manipulators in which the inverse mapping is difficult to derive analytically or in which the mappings change as a function of time.

It can be seen that the system of figure 10.14 is an implementation of equation (10.7), also known as *inverse Jacobian control.* In this type of control, the inverse-kinematics mapping from Cartesian coordinates to joint coordinates is approximated by a neural network. Such an approach was taken by Jordan and Rummelhart (1992), who showed that a coarse mapping can be obtained rather quickly, but that it is difficult to obtain a highly accurate one (see also Torras 2002). One solution to the problem is to subdivide the network so that each portion of the network is responsible for only a portion of the input space. This can be done using the *self-organizing feature map* method or by using *context-sensitive neural networks* (Yeung and Bekey 1989). Another approach is to use the neural network only to learn deviations from a nominal kinematics embedded in the original robot controller (Ruiz de Angulo and Torras 1997).

Some of the results of applying neural networks to the inverse-kinematics problem are clearly promising and produce excellent results in laboratory studies. Clearly, it would be wonderful to obtain good approximations to inverse kinematics without having to obtain and solve the actual inverse-kinematics equations. However, neural networks converge slowly, as discussed in chapter 6. Convergence may require thousands of iterations through the problem space, so that it is not clear that neural networks are useful in practice for real-time control.

10.9.2 Inverse Dynamics

As we have seen, when dynamics need to be taken into account, the robot manipulator control problem becomes even more difficult, since in addition to dealing with inverse kinematics, it is necessary to map the end effector accelerations to the required joint torques (or forces). Since the cerebellum is known to be involved in the production and learning of smooth movements in humans, several cerebellar models have been applied to the control of robot arms. The first such model was developed by Albus (1971, 1981) and is termed a *cerebellar model articulation controller* (CMAC); it was basically a table lookup method. CMAC has been combined with a least-mean-square (LMS) error minimization scheme to control a 5-dof robot (Miller et al. 1990). Excellent reviews of the major issues in this approach to the inverse kinematics problem are given by Torras (1995, 2002).

As just stated, the cerebellum is known to play a role in movement control in humans. There is increasing evidence that to learn voluntary movements, humans develop internal models (in the cerebellum) for motor control and trajectory planning (Kawato 1999). These models are in fact inverse-dynamics models, which lead to a different formulation of the control problem than we have encountered up to this point. Consider figure 10.15(a), which illustrates the standard engineering formulation of the robot control problem, in which we have assumed a simple PID controller. Kawato points out that in biological systems, including humans, feedback delays are relatively large. For example, the combined delay, including visual feedback, involved in arm movements and transmission along nerve fibers will be of the order of 150–250 ms. This delay is inherent in the biological system and cannot be reduced. Furthermore, the control system will not respond well to error signals if the controller gains are small. Because of the presence of delays, the gains cannot be increased without causing instability. Hence, biological systems use a completely different, open-loop approach to movement control, as illustrated in figure 10.15(b). Such systems use a feedforward controller, which learns the inverse dynamics of the system, so that the combination of the controller and the system now has a value of approximately one, as shown in the figure. This implies that after learning, the out-

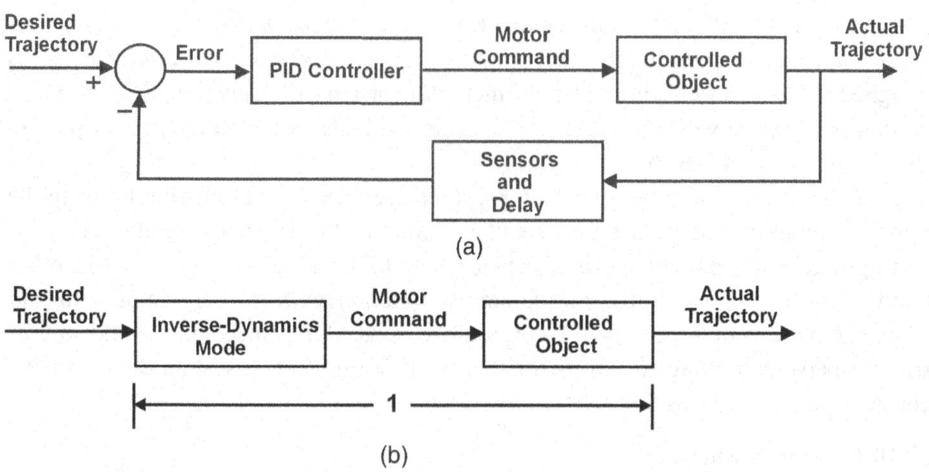

Figure 10.15
(a) Engineering feedback control system for movement; (b) biological feedfoward movement control system using an internal model of inverse dynamics

put is identical to the input. The existence of such internal models of inverse dynamics has been experimentally verified and is increasingly accepted.[26]

The question then arises, "If inverse-dynamics models work in biological systems, why not apply them to the control of robots?" In fact, inverse-dynamics control is a well-known method of control of complex systems. However, when applied to robotics, it has a somewhat different meaning than in the previous paragraph (Spong 1996). The term *inverse dynamics* is used in robotics to describe a specific method of feedback linearization. This method is used to transform a given nonlinear system into a linear system by means of nonlinear feedback. In the case of robot manipulators, nonlinear feedback is used to transform the nonlinear dynamics equation (10.8) into a double integrator. Standard techniques can then be used to control this simple linear plant. Typically, such controllers are proportional-derivative controllers with some form of feedforward acceleration control.

It should now be evident, if the reader had any doubts, that the precise control of movement of a robot manipulator is a complex subject.

10.10 Arm Prosthetics and Orthotics

In chapter 8 we saw that some principles of robotics have been applied to the development of both prosthetic and orthotic devices for the lower extremities to enable

26. According to Kawato (1999), "Internal models are neural mechanisms that can mimic the input/output characteristics, or their inverses, of the motor apparatus."

persons with disabilities to walk again. Recall that a *prosthetic* device is a replacement for a missing limb (or other body portion), whereas an *orthotic* device is designed to assist an existing but nonfunctional (or partially functional) limb. Thus, amputees are fitted with prostheses, whereas individuals with certain types of paralysis may be fitted with orthoses.

There is a major difference between prosthetic replacement of missing limbs in the upper extremities and such replacement of limbs in the lower extremities. If a leg is amputated, the patient needs a replacement to be able to walk. On the other hand, if an arm is lost, most people simply function with the remaining arm and want a replacement arm only for cosmetic reasons. (Of course, this would not be true for persons needing two arms for their professions, but it is often the case.) We consider prosthetic arms first.

10.10.1 Arm Prostheses

It is intuitively clear that a prosthetic replacement for a missing arm could be basically a robot manipulator, if it can (1) be suitably attached to the body, (2) obtain a source of power, and (3) have a means of control. Attachment to the body is the first challenge, since this implies that the prosthetic arm will not weigh much more than a living arm and that it can be attached either to a humeral stump or to the shoulder socket. Assuming that this can be done, the joints of the prosthetic arm must be powered by lightweight electric motors, batteries must be provided, a microprocessor is needed to process control signals, and a source of control information must be found. Experience has shown that if amputees have no easy way to control an artificial arm (or any artificial limb), they will not use it (Shurr and Michael 2002).

If an amputee's arm has been removed below the elbow, it is possible to use a sleeve that fits on the stump and carries an artificial hand. Most upper-extremity prostheses are limited to replacing hand function in this way. Since no arm is required in the case of such prostheses, we do not discuss them here. Artificial hands are considered in the next chapter.

Upper-extremity prostheses are either body powered or electrically powered. Body power is provided by available movement in the shoulder of the amputated arm and/or the opposite shoulder. In the case of electrical power, the signals for control of the prosthesis are usually obtained from electrodes that pick up the signals produced by muscle fibers in the stump when activated. These signals, known as *myoelectric signals*, are present even with residual muscles and can sometimes be used for control.[27] They are the ideal control signals, since they become active when the amputee wills

27. These signals are frequently termed *EMG signals*, referring to *electromyography*. In current usage, *electromyographic* has been replaced by *myoelectric*, which puts the muscle prefix "myo-" first. Nevertheless, the abbreviation EMG is still frequently used.

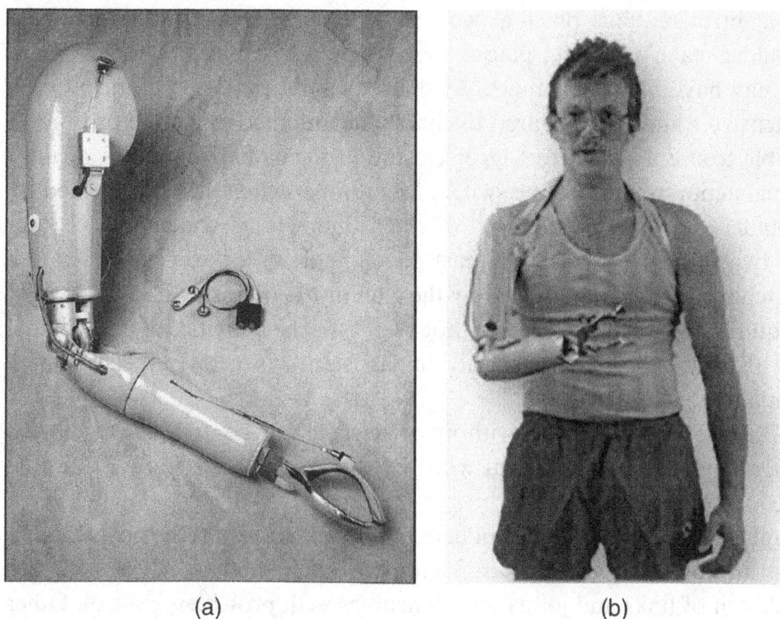

Figure 10.16
Upper-extremity prostheses: (a) prosthetic arm from Hosmer Dorrance Corporation (photograph courtesy of Hosmer Dorrance); (b) amputee fitted with Otto Bock prosthesis (photograph courtesy of Otto Bock Healthcare)

a motion. For below-elbow amputees, or even those with a long remaining upper-arm stump, the electrodes can be placed on the upper arm over the muscles. This becomes more difficult if the amputation takes place close to the shoulder, but even then, the electrodes can sometimes be placed inside of the shoulder cup of the prosthesis.

Two examples of such prostheses are shown in figure 10.16, designed for persons who have lost all or nearly all of the upper arm (known as a *transhumeral amputation*). The devices are attached to the body by straps, which are clearly visible in figure 10.16(b). Cables connect the shoulder socket to the forearm and a simple terminal device (which may be covered with a glove for cosmetic reasons). Since this device is basically a 2-dof manipulator, it requires actuation at both the elbow joint and the gripper. Frequently, myoelectric signals are used for control of the elbow and cable control for the terminal hook. For example, the cable might be controlled by movement of the stump or the shoulder itself (with a shrugging movement). The motion is transmitted to the prosthesis via a cable control system, which usually runs from the prosthetic arm across the back to a loop around the healthy shoulder. A second control signal can be obtained from the opposite shoulder.

Myoelectric control requires placing two sets of electrodes in the socket of the amputee's shoulder, each set being placed over flexor and extensor muscles. Since these muscles may have had other functions than opening or closing a grasp in the intact arm, extensive training is required to enable the amputee to use the prosthesis. It is also possible to use a switch to trigger closure of the end effector; this requires finding a residual action to depress the switch. Membrane switches can be located inside of the shoulder socket. (A membrane switch is a momentary contact device, typically made of two thin sheets, where at least one is made of a flexible material like polyester.) Switches can also be activated by the chin or the opposite limb. In view of the training requirement, such devices may not useful for the activities of daily living. Amputees will then use the prosthesis only for assistance in certain specific tasks, such as those related to their work.

Figure 10.17(a) shows a prosthesis with myoelectric elbow control; the electrodes are shown separately. The artificial arm worn by the amputee in figure 10.16(b) is cable controlled.

We have emphasized the role of control here to make it clear that the problem hindering progress in the development of artificial arms is not the lack of robotic technology in the design of links and joints, but difficulties with prosthesis control. Other control means, such as voice, have been considered but found to be impractical. The brief analysis presented here makes it clear why a person with one functional arm may have no interest in a complicated and poorly functioning artificial arm with limited usefulness.

If an individual is confined to a wheelchair, it possible to attach one or both arms to the chair (Prior 1990). This solves the attachment problem. Batteries can be mounted on the wheelchair. The control problem, however, is still present.

10.10.2 Orthotic Arms

One might logically assume that if the prosthetic-arm problem has proven difficult to solve, enabling the use of an arm with a nonfunctional limb might present even more serious problems. This is not necessarily the case, however, since an orthotic arm can simply be strapped to a flailing, nonfunctioning arm. Further, those being equipped with an orthotic arm may not have a normal second arm on which to rely for functionality, which increases their motivation for learning the difficult protocols necessary to control the orthosis. Also, they are frequently confined to wheelchairs, which can support one or two robot arms used as orthoses. In such cases, a nonfunctional or partially functional arm is used to control the robot arm by means of joysticks or keyboards.

An important application of robotics in orthotics is in assisting person with cerebral palsy, who may have uncontrollable jerking or oscillations of their arms when they attempt purposeful movements. A multi-dof orthotic arm can be strapped to

Figure 10.17
Rancho Orthotic Arm mounted on a wheelchair; note the tongue-activated control switches (photograph courtesy of Daniel Antonelli and the Rancho Los Amigos Medical Center)

the uncontrolled limb, with adjustable damping in the joints. This additional damping can reduce the severity of the uncontrolled movements and make functional movements possible.

We conclude this discussion of artificial arms with one more example, the Rancho Orthotic Arm (figure 10.17).[28] Although this arm never became a successful commercial product, the very difficulties it presented are illustrative from a robotics point of view.

It is clear from the figure that this arm is in fact a robot manipulator. It was designed for use by a quadriplegic person, who was completely paralyzed below the

28. The name comes from the fact that this arm was developed at Rancho Los Amigos Medical Center in Downey, California, in the 1970s. I was at Rancho as a consultant during this period and played a minor role in developing the control systems for the arm.

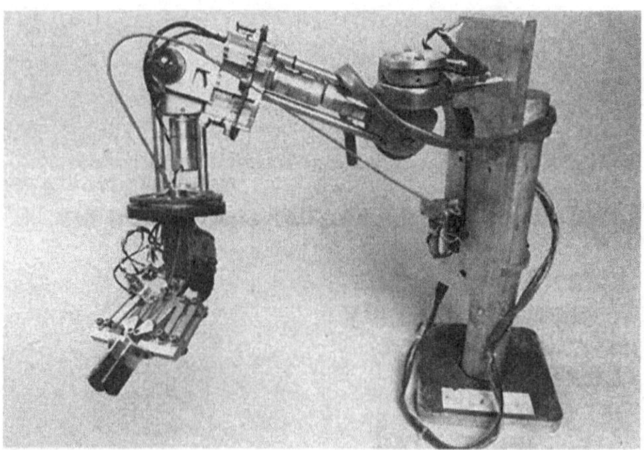

Figure 10.18
Stanford-Rancho robot manipulator (photograph courtesy of Gio Wiederhold Stanford)

neck. The arm was mounted on a wheelchair, with straps that attached to the patient's arm. It had 6 dof (2 in the shoulder, 1 in the elbow and 3 in the wrist), plus an additional degree of freedom in the end effector. Electric power for the joint motors was provided by batteries carried on the chair. Thus, the issues of attachment and power were solved. The control problem was clearly serious, since the only sources of control were above the neck. The designers of the arm (James Allen, Andrew Karchak, Vert Mooney, and others) developed a very ingenious control method: tongue-actuated switches. A bracket extending from the back of the chair, actuated by a movement of the head, brought a small fixture with seven levers in front of the patient's mouth, as can be seen in the figure. (This fixture could be retracted by a head-movement-activated switch.) The patient could move the switches up or down with her tongue; each switch controlled 1 degree of freedom of the manipulator.

Readers will note immediately why this method was doomed to failure: The subject was required to perform inverse kinematics in her head to move the robot arm to a desired location in space! Nevertheless, in spite of the difficulty associated with this task, one resident at the hospital acquired remarkable skills in controlling the arm. After an extensive period of practice, she succeeded in writing a letter to her daughter using a pen held in the robot end effector. In our opinion, this success supports the view that the subject was able to learn the inverse kinematics and use them for control.

Later modifications replaced the switches with contacts located on a dental bridge inside the mouth. Eye movement control was also tried, but the problem of inverse kinematics remained, since efficient software for solving the kinematic equations was not yet available.

The Rancho arm was in fact one of the first robot manipulators built in the United States. This fact was not lost on researchers at Stanford University and the NASA Jet Propulsion Laboratory, who modified the arm to use computer control. The resulting Stanford-Rancho arm is shown in figure 10.18.

10.11 Concluding Remarks

This chapter has summarized the state of the art in manipulators, including their biological foundations, industrial implementation, and use in rehabilitation. We believe that in the future, most mobile robots will be equipped with arms. These new arms will made of composites or other lightweight materials and will have embedded processors and new, very small, high-torque motors. The trend to add arms to mobile robots will lead to autonomous robots that invariably include manipulation as well as mobility.

11 Control of Grasping in Human and Robot Hands

Summary

To grasp and manipulate objects, robots need the mechanical equivalent of both arms and hands. Arms were discussed in the previous chapter. In this chapter we review the structure and function of the human hand and then study both simple grippers for industrial robots and multifingered robot hands developed for research purposes. The nature of the grasping process is analyzed using the principles of "opposition spaces" and virtual fingers. Several multifingered robot hands are presented. A case study in the chapter concerns the design of the Belgrade-USC robot hand, an anthropomorphic, five-fingered end effector. Prosthetic hands are discussed in some detail, since they can be considered special-purpose applications of robot hands.

11.1 Introduction to Hands

The human hand is a truly remarkable instrument, both structurally and functionally. When attempting to design a "hand" for a robot or a prosthetic hand for an amputee, we generally attempt to imitate its ability to grasp objects of arbitrary shape, as illustrated in figure 11.1. But the hand is a complex and versatile system, capable of large number of other functions, such as pointing, waving, stroking, typing, and forming a fist for hitting, to name a few. MacKenzie and Iberall (1994) list three hundred tasks that hands perform related to the activities of daily living. To accomplish the variety of tasks of which it is capable, the hand depends on control inputs from the central nervous system and numerous sensors that provide feedback on such variables as finger segment flexion, forces being applied in various locations, geometric patterns of objects being touched, slippage, and object temperature. Clearly, artificial hands cannot be expected to duplicate all these functions.

In this chapter we discuss only a small subset of the capabilities of the hand, primarily those associated with reaching and grasping, since these functions are also

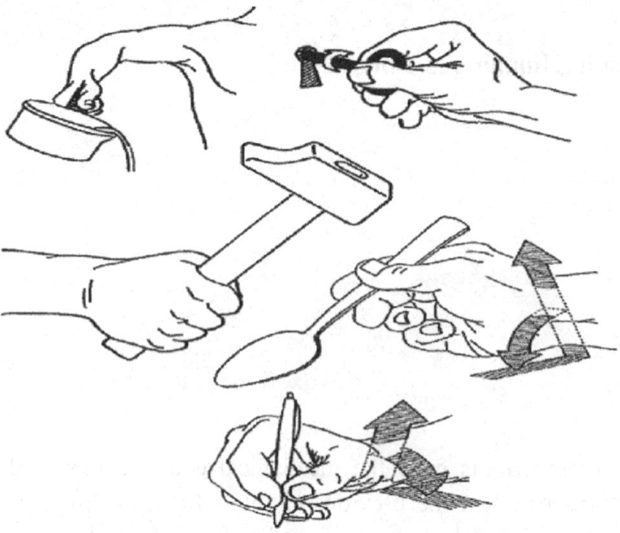

Figure 11.1
Some human grasp functions (from Kapandji 1982, reproduced with permission of Editions Maloine, Paris)

essential for robot arms. Control of grasp and manipulation depends basically on intricate eye-hand coordination and interaction processes. Control of functional motions like reaching and grasping requires integration of various reflex and learned mechanisms. The following section reviews human control of reaching and grasping and presents a model of its neural control, as developed by Arbib, Iberall, and Lyons (1985). It then presents a simplified model suitable for designing the control of robot reaching and grasping, based on the "opposition principle" (Iberall and MacKenzie 1990).

As with other aspects of robot control presented in this book, we believe that classical approaches to the control of multifingered hands are impractical. To control grasping by conventional methods would require considering each finger as a kinematic chain and designing a separate controller for it. Then, given the object geometry (from prior knowledge or a vision system) and the current hand pose, the controller would synthesize the desired finger segment trajectories for completing the grasp. In spite of the difficulty of such a control approach, this is exactly the method used for control of the Utah-MIT hand, as indicated in section 11.4. It is also possible to build grasp controllers using knowledge of the task to be performed, simplification of both hand description and object, and design of robot hands capable of a small number of behaviors (Bekey, Tomović, and Zeljković 1990; Bekey et

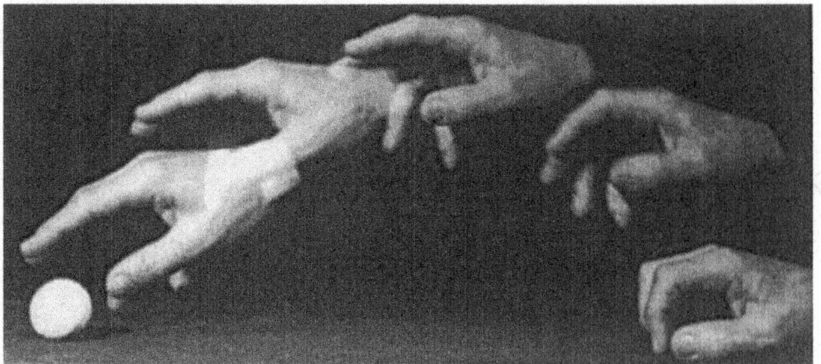

Figure 11.2
Trajectory and preshaping of hand moving toward object (photograph courtesy of Marc Jeannerod)

al. 1993). The design of such hand controllers for grasping, using neural networks and knowledge-based systems, does not require a precise mathematical representation of the finger kinematics. These two approaches to design of grasp controllers are presented in sections 11.5–11.7.

11.2 Reaching and Grasping

When a specific task needs to be accomplished, the human hand preshapes into a posture with capabilities that match task variables, thus creating a special-purpose mechanism for the desired job. For prehensile tasks, the hand preshapes into a posture suitable for grasping a *given object* for the *given task*, then encloses the object with the fingers. As we show later, the shape assumed by the hand depends both on the geometry of the object being grasped and on the intended task. For example, a pencil is grasped differently for writing than for passing it to another person. It has also been shown (Jeannerod 1981, 1989) that when reaching for an object, starting from a resting position on a table, the hand follows a parabolic trajectory in the vertical direction, while moving forward in the horizontal direction toward the object. This trajectory and the preshaping are shown in figure 11.2.

As indicated previously, the hand is capable of a large number of possible configurations, of which grasping encompasses only a small subset. To understand the physiological basis of these configurations, consider the simplified drawing of the anatomy of the human hand in figure 11.3. The wrist structure consists of eight small *carpal* bones (labeled A in the figure). The wrist joint is formed between the major bone of the forearm (the *radius*) and the carpal bones. During flexion of the wrist there is considerable movement between the first and second row of carpal bones, in the *midcarpal* joint (B in the figure). Each finger consists of a number of bone segments known as *phalanges* (two in the thumb and three in each of the remaining

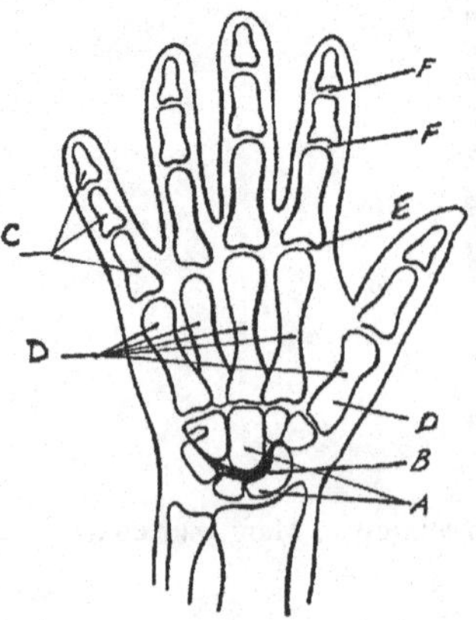

Figure 11.3
Anatomy of the hand (adapted from Murray 1969)

fingers) (C). The bone linking the first phalanx to the carpal bones is known as the *metacarpal* (D). Hence, the joint between the metacarpal and the first or proximal phalanx is called the *metacarpo-phalangeal* joint (E), and the joints between the phalanges as called *interphalangeal* joints. It should be evident that the hand has a highly complex, articulated structure; designers of robot hands generally do not attempt to replicate this structure in any detail. Rather, multifingered robot hands are designed to approximate some of the functions of the hand, such as grasping. To perform the movements of which the hand is capable, the bones depicted in figure 11.3 are actuated by a large number of muscles, connected to the bones by means of tendons and controlled by a network of nerve fibers. Since each muscle can exert force only when it contracts, the reader can imagine the complex neuromuscular control system present in the hand. It is beyond the scope of this book to examine these structures in detail; treatments can be found in any textbook of human anatomy (e.g., Taylor and Schwartz 1955; Murray 1969; Netter and Hansen 2003).

Let us now consider grasping specifically. A number of researchers have identified a small subset of the possible postures of the hand as representative of possible grasping modes (e.g., Napier 1956). The specific prehensile postures vary from discipline to discipline (e.g., rehabilitation, manufacturing). We assume here that the set of grasp modes consists of the same six elements discussed in case study 6.2 and presented in figure 6.15 (Liu, Iberall, and Bekey 1988). The six grasp modes shown in the figure are illustrated in common tasks.

Such a classification of grasp modes is convenient for the study of knowledge-based control of robot grasping (Bekey et al. 1993). To ensure stability of the grasped object in the hand, it is essential that balanced forces be applied in opposite direc-

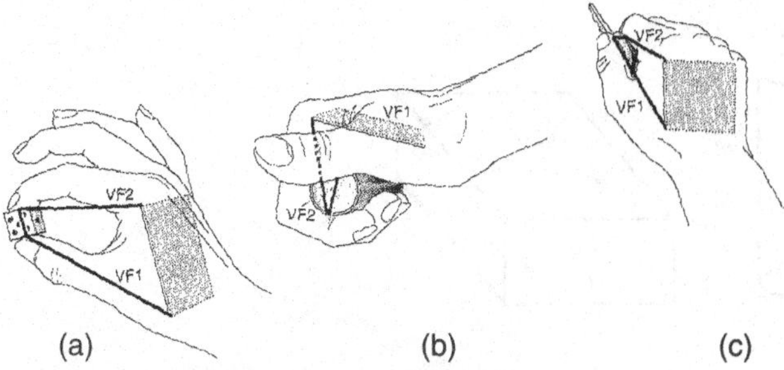

Figure 11.4
Opposition spaces and grasp postures. Virtual fingers are indicated by VF1 and VF2. (a) Pad opposition occurs along an axis x, generally parallel to the palm. (b) Palm opposition occurs along an axis z, generally perpendicular to the palm. (c) Side opposition occurs along an axis y, generally transverse to the palm (from MacKenzie and Iberall 1994, p. 34, courtesy of Christine MacKenzie and Thea Iberall, reproduced with permission of Elsevier)

tions. As shown by Iberall (1987), depending on the grasp mode, these forces may be applied by the tips or the lateral surfaces of the fingers or by the palm. When several fingers act in unison in the generation of opposition forces, they may be viewed as a *virtual finger*, as may the palm. The specific finger configurations generating these opposition forces belong to three general classes, as shown in figure 11.4.

It should be noted that the specific grasp mode (and corresponding opposition space) selected by the nervous system for the hand depends both on the geometry of the object to be grasped and on the task to be performed. Thus, a screwdriver will be grasped using a power grasp for actually driving screws, but perhaps with a lateral grasp when it is stored in a drawer. Objects come in a bewildering variety of shapes. However, it is possible to abstract them into a small number of primitive shapes. Such an abstraction appears to be consistent with the way the human visual cortex classifies object shapes (Biederman 1987). Using this principle, we assume in this discussion that objects can be considered as belonging to one of five geometric primitives, as illustrated in figure 11.5. Clearly, complex objects may require decomposition and approximation in order for this classification to be valid. For example, a screwdriver may be considered two cylinders, and an axe may be viewed as a junction of a cylinder and a box or parallelepiped.

Using the object shape simplification shown in figure 11.5 and a library of tasks in some domain, it is possible to construct a grasp selection control system (Bekey et al. 1993). Basically, the system takes as its inputs the desired task and the geometry of the object to be grasped (which may be obtained from an online vision system) and infers the most suitable grasp mode (from a library such as that depicted in figure 6.15).

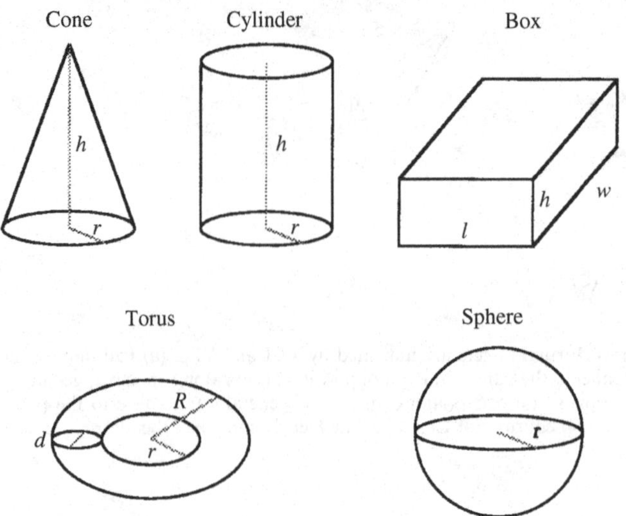

Figure 11.5
Five geometric primitives

The preceding paragraphs have indicated some of the issues involved in the control of grasping. The neural basis of many of the components of such control, and even the location of those components in the brain, are known, as shown in figure 11.6 (Arbib, Iberall, and Lyons 1985). This remarkable diagram includes all the phases of grasping in the foregoing discussion.

Figure 11.6 consists of several *schemas*, functional groups of elements responsible for particular aspects of perception or action. However, it is important to note that the schemas shown in this figure correspond to specific locations in the brain. In other words, the construction shown in the figure is not hypothetical but reflects the structure and organization of the brain and neuromuscular system. Note that the figure includes a specific schema for enclosure, as the final step for finger closure after the preshaped hand reaches the object.

11.3 Simple Robot End Effectors

Let us now take a major step away from the complexity, intricacy, and beauty of the human hand and consider some of the mechanisms used in robotics for grasping objects. These mechanisms are collectively known as *end effectors.* (We reserve the term "robot hands" for those devices that are clearly inspired by the structure of the human hand and examine them separately in the next section.)

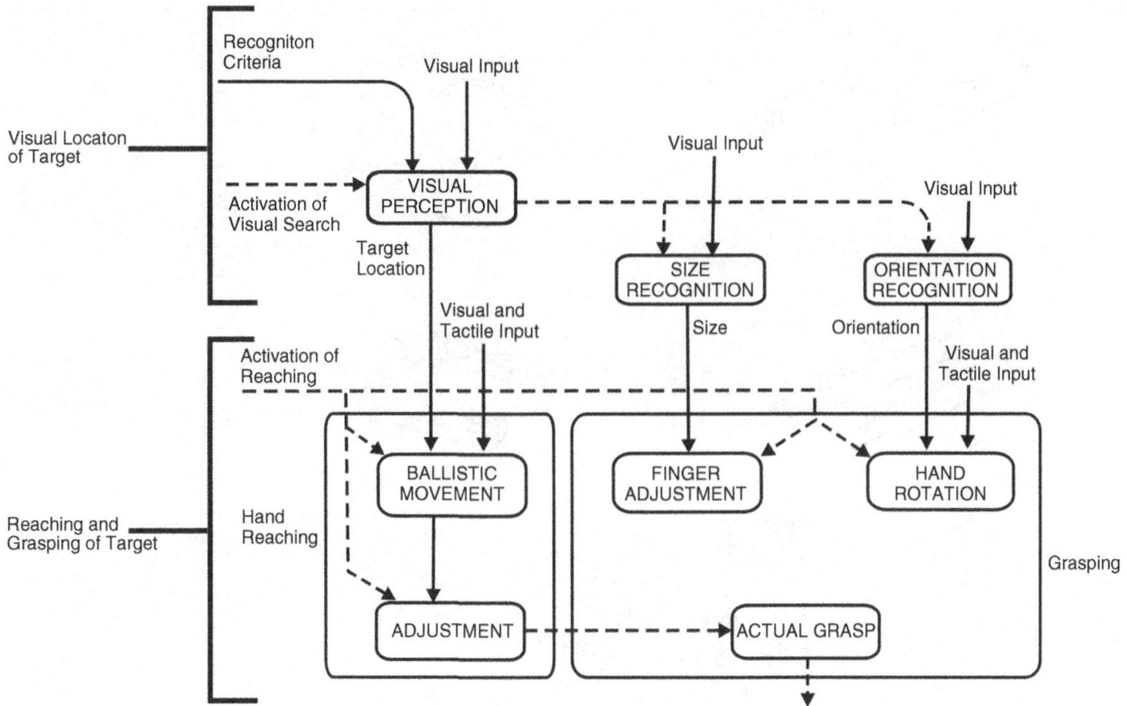

Figure 11.6
Neuromuscular control of grasping (illustration courtesy of Michael Arbib)

As with the hand, the goal of an end effector is to apply opposing forces to an object sufficient to keep it stable while a manipulator moves it from one location to another (as in pick-and-place operations) or performs an assembly operation with it. Some typical industrial end effectors are shown in figure 11.7.

It should be apparent from figure 11.7 that industrial end effectors are special-purpose devices, suited for grasping or manipulation of particular classes of objects. One end effector is better for grasping flat sheets of metal, whereas another may be better suited for manipulation of a paint spray gun. Parallel-jaw grippers are in very common use. Two-fingered pinch grippers and enclosing grippers are designed to grasp cylindrical objects. Since industrial manipulators are much more expensive than the end effectors they use, it is common to locate a collection of different end effectors near a robot so it can conveniently change from one to another. These repositories of end effectors are known as *magazines*. The use of magazines of end effectors provides an industrial robot with some (but obviously not all) of the versatility of a multifingered hand. As we show in the next section, multifingered hands

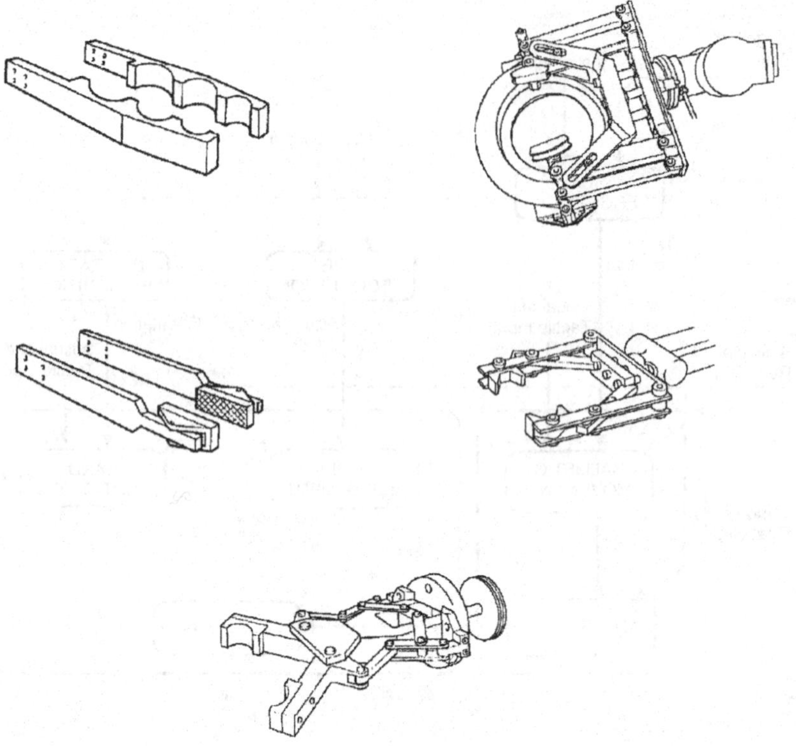

Figure 11.7
Typical industrial end effectors

for robots are more complex in structure than the simple end effectors in common use, and hence they are both more expensive and less reliable. For industrial use the combination of lower cost and higher reliability dictates the common use of simple end effectors, rather than more versatile "hands."

The end effectors shown in figure 11.7 are clearly designed for grasping specific shapes. Other types of end effectors are designed to grasp arbitrary shapes (within constraints of size and weight). One such gripper for small parts is shown in figure 11.8. The gripper consists of a number of wire fingers that are opened and closed by alternately inflating two air bags located at the fulcrum point of the fingers. The gripper can grasp odd-shaped parts that are hard to orient. Furthermore, since the gripping force is distributed among many fingers, the unit can also pick up soft or fragile workpieces.

It should be evident from these examples that industrial grippers come in a wide variety of shapes, as required by the objects they need to grasp.

Figure 11.8
Flexible gripper for small parts (photograph courtesy of Schunk GmbH)

11.4 Multifingered Robot Hands

The simple grippers described in the previous section are in common use in a variety of manufacturing applications. In contrast, more complex hands, with a structure resembling that of the human hand, have been built for research purposes, in order to better understand the nature of grasping. The goal of this chapter is to explore the potential applications of biologically based designs for such hands. In our view, these applications include grasping tasks in unstructured environments (such as those existing in space, under water, in mines, or in the presence of hazardous substances). Some experiments have attempted to demonstrate that dextrous multifingered end effectors (with three, four, or five fingers) have advantages in certain highly structured industrial environments, specifically, flexible assembly, but the evidence that has emerged from these experiments not clear. Over the past twenty years, many university robotics laboratories throughout the world have fabricated robot hands for research, whereas only a small number have been offered in the marketplace. In this section we describe multifingered hands only as research tools, without claiming any specific industrial applications for them (but see the Barrett hand shown in section 11.4.3).

Among the major books and articles devoted to this subject are Cutkosky 1985; Mason and Salisbury 1985; Venkataraman and Iberall 1990; and Bicchi and Kumar 2001.

11.4.1 General Considerations

Dextrous hands may be classified according to hardware features (size, weight, number of fingers, articulations), sensory system (proximity, tactile, vision), or control philosophy. They also may be classified as (1) research-oriented or (2) application-oriented dextrous hands.

Research-oriented robot hands are designed as universal tools for the general study of human and machine grasping in terms of control, sensors, and hardware. Consequently, their design reflects an attempt to create devices that mimic the functions of the human hand and resemble it in external appearance, without attempting to duplicate the anatomical structure. Bearing in mind that the human hand, including the wrist, represents a compact, fast, biomechanical control system with 22 dof, it is easy to understand that a mechanical replica of such a shape-adaptive device is no simple matter to reproduce. Coordination and control of grasping movements involving a large number of actuators as well as the processing and integration of abundant tactile and video information are also challenging tasks. Other performance requirements, like weight, size, compactness, and simple operation, are not of primary interest for this kind of application. The Utah-MIT dextrous hand shown in (figure 11.9(b)) is an outstanding example of research-oriented design.

Application-oriented designs of multifingered hands are meant for a specific range of grasping functions in terms of target features and manipulation tasks. Naturally, a priori specification of grasping functions facilitates greatly the design, control, and construction of such multifingered end effectors. For example, some hands have been designed specifically for small-parts assembly operations. An outstanding example of a special-purpose hand is the one being used for the NASA Robonaut (shown in figure 11.9(g)), a two-armed robot designed for servicing operations on the exterior of the International Space Station.

To be accepted in the industrial (or space) environment, multifingered robot hands must meet rather strict requirements. For tasks in which the overall size of the robot hand is approximately that of the human hand (so that it can perform tasks previously done by humans), these requirements include the following:

- *Self-contained* It is essential that all components of the hand, including the hardware, actuators, and local electronics, be an integral part of the terminal device. Except for the power supply, the hand must be self-contained so that it is easily replaceable.

• *Weight* A self-contained dextrous hand should weigh less than 2 kg so that the payload factor of the manipulator is not significantly deteriorated by the weight of the end effector.

• *Flexibility* All object shapes and sizes fitting the human hand should be graspable by the robot hand.

• *Computation* The global control system of the robot hand must not require excessive processing power and speed.

11.4.2 Sensors

It is well known that the dexterity of the human hand depends on the richness of its external (exteroceptive) and internal (proprioceptive) sensory information. These factors affect also the dexterity of robot hands. However, technology for robots cannot yet offer sensors that can provide information whose diversity and quantity approaches that supplied by biological exteroceptive and proprioceptive sensors. Characteristic of the current stage of sensor technology for dextrous hands is the great number of experimental solutions using the most varied physical phenomena.

Let us first point out the requirements for sensors that are specific to robot hands:

• *Distribution* Grasping of objects implies contact between the hand and an area of the target. Consequently, sensory feedback for grasp control must include information about both the intensity (force or pressure) of the contact and its topology. In other words, sensitive areas rather than point sensors are needed for robot hands. This can be accomplished through the use of matrix-type sensors or surface-sensitive transducers. The technological problems involved in the use of either are not easy to overcome.

• *Miniaturization* Several layers of different kinds of sensors should be distributed over the whole hand area. It is technologically difficult, however, to produce reliable, miniaturized, distributed, sensitive multiple layers of sensors fitting the measurements of human-sized robot hands.

• *Pliability* Since robot sensors must be evenly distributed over the hand and finger surface, it is desirable that the sensory layers be pliable, so that they may be fitted to anthropomorphic surfaces.

These requirements concern only the general performance features of robot sensors, without specifying the actual nature of the sensory information they should provide. In the absence of a closer exploration of task requirements, it is not possible to determine exactly what kind of sensors are needed for a particular dextrous hand. However, there is a basic spectrum of sensory information essential for reliable and stable grasping:

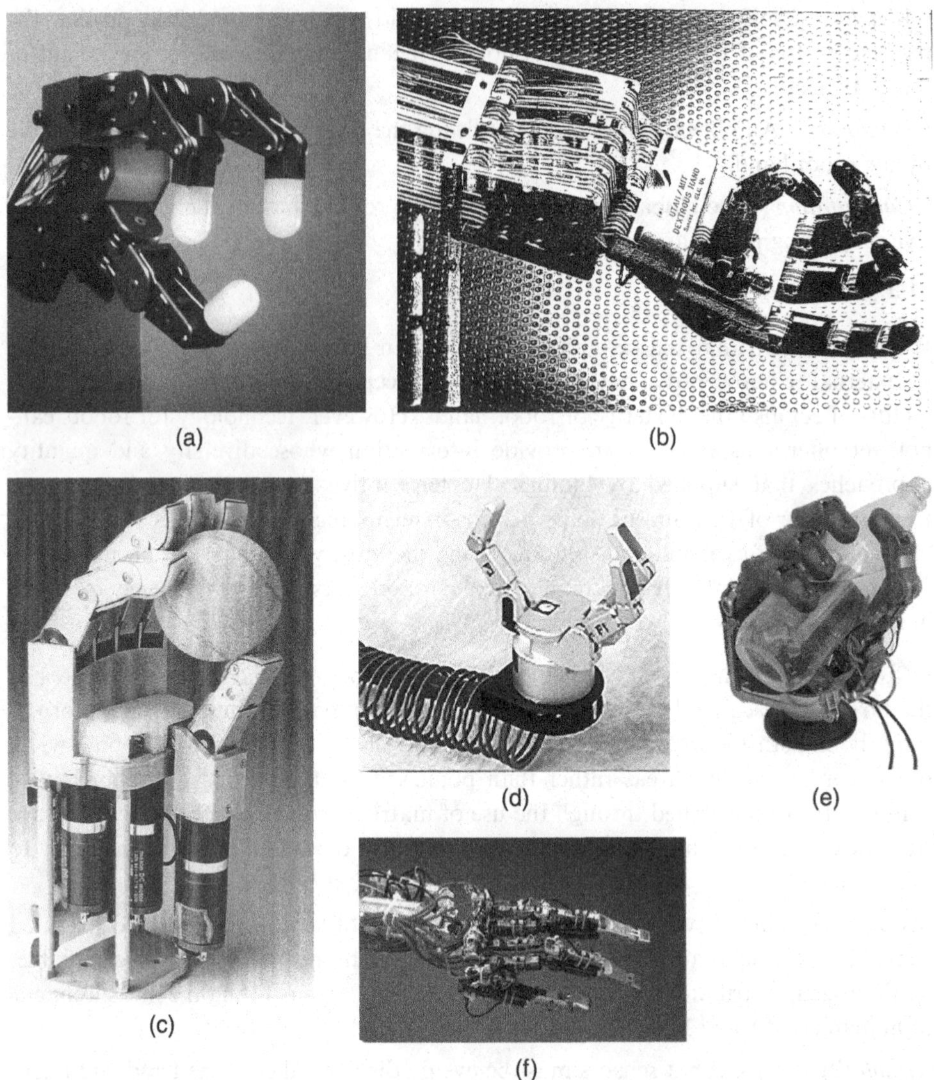

Figure 11.9
Multifingered robot hands: (a) Salisbury hand (photograph courtesy of Hank Morgan and Ken Salisbury); (b) Utah-MIT hand (illustration courtesy of Steven Jacobsen); (c) Belgrade-USC hand; (d) Barrett BH-Series hand (photograph courtesy of Barrett Techology, Inc.); (e) DLR hand (photograph courtesy of Gerd Hirzinger); (f) Omni-Hand III, copyright 1997 Ross-Hime Designs, Inc. (photograph reproduced with permission of Mark Rosheim); (g) Robonaut hand (photograph courtesy of Robert Ambrose)

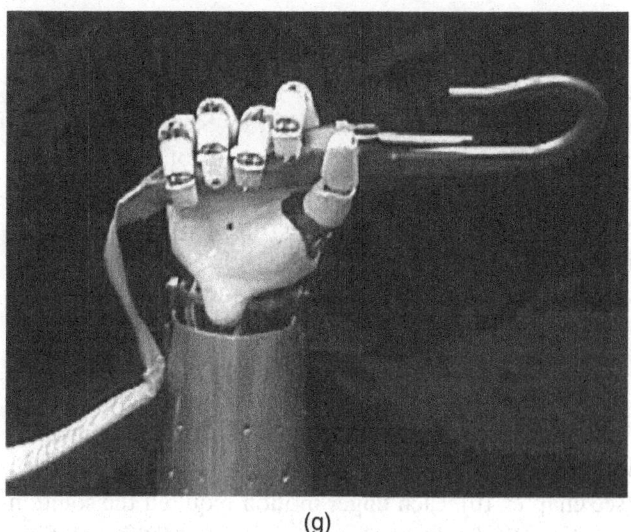

(g)

Figure 11.9
(continued)

- contact data
- finger pressure (in multiple directions)
- torques applied by the object or environment on the fingers
- finger position (joint angles)
- slippage

In addition, information about the force vector at the robot's wrist is important for certain manipulation tasks. Wrist position in space is important for robot control based on computer vision. Proximity sensors may be also needed in some instances.

11.4.3 Examples of Robot Hands

Of the many such hands designed during the past twenty years, we concentrate here on only seven models, shown in figure 11.9. All these hands have been (or are) in use; the Barrett and Omni hands are commercial products. Readers can evaluate these hands in terms of the requirements outlined in the previous section.

11.4.3.1 The Salisbury Hand

The Salisbury hand (Fearing 1986), originally designed in the laboratory of Ken Salisbury while he was on the faculty at MIT, has three fingers. It is also known as the Stanford-JPL hand. For some time in the 1990s, this hand was a commercial product, it is no longer available. A number of Salisbury hands are still in use in research laboratories.

11.4.3.2 The Utah-MIT Hand

The Utah-MIT hand (Jacobsen et al. 1986) is the most complex robot hand in existence. It was designed by Steven Jacobsen at the University of Utah and John Hollerbach, while the latter was on the faculty at MIT, and built by Sarcos, Inc. in Salt Lake City, Utah. This hand could be configured into a variety of grasp modes, including pinch. Its performance on a large number of tasks was truly remarkable. For example, it could be commanded to tap with the fingertips at 10 Hz, probably faster than most human abilities. As can be seen in figure 11.9(b), the hand's finger joints were controlled by cables, pulled by servo actuators, resulting in a controller structure of considerable weight and size that was carried by a separate robot arm. Note that there were four joints per finger, so that each could be flexed, extended, and moved laterally. Since the cable "tendons" used for actuation could only be pulled and not pushed, the resulting 16 dof required sixteen joint position encoders and thirty-two separate cables and actuators. Since each finger was controlled as if it were an RRR manipulator (see chapter 10), each finger motion required the solution of a problem in inverse kinematics. To facilitate the computation of this solution, the Utah-MIT hand was provided with a software system known as CONDOR (Narasinham, Siegel, and Hollerbach 1990). Some ten of these hands were built for various research laboratories, mostly under support of the National Science Foundation.

11.4.3.3 The Belgrade-USC Hand

Since the author was involved in the design and application of the Belgrade-USC hand (Bekey, Tomović, and Zeljković 1990), we consider it separately as a case study in section 11.5. It was a five-fingered hand. In contrast with the Utah-MIT hand, it did not have separate controls for individual finger segments. Rather, the hand was designed to close autonomously when touch sensors detected an object in contact with the fingers or palm.

11.4.3.4 The Barrett BH-Series Hand

The Barrett BH-Series hand (Townsend 2000), a three-fingered hand, has a number of unusual features:

• Like the Belgrade-USC hand, it is completely self-contained, requiring only an 8 mm cable to supply power and two-way communication with the main robot controller. The "palm" of the hand contains four brushless servomotors, several microprocessors, communications and signal processing electronics, and sensors.

• It weighs about 1.18 kg. This light weight makes it suitable for a variety of applications.

• The hand communicates with any PC using a serial port.

• Altogether, the hand has eight joints. Of its three multijointed fingers, two have 3 degrees of freedom and one has only 2 dof, supporting a large variety of grasp types.

• Each of the servomotors drives two joint axes. Torque is channeled through these axes using a proprietary TorqueSwitch mechanism. Thus, when a fingertip contacts an object and reaches the required torque, both joints are locked, and motor currents are turned off.

This hand is a commercial product; more than thirty had been sold by 2001, primarily to automotive manufacturers and suppliers in Japan and the United States and to NASA.

11.4.3.5 The DLR Hand

The DLR hand (Butterfass et al. 2001) is the second-generation robot hand developed at the Institute for Robotics and Mechatronics at the German Aerospace Center, under the direction of Gerd Hirzinger. It has a number of significant features:

• It has four identical fingers, one of which acts like a thumb in opposition to the other three. Each finger has three phalanges and three joints, like a human hand. Thus the hand has 12 dof.

• The proximal joint has 2 dof (flexion/extension and adduction/abduction).

• The motions of the middle and distant finger segments (phalanges) are not individually controllable, but are coupled (like the joints on the Belgrade-USC hand).

• There are numerous sensors on the hand, including Hall effect position sensors integrated into the motors. Each joint has a strain gauge for joint torque measurement and special potentiometers for position measurement.

• Each fingertip contains a tiny force-torque sensor, able to detect the three orthogonal components of force and the rotational components of torque during a contact with the environment.

• Twelve cables connect the hand with the external world.

11.4.3.6 The Omni Hand

The Omni hand, developed by Mark Rosheim (1989, 1994), is a three-fingered hand. It has a number of significant features:

• Each finger has 3 dof and the fingers are interchangeable.

• The hand differs from the others reviewed in the structure of its wrist and palm. The palm folds like a human palm, making a truly opposable thumb possible.

• The wrist has 3 dof (pitch, yaw, and roll). The combination of the articulated wrist, folding palm, and multiple-dof fingers makes this hand very versatile.

• The finger actuators are proprietary electromechanical devices.

11.4.3.7 The Robonaut Hand

The Robonaut hand (Ambrose et al. 2000) was specifically designed to work in space and to grasp the standard tools available to an astronaut performing extravehicular activities (EVAs). Robonaut is discussed in more detail in chapter 13. The hand has five fingers and a size and capability close to that of a gloved astronaut. It has some other interesting features:

- The tolerances on parts made of different materials are selected so that they will perform acceptably even when they expand or contract under the extreme temperature variations in the vacuum of space.
- All parts use space lubricants of extremely low volatility.
- Brushless motors are used to ensure long life in a vacuum.
- The forearm houses all fourteen motors, twelve circuit boards, and all the wiring for the hand.
- The Robonaut hand has 14 dof, divided into two groups. Group 1 (the "dextrous set," used for manipulation) includes those in the thumb, the index, and middle fingers, which have 3 dof each. Group 2 (the "grasping set") includes those in the ring and little fingers, with 1 dof each, and the single degree of freedom in the palm, for a total of 12. There are also 2 dof in the wrist.

We emphasize that the seven hands reviewed in the foregoing are only examples, chosen from a large number of robot hands developed throughout the world. Another very interesting one was the four-fingered hand developed for the humanoid Cog at the Massachusetts Institute of Technology (Matsuoka 1997). The Cog hand had four actuators (one for each finger) and thirty-six sensors. The phalanges were not individually controlled; they depended on an internal synergy comparable to that of the Belgrade-USC hand (see the following section). The sensors included force-sensing resistors, joint motion potentiometers, and motor current sensors.

There is currently a resurgence of interest in robot hands (e.g., Namiki et al. 2003), triggered in part by the availability of new lightweight materials, miniature sensors, and high-powered processors.

11.5 Case Study 11.1: The Belgrade-USC Hand

Shortly after World War II a group of engineers in Yugoslavia, under the direction of the late Rajko Tomović of the University of Belgrade, designed and fabricated a unique prosthetic hand. The motivation for the design of this hand came from the large number of Italian soldiers who had lost a hand in the war's African campaign.

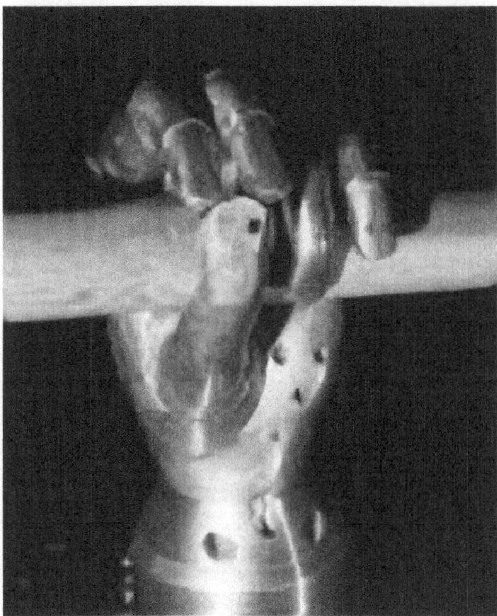

Figure 11.10
The original Belgrade hand (photo courtesy of the late Rajko Tomović)

The so-called Belgrade hand (figure 11.10) was anthropomorphic, had five fingers, was approximately the size of a human hand, and was made of aluminun. The fingertips were equipped with pressure sensors. Its control philosophy was very simple: When the fingertips contacted an object, a single actuator began to close all five fingers until the pressure was approximately equalized among them. This made the hand shape adaptive without any effort on the part of the wearer (Tomović and Boni 1962). Unfortunately, the control and sensor technologies of the day were not adequate for the task, and Belgrade hands were not sufficiently reliable to be used. (There were also attempts at NASA to use this hand as a robot end effector; these efforts were also not successful.) Nevertheless, the Belgrade hand, as the first five-fingered hand, was a significant accomplishment.

In the 1980s there was a rebirth of interest in multifingered hands for robots and a cooperation between the University of Southern California and the University of Belgrade was initiated to produce a new version of the Belgrade hand, which became known as the Belgrade-USC hand. The physical hand was designed and fabricated in Yugoslavia (at the University of Novi Sad), and the sensing and control algorithms were developed at USC. It is illustrated in figure 11.11.

This case study discusses the design principles of the Belgrade-USC hand, in order to highlight a number of more general issues in the design of multifingered robot

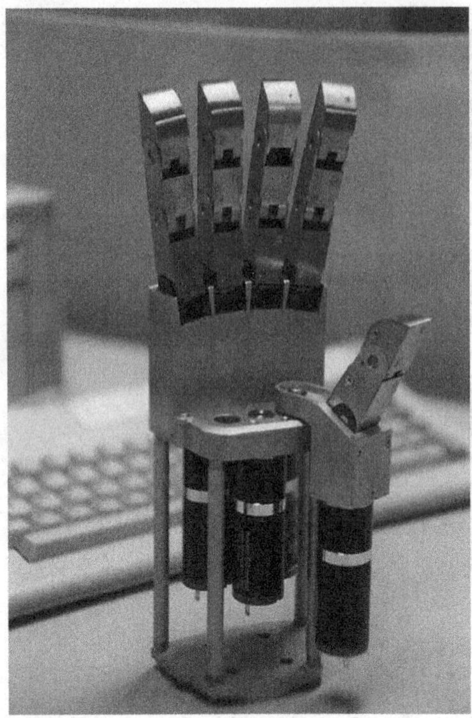

Figure 11.11
Belgrade-USC hand, Model II

hands. Using this grasping device, we show here how the study of human expertise in motor control can serve as a useful tool in the design and control of multifingered robot hands. Nonnumerical approaches to grasp synthesis, combined with other methods, can produce excellent results in the development of versatile end effectors. An important lesson to be drawn from this example is that prior to developing a hand design, it is necessary to formulate a philosophy of control for an intelligent robotic system. In other words, the control strategy must precede the mechanical design. Some five or six Belgrade-USC hands were fabricated and used for many years in university research laboratories in the United States, Germany, and Yugoslavia. In principle, the Belgrade-USC hand was an end effector suitable for use in an industrial environment. However, it was not adopted for industrial purposes in spite of its versatility and potential for replacement of the human hand in certain assembly operations, since it could not compete with a magazine of simple grippers in overall reliability. The following sections describe various aspects of the design of this hand (Bekey, Tomović, and Zeljković 1990).

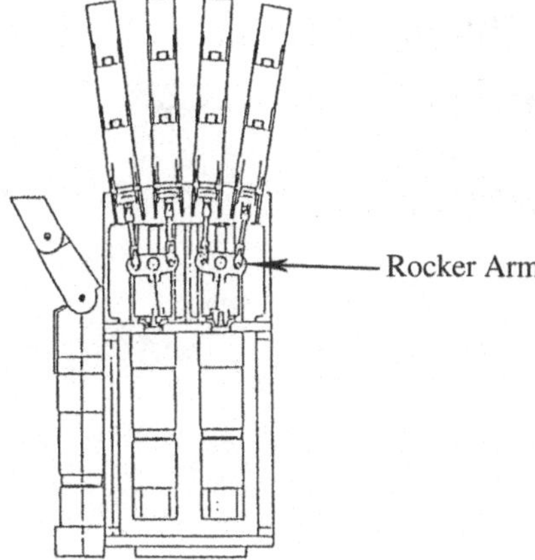

Figure 11.12
Belgrade-USC hand showing finger spreading, thumb articulation, and rocker arm for control of a pair of fingers

11.5.1 Hardware

The physical design of the hand was governed by the following requirements:

- completely self-contained, except for power supplies
- self-adaptive and capable of autonomous grasping, like the original Belgrade hand
- anthropomorphic and approximately human-sized, with five fingers
- capable of at least four grasp modes
- controllable with a standard PC

The design that followed from these requirements included four identical three-jointed fingers and a two-joint thumb, as illustrated in figure 11.11 and the schematic of figure 11.12. The thumb had a third degree of freedom, since the structure that supported it could rotate 120° about an axis parallel to the wrist structure major axis, as shown in figure 11.12. This rotation enabled the thumb to move into opposition with the second, third, and fourth fingers. Note that although the fingers had multiple joints, these did *not* confer degrees of freedom, since they were not individually controlled. All the finger joints were supported by miniature ball bearings and were driven by DC servomotors located within the wrist structure. The motors (shown clearly in figure 11.11) drove the fingers directly (through reduction gears), so there were no cables or tendons.

A number of the control features of the hand were unique. First, the motions of the finger segments were not individually controllable; the segments were connected by linkages to display motions similar to those of human fingers during grasping.

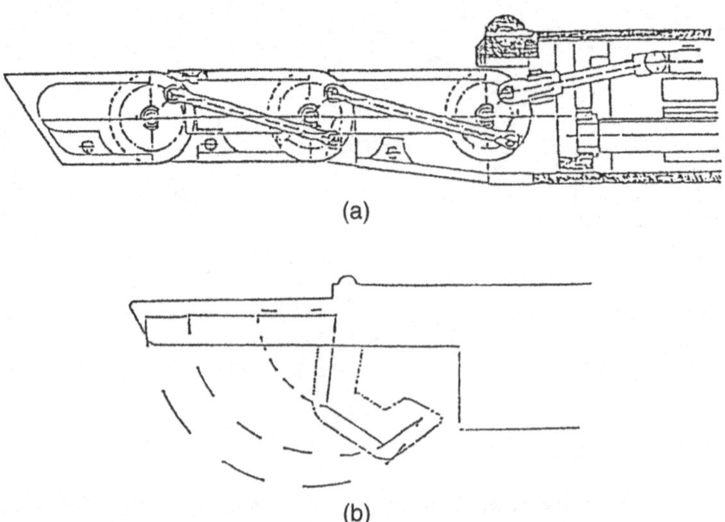

Figure 11.13
Kinematics of finger movement in the Belgrade-USC hand: (a) linkages (b) motion (from Venkataraman and Iberall 1990, reproduced with permission of Springer-Verlag)

This relationship (which Tomović called "internal synergy") led to fixed lever arm relations between the phalanges, as shown in figure 11.13(a). As is evident from the figure, rotation of the knuckle joint would produce rotation of the two additional finger joints. The synergistic motion of the finger segments is shown in figure 11.13(b), which illustrates corresponding points on the joint trajectories.

The second unique aspect of the design of the hand was its autonomous shape adaptation during grasping. There were four motors in the wrist structure (as can be seen in figure 11.12). Two motors were used to position and flex the thumb. The remaining motors were used to flex two fingers each. Contact with an object initiated closure of the four fingers. The motor drive to a pair of fingers was applied through a rocker arm (shown in figure 11.12) designed in such a way that if the motion of one finger of a driven pair was inhibited, the second finger continued to move, thus achieving passive shape adaptation without external control. An illustration of this effect on grasping a circular contour is shown in figure 11.14, in which the broken line shows the finger position without the rocker arm adaptation, and the solid line shows the actual finger position. The autonomous shape adaptation feature provided the equivalent of additional degrees of freedom to the hand, without active control. Finger closure terminated when sensed forces at the fingertips were approximately equal to a preset level.

Even though there were fifteen joints in the hand, there were only 4 dof. The remaining eleven motions were produced through synergies with actuated motions.

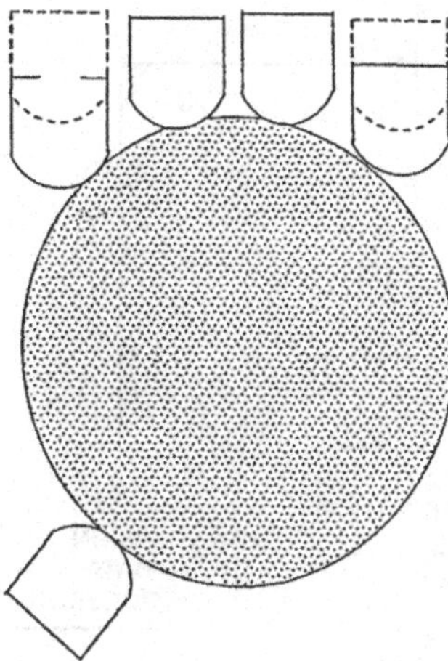

Figure 11.14
Finger position in shape adaptation

11.5.2 Sensors

Basic preshaping decisions for the hand were made using an external vision system (Rao et al. 1988). The hand itself was equipped with two sets of sensors:

- position sensors, to indicate the rotation of the finger base with respect to the palm plane;
- touch-pressure sensors in the fingertips and on the palm, to detect contact with an object and the force being exerted upon it.

Rather than using shaft encoders for sensing finger rotation, we selected simple and inexpensive potentiometers. The high precision possible with encoders was not necessary, since the nature of the grasp control called for only two signals: contact and contact force.

Force sensing was provided using Interlink Electronics force-sensing resistor (FSR) elements. The electrical resistance of these devices changes as a result of applied force. The sensors used in the hand were capable of reading forces in the range of 20 g to 5 kg. The FSR was obtained in thin sheets that were mounted on the fingertips and on the palm. The properties of this material are shown in Bekey, Tomović, and Zeljković 1990.

11.5.3 Control System Architecture

Earlier in this chapter we noted that humans adjust the shape of their hands to match the shape of objects to be grasped. In the Belgrade-USC system, the preshape was

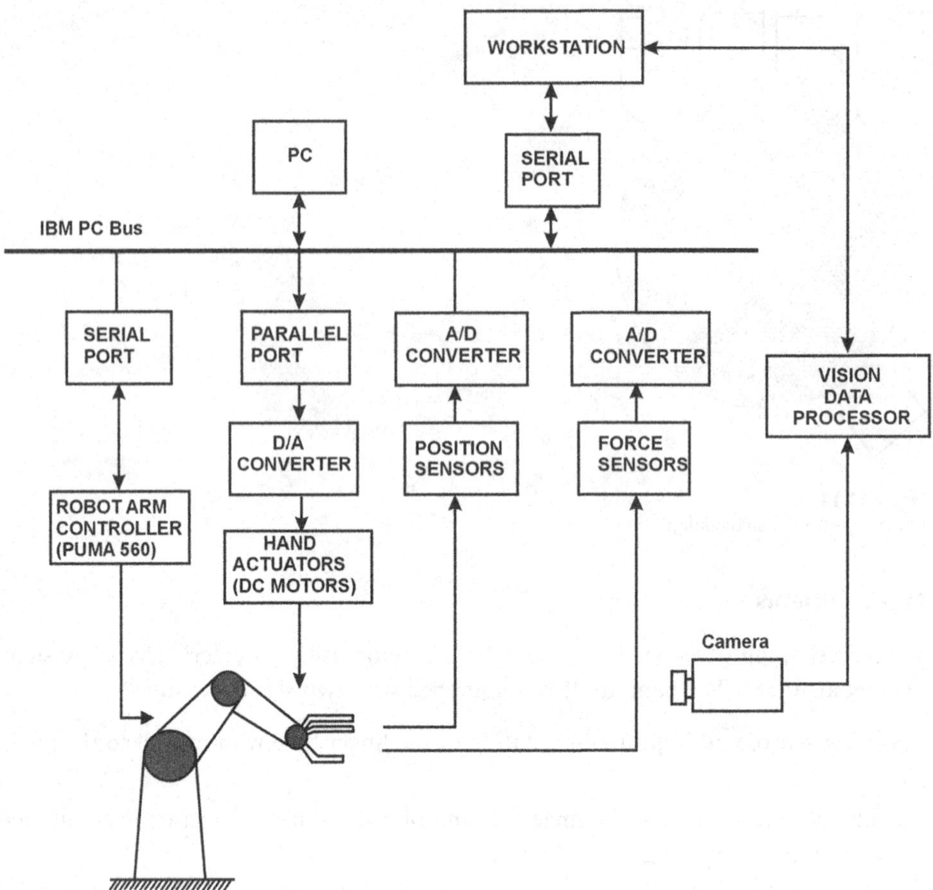

Figure 11.15
Control system architecture for Belgrade-USC hand

controlled using a knowledge-based system. The object to be grasped was observed using a computer vision system. A data structure received from this system provided information on the location, orientation, and geometry of the target object. The geometric information was combined with task information using a high-level computer program to produce the most desirable grasping mode (Rao et al. 1988).

Given the desired preshape, the hand motors were driven to achieve it. Beyond the basic preshaping, the control of hand shape was based on shape adaptation and sensor feedback. This approach to control implies that neither the kinematic nor the dynamic equations of finger motion needed to be solved. The overall control architecture for the hand is shown in figure 11.15.

This brief description has omitted the details of the electronics, the motor drivers, the method of addressing various sensors, and the software, because the approach to all of these elements is now dated, and they would be handled differently today. However, the design issues outlined in this case study may prove useful in future designs of biologically inspired robot hands.

11.6 Prosthetic Hands

The complex nature of the human hand, under the control of the central nervous system, is brought into focus when the hand is lost in battle, in an accident, or (in some cultures) as a result of legal punishment. When people lose a hand, they are losing an intimate partner of their minds that combines fine coordinated movement, tactile sensation, proprioceptive feedback, and aesthetic appearance. Amputees look to engineers for a prosthetic device that captures the beauty and versatility of the original hand.

An artificial hand *is* a robotic hand, but as we shall see, there are both similarities and differences between them. Clearly, both are designed to be *functional*, in the sense of accomplishing the tasks for which they were designed. They differ from one another in the following ways:

- The cosmetic aspects of prosthetic hands are very important; they must closely approximate the appearance of living hands. Robotic hands need not appear humanlike; they must provide controllable and secure grasps.
- Prosthetic hands must be lightweight, comparable in weight and size to human hands. There is no such weight restriction on robot hands.
- Unlike a natural hand with its remarkably versatile attachment structure, secure attachment of a prosthetic hand to the forearm stump is difficult. Robot hands attach to mechanical structures using conventional means like threads or clamps or bolts.
- Control of prosthetic hands is more difficult than for robot hands, because the sources of control information are much more limited. (We encountered this problem previously in connection with artificial arms.)
- The function of a prosthetic hand is expected to be much more limited than that of a living hand. It is generally expected to provide only pulp pinch and/or power grasp. A robot hand is also limited to only one or a few grasp modes, which may not resemble human grasping.

We shall encounter other differences as we review several artificial hands.

The design of a functional and cosmetic replacement hand for amputees has been an elusive goal for engineers for centuries. The first recorded instance of an artificial

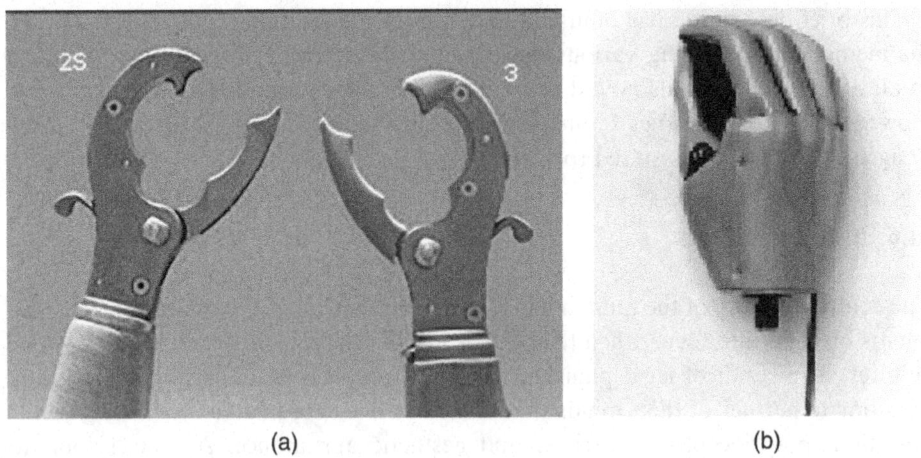

(a) (b)

Figure 11.16
Hook prosthetic hands: (a) prehensors from TRS (reproduced with permission of TRS Inc.); (b) glove-covered hook with two articulated fingers from Hosmer Dorrance (reproduced with permission of Hosmer Dorrance Corporation)

hand was in 200 BC when a Roman general lost his hand during a war and was fitted with an iron hand. One of oldest mechanical prosthetic devices is the hook, as exemplified by the fictional pirate Captain Hook. Even today, the most common hand prosthesis is the parallel hook, shown in figure 11.16. The two hooks on the prehensors in figure 11.16(a) are pressed together by a spring; they are opened for grasping by such methods as pulling on a cable by motion of the opposite shoulder. The hook shape of the prosthesis means that picking up objects with handles requires no adjustment, and the presence of two hooks makes it possible to grasp such objects as a pencil. Hook prostheses appear to be quite primitive, but their very simplicity makes them both useful and inexpensive.

In the past, realistic appearing artificial hands were available only for cosmetic uses. Now, the availability of lightweight alloys and composite materials, integrated circuits, reduced voltage requirements, battery-saving features, reinforced silicones for cosmetic gloves, and space age miniaturized motors and circuits have led to lightweight electrically powered prosthetic hands for adults and children that are both functional and cosmetically appealing (e.g., the Steeper Electric Hand and the Otto Bock SensorHand, shown in figure 11.17). Users can comfortably wear such devices during an eight-hour day.

The first electrically powered hands derived their signals from external switches, touch pads, or other body motions. In recent years a number of myoelectrically controlled hands have become commercially available. When muscles contract, they produce electrical signals associated with the polarization and depolarization of indi-

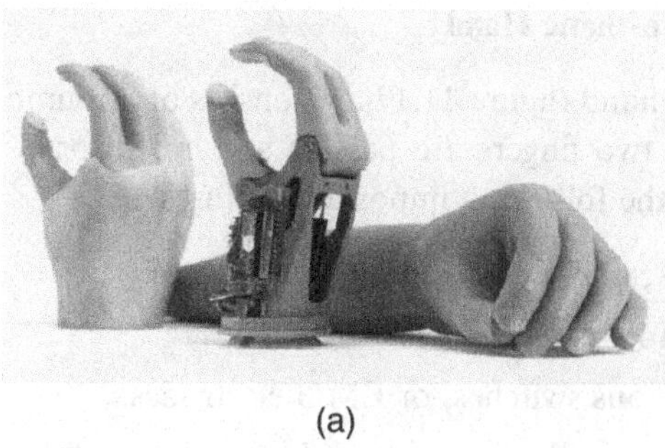

(a)

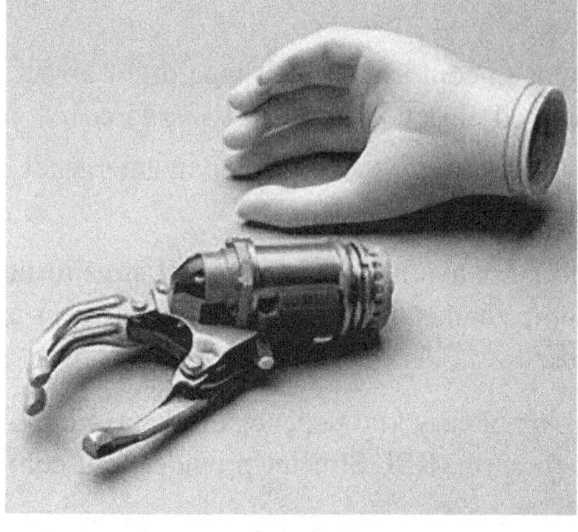

(b)

Figure 11.17
Myoelectric hands and covering gloves: (a) RSL Streeter MultiControl electric hand (photograph courtesy of Liberating Technologies, Inc.); (b) Otto Bock SensorHand (photograph courtesy of Otto Bock HealthCare)

vidual muscle fibers. These signals are conducted by the surrounding tissue, distorted by the tissue impedance, and contaminated by signals from adjacent muscles. When they are detected by electrodes they are known as electromyographic (EMG) or myoelectric signals. EMG electrodes can be located on the surface of the skin above the muscle of interest (or implanted on or near the muscle). Since a number of the muscles responsible for finger contraction are located in the forearm, it is theoretically possible (but difficult) to detect the myoelectric signals produced by these muscles and use them for control of an artificial hand. Figure 11.17 shows two such EMG-controlled hands, the RSL Streeter MultiControl hand and Otto Bock SensorHand. Frequently, such hands can be actuated either by myoelectric signals or by conventional means such as touch switches.

11.6.1 The RSL Streeter MultiControl Prosthetic Hand

The RSL Streeter MultiControl prosthetic hand (figure 11.17(a)) consists of a thumb and two activated fingers. The remaining two fingers are passive and are incorporated in the covering glove. The hand has the following important features:

- The hand is capable of rapid response: it can close or open in 0.9 seconds.
- It is light, with a total weight of about 214 g (7.5 oz).
- It can accept inputs from touch pads, various switches, or EMG electrodes.
- The hand has a power management system that provides additional grip force when needed (like an automatic transmission).
- It has soft, resilient finger pads that allow it to conform to the shape of an object, thus making it easy to grasp an irregular-shaped object without excessive force.
- It has a breakaway thumb that allows the user to release the grip in an emergency.
- The hand has a number of control modes.
- Like other commercial hands, these devices are available in a range of sizes to fit both children and adults with different stump diameters. Gloves are available in a range of colors to match the user's skin tone.

Control of this (and other similar hands) is not easy and depends on the availability of suitable control sites on the subject's forearm. RSL Streeter provides four control strategies:

1. *Single-site voluntary opening with threshold control* This implies the use of a suitable myoelectric site. A single pulse (from a brief muscle contraction) will cause the hand to open; a sustained signal above a threshold will cause it to close at a constant speed. Note that the pulse to open could also be obtained from closure of a switch by means of motion somewhere on the body.
2. *Single-site pulse to open, sustained signal to close, proportional closing speed* This control mode implies that the user is able to produce a range of myoelectric signal amplitudes, thus enabling the hand to close over a range of speeds. Note that the myoelectric signal is a random-appearing complex waveform that must be rectified and filtered for use as a control signal.
3. *Two-site threshold control* This strategy requires the availability of two suitable myoelectric sites (i.e., two controllable muscles, preferably those used for finger flexion and extension in an intact hand). One signal is used for opening and one for closing the hand.
4. *Two-site proportional control* This is an extension of mode 2 with two myoelectric sites. Proportional control may extend to grip strength as well as speed.

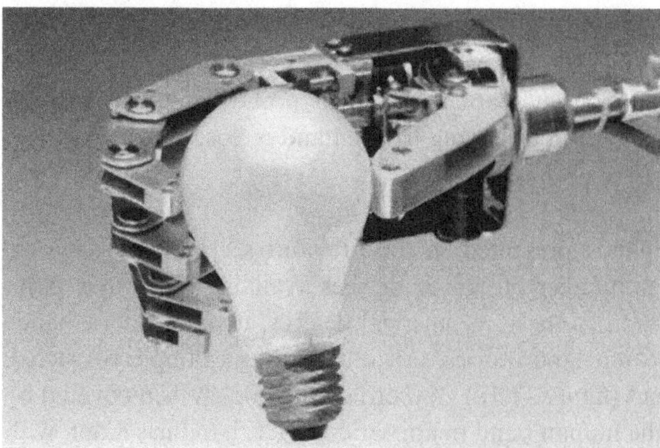

Figure 11.18
Southampton hand (photograph courtesy of Peter Kyberd)

To these four modes it may be possible to add an additional degree of freedom by using a wrist rotation device, available from manufacturers such as Otto Bock. The controllers are more complex when the wrist is involved and require two channels. The battery power requirements also increase when both hand and wrist are controlled.

11.6.2 The Otto Bock SensorHand

The Otto Bock SensorHand, shown in figure 11.7(b), is advertised by the manufacturer as "the most advanced hand in the world." Its major features include the following:

- It is controlled by a microprocessor, which allows for more control options.
- It is equipped with slippage sensors. When an object being grasped is about to slip, sensors in the thumb and finger lever detect changes in the object's weight or center of gravity. The data are then transmitted automatically to the microprocessor that controls grip force, which adjusts the grip force accordingly. This feature means that the wearer need not keep constant watch over the objects being held.
- The structure of the hand and glove, along with the embedded sensors, enable it to grasp any object securely, even fragile items and liquid-filled containers.
- Like the Streeter hand, the Otto Bock hand has a variety of control options using one or two EMG electrodes. A brief myoelectric opening signal stops the Sensor Hand's auto grasp response. A longer myoelectric signal opens the hand. Two independent measurement and regulation systems proportionally control grip speed and

grip force. An unusual control mode allows for proportional opening and proportional closing from a single electrode.

• The so-called Flexi-Grip function lets patients use their sound hand to reposition an object held by the prosthesis, without having to open and reclose it.

11.6.3 Other Prosthetic Hands

The commercial prosthetic hands presented in the previous sections are just two examples from among a large number of similar devices available in Europe, Asia, and North America. In addition to these commercial devices, many artificial hands have been developed in research laboratories. An outstanding example of such a hand is the Southampton hand (figure 11.18) (Kyberd et al. 2001). When covered by a glove, it closely resembles the human hand in appearance. The hand has 4 dof, with four motors and three types of sensors. Note that it appears to be a robot hand, with some resemblance to the Belgrade-USC hand discussed previously. The emphasis in the design of this hand was on autonomy, in order to reduce effort on the part of the user. Control is achieved by means of myoelectric sources, force sensors, and slippage sensors. A sensor-based microprocessor controller adjusts the grasping force as required, to ensure a secure grasp without slippage. To the best of our knowledge, this hand is not yet commercially available.

Despite the advances described above, current prosthetic devices make no pretense of approaching the versatility and functionality of the human hand. With the exceptions cited above, commercially available prosthetic devices, whether hands, hooks, or non-standard devices, generally have only one or two degrees of freedom and limited manipulative ability.

11.7 Concluding Remarks

As with other chapters in this book, we have only introduced, in this chapter, a complex and interesting subject. There are numerous papers dealing with mathematical models of prehension in robot hands (e.g., Li and Sastry 1988). Neural networks have been applied to the translation of task and object geometry to grasp modes (e.g., Liu, Iberall, and Bekey 1989); a simple example of such an application was discussed in section 6.8. Knowledge-based systems have been used to predict grasp modes (e.g., Stansfield 1991; Bekey et al. 1993). We anticipate that both research into and applications of robotic and prosthetic hands will continue for some time, since the gap between the capabilities of human hands and robotic hands remains so large.

12 Control of Multiple Robots

Summary

In this chapter we discuss the control of multiple robots, with emphasis on achieving cooperative behaviors. After describing some of the major issues and some of the history of multiple-robot systems, we consider a special case involving centralized control. We then review several multiple-robot system architectures. Swarm and cellular robots are reviewed, followed by a discussion of communication problems among robots in a group. The organization and the control of robots in formations, including both formations in 2-D and 3-D, are covered in a major section. We then discuss task assignment to robots in a group. The closing section summarizes the major design issues in multiple-robot systems.

12.1 Principles and Problems of Multiple-Robot Systems

There is increasing interest in the development and deployment of groups of autonomous robots to perform tasks that may be difficult, undesirable, or impossible for a single robot. Such tasks may include the following:

• Explorations in hazardous environments (e.g., under water, on distant planets, or inside of earthquake-damaged structures) where failure of one robot should not lead to failure of the entire mission and where redundancy may increase the fault tolerance of the colony.

• Tasks beyond the limits of single robots (e.g., cooperative lifting or pushing of large and heavy objects or assembly of large, complex structures, say, in space or on a planetary surface).

• Tasks that can be completed more rapidly by multiple robots than possible is for a single robot (e.g., collecting trash in a stadium after a football game or searching for unexploded land mines).

• Complex tasks (e.g., exploration, excavation, communication, and object retrieval on a planetary surface) that may be less expensive with a group of specialized, simpler vehicles than with a single, multipurpose robot.

• Highly distributed sensing, in which large colonies of simple and inexpensive robots are used as mobile, communicating sensors. We visualize such colonies with thousands or tens of thousands of members, distributed over large areas, capable of chemical sensing (e.g., to detect the presence of explosives, hazardous chemical substances, or chemical signals); motion sensing (e.g., to detect small earth movements from seismic activity or movements of large numbers of people or animals); or temperature sensing (e.g., to detect survivors in rubble or changes in water temperature in the ocean).

As we have seen in previous chapters, control of biologically inspired autonomous robots requires integration of principles from biology, control theory, kinematics, dynamics, computer engineering, and other disciplines. Applications such as those just listed may require the additional consideration of issues from animal ethology, social psychology, organization theory, and even economics.

From a control point of view, the basic question addressed in this chapter is "How do we control multiple robots so as to achieve collective tasks?" Clearly, there are two basic approaches: centralized control and distributed control. Centralized control can be a highly efficient (and inexpensive) method of controlling small groups of robots, since the control software and hardware resides only in the supervisory robot, and the "workers" need have only the actuators needed to perform a given task. Even much of the sensing can be centralized in such a structure, as we illustrate with a specific colony design in section 12.5. The centralized approach is used by many military and industrial organizations, which are fundamentally hierarchical in structure. The central controller (the "general") exercises control over a small number of high-level officers, each of whom, in turn, controls junior officers, who control noncommissioned officers. The latter then command the front-line soldiers. To accomplish a global goal with a colony of robots under this model, the colony may require supervision and subtask assignment by a hierarchy of "supervisor robots." The distributed-control model, inspired by insect societies, is a generalization of behavior-based control to multiple robots in which it is assumed that close coupling of perception with action among members of a colony, each working on local goals, can accomplish a global task.

Control of groups of robots is also more difficult than control of single robots for other reasons. By definition, multiple robots are able to operate in each other's reachable spaces. In effect, each robot in a colony becomes a moving obstacle to other robots, so that obstacle avoidance becomes a necessity. As we have seen, obstacle avoidance can be simply one of the behaviors in a behavior-based system. There are major research programs related to highly distributed networks of mobile sensors; these programs can be viewed as another instance of multiple, cooperative robots. A colony may require the ability to reprogram robots to different tasks in the event of failure of a member of the colony, and it may require explicit or implicit communication among the members. The issues of communication and organization are complex and are considered in later sections of this chapter. We can look for examples in the behavior of colonies of insects, such as ants, bees, or termites.

A further issue in robot colonies concerns the nature of the group interactions. The preceding paragraphs have discussed *cooperative* behaviors among multiple robots. However, as indicated by Wilson (1971, 2000), in animal societies group behaviors may include competition as well as cooperation, aggressiveness as well as altruism. Some attempts have been made to produce some of these alternative behaviors among multiple robots as well.

A final important issue in the design of a multiple-robot system is scalability, that is, the ability to apply a design to larger and larger numbers of robots. Whereas early work in this field emphasized cooperation between, say, two or three robots, in the future it will be necessary to design systems with hundreds, thousands, or perhaps millions of members. Can a given design be scaled to such large numbers?

Research in cooperative robotics also benefits greatly from research in multiple intelligent agents. A robot can be viewed as an agent implemented in hardware, so these two areas clearly overlap. We consider all these aspects of robot colonies in the following sections.

Excellent reviews of the state of the art in multiple mobile cooperative robotics up to the late 1990s have been published by Cao, Fukunaga, and Kahng (1997), Arkin and Balch (1998), Agah (1996), and Balch and Parker (2002). However, this is an extremely active area of research in robotics, and no published review will be completely up to date even when it is published.

12.2 Biological Inspiration: Sociobiology

In the animal kingdom we can observe cooperative behavior on both a large and a small scale. Organization of behavior in large groups is observed in insect societies, such as bees or ants or termites; these are among the best known of the twelve thousand species of social insects. Smaller groups of animals, such as flocks of birds, herds of wildebeest, and prides of lions also exhibit group behavior. We owe much of our understanding of such behavior to the work of Edward O. Wilson (1971, 2000). Wilson coined the word "sociobiology" to describe this field. One need but browse through the nearly seven hundred pages of the Wilson's (2000) volume with the same title to get an impression of the complexity and vastness of this field. Clearly, we cannot do justice to sociobiology by summarizing it in a few paragraphs. We can only hope to hint at a few of its salient points, so that interested readers can delve further into this fascinating field.

Insect societies are governed by rigid instincts and appear to have a small number of built-in rules for behavior. As stated by Wilson (2000, v): "Each insect colony is an assemblage of related organisms that grows, competes, and eventually dies in patterns that are consequences of the birth and death schedules of its members." In such a society there is little if any real autonomy or learning. Yet such societies display remarkable collaborative behaviors. It is important to note that insect societies are *self-controlled* or *self-regulating* systems, in the sense that there is no king that issues orders to viceroys who in turn issue orders to lieutenants, and so forth. (The queen in some insect societies is in fact an egg-bearing machine, with no real authority over members of the society.)

Insects in societies display at least three common characteristics:

1. Members of the society collaborate in caring for the young.
2. Those members with the highest reproductive potential have a higher standing in the society, so workers that are sterile tend to work for the benefit of their more fertile fellows.
3. There is an overlap of at least two generations in the work being done for the colony, so that offspring help their parents during some time in their lives.

Clearly, these aspects of behavior are not directly applicable to robot societies. There are, however, two additional aspects of organization of insect societies that can serve as models for robot societies as well:

4. Insect societies have evolved a large degree of specialization among their workers; there may be as many as ten distinct specializations.
5. To coordinate their activities, insects have developed a surprisingly rich repertoire of communication methods.

The issue of specialization is an important one for group robotics. As we show in section 12.4, if robots are to perform a variety of tasks, it is possible to deploy either identical (but highly versatile) systems or a variety of specialized ones. Generally speaking, general-purpose robots will be significantly more expensive than those designed for a special task. If we take our cue from the biological example presented by insect societies, then specialization will be the preferred choice.

Communication is another fascinating characteristic of insects in societies. For example, in order to communicate, insects may tap each other, stroke each other, contact each other's antennae, or exchange liquid or food particles, and they may release a variety of chemicals to assist in recruitment for completion of a task or to announce danger. This is only a partial list of communication modalities, but it illustrates the importance of communication for a self-regulating colony.

The preceding list of communication modalities raises a corresponding question for robot societies: Do robot colonies require a variety of communication methods if they are to be successful in the absence of central control? We discuss this question in section 12.8. Agah (1994) coined the term "sociorobotics" to acknowledge biological inspiration for his study of multiple-robot systems.

Tinbergen (1996) discusses a variety of social behaviors among animals, including simple imitative cooperation, varieties of mating behaviors, family and group behaviors (such as flocking, communal attack, herding, fighting, and mutual hostility among members of the same species). These are simply examples of the complexity of group behaviors in animals. A relatively recent example of this complexity in the behavior of lionesses protecting their territory is given by Heinsohn and Packer (1995). Apparently some lionesses are much more aggressive than others in repelling intruders. Some lionesses lag behind, effectively allowing the others to fight for them. Are the more aggressive members of the pride displaying an altruistic behavior and protecting those who are more "cowardly"? The answer is not yet clear, but it illustrates that multiple-animal behaviors are not just the result of simple aggregations of individuals. Some attempts have been made to produce some of these behaviors among multiple robots as well.

12.3 A Brief History of Multiple Robots

Work on multiple mobile robots (either autonomous or teleoperated) has lagged behind research on single robots. This may be due primarily to the fact that for many years robot hardware (and software) was so unreliable that it required an enormous amount of effort to keep a single robot working, to say nothing of multiple machines. This state of affairs began to change in the mid-1990s as reasonably priced commercial mobile robots became available and reliability gradually increased.

One of the earliest research efforts in this field is due to the biologist W. Grey Walter (1950), who constructed two electromechanical "tortoises" (introduced in chapter 5) more than fifty years ago. These robots, named Elmer and Elsi (see figure 12.1), were used in what may have been the first experiments on autonomous robots. Each robot's controller contained vacuum tubes (since this was before the invention of the transistor), a photocell, a touch sensor, a motor for movement, a motor for steering, and batteries. When no light was present, the robots moved randomly in the environment. When a light source was present, the photocell circuit caused the robots to move toward the light when they were far away and move away from it when they were too close. Placing lights on the robots allowed for study of the interaction between the robots and their reflection in a mirror, which Walter termed "self-recognition." It is interesting that even in these early experiments, more than

Figure 12.1
One of Grey Walter's tortoises (photograph courtesy of Owen Holland)

fifty years ago, robots were described in biological (and indeed, anthropomorphic) terms. In Walter's (1950) own words:

> These machines are perhaps the simplest that can said to resemble animals. Crude though they are, they give an eerie impression of purposefulness, independence and spontaneity.

In the 1980s there was significant interest in the control of multiple manipulators (e.g., Zheng and Luh 1985; Hayati 1986; Koivo and Bekey 1987). In multiple manipulators, two robot arms grasp the same object, such as a large panel for automobile assembly. The main feature of systems of multiple manipulators is that they experience a *decrease* in their total degrees of freedom at the moment of grasping. Two 6-dof arms grasping the same object, for example, do not have 12 degrees of freedom, since the arms constrain one another's movements. More recently various pictures of cooperative manipulation have appeared in the literature, such as figure 12.2 from our own laboratory at USC. However, the activity depicted in this figure (and other similar ones) is somewhat artificial. The two robots were able to move holding the pipe, but they required assistance in grasping it.

A more interesting approach to dual-arm manipulation, demonstrated at the University of Pennsylvania (Yun 1993; Desai, Zefran, and Kumar 1998) and at Stanford University (Chang, Holmberg, and Khatib 2000), involves the use of force sensors in

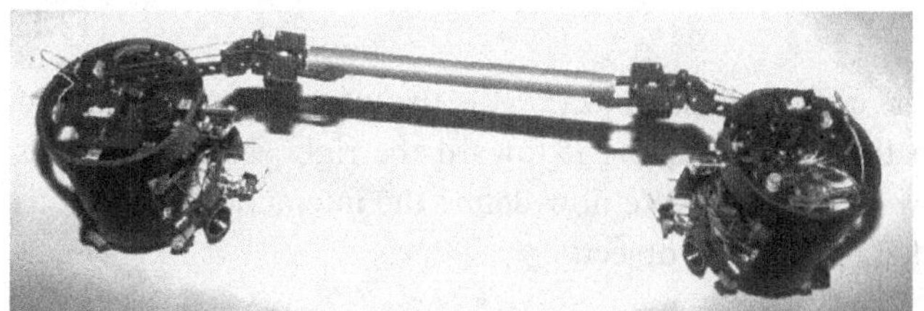

Figure 12.2
Cooperative manipulation (from Agah 1994, used with permission of author)

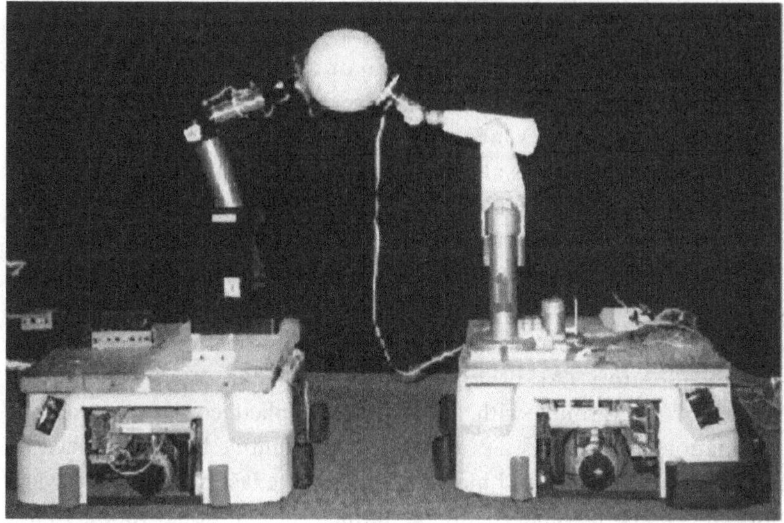

Figure 12.3
Cooperative manipulation without grasping (photograph courtesy of Vijay Kumar)

the end effectors (hands) while they hold an object by pressure alone, without actual grasping. In the most interesting aspect of the projects involving this approach, robot arms were mounted on mobile platforms, so that both the platforms and the arms were able to move while holding an object. To illustrate the control complexity associated with this task, consider the situation depicted in figure 12.3. Each robot manipulator is equipped with a flat "palm" but no gripper, so that they only push (but not pull) an object.

Consider movement in only one dimension. The robots are required to move the object back and forth, say, in sinusoidal motion. A problem arises here as a result of the restriction to pushing and the need to coordinate the two forces so as to hold the object securely. From Newton's law of motion, assuming that both the object and the robot palms are rigid, we can write

$$m_0\ddot{x}_0 = F_1 + F_2, \tag{12.1}$$

where x_0 is the position of the center of the object, m_0 is the mass of the object, and F_1 and F_2 are the forces on the object pushing it toward the right. Note that F_1 is always positive and F_2 is always negative. We now define the interaction force F_I as the minimum net force needed to hold the object:

$$F_I = \min\{F_1 - F_2\} = \frac{F_1 - F_2 - |F_1 + F_2|}{2}. \tag{12.2}$$

Clearly, this force will depend on the weight of the object (to cancel the gravitational force) and the coefficient of friction between the palms and the object. If the interaction force is given by the designer, it is now necessary to design a controller that maintains it while also regulating the desired motion of the object. This is a complex problem in control theory because of the constraints in (12.1); see Yun 1993 for a complete analysis of the control issues involved.

Cooperative work by two robots has also been used in such tasks as moving a sofa (Rus, Donald, and Jennnings 1995). Kube and Zhang (1994) have studied cooperative multirobot systems, both in simulation and in hardware, in such tasks as box pushing. Like other investigators, Kube and Zhang found that the performance of a colony improves as more robots are added, up to a critical value. Beyond this value, increases in the population may actually result in a decrease in performance.

Arkin and his collaborators (Arkin 1993, 1998; Arkin and Balch 1998) have studied schema-based robot groups with and without explicit communication. In schema-based navigation each goal is decomposed into primitive behaviors, such as move-to-goal, avoid-static-obstacle, and move-randomly (the last being used in wandering). Cooperation is made possible by the recruitment of additional robots to accomplish a task.

Two classes of robot communication, active and nonactive, were introduced in Premvuti and Yuta 1991. In this classical paper, robots can switch between autonomous modes and modest-cooperation modes. Robots can work for their own purposes, or they can work for a common goal jointly with others. In the latter case, a leader is chosen, thus making these into Type I groups, as discussed in the following section.

Pioneering work in behavior-based multiple-robot systems was done by Matarić (1994) and is discussed in section 12.6.1. The tropism architecture of Agah was discussed in sections 5.8 and 6.11, so it is mentioned only briefly in this chapter.

The works cited in the preceding discussion are representative of the large amount of work in the field of multiple robots. Interested readers are referred to the *Proceedings of the Conferences on Distributed Autonomous Robotic Systems* (DARS) (e.g., Parker, Bekey, and Barhen 2000) and the *Proceedings of the NRL Workshops on Multi-Robot Systems* (e.g., Schultz and Parker 2002). Our group at USC has been active in multiple-robot system studies for a number of years, including such projects as reconnaissance using teams of heterogenous mobile robots (Sukhatme, Montgomery, and Matarić 1999), using multiple robots for interior mapping (Dedeoglu, Matarić, and Sukhatme 1999), autonomous task assignment among multiple robots (Fontan and Matarić 1998), and large sensor networks in which some of the sensors are carried by robots (Howard, Matarić, and Sukhatme 2002a). Clearly, this is a very active area of research in robotics.

12.4 Control Issues in Autonomous-Robot Colonies

The question of control in robot colonies requires a wholly different perspective from that of individual robot control. In general, control of each member of a colony by an external controller is possible only with small groups; it is difficult or impossible as the colony grows in size, just as it is difficult or impossible for a general to control the actions of each individual soldier on a battlefield. Hence, we see the emergence of three major types of control strategies:

• *Type I—Centralized and hierarchical control* By analogy with control of an army, factory, or a construction crew, each individual in this type of control strategy is responsible for a small number of other individuals, who in turn control a like number of others, to the bottom level where the actual physical work is done. In this architecture the global goal of the entire team may be known in detail only to the top level or levels of the hierarchy. For example, a man carrying supplies to a construction crew may not know whether he is working on an apartment building or an office building and may not know any details of its size, construction features, and so on. If the group is small enough, there may be a single controller or supervisor for the entire group. There is clearly a problem if a high-level supervisor is ill or disabled. A lower-level individual must in such a case take the place of the disabled leader. This happens in military situations: If a sergeant is killed, one of his men takes over. In group robotics, this implies that each robot has an internal model of a supervisor sufficient to allow it to take over the work of the supervisor who has failed. At present, there is no theoretical basis for enabling a robot to take over the work of another using internal models.

• *Type II—Decentralized and local control* A completely different approach is based on allowing each individual to operate on local information while accomplishing global goals. In effect, this strategy requires that these goals be contained implicitly within the rules of behavior governing each individual. In effect, cooperation is an emergent property of the robot group; it is a consequence of the way in which the robots interact with the environment. As a biological example, consider a colony of African termites building one of the huge mounds of sand for which they are famous, within which the colony has its home. It is clear that the termites carrying grains of sand to the mound do not have a global blueprint of the structure in their brains. Rather, they follow simple rules, such as "Move toward home, go as high as you can, and deposit your load." Note that this is very different from a group of workmen building Notre Dame cathedral, but even here, individuals do not have a complete blueprint. Only the architect and the chief contractor have complete blueprints. The workers digging a foundation, building a wall, or installing windows have only a limited view of the global goal.

• *Type III—Hybrid structures* Some organizations may incorporate both Type I and Type II structures, in which some groups are highly autonomous and decentralized and others operate under central authority. For example, Parker (1994) suggests four architectures for multiple-robot control: (1) local control alone (Type I), (2) local control augmented by a global goal, (3) local control augmented by a global goal and partial global information, and (4) more complete global information, resulting in a control structure approaching Type II.

A great deal of information can be gained from a study of human work structures, such as factories. The organization and control of such groups is the subject of numerous publications (e.g., Hackman 1990; Majchrzak and Gasser 1992). In human organizations, the success of workgroups is related to such factors as reward systems, information access, training programs, and task overlap. Majchrzak and Gasser (1992) also describe the importance and effects of various levels of autonomy and self-regulation in groups.

12.5 Case Study 12.1: Centralized Control of Very Simple Robots

A centralized, single-supervisor architecture is particularly applicable to situations in which a central controller can be physically located in a position from which it has line-of-sight communication with a colony of simple, control-free "slave" robots (Khoshnevis and Bekey 1998). Such a situation may arise where a controller can be mounted in the ceiling of a warehouse and then used to control a group of forklifts or tractors that move boxes from and to desired locations. We describe here such a system as developed at USC and implemented with a group of simple robots.

The system is designed to overcome the following limitations of individual autonomous robots:

• The individual robot structure is larger than desirable as a result of accommodation of sensors and controllers. A considerable amount of on-board power is consumed in sensing and in control data processing (by the on-board computer) for each robot. The sophistication and intelligence of control is limited because of the relatively small scale of the on-board computers.

• Overall positional sensing is limited by the precision of the sensors (e.g., ultrasound) and slow because of the need for frequent localization (resulting from possible changes in the environment).

• Manufacturing costs are added because of the desire for compactness in the packaging of the structure, actuators, power supply, sensors and transducers, and computers.

• Unit cost is added because of the complexity and large number of expensive and compact components. The unit cost becomes more significant in the case of collaborative multiple robots.

These limitations become particularly acute when one attempts to build insect-sized robots with some on-board intelligence.

As an illustration of the alternative, centralized approach, consider the situation illustrated in figure 12.4. The figure shows a swarm of small robots performing electronic assembly. The conveyor at the top of the figure carries circuit boards. The robots pick up components from the store at the bottom of the figure and deliver them to appropriate locations on the boards. Each robot is extremely simple, being equipped with only a gripper and inputs to cause it to move in a desired direction, stop, grasp an object, place it in a desired location, and release it. Note that the robots do not require navigation or obstacle avoidance features, since their movements are controlled by a central computer with a vision system that tracks and controls their movements.

A system of the form of that in figure 12.4 was fabricated at USC and used to control several simple robots, as illustrated in figure 12.5. As figure 12.5(b) shows, the robots (about $4 \times 6 \times 4$ cm in size) in this system are indeed very simple, since

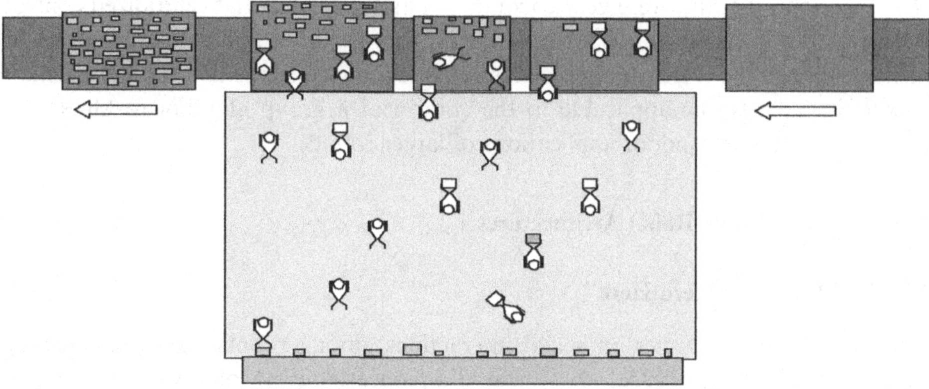

Figure 12.4
Centralized control of multiple robots (illustration courtesy of Behrokh Khoshnevis)

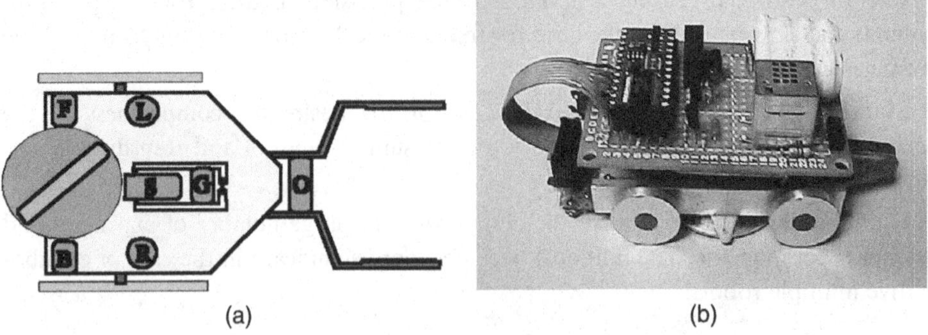

Figure 12.5
Simple robots for centralized control scenario: (a) schematic: B-backward, F-forward, R-right, L-left, G-gripper; (b) physical prototype robot (illustration and photograph courtesy of Behrokh Khoshnevis)

sensing, control, and task allocation are centrally determined. They do not require vision or ultrasound collision avoidance systems, since a ceiling-mounted computer with a sophisticated vision system controls their movements. Each robot needs an appropriate identification or signature to enable the central computer to control its desired position and orientation. The central station is equipped with a high-speed digital camera, an infrared light source, and image-processing software. The schematic in figure 12.5(a) indicates a number of light-sensitive spots to control direction and gripper closure. In the practical realization of figure 12.5(b), each robot carries a single photodetector and decoder circuitry. Each is assigned a code so that each control command can select a particular robot, then send encoded commands for motion direction and grasping.

Preliminary results from this system, using a small number of robots of the type shown in figure 12.5(a), were very encouraging and indicated that centralized control is indeed a reasonable option for multiple robots performing a restricted range of tasks. Although the scenario discussed in the foregoing concerns the use of very small robots, it may also be applicable to the control of a group of robot forklifts in a warehouse or similar indoor applications of larger robots.

12.6 Some Multiple-Robot Architectures

12.6.1 Matarić's "Nerd Herd"

One of the earliest studies on social interactions among robots was conducted by Maja Matarić (1992b, 1992c, 1993), who did pioneering work on a variety of group behaviors. Her work was both theoretical and experimental; her experimental work

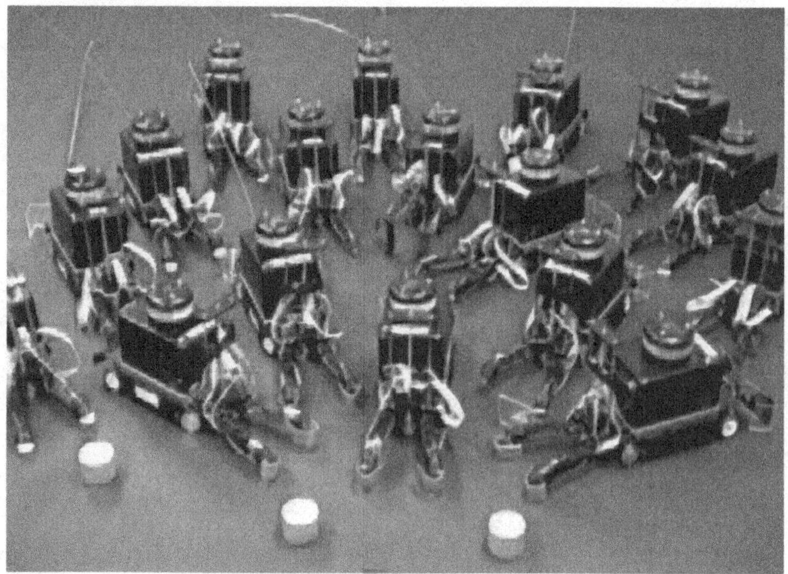

Figure 12.6
The "Nerd Herd" colony of twenty robots (photograph courtesy of Maja Matarić)

was performed on a collection of twenty identical mobile robots. Since these robots did not possess a great deal of intelligence, they were sometimes referred to as the "Nerd Herd." The robots (illustrated in figure 12.6) were about 12 in tall, moved on four wheels, were equipped with piezoelectric bump sensors on the sides and rear and with a gripper suitable for grasping cylindrical pucks.

The interaction primitives used in Matarić's experiments were

- collision avoidance;
- following (staying behind or alongside another robot);
- dispersion (having a group of robots spread out over an area until they achieve a desired separation);
- aggregation (the inverse of dispersion: having a group of robots gather until they achieve a desired proximity);
- homing (ability to reach a desired goal region or location);
- flocking (having a group move in a coherent way without a specified leader).

Note that flocking includes components of collision avoidance, following, dispersion, and aggregation, so it may be considered a composite group behavior. A unique feature of Matarić's work was the use of animal models to develop very simple algorithms for the above behaviors. Thus, collision avoidance with other robots was determined by the following algorithm:

```
Avoiding Other Agents
If another robot is on the right
      turn left
      otherwise turn right
```

This simple strategy was successful for Matarić's Nerd Herd because all the robots were alike; it might require modification for colonies of heterogeneous robots. If a robot fails to recognize another, it will then use an obstacle avoidance program. As one more example of the software structure of Matarić's robot groups, consider the behavior known as following, which depends on the output of IR sensors:

```
If an object is on the right only,
      turn right.
If an object is on the left only,
      turn left.
If an object is on the left and right,
      keep going and count time.
If an object is on the left and right,
And has been there for some time,
      stop.
```

In later work Matarić combined some of the behaviors given in the foregoing list into other composites, such as foraging. In foraging behavior the robots begin to search for food by dispersion. Once an agent has located "food" (say, one of the pucks), it begins homing. During the homing phase, it may avoid other robots not carrying food but follow others who have it. The group of food-carrying robots then forms a flock. In this way, primitive behaviors can be combined to produce more complex ones.

Similar basic behaviors for foraging, acquisition, and delivery were studied by Arkin and his students (Arkin 1998).

12.6.2 The MissionLab Architecture

The Autonomous Robot Architecture (AuRA) developed in the Mobile Robot Laboratory at Georgia Institute of Technology (see chapter 5) is a hybrid reactive-deliberative architecture. It has been used by the Georgia Tech group as a basis for development of a set of software tools (known as MissionLab) that allow a non-expert user to specify robot missions and automatically configure the software needed for coordination and control of a group of robots, either in simulation or in hardware. MissionLab also allows for formation control of teams of robots under teleoperation.

Although the full description of the software developed in this research program is beyond the space available in this book, we summarize here some of its major features (Mackenzie, Arkin, and Cameron 1997); the system is described at length in Arkin 1998. The work begins with decomposition of the control program of a reactive, behavior-based robot into a collection of behaviors and coordination mechanisms for these behaviors. Then, a multiagent robot configuration is created in three steps: (1) identifying an appropriate set of skills for each robot; (2) translating those mission-oriented skills into sets of suitable behaviors (called *assemblages*); and (3) construction or selection of suitable coordination mechanisms to ensure that the correct assemblages are used throughout a given mission. Use of the architecture and software tools assumes that users are familiar with behavior-based robot control and programming languages such as C++.

The software primitives used to construct the behavioral agents are spcified using a configuration description language (CDL). For example, a partial CDL description of a multiagent system for trash collection, called Janitor, is shown in figure 12.7(b). Temporal sequencing of behaviors is accomplished by means of finite-state automata (Arkin and Mackenzie 1994; Mackenzie, Arkin, and Cameron 1997b).

The MissionLab set of software tools was built using the CDL. It includes a graphical configuration editor (CfgEdit), a multiagent simulation system, and two different architectural code generators. To illustrate the application of these tools, consider the janitorial robot system just mentioned, as described by Mackenzie, Arkin, and Cameron (1997). The goal of the project was the creation of a robot system (of one or more robots) that would wander around looking for and picking up empty soda cans, find recycling baskets, and place the cans in the baskets. The operating states for each robot were Start, Look_for_can, Pick_up_can, Look_for_basket, and Put_can. Figure 12.8 shows the state diagram implementing the trash-collecting robot; the circles in the figure represent the operating states within the finite-state machine, and the rectangle in the center of each circle lists the behavior active in that state.

The system prompts the user to bind the architecture to a specific robot, and multiple robots can be included simply by using the copy feature in CfgEdit. MissionLab includes an operator console allowing for execution of a mission in simulation or in hardware. A typical simulation run is shown in figure 12.9. Hardware missions were performed in the laboratory using Denning robots (cylindrical robots surrounded by a ring of sonars, manufactured by Denning Branch International in Australia). The system has also been used for controlling U.S. Army autonomous vehicles in the field.

(a)

```
   /* Define cleanup behavior as a prototype */
1. defProto movement cleanup();

   /* Instantiate three cleanup agents */
2. instAgent Io from cleanup();
3. instAgent Ganymede from cleanup();
4. instAgent Callisto from cleanup();

   /* Create an uncoordinated janitor society */
5. instAgent janitor from IndependentSociety(
           Agent[A]=Io,
           Agent[B]=Ganymede,
           Agent[C]=Callisto);
   /* janitor agent is basis of configuration */
6. janitor;
```

(b)

Figure 12.7
(a) Three trash-collecting robots, named Io, Ganymede, and Callisto; (b) partial CDL description of multi-agent Janitor configuration. Comments are bounded by /* */ (from MacKenzie, Arkin, and Cameron 1997, 37, reproduced with permission of Kluwer Academic Publishers)

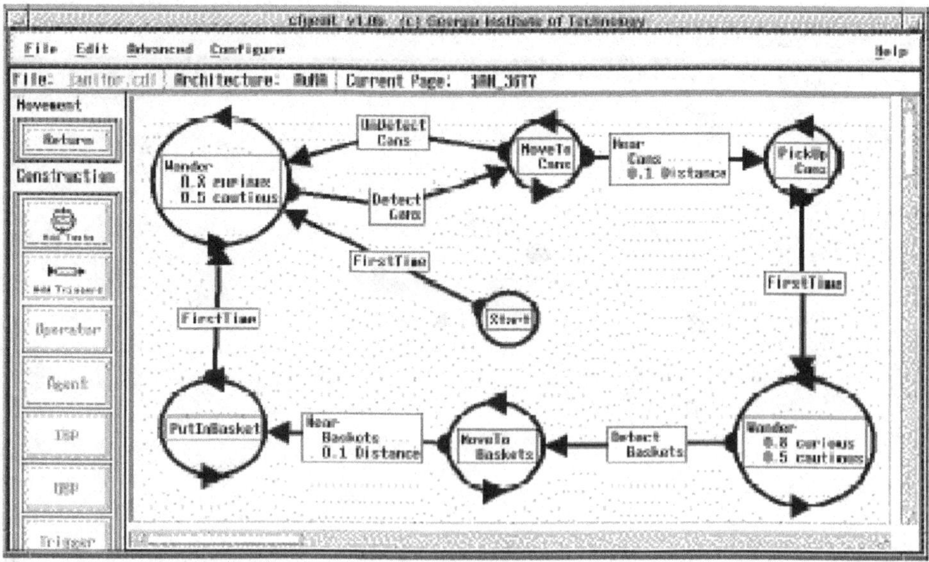

Figure 12.8
State diagram defining the finite state automaton for the trash collecting robot (from MacKenzie, Arkin, and Cameron 1997, 42, reproduced with permission of Kluwer Academic Publishers)

12.6.3 The ALLIANCE Architecture and the Problem of Cooperative Multirobot Observation of Multiple Moving Targets (CMOMMT)

The goal of the ALLIANCE architecture (Parker 1994, 1997, 1998, 1999; Parker and Emmons 1997) was the development of a fault-tolerant, adaptive, distributed, behavior-based software system for control of teams of robots. The robots were to accomplish a mission (such as hazardous-waste cleanup) in a dynamic environment and in the presence of failures in their action selection mechanisms and with noisy data from sensors and erratic (noisy) end effectors. Clearly, this is a tall order for a robot team, but ALLIANCE has been very successful at accomplishing its goals and has inspired other approaches to the control of multiple robots. The major features of the architecture are the following:

- No centralized control is utilized. All the robots are fully autonomous and have the ability to perform useful actions even in the presence of failures of other robots.
- The robots on the team can detect, with some finite probability, the effects of their own actions and those of other members of the team. Note that being able to detect their own actions implies the presence of sensors and feedback control on the robots. The actions of others are detected via an explicit communication mechanism.

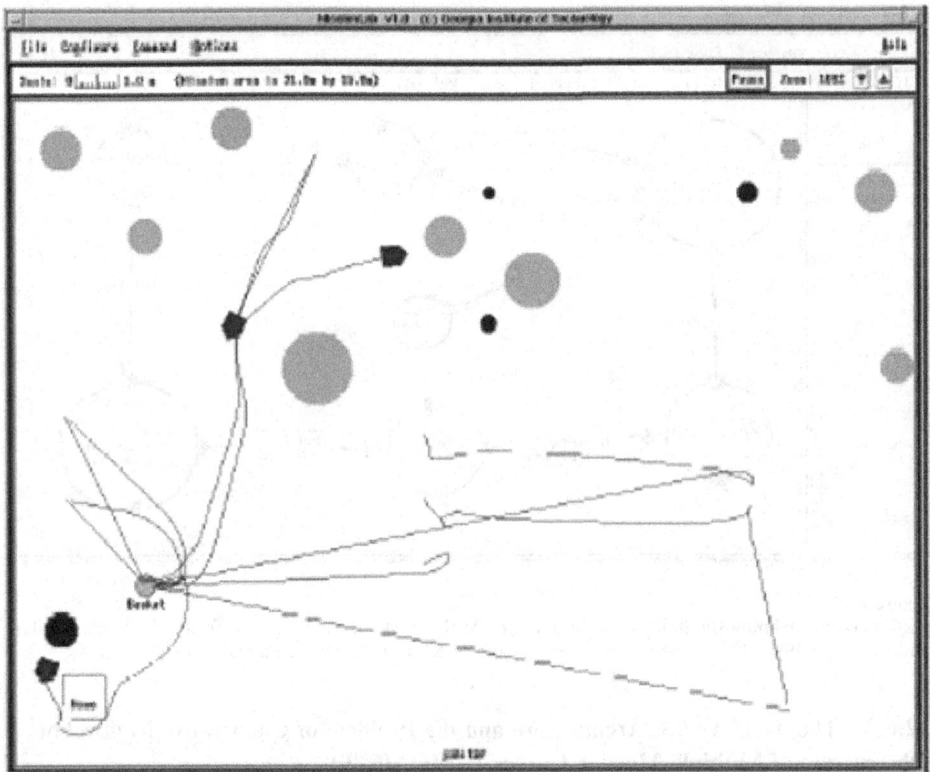

Figure 12.9
The Trashbot configuration executing in simulation. The cans have all been gathered and placed in the circle labeled "Basket." The lines show the paths the robots took completing the mission. The remaining circles represent obstacles of various types (from MacKenzie, Arkin, and Cameron 1997, 43, reproduced with permission of Kluwer Academic Publishers)

• The robots are able to select appropriate actions throughout a mission, taking into account the environment, their own internal state, and the actions of other robots.

• A particularly interesting aspect of this architecture is that the robots are endowed with *motivations* that enable them to perform tasks only when those actions are expected to have the effect on the world that the robot desires. Specifically, *robot impatience* and *robot acquiescence* are modeled. For example, if robot A is not accomplishing some actions needed by robot B (say, due to a sensor failure), then B becomes increasingly *impatient* to take over the task of A, and eventually does so in order to complete its assigned task (i.e., to have its desired effect on the world). Similarly, when C becomes aware (from feedback information) that it is not completing its tasks adequately, it will try to find other actions it can take until it becomes aware that other robots have take over the task. Then it *acquiesces* to the situation.

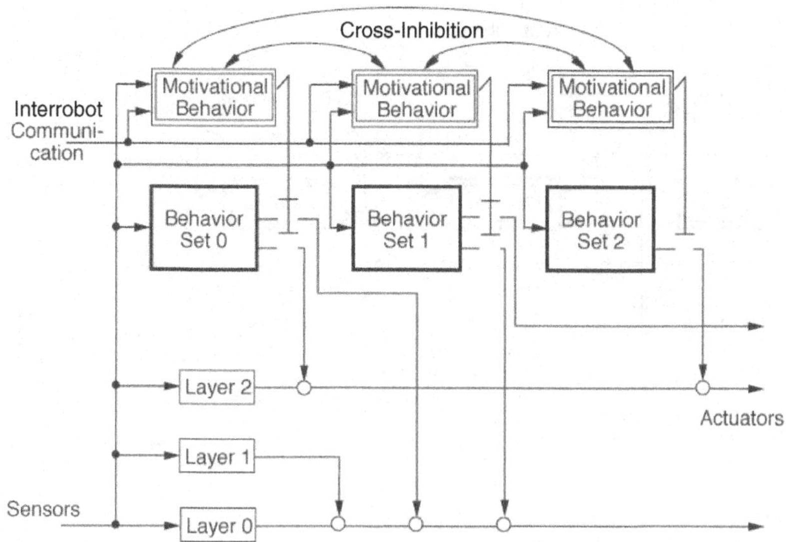

Figure 12.10
The ALLIANCE architecture for control of robot teams (from Parker 1998, copyright 1998 by the Institute of Electrical and Electronics Engineers (IEEE), reproduced with permission of IEEE and Lynne Parker)

The ALLIANCE architecture is illustrated in figure 12.10. Note that this architecture is implemented on each robot in the team. Each "behavior set" corresponds to some high-level task-achieving function, the selection of which is dependent on the motivational behaviors shown at the top of the figure. Note also that the outputs of the "motivational behavior" boxes are not explicitly connected to behavior sets. Thus, the behavior set corresponding to a particular motivation may or may not result in outputs to the actuators. Note also that (as in other behavior-based architectures) the task-achieving behaviors receive inputs from sensors and control some actuators. Lower-level behaviors correspond to survival behaviors; higher-order behaviors may correspond to exploration or map building. Layer 0 represents reactive survival actions and is active all the time. Layer 1 may correspond to collision avoidance and also be active continually, whereas Layer 2 corresponds to higher-level competences that are turned on or off by the appropriate behavior sets.

To reduce the need for parameter tuning, ALLIANCE has been extended to include a learning mechanism, and the modified version is referred to as L-ALLIANCE (Parker 1998). The learning mechanism employed requires the robots to monitor and evaluate their performance in the changing environment and then update parameters as required. This extension is illustrated in figure 12.11.

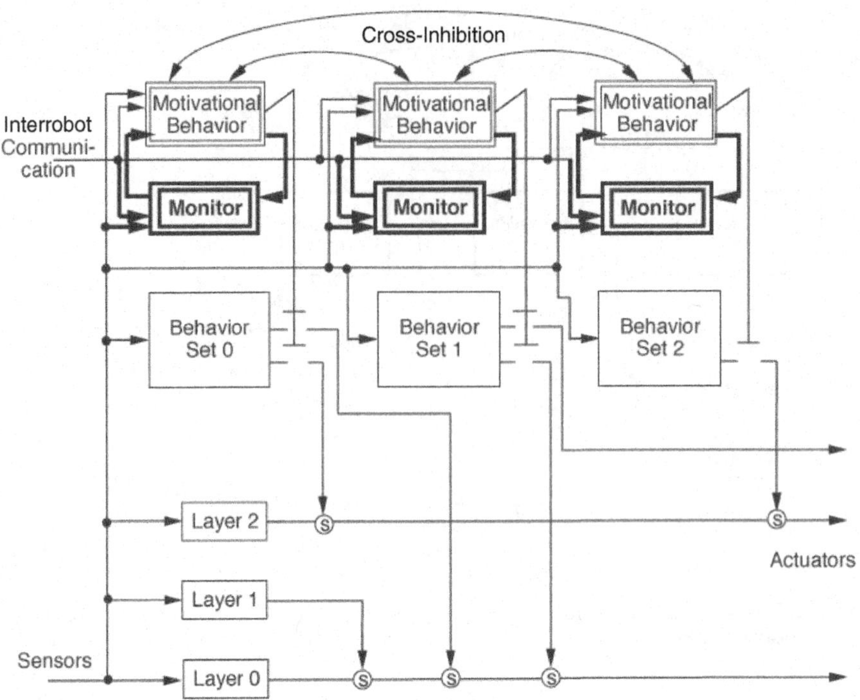

Figure 12.11
L-ALLIANCE architecture illustrating monitor elements (from Parker 1998, copyright 1998 by IEEE, reproduced with permission of IEEE and Lynne Parker)

The ALLIANCE architecture has been used to coordinate robots in a task known as *cooperative multirobot observation of multiple moving targets* (CMOMMT). This task is clearly relevant to such applications as reconnaissance and security patrolling. The use of robots for observation activities can reduce the number of fixed sensors needed to cover a given area, and robots can follow a target if necessary to areas where no other sensors are present.

The CMOMMT problem can be defined as follows (Jung and Sukhatme 2001): Given a bounded, enclosed region S, a team of m robots R, a set of targets $O(t)$, and a binary variable $In(o_j(t), S)$ defined to be TRUE when a target $o_j(t)$ is located within region S at time t, an $m \times n$ matrix $\mathbf{A(x)}$ defined with elements

$$a_{ij}(t) = \begin{cases} 1, & \text{if a robot } r_i \text{ is monitoring target } o_j(t) \text{ in } S \text{ at time } t, \\ 0, & \text{otherwise,} \end{cases}$$

and the *LOGICAL-OR* operator defined as

$$\vee_{i-1}^{k} h_i = \begin{cases} 1, & \text{if there exists an } i \text{ such that } h_i = 1, \\ 0, & \text{otherwise,} \end{cases}$$

then the goal of CMOMMT is to maximize the quantity

$$Observation = \frac{\sum_{t=0}^{T} \sum_{j=1}^{m} \vee_{i=1}^{k} a_{ij}(t)}{t \times m},$$

that is, to count the number of robots actually monitoring targets over the entire time interval in question. In defining the ALLIANCE architecture in which robots were coordinated in the performance of the CMOMMT task, Parker assumed that a global coordinate system was known and that the observation sensors had a perfect field of view; roles were assigned among mobile robots using one-way communication. Experiments were performed in a bounded, enclosed region. This architecture has been very successful and paved the way for its use to explore other related problems. For example, Jung and Sukhatme (2001) worked on the CMOMMT problem in an indoor office environment with multiple corridors.

12.6.4 The Pheromone Architecture

Pheromones are chemical markers used by insects for communication, coordination, and sexual attraction; they are detectable in infinitesimal quantities (consisting of a few molecules) by opposite-sex members of the same species. Insects follow a pheromone trail in a zigzag fashion, moving across the trail in one direction and then another. David Payton and his associates have developed an approach to mobile robot coordination inspired by pheromone trail following (Payton et al. 2001; Payton, Estkowski, and Howard 2002). Their approach exploits the notion of a "virtual pheromone," implemented using simple beacons and directional sensors mounted on top of each robot. Virtual pheromones facilitate simple communication and coordination and require little on-board processing. Collections of robots coordinated via virtual pheromones are able to perform complex tasks such as leading the way through a building to a hidden intruder or locating critical choke points. The robot collective becomes a computing grid embedded within the environment while acting as a physical embodiment of the user interface.

Over the past few decades, the literature on path planning and terrain analysis has dealt primarily with algorithms operating on an internal map including terrain features. The virtual pheromone approach externalizes the map, spreading it across a collection of simple processors, each of which determines the terrain features in its locality. The terrain-processing algorithms of interest are then spread over the population of simple processors. The user interface to this distributed robot collective is itself distributed. Instead of asking the user to communicate with each robot

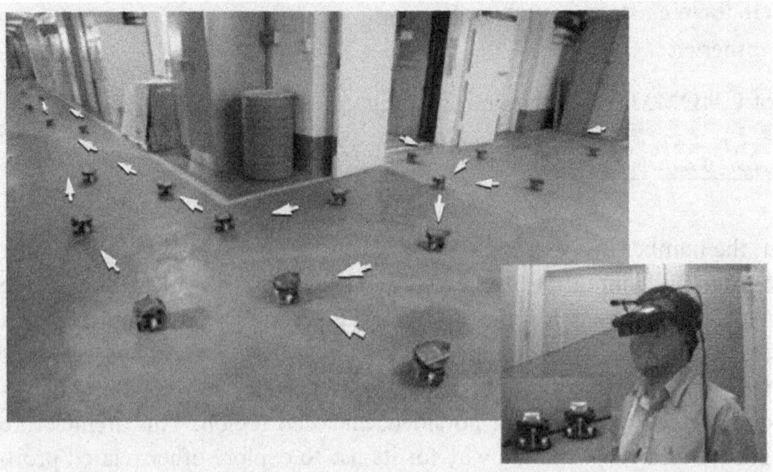

Figure 12.12
Pheromone robots acting as a virtual display (photograph courtesy of David Payton)

individually, the entire collective works cooperatively to provide a unified display embedded in the environment, as illustrated in figure 12.12.

12.6.5 The Ranger-Scout Architecture

A different type of multiple-robot system involves robots of two or more drastically different sizes (and perhaps weights and shapes). So-called marsupial robots, in which, by analogy with kangaroos and other marsupials, a larger robot will carry one or more smaller robots and then deploy them as appropriate, fit this classification. The marsupial approach to deployment of multiple robots was used by Murphy (2002), as well as others in the late 1990s. A unique approach to marsupial robotics for reconnaissance has been pioneered by researchers at the University of Minnesota (Rybski et al. 2000; Rybski et al. 2002), who employ a relatively large robot named Ranger to carry and then deploy perhaps a dozen much smaller Scouts to explore actual terrain (figure 12.13). The Scouts are sufficiently small as to make them difficult to detect. They are roughly cylindrical in shape, with a diameter of about 4 cm and a length of about 12.5 cm, and are capable of jumping over obstacles by means of an internal spring.

12.7 Swarm and Cellular Robotics

Cellular robotics refers to an organization of simple robots located in "cells" or discrete locations within a given space. The robots can achieve a particular spatial orga-

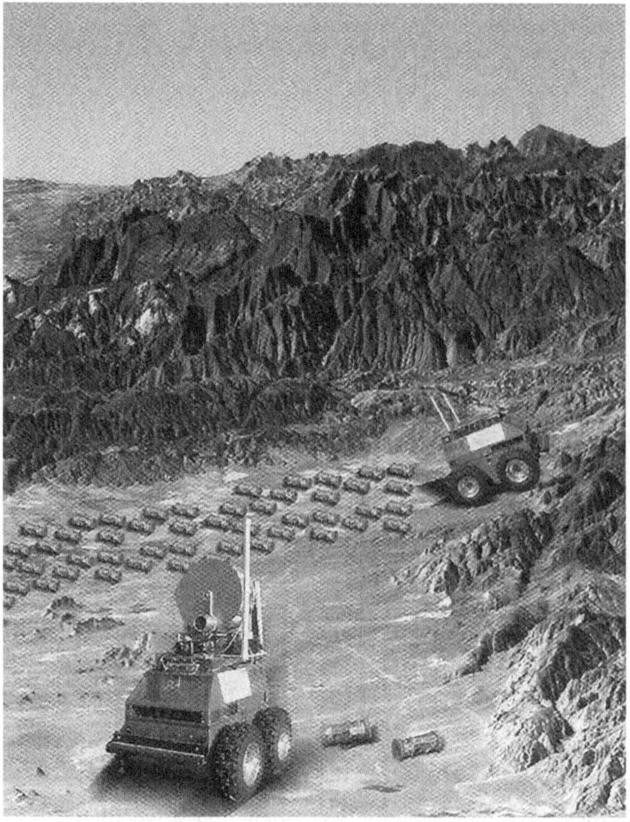

Figure 12.13
Heterogeneous Ranger and Scout robots (from Rybski et al. 2000, copyright 2000 by IEEE, reproduced with permission of IEEE and Nikolaos Papanikolopoulos)

nization by moving to unoccupied cells. They can pass objects among themselves and reconfigure to achieve desired objectives. Thus, *cellular robotic systems* (CRSs) may be viewed as a robotic implementation of distributed computing. The theory of CRS has been developed extensively by Beni, Hackwood, and Wang (Beni 1988; Beni and Wang 1991; Beni and Hackwood 1992).

Cellular robotic systems employ large numbers (swarms) of autonomous robots of low complexity that cooperate to perform a given global task. They have no centralized control (thus they belong to the Type II control architecture described in section 12.4). CRSs can be used as sensing systems if each basic unit is capable of sensing some external variable and can respond in a simple way. Hackwood and Wang (1988) give an example of a large number of robots equipped with cameras, sonars, or radars that respond to a sensory stimulus only by arrangement of their

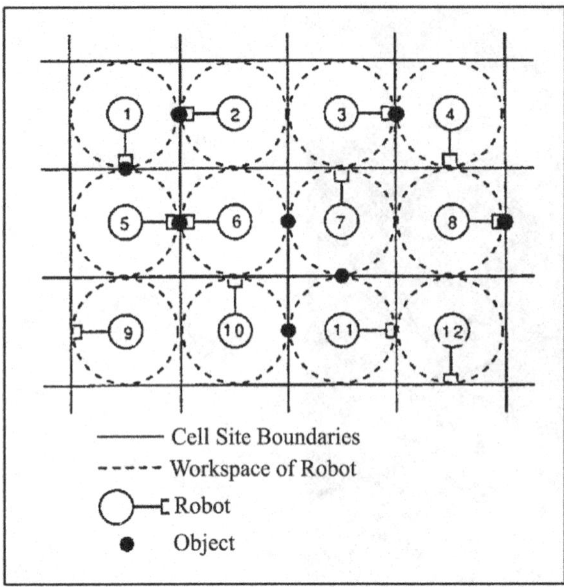

Figure 12.14
Cellular robot system for intercellular material transfer (from Hackwood and Wang 1998, reproduced with permission of Susan Hackwood)

configuration. For example, if all the robots are arranged in a ring, and all detect the same object, they might all point in the same direction. If the sensors on any of the robots are destroyed or fail, the remainder of the robots can rearrange themselves and continue to perform the task at hand. CRSs can also be used to transfer material between cells, as illustrated in figure 12.14.

In the system shown in the figure, objects can be passed from one unit to the next using local rules. Such a system might sort the objects into specific patterns or assemble them into complex objects. Since no central controller is present, a failed element can be replaced by another without aborting the global mission. Hackwood and Wang suggest that such structures could provide a new approach to manufacturing.

Another version of a "swarm" of robots is "smart dust," a very large collection of very simple devices containing a sensor, a transmitter, and a battery, each perhaps a few millimeters in size, that can be dropped from an airplane, each device being equipped with a means to slow its descent (Kahn, Katz, and Pister 1999; Doherty et al. 2001). Assume that each of these tiny devices can sense the presence of a particular chemical compound and that each is programmed to transmit a "yes" or "no" signal at a particular time. The combined signal from a large number of very weak transmitters may be sufficient to reach a distant receiver and thus provide an alarm signal indicating the presence of the chemical. Such an alarm system could be virtually undetectable and thus of great help to the military. Note that since each of these tiny devices contains a sensor, an actuator (that turns on the transmitter), and a decision element, they can be considered robots.

Fukuda and his colleagues have further developed the idea of cellular robotics by allowing each robot in the group to be dynamically reconfigurable. The theory and applications of these CEBOTs (cellular robotic systems) have been published in a number of papers and a book (Fukuda and Nakagawa 1987; Fukuda, Ueyama, and Arai 1992; Fukuda, Kawauchi, and Hara 1993; Kawauchi, Inaba, and Fukuda 1993). Rather than being confined to a particular cell, CEBOT elements are autonomous and mobile, which makes them able to seek and physically join other appropriate cells. In contrast with the CRSs previously discussed, CEBOTs may include some control elements, embodied in "master cells" capable of communicating with and controlling other cells in their network. Genetic algorithms have been used to enable self-organization of CEBOTs. An interesting application discussed by Kawauchi, Inaba, and Fukuda (1992) involves the carrying of an object by a group of mobile cellular robots, each equipped with sensors (for collision avoidance and position detection), CPU, and communication devices. The robots initially move in random directions but eventually become synchronized to enable them, collectively, to move the object in question.

The concept of *swarm intelligence* was first employed by Beni and Wang (1989) in connection with the control of cellular robotic systems. It was broadened by Bonabeau, Dorigo, and Theraulaz (1999) to cover any algorithms or distributed problem-solving systems inspired by insect colonies and other animal societies. Once so defined, swarm robotics was no longer restricted to cellular systems, but rather became a paradigm for solving problems involving large numbers of robots (swarms) with emphasis on control and communication. Of course, other papers (and books) using biological inspiration for collective robotics had already been published, such as Martinoli and Mondada 1995. More recently the term *SwarmBot* has been used to describe a swarm of mobile robots with the capability of connecting and disconnecting from one another (Mondada et al. 2002). Figure 12.15 shows a Swarm Bot, consisting of a group of simulated robots (*s-bots*) performing a cooperative task.

The study of swarms using evolutionary algorithms (see chapter 6) has also become a topic of major research interest (Nolfi and Floreano 2000).

12.8 Communication among Multiple Robots

The issue of communication among multiple robots is complex. Clearly, if an entire group is to accomplish some global goal, some form of communication among the

Figure 12.15
A simulated Swarm Bot performing a task (photograph courtesy of Marco Dorigo)

group's members is necessary, but what form this communication should take is not clear. Most researchers in this field describe three methods for such communication:

- Point-to-point communication, in which individual robots can communicate with one another, passing on information such as goal locations and locations of obstacles.
- Broadcast methods, in which a human or robot supervisor can send messages to all members of the group.
- Communication via the environment, in which messages are implicit in the effect of individual robots on their environment. Among insects, the use of chemical trails by ants is an example of such a form of communication (Holldobler and Wilson 1990). Animals frequently mark the boundaries of their territory with urine or other chemical markers.

Sometimes, the high-level goals of a group or colony are implicit in the local rules being used by individual members to achieve the colony goals, and no explicit form of communication among the members exists. African termites are known to build large mounds with no apparent communication among them. Each termite apparently operates on local rules causing it to pick up a grain of sand, carry it to the top of the mound, drop it, and return for more.

12.8.1 Cooperation without Communication

We consider first the accomplishment of a global task by a colony of robots without explicit communication among the members of the colony. Arkin has shown that a *foraging* task can be accomplished by robots without any communication among

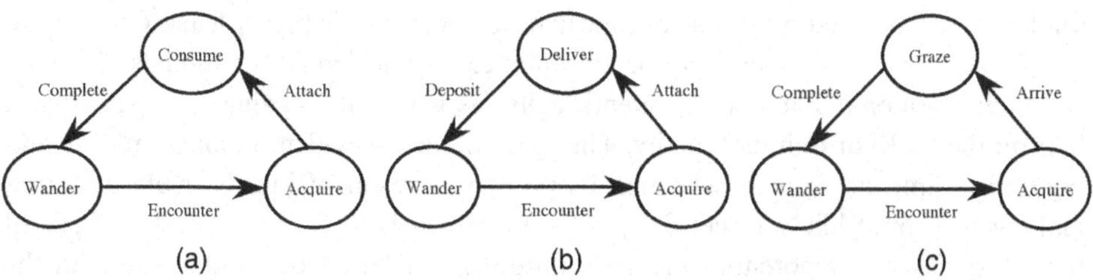

Figure 12.16
Finite-state models of three multiple-robot tasks: (a) foraging, (b) consuming, and (c) grazing (adapted from Balch and Arkin 1994)

them and that if each robot simply acts upon its own perception of the world, a form of cooperation among the agents emerges, in the sense that multiple agents converge to work on the task (Arkin and Hobbs 1992).

Balch and Arkin (1994) have extended these results to a more complex situation in which the tasks include *consuming* and *grazing* as well as foraging. The tasks involve attractors (items of interest in the foraging task, and items of food in the consuming task). The attractor in the grazing task is the presence of an ungrazed area. The robots then move toward it, and then move over it until it is grazed (completely covered with tracks). The number of attractors and their mass are parameters for the foraging and consuming tasks. Each task is represented by finite-state acceptors (FSAs), consisting of several states, in which transitions between states are triggered by perceptions. The FSAs for these three tasks are shown in figure 12.16. If no attractors are present in a robot's field of view, it wanders until one is found.

Balch and Arkin considered teams of simulated robots engaging in three forms of communication:

• *No communication* Each robot is able to use its own sensors to identify other robots, obstacles, or attractors, but it receives no information on the location of these entities from other robots. It must attempt to complete its assigned task relying only on its own perceptions.

• *State communication* Robots are able to communicate their internal state (i.e., whether they are in the foraging, grazing, or wandering mode) to other robots. In an even simpler one-bit commucation mode, the robots display a zero when they are in the wander mode and a one otherwise.

It should be noted that state communication can be either explicit (i.e., intentional) or implicit. In implicit state communication, the receiver of a message can infer the sender's state without any intentional transmission of that state on the part of the sender. Humans can frequently infer the emotional state of another person with no intention on the sender's part to communicate that information (e.g., when the sender is embarrassed and blushes or when he perspires visibly in fear). Other emotional states, such as anger, may be communicated intentionally. Animals also convey their emotional state, either intentionally (as with tail wagging or the raising of hair on the neck) or unintentionally. The latter might be evident in an animal's walking with a limp or more slowly than its peers as a result of injury or old age, thus making it a more likely target for predators, which look for such implicit signs of the internal state of potential prey. Communication based on modifications in the environment rather than direct message passing is sometimes termed *stigmergic* (Matarić 1993); sensing the external state of other agents is a form of stigmergic communication.

• *Goal communication* Information concerning goal locations (or, in other tasks, the type of goal or the presence of predators) is transmitted to other agents intentionally. We have seen that ants use scent to provide communication via the environment. By contrast, bees use explicit goal communication: a bee returning from a search for flowers will execute a "dance" at the entrance to the hive. The parameters of the dance provide information on the direction and distance from the hive of a source of nectar it has discovered.

The results, obtained from thousands of simulations of multiple robots in foraging, grazing, and consuming tasks, with and without communication, with randomly chosen initial conditions, are shown in table 12.1. Since the results have a random

Table 12.1
Summary of performance results for different levels of communication

Task	Average improvement	Best	Worst
Forage			
State versus No Communication	16%	66%	−5%
Goal versus No Communication	19%	59%	−7%
Goal versus State Communication	3%	34%	−19%
Consume			
State versus No Communication	10%	46%	−9%
Goal versus No Communication	6%	44%	−16%
Goal versus State Communication	−4%	5%	−30%
Goal versus State (low mass attractors)	−1%	23%	−19%
Graze			
State versus No Communication	1%	19%	0%
Goal versus No Communication	1%	19%	0%
Goal versus State Communication	0%	0%	0%

Source: From Balch and Arkin 1994, reproduced with permission of Kluwer Academic Publishers.

element it is necessary to average the results. Several conclusions can be drawn from the results:

1. Communication improves performance significantly in tasks with little implicit communication, such as foraging and consuming.
2. Communication appears unnecessary in tasks for which implicit communication exists (such as grazing).
3. More complex communication strategies, such as goal communication, may offer little or no benefit for tasks that involve implicit communication.

Other studies (e.g., Arkin, Balch, and Nitz 1993) have also investigated systems of multiple robots in which each robot communicates its state and location using a broadcast system. These studies have shown that the performance of a group of robots is enhanced by communication among the robots.

12.8.2 Communication among Embedded Robots

When robots are embedded into the Internet, communication among them becomes critical for the accomplishment of complex tasks, such as search and rescue (Sukhatme and Matarić 2000). When mobile robots become nodes in a communication network (which may also include fixed nodes), it is clear that communication is also not a luxury, but a necessity. Communication can help to reduce any uncertainty over the robots' location and facilitate their interaction with both humans and other robots in the completion of a complex and difficult task, say, search and rescue following an earthquake or other catastrophic event. If robots can search autonomously for survivors in the rubble, communication with one another and with outside supervisors becomes critical. Otherwise, if one robot finds a survivor, it cannot alert humans or call for help.

An unusual approach to inter robot communication (and human-robot communication) was reported by Sekmen, Koku, and Sabatto (2001). In the experiment conducted by these researchers, a supervisor robot and three subordinate, smaller robots were equipped with both voice synthesizers and voice recognition systems. A human could command the supervisor robot with a verbal message. The supervisor, in turn, could send a command to all the subordinate robots by means of a simple phrase. If the phrase was in the robots' library, they would execute the command; otherwise they would ignore it. Since each repetition of the supervisor's synthetic voice was nearly identical (in contrast with the variability of humans repeating a phrase), the subordinate robots' recognition accuracy was 100% for a small number of phrases. In response to a voice command, subordinates could use their on-board cameras to recognize targets and to perform tasks such as lining up single file behind the supervisor. The major potential advantage of this system seems to be that other human users could intervene and present verbal commands to either the supervisor or the workers.

Communication in simulated robot colonies using the tropism architecture (section 6.11) produced some interesting results. Member robots in these colonies that were unable to move large objects alone could broadcast a call to other member robots for help. The radius for successful communication was limited, and robots outside that radius could not hear a call for help from another robot. When the communication radius was relatively small, say, 10% of the field size, it was found that the robots had a tendency to cluster within that radius. This was an unexpected, emergent result (Agah and Bekey 1997b).

12.9 Formation Control

Organized patterns of movement occur frequently in the animal world. For example, birds (such as Canada geese) sometimes fly in inverted V-formations, which provide them with aerodynamic advantages that enable them to fly long distances with less fatigue (figure 12.17). It has now been shown that when white pelicans are trained to fly in formation with proper spacing, their heart rates (and hence energy expenditures) drop by about 30% (Weimerskirch et al. 2001). When the lead bird in the formation becomes fatigued, another one replaces it. Fish may swim in schools in which they maintain approximately constant distances from one another. In some cases

Figure 12.17
Birds flying in inverted V-formation

they organize in specific formations to surround their prey. Human beings also organize in specific formations for particular purposes (e.g., in sports, hunting, or military operations). Armies have organized in specific formations for hundreds of years. Groups of aircraft generally fly in formation. Recently formations of spacecraft have been used for scientific purposes, such as the creation of a huge interferometer with kilometer spacing between mirrors (Lawton, Beard, and Hadaegh 1999). In such formations it may be necessary to maintain intervehicle spacing to accuracies of centimeters in total ranges of kilometers. Clearly, such accuracies require new approaches to control. In the following sections, we consider first formation control in small groups of mobile robots moving in two dimensions using only local information. Then we look at formation control architectures in general and finally at formations of space vehicles.

12.9.1 Formation Control Using Only Local Information

Most early work in formation control of robots has assumed global knowledge of some form, such as a global coordinate system or knowledge of all other robots' positions and headings. Balch and Arkin (1998) identified three approaches to formation control: *unit center referenced*, *leader referenced*, and *neighbor referenced*. In the first, each robot decides its position relative to the centroid of all robots; in the second, each robot uses a leader's known position as a reference for its own position; and in the third, a neighbor's position is used as a reference for the robot's own position.

It has been shown that if each mobile robot in a formation has its own coordinate system (say, from GPS measurements), sensors that enable it to measure its distance from and orientation with respect to its neighbors, and actuators that enable it to change its position and orientation, then the formation can be maintained without a supervisor (Yamaguchi, Arai, and Beni 2001). Consider the situation shown in figure 12.18.

Each robot shown in the figure has its own coordinate system and is able to measure and adjust its position and orientation relative to its neighbors. Since each robot interacts only with its neighbors and determines its position autonomously without a supervisor, it is evident that a formation can be maintained autonomously. Clearly, this strategy is neighbor referenced, in Balch and Arkin's terminology. As with other group behaviors discussed in this chapter, it is necessary to predict the global (group) behavior from the local specified group behavior.

Another local algorithm for formation control has been presented by Fredslund and Matarić (2002a, 2002b, 2002c), who report on the ability of n robots to establish and maintain a formation by enabling each robot to keep one other robot (a *friend*) at a desired angle and distance. Each robot is assumed to have a sensor for this purpose (the *friend sensor*). The friend sensor is assumed to be pointing forward

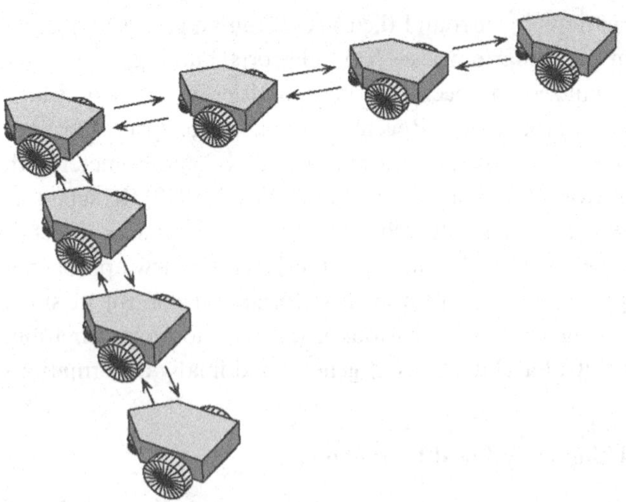

Figure 12.18
Mobile robots in formation (illustration courtesy of Hiroaki Yamaguchi)

with a $\pm 90°$ field of view and to have limited visibility. Each robot has a unique identifier, detectable by other robots via their friend sensors. The identifier is broadcast regularly by each robot, as a "heartbeat" message to signal to other robots that it is "alive and well." Hence, each robot knows how many other robots are in the formation. The robots are organized in a *chain of friendships* by the order of their identifiers. One robot is the leader (called the *conductor*); this robot does not follow any friend. The leader broadcasts a message indicating the current formation but does not broadcast any heading. Any one of the robots can be the conductor.

There are some limitations to this approach, which applies only to formations that can be formed from a chain of friendships, and so can have at most two free ends. Other restrictions are discussed in Fredslund and Matarić 2002a. Three examples of formations formed from chains of friendships are shown in figure 12.19. In the *centered formation* of figure 12.19(a), the conductor is in the center. In this formation, robots with identifiers smaller than the identifier of the middle robot find a friend with an identifier immediately greater than their own; robots with identifiers larger than the identifier of the middle robot find a friend with an identifier immediately less than their own. In *noncentered formations*, all robots find a friend with an identifier less than their own, leading to the structure of figure 12.19(b). For other formations, like the diamond in figure 12.19(c), given the number of robots n, the formation code, and its own identifier, each robot can determine its angle in the chain of friendships, as shown in the figure. Note that figure 12.19 illustrates formations of simulated robots.

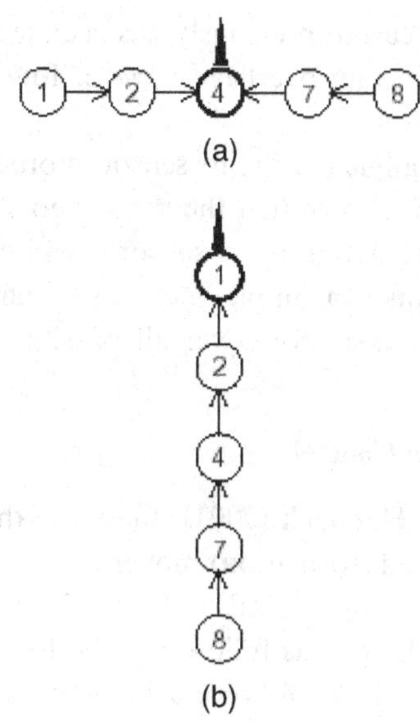
1
2
4
7
8
(a)
1
2
4
7
8
(b)

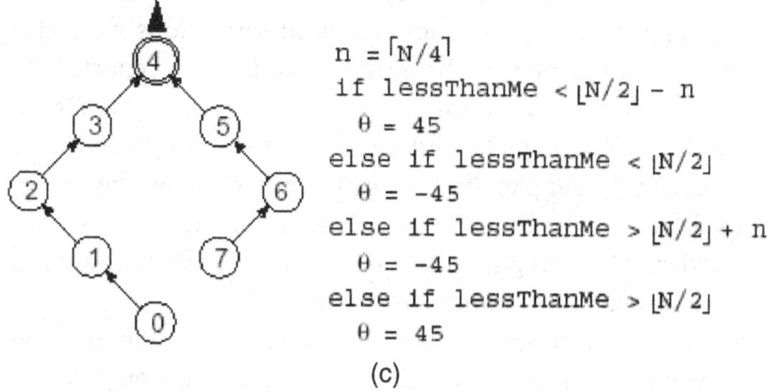
4
3
5
2
6
1
7
0
n = ⌈N/4⌉
if lessThanMe < ⌊N/2⌋ - n
θ = 45
else if lessThanMe < ⌊N/2⌋
θ = -45
else if lessThanMe > ⌊N/2⌋ + n
θ = -45
else if lessThanMe > ⌊N/2⌋
θ = 45
(c)

Figure 12.19
(a) Centered, (b) line, and (c) diamond formations, indicating the chain of friendships. The dark triangles indicate the conductor (from Fredslund and Matarić 2002a, copyright 2002 by IEEE, reproduced with permission of IEEE and Maja Matarić)

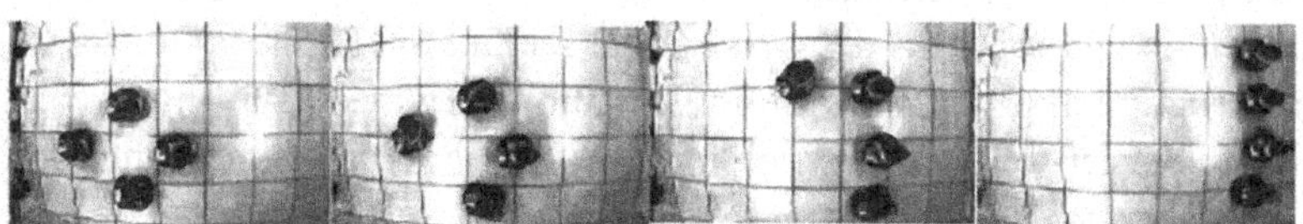

Figure 12.20
Overhead view of formation change from diamond to line (from Fredslund and Matarić 2002a, copyright 2002 by IEEE, reproduced with permission of IEEE and Maja Matarić)

The chain of friendships algorithm discussed previously has been tested experimentally with real robots. Figure 12.20 shows an overhead view of four mobile robots changing from a wedge to a line formation.

Each robot must keep its friend at an angle θ using its sensor in order to create and maintain the desired formation. It can be shown that the robot's goal for all types of formations is simply to keep its friend centered in the sensor's field of view by panning the sensor by θ degrees. Clearly, this is a simple, local algorithm that does not rely on global information. Although it does not cover all possible formations, its simplicity recommends it highly.

12.9.2 Global Approaches to Formation Control

As pointed out by Beard, Lawton, and Hadaegh (2001), there are three fundamental approaches to the study of coordinated (formation) movement in robotics: leader following, behavior based, and virtual structure based. In leader following, one of the robots is designated as the leader and the rest as followers. The followers track the position and orientation of the leader. If the followers in figure 12.19 are given the global coordinates of the conductor, the strategy changes from a local to a global, follow-the-leader strategy. Leader-following-referenced formations were also discussed by Balch and Arkin (1998); one of the earliest studies was reported by Wang (1991). Control of follow-the-leader formations is also discussed in Desai, Ostrowski, and Kumar 1998 and Yamaguchi and Burdick 1998. An architecture based on leader following is clearly simple, since only the behavior of the leader needs to be specified. This is also its weakness, since there is no explicit feedback from the followers to the leader, which might go so fast as to "lose the followers in the dust." Also, failure of the leader may doom the formation.

Behavior-based architectures prescribe several desired behaviors for each agent in a formation, such as obstacle avoidance, goal seeking, and formation keeping. Then, the control action for each agent is some form of weighted average of the control for each behavior, as described, for example, in Balch and Arkin 1998. One of the advantages of a behavior-based approach is that it provides feedback to the formation, since each agent reacts to the positions of its neighbors. However, the group behavior is difficult to specify mathematically. A behavior-based approach that grows formations from individual robots is described in Naffin and Sukhatme 2004.

In the virtual-structure approach, the entire formation is treated as a single structure, which makes it easier to define control laws for it. The motions and control of the individual robots are then derived from the motion of the virtual structure (Lewis and Tan 1997). Such an approach is illustrated in figure 12.21. In step 1 of the figure, the robots are situated in a triangular virtual structure. A virtual force field moves the structure (step 2), and then the robots reposition themselves with respect to the structure (step 3). The flow of control is actually bidirectional: Movement of the vir-

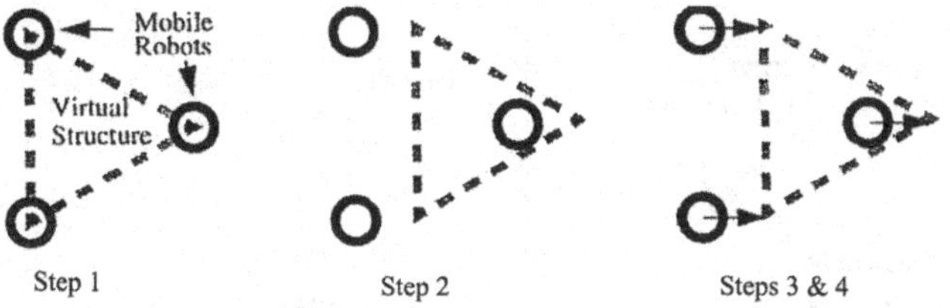

Figure 12.21
Steps in the virtual-structure control algorithm for moving in formation (adapted from Lewis and Tan 1997, illustration courtesy of M. A. Lewis)

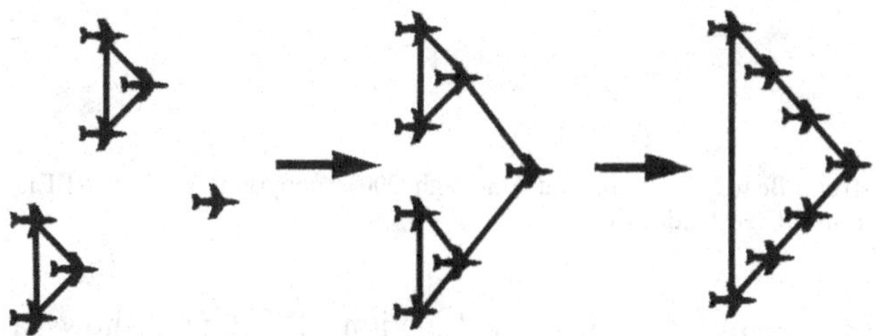

Figure 12.22
Reconfiguration of virtual structure (adapted from Lewis and Tan 1997, illustration courtesy of M. A. Lewis)

tual structure causes the robots to reposition themselves; movement of the robots can cause the virtual structure to reposition. Virtual structures can be dynamically reconfigured as robots are added or removed or the formation needs to be changed, as shown in figure 12.22.

12.9.3 Spacecraft Formation Architectures

The control of formations in three dimensions is clearly a much more complex problem than those reviewed up to this point in the chapter. In recent years there has been increasing interest in flying groups of autonomous spacecraft in formation for a variety of applications, including space-based interferometry and military surveillance. We describe briefly first the interferometry task, and then the high-precision formation control architecture devised for its implementation (Beard, Lawton, and Hadaegh 2001).

An interferometer (figure 12.23) works by collecting two beams of light that have traveled different paths from the same source, then combining the two beams to create an interference fringe pattern. The width, angle, and intensity of the fringe pattern determine the mutual coherence function of the light source, $\mu(u, v)$, where u and v are frequency variables. The intensity map or image is obtained through an

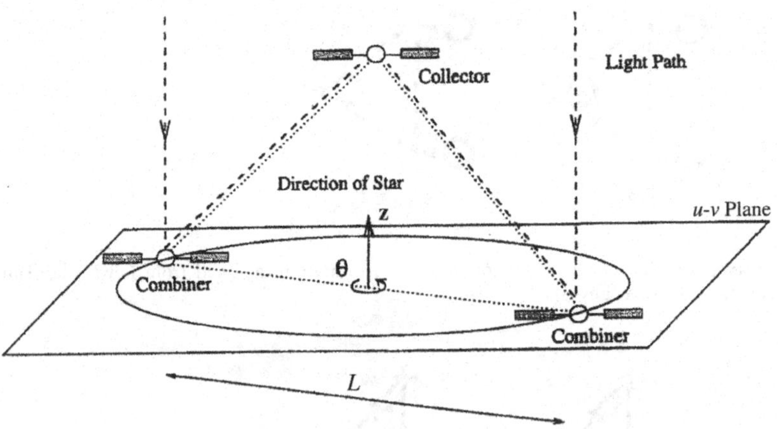

Figure 12.23
A three-spacecraft interferometer (from Beard, Lawton, and Hadaegh 2001, copyright 2001 by IEEE, reproduced with permission of IEEE and Fred Hadaegh)

inverse Fourier transform of the mutual coherence function. The figure shows a three-spacecraft interferometer, configured as an equilateral triangle. The vector **z** points in the direction of a star, the light source. To image a star, the entire formation must undergo a sequence of maneuvers, each ending at specific values of (L_i, θ_i), the distance between the "combiner" spacecraft and the rotation angle, as shown in the figure. During the collection process, the relative distance between the spacecraft (which may be kilometers) must be controlled with errors on the order of centimeters or better. The details of the control system used to accomplish this task are given in Wang, Hadaegh, and Lau 1999 and Beard, Lawton, and Hadaegh 2000 and 2001.

The architecture for formation control of these spacecraft is a synthesis of follow-the-leader, behavioral, and virtual-structure approaches. There are several levels of control in formation flying. At the highest level of abstraction is the dynamic transition from one subtask to another. In the interferometer task, this implies that the entire formation must rotate to a new pose, and then maintain it with extremely high precision. At the next level, there must be a mechanism to coordinate the motion of each spacecraft to synthesize the desired behavior of the entire formation. At the lowest level, each individual spacecraft must be controlled to be consistent with the coordination mechanism. A block diagram of the architecture supporting this work is shown in figure 12.24. In this figure, $\mathbf{S}_i$ represents the ith spacecraft, with control vector $\mathbf{u}_i$ representing the control forces and torques and output vector $\mathbf{y}_i$ representing the measurable output variables, such as position and attitude. $\mathbf{K}_i$ is a matrix representing the local controller for the ith spacecraft. The inputs to $\mathbf{K}_i$ are the outputs of the ith spacecraft $\mathbf{y}_i$ and the coordination variable ξ. The outputs of $\mathbf{K}_i$ are the control vector $\mathbf{u}_i$ and the performance variable $\mathbf{z}_i$. **F** is the formation controller, the

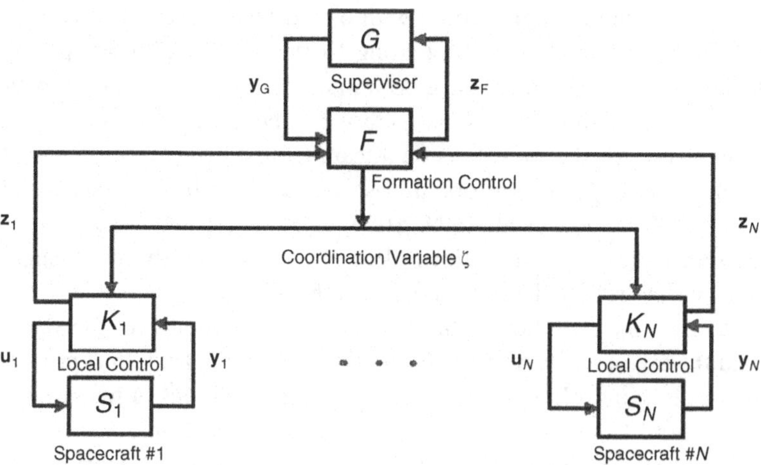

Figure 12.24
Architecture for formation flying of spacecraft (from Beard, Lawton and Hadaegh 2001, copyright 2001 by IEEE, reproduced with permission of IEEE and Fred Hadaegh)

primary coordination mechanism in the system. Its output is the coordination variable ξ, which is broadcast to all the spacecraft. In addition, the formation controller outputs a vector variable $\mathbf{z}_F$ that encapsulates the performance of the formation to the supervisor **G**. The inputs to **F** are the performance variables from each spacecraft $\mathbf{z}_i$ and the output of the supervisor $\mathbf{y}_G$. Finally, the system **G** is a discrete-event supervisor that uses the performance vector $\mathbf{z}_F$ to determine the input to the formation controller $\mathbf{y}_G$.

Of course, the preceding is only a high-level description of the block diagram. The detailed equations for each block require representation of the spacecraft dynamics, the rule-based supervisor, the dynamics of the formation controller, and the local control laws for each spacecraft for various modes of operation. These complex equations are derived in Beard, Lawton, and Hadaegh 2001. We have included this high-level presentation here only to indicate that the principles of robot formation control can be extended to formations of spacecraft, albeit with significantly greater complexity.

12.10 Robot Soccer

The use of multiple autonomous robots in teams to play a competitive game such as soccer is major challenge. The robots have to accomplish specific goals (such as moving the ball into the opponent team's goal) in real time in a highly dynamic environment. To accomplish this task they require closely integrated perception, reasoning, and action, while maintaining communication with one another. They must be able to adapt the strategy on which they are operating to the actions of the opposing team. During the last several years there have been international RoboCup competitions in which teams of miniature robots compete with one another under specific rules governing robot size, number of robots on a team (generally five), playing field size, allowable number and type of sensors, and so on, as described in Asada and Kitano 1999a and 1999b, Asada et al. 1999, Stone et al. 2001, and Veloso et al. 2002. The competition takes place in various "leagues," including those for small and medium-sized wheeled robots, legged robots, and the like.

It is interesting to note the evolution of the RoboCup competition over the years. Early rules allowed the use of a camera suspended over the playing field, thus providing a global view for the team robots. However, tracking the positions of ten moving robots (five on each team) as well as the orange-colored ball in real time was a nontrivial task. Each robot needed some identifying marker. Furthermore, the results of the image processing then required additional computation to enable the controller to determine the appropriate commands for each robot. The translation from global perception to individual action is not equivalent to the centralized control of multiple robots, described in section 12.5, since the global view is available to each member of the team. The robots were able to work as a team by integrating the perceptual information with their respective roles, such as attacker, defender, or goalie, to produce appropriate behaviors. For example, the Carnegie Mellon team used an optimization function to minimize the distance between a robot and the ball and to maximize robots' distance from other robots. A typical wheeled robot team in the competition is shown in figure 12.25.

In more recent years of the competition, as the on-board capabilities of the robots has increased (in terms of both perception and computing power), they have become

Figure 12.25
The Roobots, a RoboCup small-sized robot league team from the University of Melbourne (photograph courtesy of Nick Barnes)

increasingly autonomous and less reliant on remote commands. Instead, they use their on-board cameras and compute their position and orientation in the field (also known as localization, see chapter 14). This process is facilitated by the presence of colored landmarks on the field. Nevertheless, the data are clearly noisy, and some form of probabilistic computation is required.

Sony AIBO robots have been used in legged-robot competitions. In these contests, the robots are fully autonomous. Since both teams in a legged-robot contest use the same type of robots, the major differences between the teams in such contests relates to their cognitive ability, including strategy and teamwork. A typical group of legged competitors is shown in figure 12.26(a), and a goalie in defensive position is shown in figure 12.26(b).

We expect that in the future, robot soccer competitions will make use of two-legged humanoid robots, such as ASIMO (see chapter 13). In twenty years we may see full-size humanoids competing with humans on a soccer field, but much progress in all aspects of robotics will be required before that is possible.

12.11 Heterogeneous Robot Teams

With the exception of the Ranger-Scout teams described in section 12.6.5, the groups of robots we have considered so far consist of identical members, such as Kheperas, Pioneers, or AIBOs. There is increasing interest, however, in teams consisting of two or more classes of robots that may differ in size and capability.

There are two common varieties of heterogeneous robot systems. In one type, the robots are designed for different tasks and may be drastically different in structure and function. An example of an approach of this type is the reconnaissance team designed by Sukhatme and his students (Sukhatme, Montgomery, and Matarić 1999; Sukhatme, Montgomery, and Vaughan 2001), consisting of a robot helicopter and one or more ground vehicles, such as Pioneer robots (figure 12.27). Clearly, such teams require communication, either point to point (robot to robot) or as a broadcast from the flying robot. In the team described by Sukhatme, the helicopter was able to transmit GPS coordinates of goal locations to the ground vehicles, which then navigated toward them. In another mode, the ground vehicles could track the position of the helicopter (using radio and/or vision) and follow it on the ground.

The second approach to multiple heterogeneous robots working as a team has been termed marsupial robotics (Murphy 2002). This is the approach used by Papanikolopoulos and his associates in the Ranger-Scout example discussed in section 12.6.5. Drawing its inspiration from kangaroos, this approach assumes that a larger ("mother") robot will carry a number of much smaller robots to be deployed by the mother robot at the appropriate time.

(a)

(b)

Figure 12.26
(a) A legged RoboCup team (of Sony AIBO robots) from Carnegie Mellon University; (b) the goalie (photographs courtesy of Manuela Veloso)

Figure 12.27
Heterogeneous robot team (photograph courtesy of Stefan Hrabar and Gaurav Sukhatme)

12.12 Task Assignment

Another major research area in multiple-robot systems concerns the assignment of specific tasks to individual robots on a team. Whether the teams are homogeneous or heterogeneous, different robots will generally be required to perform different tasks; for example, one may be a scout, reporting on observed events, one may be carrying supplies, and a third could be a beacon to transmit goal information to other robots on the team. Clearly, the designer can assign the various tasks required to accomplish a particular goal to various robots. More interesting is the situation in which the robots are able to select the most appropriate tasks themselves, given some knowledge of their own abilities and limitations and those of other robots on the team. For example, only one robot on a team may be equipped with a vision system, and another may be the only one with dual manipulators. The question is: How do the robots pass this information to one another and select the appropriate role for themselves? Furthermore, can they do this dynamically, so that each robot selects the appropriate action at each instant of time, ensuring completion of a global task by the entire team? Further, membership on a team implies some awareness of other team members. Working in teams is common for humans, but largely unexplored territory for robots. For example, what if a robot on a team becomes inoperative; can another robot step in and take its place so that the global effort is not brought to a halt? Would such an intervention require each robot to have internal models of the other robots, so all the robots on the team would know the appropriate behaviors for a given sensory input?

A behavior-based approach to these issues was formulated by Parker (1998) in her ALLIANCE architecture, as described in section 12.6.3. In this architecture, robots have "motivations" that increase or decrease as tasks for which they are suited become available.

Another approach to task selection is based on negotiation among the robots. We describe here two projects using this approach, one at Carnegie Mellon University and at the other at the University of Southern California. The CMU project was concerned with developing a system for large-scale assembly by heterogeneous robots, such as that which might be needed for autonomous construction of planetary habitats (Simmons et al. 2000a, 2000b). Such construction requires explicit and tight coordination among robots with widely differing capabilities. The solution proposed by Simmons and colleagues involves both negotiation among the robots and some control by a foreman. In this scheme, individual robots may be able to solve certain problems on their own, or they can negotiate task selection with other robots. The robots execute plans dynamically through the construction of *task trees*. Nodes in these trees represent primitive behaviors executed by a given robot under appropriate commands. The nodes are further decomposed into subgoals. The tasks may be constrained or time-ordered. Temporal constraints and goal decomposition are encoded using the authors' Task Description Language (TDL) (Simmons and Apfelbaum 1998). The task tree representation is distributed, so that each robot keeps only that part of the tree that deals with its own goals and actions. The multiagent version of TDL then forms a basis for coordination, since it takes care of the required synchronization of the robots in performance of a given task. The example task considered in Simmons et al. (2000) concerned the placement and connection of a beam in a construction project using three robots: a robot crane (Robocrane) constructed at NIST (Albus, Bostelman, and Dagalakis 1992); a mobile robot equipped with stereo cameras on a pan-tilt head, referred to as a "roving-eye" robot; and a mobile manipulator. For this task, Robocrane is equipped with cables that hold an 8 ft long beam. All three robots are in communication with each other using Ethernet. The system alerts an agent known as a "foreman" to manage the task; the foreman has access to all three of the above robots (figure 12.28). The foreman calls upon Robocrane to move the beam into the general goal area, at which time the roving-eye robot can see fiducial markers on the beam. The roving-eye robot and the mobile manipulator then coordinate their positions and behaviors, until the manipulator has correctly grasped and placed the beam. If the position of the beam is not appropriate, the mobile manipulator negotiates directly with Robocrane to adjust the position.

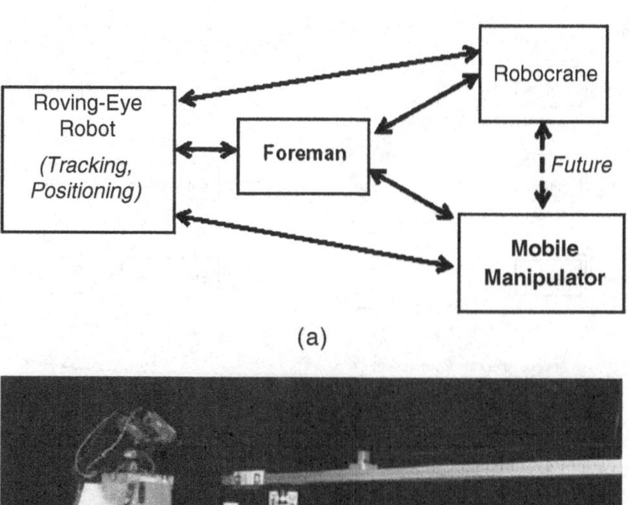

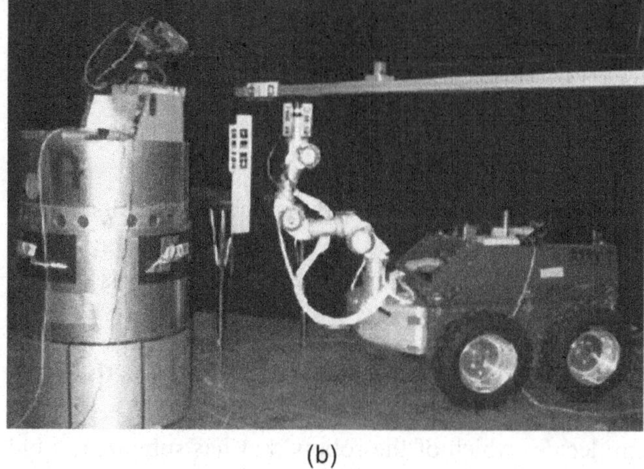

Figure 12.28
Relation between foreman and three robots in task assignment: (a) block diagram; (b) view of the two robots and beam (from Simmons et al. 2000, reproduced with permission of the authors)

The scenario described here makes use of a supervisor to assign jobs to robots with fixed capabilities. In principle, the robots should be able to negotiate among themselves which robot will perform which jobs, by posting bids on a "blackboard" or some analogous storage unit on a computer in which their capabilities are described and matched to a task to be performed. This is the approach taken in the USC project (Matarić, Sukhatme, and Østergaard 2003). Matarić and her colleagues were concerned with dynamically allocating tasks to robots in the presence of uncertainty, clearly a difficult problem that does not yet have a general solution. Global action selection can be viewed as a mapping of each robot's state space to an action, or globally, as a mapping from the combined state space to the combined action space of the team, that is,

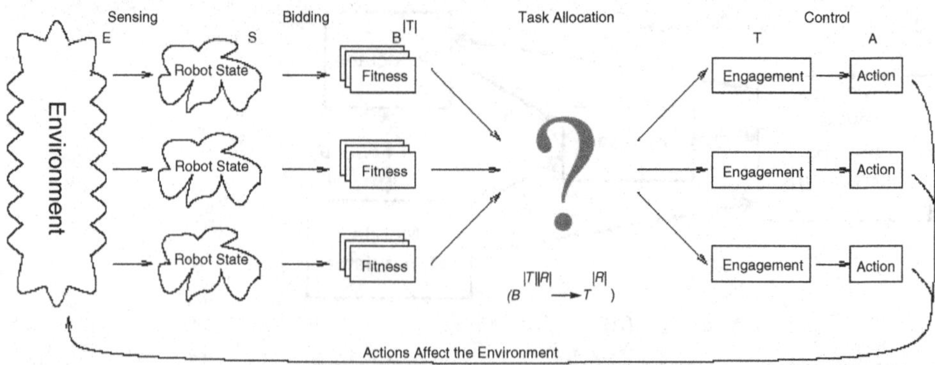

Figure 12.29
Multirobot task allocation based on bidding (from Matarić, Sukhatme, and Østergaard 2003, 257, reproduced with permission of Kluwer Academic Publishers)

$$S^{|R|} \rightarrow A^{|R|}, \tag{12.3}$$

where S is the state space space of a robot, R refers to the number of robots, and A is the set of actions available to the robots (Matarić 1994). Following the approach introduced in Gerkey and Matarić 2001, the task allocation problem was divided into three phases:

1. Each robot bids on a particular task based on its perceived fitness to perform the task, the bids are recorded in a proper format in computer memory.
2. An auction-like mechanism decides which of the robots that has submitted a bid gets the task.
3. The winning robot's controller performs one or more actions to execute the task.

This decomposition was used to construct a general formulation of the multirobot coordination problem, in which a *bidding function* determines each robot's ability to perform the task, a *task allocation mechanism* selects the proper robot for a given task from the bids, and the robot controllers select the appropriate actions for each robot. The process is illustrated in figure 12.29, and the task allocation strategy is illustrated in table 12.2. However, rather than proceeding on the basis of the mapping in equation (12.3), the strategy is based on the mapping

$$B^{|T||R|} \rightarrow T^{|R|}, \tag{12.4}$$

which indicates that the robots' bids B for a given task T are mapped to the task assignment. Matarić and her colleagues investigated four task allocation strategies, as illustrated in table 12.2.

Table 12.2

Strategies	Coordination		
		Individualistic	Mutual Exclusion
Commitment	Commitment	strategy 1	strategy 2
	Opportunistic	strategy 3	strategy 4

The table shows the four tasks arranged along the separate axes of *coordination* and *commitment*. As shown in the table, there are two levels of commitment: full commitment or opportunistic, no-commitment strategies. A fully committed robot will complete its current task before considering any new possibilities, whereas the opportunistic strategy allows a robot to drop a current task in favor of a new one that it views as more appropriate. There are also two levels of coordination: an individualistic or uncoordinated strategy implies that each robot works based only on local information, whereas a coordinated strategy ensures that only one robot is assigned to a given task (i.e., mutual exclusion). The four strategies were studied both in simulation an in an experimental study with several real robots, including various levels of noise. Although the results of the study are not sufficiently general to be included here, they indicate a principled way of approaching the task assignment problem.

A related problem, also studied by Gaurav Sukhatme and Maja Matarić at USC, involves leadership among robots. Clearly, in a large team of, say, a hundred robots, not all of them need to have the same degree of sophistication, so the issues of task assignment and communication become increasingly important.

12.13 Design Issues in Multiple-Robot Systems

We have shown that certain themes, like degree of centralization, heterogeneity, and system robustness, recur again and again in the design of multiple-robot systems. This section attempts to identify some of the issues that must be considered in the design of a multiple-robot system, including (in arbitrary order) the following:

- interrobot communication
- homogeneity versus heterogeneity
- task assignment and specialization
- degree of autonomy
- hierarchy and organization
- reliability, self-repair, and robustness

- control architectures
- mutual recognition
- deployment
- localization
- formation control
- scalability

Many of these issues have been discussed earlier in this chapter and are included here purely for the sake of completeness. The following sections consider each issue in turn, except for homogeneity vs. heterogeneity and degree of autonomy, which were discussed earlier.

12.13.1 Interrobot Communication

Interrobot communication is a surprisingly complex issue the resolution of which is still not completely clear at the time of this writing. Obviously, some form of communication among members of a multiple-robot system is essential, but how should it be accomplished? Explicit, broadcast-type communication may not be practical if the robots are used for surveillance or reconnaissance work. Further, communication requires power, and this may be a problem in a system with large numbers of very small robots. Cao, Fukunaga, and Kahng (1997) suggest three possible structures:

1. Communication via the environment, in which no explicit communication exists directly between the robots. This has also been called "cooperation without communication" (Arkin and Hobbs 1992). Animals may leave trails or traces to be used by other members of the species as signals.
2. Interaction via sensing. In this approach the robots can sense each other but do not communicate explicitly. The sensing could be by means of touch (as with the antennae of insects), or it could take place by means of vision, sound, or smell, which enable animals to recognize members of their own species. Such communication may be essential for some forms of formation control (e.g., Matarić 1994).
3. Explicit communication. In the past such communication frequently used the broadcasting of RF signals. There is currently considerable interest in wireless networks for communication, with a variety of protocols to enable selective communication among members of a group (e.g., Sukhatme and Matarić 2000).

12.13.2 Task Assignment and Specialization

Task assignment was discussed in the section 12.12. There are at least two issues here: first, how to provide the needed high-level instruction to the entire group ("clear mines from Sector A-3," "look for survivors in the vicinity of latitude *xx*, longitude *yy*"). Second, instructions must be provided to individual robots (or the robots must determine on their own what task they should perform). There are a number of options here, some related to other issues discussed in subsequent sections. If a group is organized hierarchically, then there may be supervisor robots that make specific task assignments. If the robots are heterogeous and specialized in their abilities, then the assignment may flow automatically from their specialization (e.g., cleanup robots will clean, while carpenter robots will build, and rock-gathering robots will gather rocks). A more likely scenario may require the use of robots for tasks for which they may not be the first choice, and more complex decision processes will be required in such a case (see, for example, Gerkey and Matarić 2000).

A particularly interesting issue arises in cases in which a group of robots is tasked with a global goal, but the robots have only local information (as with colonies of termites building large mounds of sand) and/or in which the robots have multiple capabilities among which they must select the appropriate ones autonomously.

12.13.3 Hierarchy and Organization

This issue has been considered earlier in the chapter. There is clearly a relation between the hierarchical organization of a group and the degree of autonomy of its individual robots.

12.13.4 Reliability, Self-Repair, and Robustness

One of the reasons for using multiple robots for a mission is simply the need to increase the probability of completing the mission in the presence of failure of one or more of the team's members. The Prospector mission to Mars was dependent on the single robot Sojourner; had it failed, no exploration would have been possible. But there are related issues here: If the robots on a team are specialized, can robot B in fact substitute for a failed robot A? Can a drone substitute for a missing queen in a bee colony? Or is some form of automatic reprogramming required? Must B have an internal model of A to enable it to reconfigure itself as required? A related issue in large colonies is the need to remove damaged or inoperative robots so they do not interfere with the team's tasks (by becoming obstacles) or perhaps to assign specialized robots to perform repair or part substitution in defective robots.

12.13.5 Control Architectures

Earlier in this chapter we provided examples of control architectures for multiple-robot systems, including the pheromone model, the ALLIANCE architecture, and the Ranger-Scout architecture. Of course, there are many more possible software structures for control of such systems, and we expect that new ones will emerge in the future as multiple-robot systems become more prevalent.

12.13.6 Mutual Recognition

Animals have an almost uncanny ability to distinguish members of their own species from others. Birds of nearly identical plumage seem to be able to find their own with no difficulty. Will such a recognition of membership in a group be important for large colonies of robots? We expect that some form of "friend or foe" recognition will be essential, as will be recognition of group members for purposes of support or repair.

12.13.7 Deployment

The method of deployment of large numbers of small or very small robots is not obvious. The tiny smart-dust robots discussed in section 12.7 are designed to be deployed from aircraft. The Scouts discussed in section 12.6.5 are brought to a location by Rangers, and then dropped; they use their jumping ability to scatter throughout the terrain thereafter. Tiny waterborne robots could be dropped into the water en masse, with the assumption that currents will disperse them as required. Robots could be dispersed by explosive forces, by wind, by systematic drops in desired locations, or even by programmed movement from a start location with either random or specified trajectories. In any case, deployment is an issue that will have to be addressed in the design of any multiple-robot system, especially if a particular distribution of robots is desired.

12.13.8 Localization

We have shown that the ability of a robot to determine its location in space is essential for navigation, goal seeking, and other activities, and we discuss this issue at even greater length in chapter 14. Clearly, the issue is different in multiple-robot systems, since now the localization of many robots must be addressed. If the numbers of robots are large, and the dispersion is small, it may be sufficient to localize the center of gravity of the swarm. On the other hand, if the numbers are small, it may be possible to use other members of the group as beacons or reference points and have each robot determine its location accordingly (e.g., Roumeliotis and Bekey 2000a; Howard, Matarić, and Sukhatme 2002b).

12.13.9 Formation Control

If a group of robots is to move in a coordinated manner, the control of the formation needs to be considered. Such control could be obtained from the environment, from communication among the robots, or perhaps by simple rule-based behaviors.

12.13.10 Scalability

Although we have not discussed the issue at length in this chapter, the question of scaling is of enormous importance in the design of robot teams. It is reasonable to expect that control strategies that work for small teams of three or four robots could be scaled up for implementation for teams of ten or twenty robots. However, it is unlikely that they can be scaled to groups of hundreds or thousands of robots. Entirely different approaches will be required for the control of very large groups. Swarm robotics may provide some of the answers, but what if the robots in the group are heterogeneous?

12.14 Conclusions

This chapter has addressed some of the control issues that arise in connection with groups of robots. It should be evident that ensuring cooperation among robots while they are carrying out complex tasks is a major challenge for robot designers. We have described a number of approaches used in the past and summarized some of the major design issues involved. We expect that research on methods of obtaining collaborative behavior in large groups of robots will continue to be a major challenge for the next decade or two, particularly for cases in which the groups include humans as well as robots.

13 Humanoid Robots

Summary

Humanoid robots are machines whose structure and/or behavior imitates that of humans. In this chapter we survey some notable humanoids and discuss their features, advantages, and disadvantages. We begin with some historically interesting humanoids, both real and fictional. Some current humanoids are then discussed, with emphasis on structure, locomotion, intelligence, and emotions. The objective of the chapter is to stimulate thinking about those aspects of humanoids that are truly essential as models of human behavior and those that are primarily cosmetic in nature.

13.1 Introduction: Why Humanoids?

In the minds of many people, the word "robot" is associated with images of humanlike machines with superhuman intelligence, strength, or abilities. Certainly, that has been the popular view in science fiction for many years. Sometimes the robot may have a somewhat human appearance, or not look human but possess outstanding memory and intelligence, like C3PO and R2-D2, respectively, in the *Star Wars* films (see figure 13.1). The robot in the movie *Short Circuit* had a vaguely humanlike appearance (though it moved on wheels rather than legs) but could read the entire *Encyclopedia Britannica* in a matter of seconds. It is attributes, such as memory and intelligence, that determine whether we call a machine a humanoid or not. Thus, from our point of view, a humanoid robot must include

1. A head equipped with at least some of the sensors in the human head, such as vision or audition. This head should, at least in some general sense, approximate the shape of the human head.

2. Some cognitive ability, enabling it to process the information received from the sensors.

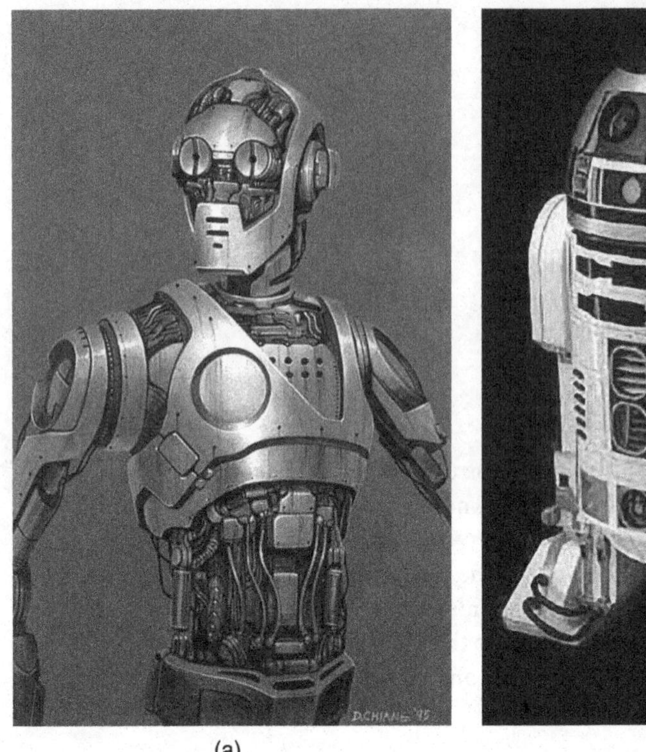

(a)

(b)

Figure 13.1
Star Wars robots (a) C3PO and (b) R2-D2 (reproduced with permission of Lucas Films)

3. Some ability to interact with humans.

These are the essential ingredients. In addition, many so-called humanoids also have

4. Anthropomorphic appearance (resembling an upright human body), sometimes with artificial muscles and humanlike legs and arms.
5. Mobility, using legs or wheels.
6. Ability to express emotions and to respond to human emotions. Robots that have this ability are termed *anthropopathic* (Swinson and Bruemer 2000).

Thus, we suggest that the primary goal of humanoid robot design is to build a machine with the essential ingredients in the preceding list and as many of the additional qualities as possible.

A quadriplegic confined to a wheelchair, paralyzed from the neck down, still retains the qualities that make him or her essentially human. For this reason we intentionally did not make biped locomotion an *essential* requirement in the preceding list, even though some two-legged machines with no cognitive ability are termed "humanoid robots" (see chapter 8). Autonomy is also not listed, even though it is clearly a desirable attribute. Certainly, some (or all) of the interaction of the robot with humans should be autonomous, if possible. However, a number of interesting humanoid robots are teleoperated, including the Honda P-series robots, introduced in section 8.6.4 and described in detail in section 13.3.

Thus, from this point of view, the robot HAL in the movie *2001* should properly be called a humanoid, even though it was not embodied in humanlike form. Nevertheless, it indeed had sensors (including vision and audition) as well as access to numerous diagnostic sensors throughout the spaceship that housed it. It also had actuators at its disposal, ranging from those for opening and closing doors to those that controlled the environment for astronauts in suspended animation.

13.1.1 Other Motivations for the Design of Humanoids

The second major goal of designers of humanoid robots is to build machines to assist humans in such tasks as operation of power plants, disaster relief, and construction. Application of robots in this way will be even more significant when we can communicate with the robots using voice and gestures and when they can learn from humans simply by imitation rather than requiring programming. The third motivation is to provide humanlike robots for entertainment, not only for the movie industry, but perhaps ultimately as playmates for children.

Finally, the greatest motivation for the development of humanoid robots is fascination. Human-appearing robots fascinate us for all sorts of reasons, the most basic of which is because of the deep human desire to understand who we are. They are imitations of some aspects of ourselves, and thus they cause us to wonder what it is that makes us characteristically human and different from machines. We deeply identify with our human form, even though some consider it to be simply a mortal vessel containing the soul. This identification even leads us to create models of extraterrestrials that are surprisingly humanlike in structure, even if their heads are larger or their arms shorter and weaker than ours. Philosophers have long been concerned with the duality between mind and body and have debated whether human intelligence requires a human body for its realization. It is certainly true that evolution has led to the human brain in the particular embodiment we have.

Our fascination with humanoid robots appears to exist even when they do not exhibit superior intelligence. The Honda robot P-3 never fails to draw a crowd, especially when it demonstrates its ability to go up and down stairs. Perhaps for this reason research in humanoids continues to attract attention throughout the world, even when there are no clear practical applications for them.

13.1.2 Humanoid Dynamics and Stability

If a biped robot is to move in a manner that approximates that of human beings, its designers must pay attention not only to the kinematics of its joints and structure, but to dynamics as well. As noted at several points earlier in the book, a walking humanoid is basically an inverted pendulum that, like its human prototypes, needs some form of active control to remain upright. As was discussed in chapter 8, vertical stability in humans requires the active participation of muscle spindles and reflex control arcs closing through the spinal cord. Small stretches of flexor and extensor muscles in the legs are sensed by the muscle spindles and initiate the control response, leading to a small amount of sway. The "inertial platform" of humans, located in the inner ear, becomes effective only for larger angular movements, such as those associated with falling. To maintain stability, robots require sensors to indicate the deviation from the local vertical. The issues associated with the design of such controllers were discussed in chapter 8. In this chapter the emphasis is on the overall design of humanoid robots.

13.2 Historical Background

Historically, we date humanoids to three sources. The first are creatures of myth and imagination. Perhaps the best known of these was the Golem, a creature of Jewish medieval folklore. According the legend, certain wise men were able to make an effigy, a sculpture of a man, come to life by placing in its mouth or on its head a charm, frequently a group of letters indicating one of the names of God. Removal of the charm returned the Golem to its original inanimate state. Rabbi Löw in Prague created one of the most famous Golems out of mud taken from a riverbank (Swinson and Bruemer 2000). The creature was given life by virtue of the rabbi's wisdom and served him well, until.... This theme, that a good robot may become malevolent, appears again and again in literature. We place the Frankenstein monster, the semimythical Big Foot creature from the forests of northern California, and the innumerable robots from science fiction films into the same class. (The Frankenstein monster was probably inspired by a 1920 German silent film on the Golem.)

The second source of inspiration for humanoids comes from the "automatons" that became quite popular in the eighteenth century. These machines were mannequins that appeared to be alive, since they could answer questions and perform some limited movements. They were actually controlled by a person who might have hidden under a desk where he pulled levers or strings, or sometimes even by a midget or small person actually hidden inside of the mannequin. A typical automa-

Figure 13.2
Typical automaton from 1890 (photograph courtesy of the Carol and Dallas Morris Collection)

ton is shown in figure 13.2. This French piece, made in 1890, is called the "Russian Tea Server." She opens and closes her eyes, turns her head, moves the tray, pours tea, and lifts the teapot, while music plays.

The third source of our ideas about humanoids probably dates to the original "robots," the artificial humans in the 1921 play *RUR* (*Rossum's Universal Robots*) by Karel Čapek. The word "robot" was derived from the Czech word *robota*, meaning slave labor or drudgery, which is precisely what the robots in the play were intended for. It is interesting to note that in the play, Rossum had tried to create an artificial human for many years with no success. His son pointed out that the reason for the failure was that his father had tried to imbue his creations with all human qualities, including feelings, appreciation for truth and beauty, and fear of death. When his son took over the project, he left out these "nonessential" features and created human-appearing "robots" that obeyed orders and performed manual labor, mostly in factories, until they wore out. Then, with no hesitation, they would throw themselves into ovens to be recycled. The distinction between human-appearing robots and robots with cognitive and emotional attributes continues to haunt the field to the present day. Of course, humanoid robots are frequently both the heroes and the villains of science fiction.

As we move from the humanoid robots of science fiction to actual implemented robots, a place of honor goes to the late Ichiro Kato of Waseda University in Japan. In many ways, Kato can be considered the father of contemporary humanoid robotics. In 1985, at a science exposition in Tsukuba, Japan, a biped robot designed and built in Kato's laboratory walked about for many days without a failure, and the Waseda piano-playing robot, WABOT-2, began a remarkable career. As illustrated in figure 13.3, WABOT-2 was indeed anthropomorphic. It sat on a regular piano bench, used a television camera to read sheet music, moved its two five-fingered hands to press the piano's keys, and used its feet to press the instrument's pedals. This was indeed a humanoid robot in general body structure and in its ability to move hands, fingers, and feet. On the other hand, it was basically a single-function robot, capable of only this one task. Furthermore, although it could indeed read sheet music, it had to be positioned in a precise location on the bench to hit the correct keys in response to the music it was reading.

Two developments in Yugoslavia in the 1970s provided important background for later development of humanoid robots. A historically interesting robot, though not a full-body humanoid, was the walking lower body designed by Miomir Vukobratović at the Pupin Institute. This machine consisted of a powered set of hollow legs that wrapped around the nonfunctional legs of a person with lower-body paralysis. The robot then provided walking mobility to the paraplegic, while he held on to horizontal bars for stability (Vukobratović and Juricić 1968; Vukobratović, Hristić, and Stojiljković 1974).[29] At approximately the same time, Rajko Tomović and his associates at the University of Belgrade developed a five-fingered hand, intended as a prosthetic device for veterans who had lost their hands in the African campaigns of World War II. This hand (described in more detail in chapter 11) was shape adaptive. A single motor, when activated by a pressure sensor on the hand, would cause

29. I saw this system demonstrated in Vukobratović's laboratory in the 1970s. The exoskeleton did indeed walk, but the patient, captured within the moving frame, had no control over its movement. I have never seen a person with a more frightened expression than this patient!

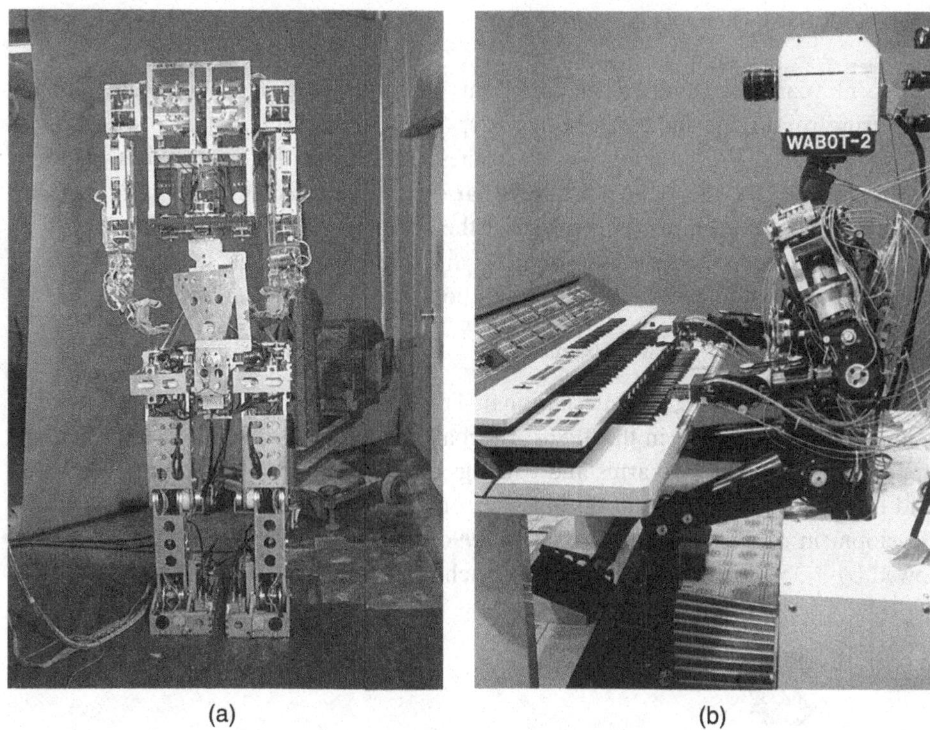

Figure 13.3
Humanoid robots from Waseda University in the 1980s: (a) biped walker WL-12 (1986) and (b) WABOT-2 piano-playing robot (photographs reproduced with permission of Atsuo Takanishi)

all the fingers to flex until they touched the object being grasped. The hand was not successful as a prosthetic device, but it provided the background for a robotic hand developed and marketed in the late 1980s and early 1990s. Development of robot manipulators for industry (see chapter 10), first in the United States and then in Japan and Europe, provided much of the needed background for humanoid arms, both in methods of analysis and in design and fabrication.

As indicated earlier, we consider at least a minimum of cognitive ability to be an essential ingredient of a humanoid robot. Hence, it is clear that another major source of background for current work in robotics has come from the artificial intelligence community. Clearly, computational intelligence is an attempt to synthesize some aspects of human intelligence. Swinson and Bruemer (2000) suggest that it is humanoid robots' cognitive architecture, including learning ability, that distinguishes them from other robots. But then, adding a brain to a robot body is clearly not an easy task (Brooks and Stein 1994).

13.3 Full-Body Humanoids

In recent years there have been numerous attempts to design and build human-appearing robots. As indicated previously, this is a very active area of research at the present time. We describe several of these projects in this section, but it should be understood that this does not imply that other humanoids not discussed here are less interesting. We simply had to restrict the coverage since this book is intended to cover a broad set of issues in robotics, and the robots discussed here represent broader classes of machines. Some partial-body robots are discussed in section 13.5.

13.3.1 The Honda P-Series Robots

The Japanese Ministry of International Trade and Industry (MITI) launched a humanoid robot project in the 1980s. Its goal was to design and fabricate a human-appearing robot with two arms and two legs, capable of biped locomotion, going up and down stairs, and picking up objects from the floor. Honda began research and development on this project in 1986. In 1996 the prototype P-2 made its debut, followed by P-3 in 1997. This remarkable machine is shown in figure 13.4.

Figure 13.4
Honda P-3 robot (photograph courtesy of Honda Motor Company Ltd.)

The Honda P-2 robot was quite large and heavy, being 1.8 m tall and weighing 210 kg. Nevertheless, it was able to walk up and down stairs while maintaing vertical stability. The P-3 robot was somewhat shorter, being 1.6 m tall and significantly lighter, weighing 130 kg including the batteries, which allowed for about thirty minutes of operation. The following description concerns the P3. The helmetlike head contained a pair of stereo cameras. The robot had a large number of degrees of freedom, including 7 in the arms and hands, 6 in the legs and 1 in the neck. In addition, the cameras could pan and tilt. The robot was designed to perform the following functions:

- biped walking, forward, backward, or sideways (on terrain with up to 2 cm height variations)
- turning in place
- walking up and down a staircase (with all steps of the same height, known in advance)
- carrying a load of up to 10 kg
- operating a lever
- grasping an object weighing up to 2 kg with one hand

The P-2 and P-3 were not entirely autonomous, being controlled from a specially designed cockpit. Given an initial position of the robot, the operator could specify the (x, y) coordinates of a desired goal position, and the robot walked there autonomously. Similarly, given a location at the base (or top) of a staircase and the number and height of the steps, the robot could ascend or descend autonomously. On the other hand, teleoperation from the cockpit, using a master-slave arrangement, was required for grasping, lifting a load or operating a lever. The operator was provided with a head-mounted display as well as a number of surrounding screens to enable her to see the world "through the robot's eyes."

Clearly, the Honda P-series robots were an impressive achievement. In view of the public's fascination with human-appearing machines, they were used in exhibits of Honda automobiles and always attracted crowds. On the other hand, these machines were not autonomous and exhibited little intelligence. As of this writing, the Japanese Humanoid Robotics Program was considering a variety of industrial applications for these robots.

13.3.2 ASIMO

A more recent member of the Honda family of humanoids is ASIMO (Advanced Step in Innovative Mobility), shown in figure 13.5 next to its larger "brother" P-3. ASIMO was discussed briefly in section 1.8.9; here it is examined in greater detail.

Figure 13.5
ASIMO (left) and P-3 (right) (photograph courtesy of Honda Motor Company Ltd.)

ASIMO is a much smaller and lighter robot, being about 120 cm in height (as compared to 160 cm for P-3) and weighing 52 kg (as compared to 130 kg for P-3). The latest version of ASIMO walks more smoothly, more flexibly, and more naturally than the earlier robots of the Honda family and is able to move more freely in ordinary environments, including climbing and descending stairways and slopes. Its arms have greater range and freedom of movement than those of the P-3: Its range of vertical arm movement is 105°, as compared to P-3's 90°. The robot's height of 120 cm was chosen because it was considered the optimum for operating household switches, reaching doorknobs in a human living space, and performing tasks at tables and benches. The robot can be programmed to perform a variety of specialized movements. It is still basically teleoperated but can be operated from a portable controller (as opposed to the cockpit required for operation of Honda's P-series robots). ASIMO can also be operated by voice and hand signals. Current models also include face recognition ability. The specifications of ASIMO are summarized in table 13.1. The Honda Corporation is leasing ASIMO robots to other companies and to museums, where it can greet visitors and perform other publicity functions.

13.3.3 QRIO (The Sony Dream Robot)

The Sony QRIO (figure 13.6) (formerly known as the Sony Dream Robot SDR-3X) is a full-body robot, even smaller than ASIMO, standing only about 60 cm high,

Table 13.1
Specifications of ASIMO humanoid robot

Weight	52 kg
Height	1,200 mm
Depth	440 mm
Width	450 mm
Walking speed	0–1.6 km/h
Operating degrees of freedom:	
Head	2 dof
Arm	5 × 2 = 10 dof
Hand	1 × 2 = 2 dof
Leg	6 × 2 = 12 dof
Total	26 dof
Actuators	Servomotor + harmonic decelerator + drive
ECU Controller	Walking/operation control ECU, wireless transmission ECU
Sensors	
Foot	6-axis force sensors
Torso	Gyroscope and acceleration sensors
Power Source	38.4V/10AH (Ni-MN)
Operation	Work station, portable controller, voice and hand signals

Figure 13.6
Sony QRIO (photograph courtesy of Sony Digital Creatures Laboratory)

with a total weight of 5 kg (Kuroki et al. 2001). Yet in this small size, it has 42 dof, a vision system, the ability to perform a variety of movements, and speech synthesis and recognition. The name QRIO is not an acronym; rather, it is derived from Quest for Curiosity and pronounced "Curio."

QRIO shares some design features with the Sony AIBO robot pet (see chapter 9). It uses an architecture based on the AIBO OPEN-R, which provides interfaces that make it possible to use a variety of AIBO input and output devices, such as the color camera and IR.

QRIO is capable of a great variety of movements, including the following:

- walking: forward, backward and sideways on a variety of surfaces
- turning while walking (up to 90° in one step)
- swinging of the arms while walking to counteract the yaw moment caused by biped locomotion
- standing up from a sitting position on a bench
- sitting down on the floor and standing up from this position
- bending down from the upper body
- locating and kicking a ball of a given color with one foot
- throwing a ball
- dancing
- jogging, in which both feet are off the ground for brief periods

The repertoire of movements grows as new applications are developed. These movements are possible because of three major developments:

1. *Actuators* QRIO's joints feature novel actuators, known as *intelligent servo actuators*, capable of producing varying levels of torque at different rates of movement.
2. *Sensors* As may be imagined, the robot is equipped with a multitude of sensors, including four on the soles of each foot. The foot sensors enable it to adjust its body attitude to compensate for variations in the floor.
3. *Walking with dynamic stability* It is clear that to walk rapidly, dance, and jog with relatively small feet, this robot does not rely on static stability. Instead, its onboard computer constantly adjusts the robot's body position based on the location of the zero-moment point (Vukobratović et al. 1989), the point at which the gravity vector and the inertial vector of the moving robot intersect the ground (see chapter 8). QRIO's control system adjusts the robot's body during movement so that the ZMP always stays within the zone of stability. (Dynamic stability based on the ZMP is also employed in control of the Honda ASIMO and the Waseda WABIAN robots.)

It is clear that QRIO is a highly sophisticated entertainment robot (Kuroki et al. 2000).

13.3.4 The Sarcos-Kawato Humanoid

The robot DB (Dynamic Brain), first mentioned in the discussion of arms in chapter 10, was designed by the Kawato Dynamic Brain Project in Japan and Sarcos, Inc. in Salt Lake City, Utah, and built by Sarcos for use in the Dynamic Brain Project, sponsored by ERATO, a research effort funded by the Japan Science and Technology Agency. Its purpose is to study neuromuscular aspects of behavior, both in man and robot, primarily in the upper body. As illustrated in figure 13.7, DB is an anthropomorphic system. At present it is mounted at the pelvis, thus eliminating concern about balance and providing access to its power supplies. DB has 30 degrees of freedom; it weighs 80 kg and is about 1.85 m tall. A great deal of effort has gone into providing DB with "muscles" and joints that enable it to achieve smooth and natural-appearing movements. This smoothness of movement was accomplished by

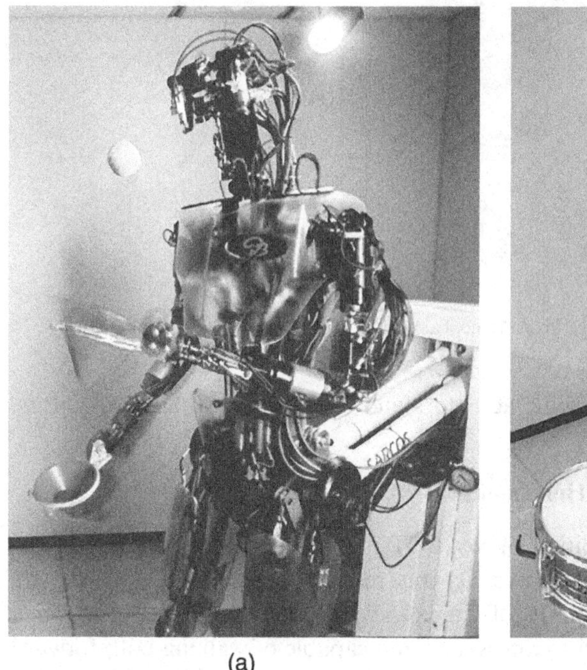
(a)

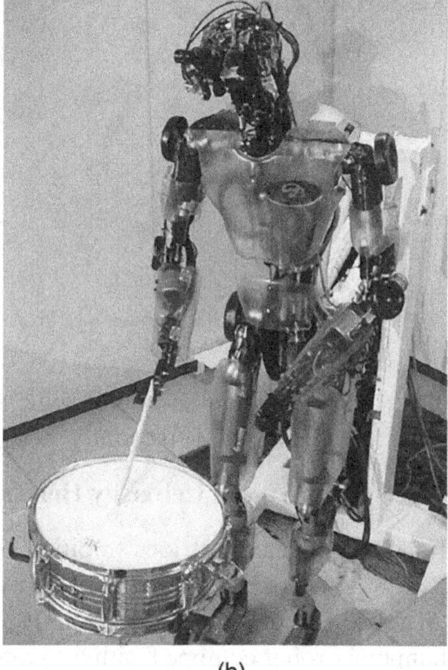
(b)

Figure 13.7
The humanoid robot DB: (a) DB juggling a ball; (b) DB playing a drum (manufactured by Sarcos Inc., copyright ATR Computational Neuroscience Laboratories, photograph courtesy of Mitsuo Kawato and Stefan Schaal)

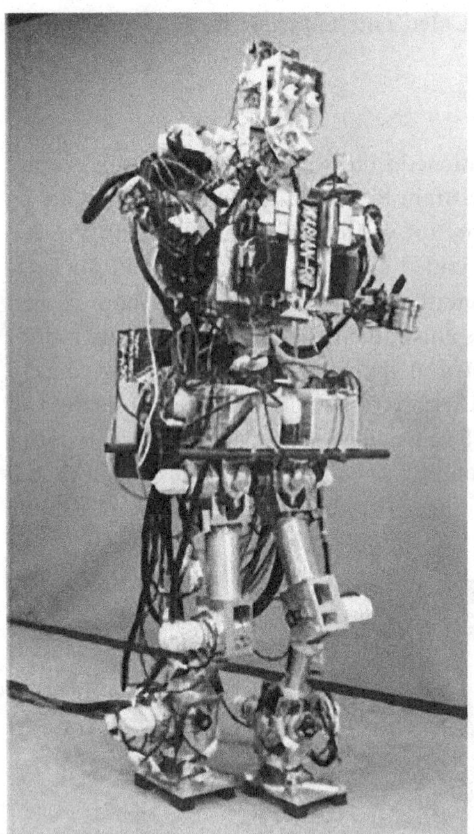

Figure 13.8
WABIAN (reproduced with permission of Prof. Atsuo Takanishi)

means of twenty-five linear hydraulic actuators and five rotary hydraulic actuators, one for each degree of freedom.

13.3.5 The Waseda University Humanoids

Following in the illustrious tradition of WABOT-1 and WABOT-2, the Humanoid Research Laboratory at Waseda University has recently developed the WABIAN (Waseda Bipedal humanoid) family of humanoids. WABIAN (figure 13.8) is a humanoid robot equipped with a vision system and capable of walking both forward and backward on its two legs. At the time of this writing the addition of speech and speech recognition was planned to allow for communication with humans (Hashimoto et al. 2000). WABIAN is relatively small (about 1.66 m tall), being comparable in size to an average Japanese female, but weighing 107 kg. The robot is capable of

Figure 13.9
Humanoid robot PINO (photograph courtesy of Hiroaki Kitano)

carrying 30 kg on its shoulders and 1.5 kg in its hand. It is entirely electrical, using household electrical current (through an umbilical) for its power supply. As it walks it swings its arms to aid in stability. In forward locomotion, WABIAN can walk at approximately the speed of a human.

WABIAN has 7-dof arms, 3 additional dof in the hand, 2 dof in the eyeballs, 2 dof in the neck, 3 dof in the trunk (enabling it to move in pitch, yaw, and roll), 1 dof each in the ankle, knee, and hip joints, and 4 passive dof in each foot, for a total of 43 degrees of freedom. Clearly, the design of this robot was a major electromechanical engineering achievement. Waseda University researchers have built a number of other humanoids in addition to WABIAN; we will encounter some of them later in this chapter.

The original WABIAN was followed by WABIAN RII and WABIAN RIII (Refined, Version III), each with additional features and versatility.

13.3.6 PINO

PINO (figure 13.9) was developed by Japan's ERATO program, which also sponsored the research that led to the development of DB, which was discussed in section 13.3.4. It is 70 cm high and weighs 4.5 kg and has twenty-six joints that are driven by motors and potentiometers. The motors are locally controlled by microcontrollers or electric circuitry and are centrally controlled by a 32-bit SH2 microcontroller. Dynamic walking is made possible by the use of a fast and inexpensive CPU that incorporates the bipedal walking algorithm (Yamasaki et al. 2000). PINO has a sophisticated control system that considers the kinematics of forward motion as well as body sway.

The design of PINO was deliberately anthropomorphic. In fact, its designers studied the human form all the way back to Greek sculptures to ensure that the robot was not only functional, but also esthetically pleasing (Kitano 2002). Furthermore, it was designed to facilitate research in humanoids, and all the software is available in open source.

13.3.7 HOAP

Fujitsu Laboratories introduced the miniature humanoid HOAP-1 in 2001. This robot stood only 48 cm high and weighed 6 kg. It had 20 dof. Its successor, HOAP-2 (figure 13.10) is 50 cm tall and weighs 7 kg. HOAP-2 has articulated fingers, allowing it to grip a variety of objects; it has 25 dof. The HOAP series uses RT-Linux

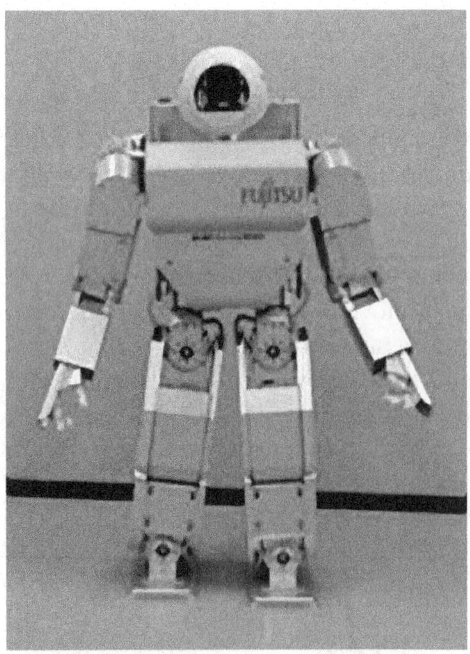

Figure 13.10
Miniature robot HOAP-2 (photograph courtesy of Fujitsu Automation Ltd.)

operating systems. A simulation program and information on its software architecture are provided with the robot, to enable users to develop their own programs, under Linux, on a PC that communicates with the robot through a USB port via an umbilical cable. There is clearly some competition among Japanese humanoids. Whereas QRIO can sing and dance, apparently HOAP-2 can perform a variety of martial-arts movements as well as assume a number of sumo wrestling stances.

13.3.8 Other Humanoid Projects

It is evident from the foregoing survey that much of the work on full-body humanoid robots is being carried out in Japan, where there is a long history of interest and government funding for this type of work. There are other Japanese humanoid robots not listed here. Of course, humanoid projects also exist in other countries, including Korea, Germany, Sweden, China, Russia, England, Italy, Iran, and the United States.[30] These humanoid projects generally emphasize some aspect of humanoid design, whether arms or legs or cognitive ability or sensors or emotions or ability to interact with humans.

13.4 Interaction with Humans

The ability of robots to interact with human beings has been viewed by many as the cornerstone of "humanoid" robotics. Human-robot interaction requires that the robot and human be able to sense one another through vision, audition, touch, and perhaps even smell and interpret these perceptions appropriately. Thus, humanoid robots are almost always equipped with vision systems, microphones, and perhaps speech recogntion. If touch is a communication modality, the robot will require synthetic haptic sensors (such as fingers equipped with pressure sensors). As noted in chapter 3, these types of sensors are available, either commercially or in research laboratories. The challenge for designers of humanoids lies in enabling the robot to interpret the perceptions the sensors provide.

Consider the variety of ways in which humans use vision to interact with one another: We observe one another's gestures, body movements, and facial expressions. Interpretation of these observations provides an indication of another's intentions (such as "come here" with movement of a flexed index finger, or "go there" with pointing), it may be a part of a sign language, and it may contain clues to emotional states. Communication by means of gestures is clearly a desirable attribute for human-robot interaction. The robot must be able to perform pattern recognition and

30. An outstanding source of information on more than one hundred humanoids and related projects worldwide is the Web site http://www.androidworld.com/prod01.htm.

matching to determine the meaning of a particular gesture. It may also be necessary to provide an on-off switch for gesture recognition, to prevent the robot from misinterpreting human movements and considering all of them signals. Note that communication via visual cues should be two-way; that is, the robot should be able to use gestures to communicate with a human or perhaps to use gestures to supplement voice signals. The humanoid Hadaly developed by Waseda University (see section 13.4.2) is able to recognize seven gestures: those that signify "thank you," "hello," "good bye," "yes," "no," "come here," and "go away" (Kobayashi and Haruyama 1997). To interpret the gestures, the robot's vision system captures them in a sequence of images. The system then identifies the right hand and face positions by color: It assumes that the largest skin color region is the face and the largest moving skin area represents the hand. The gesture is then expressed as a time series of hand movements with respect to the head. The values of this time series are used as inputs to a hidden Markov model that performs pattern matching by deciding which of the seven stored patterns is closest to the unknown gesture just captured. The authors claim that the robot has achieved 90% accuracy in recognizing and classifying human gestures among the group of seven that the robot is programmed to comprehend. A robot developed at Carnegie Mellon University can recognize a variety of gestures (Voyles, Morrow, and Khosla 1999). The tropism architecture described in chapters 5 and 6 has also been used for the interpretation of context-dependent gestures by robots (Voyles et al. 1997).

Facial expressions are charged with information, but since many of the clues they provide are complex, it is difficult to design robot vision systems to interpret all possible expressions. Humans use the shape of the mouth to communicate happiness or disappointment; we move the eyebrows to indicate puzzlement; we move our eyes, narrow them, or open them wide; and we even wrinkle our noses to indicate emotional states. We also move our head, using multiple muscles in the neck. Ideally, a robot should be able to interpret these signals.

In addition, we would expect the robot to modify its "facial features" in a like manner to indicate its emotional state. The issue of "robot emotions" is an emotional one for some people, who suggest that all a robot can do is to simulate facial expressions, tone of voice, and/or hand gestures that express emotions in people, but that this is a sham, since the robot does not have a nervous system and in fact does not feel any emotion. We believe that the issue is more complex than this simple criticism implies. For example, assume that Bob insults Jack, and Jack gets angry and hits Bob. Clearly, a robot could be built that reacts in exactly the same way. Does that mean that the robot actually feels anger? Is the release of adrenalin and other chemicals in the body a requirement for genuine emotion? We leave that decision to the reader. The following sections examine several implementations of robot emotion.

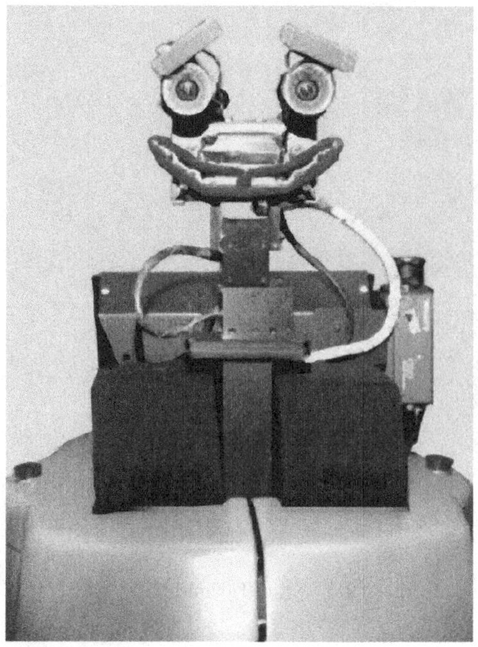

Figure 13.11
Robot face on the museum guide Minerva (from Thrun et al. 2000, photograph courtesy of Sebastian Thrun)

13.4.1 Minerva

Among the robots that have included variation of facial expression was Minerva (Thrun et al. 2000), a museum guide robot that was able to adjust the position of its mouth and eyebrows, as shown in figure 13.11. Although these movements were exaggerated by human standards, they clearly communicated such emotions as happiness and sadness. It is interesting to note that although Minerva had a humanoid head (with movable eyebrows, eyes, and mouth), its body was a wheeled, nonanthropomorphic structure. Nevertheless, children would follow it about the museum where it worked, fascinated by its humanlike interactions.

13.4.2 Hadaly

Hadaly was a synthetic beautiful woman in a French novel entitled *Eve of the Future*. The Waseda group has given name Hadaly-2 to a robot (figure 13.12) designed for interaction with humans; the robot has visual and auditory communication capabilities, capacity for eye movements, and arms with four fingers each (Hashimoto et al. 2000). As of this writing Hadaly is not yet able to recognize human facial expressions,

Figure 13.12
Hadaly-2 interactive robot (photograph reproduced with permission of Atsuo Takanishi)

but it has expressive eye movements for communication with humans. The eye movement system (known as Waseda Eye No. 3 Refined, WE-3R) allows the robot to move its eyes in 2 dof (up and down and sideways), as well as to narrow the pupil and open and close the eyelids. The pupil diameter adjusts automatically to brightness. The robot also has 2 degrees of freedom in the neck, allowing for movement of the head in pitch and yaw. It is particularly interesting that this robot communicates not only by means of eye movements, but through arm movements as well.

13.4.3 Kismet

Kismet is a robot developed in the Artificial Intelligence Laboratory of MIT, specifically to study emotional aspects of human-robot interaction. Figure 13.13 shows Kismet interacting with its creator, Cynthia Breazeal, and figure 13.14 shows a variety of emotional expressions of which the robot is capable (Breazeal 2002).

Kismet's sensors include vision (using four-color charge-coupled device cameras, two of which can move with respect to the head) and audition (using speech recognition software). Kismet also has some ability to recognize human expressions, and its eyes can follow a colored toy or similar object. Note that Kismet has a number of adjustable features for expressiveness, including movements of its eyelids, eyebrows, ears, jaw, lips, neck, and head orientation. Its auditory system can recognize some of the affective features in the human voice, and its vision system is capable of recognizing the affective aspect of some human facial expressions, including motion of the

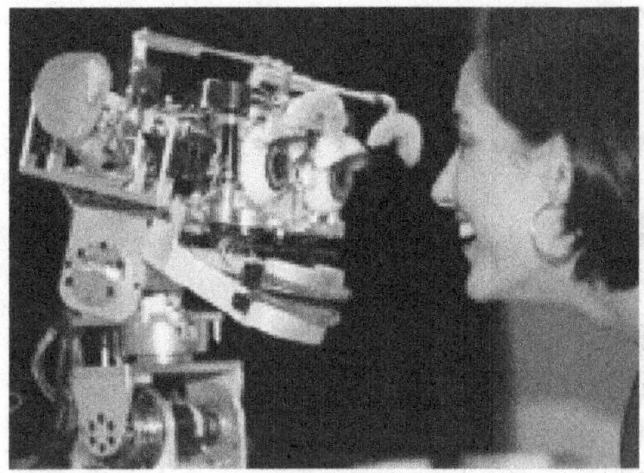

Figure 13.13
Kismet interacting with its creator, Cynthia Breazeal (photograph courtesy of Donna Coveney)

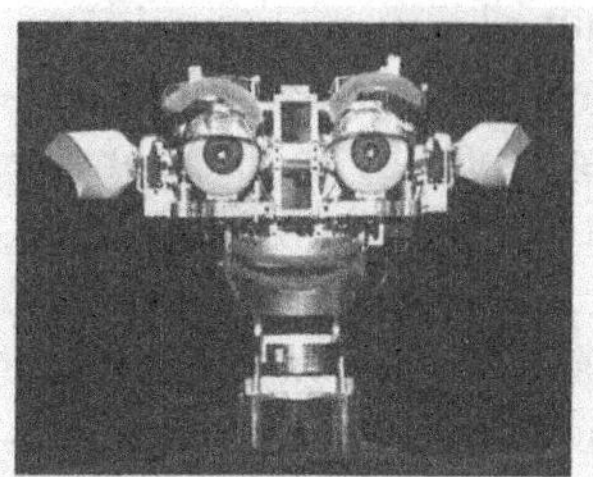 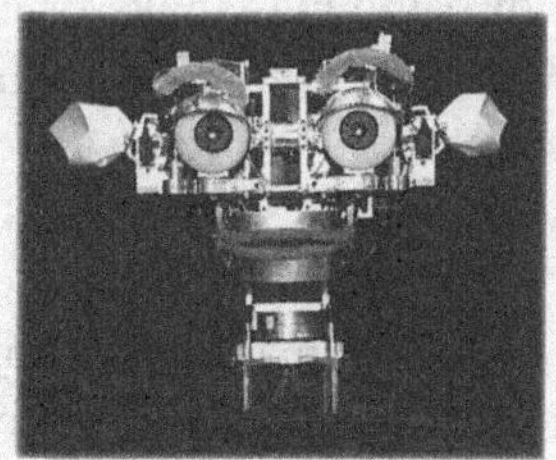 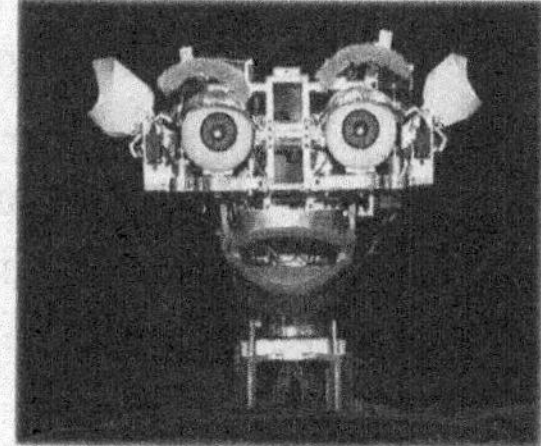

Figure 13.14
Three emotional expressions of Kismet (Courtesy of Rodney Brooks and Annika Pfluger)

head toward or away from the robot (Breazeal 2002). A great deal of attention was paid to the development of the robot's perceptual system and to providing it with an ability to recognize social cues, so that it produces the proper "emotive" response. It also has some vocalization ability. Kismet's ears move up to indicate attention or surprise, and they droop to indicate sadness or disappointment; its mouth opens or closes and curves to express attention or boredom or surprise. The ears are not at all anthropomorphic, again illustrating the principle that some aspects of function are more important than actual form. Clearly, Kismet was not a simple robot. It was capable of complex perceptual and behavioral interactions with humans. It is no longer in active use or development.

13.4.4 Leonardo

A current project involving cooperation between Breazeal (currently at the MIT Media Laboratory) and the Stan Winston Studio in Hollywood involves the development of an animated pet robot named Leonardo, shown in figure 13.15. Note that

Figure 13.15
The expressive robot Leonardo, created by Stan Winston Studio in cooperation with the MIT Media Laboratory (photograph copyright 2002 by MIT Media Laboratory, courtesy of Cynthia Breazeal)

this robot makes no pretense of being humanlike or of mimicking any animal. Leonardo has 61 degrees of freedom: 32 of those are in the face alone. As a result, Leonardo is capable of near-human facial expressions (constrained by its creaturelike appearance). Although highly articulated, Leonardo is not designed to walk. Instead, its degrees of freedom have been selected for their expressive and communicative functions. It can gesture and is able to manipulate objects in simple ways. The Stan Winston Studio is known for creating special-effect creatures, monsters, and animatronic effects for such movies as *Jurassic Park*, *The Terminator*, *Aliens*, and *Edward Scissorhands*. Leonardo, standing at about 2.5 feet tall, is the most complex robot the Stan Winston Studio has attempted at the time of publication. Leonardo is the most expressive robot in the world today. Although it is not humanlike, the developers have intentionally given Leonardo a youthful appearance to encourage people to playfully interact with it, much as one might with a young child.

The engineering challenge in designing this robot was considerable, since some 60 motors had to be incorporated in a very small volume. The studio designed a small sixteen-channel control module, with a number of innovations (such as eliminating any hum or interference signals).

Leonardo's expressiveness arises from its ability to move its eyes, mouth, ears, and arms to express a wide range of emotions. The robot has a suite of visual capabilities including a collection of visual feature detectors for objects (e.g., color, shape, and motion) and people (e.g., skin tone, eye detection, gesture, and facial-feature tracking), the ability to specify a target of attention and track it, and stereo depth estimation. Active vision behaviors include the ability to saccade to the locus of attention, smooth pursuit of a moving object, establishing and maintaining eye contact, and vergence to objects of varying depth. In the future it will also have auditory capability, including speech recognition.

Since interaction with humans involves touch as well as vision and hearing, Leonardo has a synthetic silicone skin capable of detecting pressure and location with acceptable resolution over the entire body, while still retaining the look and feel of soft skin. Working closely with the artists at Stan Winston Studio, the Media Lab staff has developed a tactile sensing system in which force-sensing resistors are placed over the robot's core and under the silicone skin and fur. The density of sensors varies according to area so that the robot has greater detection resolution in areas that are frequently in contact with objects or people.

A distributed network of tiny processing elements is also being developed to lie underneath the skin to acquire and process the signals received from the sensors. Capacitive sensing technology is used near the ear, and force-resistive sensing is used in the hand. Touch is particularly important in the robot's hand. Leonardo's articulated hand consists of integrated tactile sensor circuit boards on the palm, back, and side of the hand as well as a PIC-based 64-channel analog/digital (A/D) converter board housed inside the hand. Algorithms treat clusters of FSRs as receptive fields for higher-level processing. These "cortical-level" fields are capable of processing motion, direction, and orientation, as well determining the centroid of an object placed on the skin. This framework for processing motion and other data was designed using the hands as a test case but the final design will include a full-body sense of touch.

In our opinion, when completed, Leonardo will be the most sophisticated interactive robot in existence, making it possible to investigate a wide range of perceptual and emotion-based interactions between robots and humans. We expect that questions of the "reality" of robot emotions will need to be completely reformulated as a result of Leonardo's capabilities. In the human brain the seat of emotions is in the amygdala, and computational models of emotional responses are being developed (LeDoux and Fellous 1995). Leonardo's "brain" will have no structural resemblance to the human brain, but if it responds in a humanlike fashion to affective inputs, the boundary between robot and human will have shifted once again.

Current research projects with Leonardo concern learning by imitation and learning from social interaction with human tutors.

13.5 Special-Purpose Humanoids

Humanoid research is a very active area. In addition to the selected examples of what we have termed "whole-body humanoids" reviewed in the foregoing, a number of investigators have developed special-purpose humanoids for particular applications. As examples, consider the NASA humanoid known as Robonaut (whose arms and hand were discussed in chapters 10 and 11, respectively), designed to service the International Space Station and the robot Cog (introduced in chapter 1 and mentioned as well in the discussion of robot hands in chapter 11), designed to study the development of intelligence through interaction with the world. In the concluding part of this section, we look briefly at other special-purpose, partial-body humanoids, including ISAC, a robot designed for assistance to persons with disabilities, and the UMass Torso, a robot designed for the study of motor control and cognition. The label "special purpose" does not imply that these robots are inferior to the whole-body machines, but simply that they do not try to do everything, from cognition to mobility. On the other hand, what they do, they do extremely well.

13.5.1 Robonaut

The NASA Johnson Space Center in the United States has developed a humanoid robot to assist astronauts during extravehicular activities, when they must work in full space suits that greatly limit their dexterity. Robonaut (figure 13.16) is an upper-body humanoid that attaches to the International Space Station with its tentacle-like lower body, through which it receives power and control signals. The upper body has a head (with a vision system) and anthropomorphic arms and hands. Potentially, Robonaut will reduce response time, operating costs, and most important, some of the dangers to human astronauts working in space suits in zero gravity

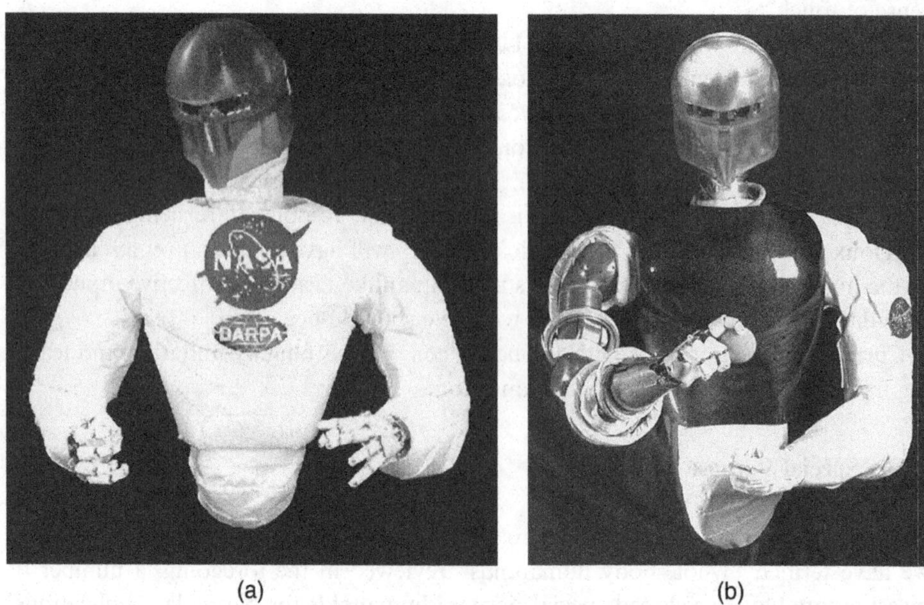

Figure 13.16
Two views of Robonaut: (a) Robonaut with fully covered arms plus torso; (b) Robonaut showing arm structure (photographs courtesy of Robert Ambrose)

(Ambrose, Aldridge et al. 2000; Ambrose, Askew et al. 2000; Bluethmann et al. 2000; Bluethmann et al. 2003; Diftler et al. 2003). Robonaut is teleoperated but has some limited autonomy.

The body of the robot is covered in Kevlar body armor and coated with Teflon. It has a 3-dof waist, a 2-dof neck, and arms with 7 dof each, thus giving it a great deal of flexible mobility. The hands of Robonaut are highly anthropomorphic, with 12 dof, enabling it to grasp tools designed for human use, such as electric drills or screwdrivers. (A further discussion of Robonaut hands can be found in chapter 11.) One of the goals of the Robonaut project is to enable the robot to share tools with astronauts. Each of Robonaut's arms is equipped with some 150 sensors (thermal, position, tactile, force, and torque), enabling it to perform a variety of complex manipulation tasks. The robot can also lift weights of over 20 lb, thus providing it with humanlike strength as well as dexterity. The back of the robot is equipped with an attachment enabling the Space Shuttle manipulator arm, also known as the remote manipulator system (RMS), to lift it and move it to a new location.

The robot's torso is approximately human-sized, as are the arms and hands. The helmetlike head conceals and protects a dual-camera vision system. Of particular interest are Robonaut's various autonomous control systems (Diftler et al. 2003). For example, the robot's hand has a reflexive grasp mode that causes the fingers to close if sensors on the palm are activated (like the Belgrade-USC hand described in chapter 11). The hand is also capable of performing haptic exploration of objects to be grasped and selecting its approach and orientation accordingly. We anticipate that future developments of Robonaut will move increasingly in the direction of autonomy, aided by learning, and away from the current teleoperated control.

13.5.2 Cog

Cog is a humanoid robot designed and constructed in the Artificial Intelligence Laboratory at MIT under the direction of Rodney Brooks. It consists of a humanoid torso equipped with anthropomorphic arms, a head with an articulated neck, and variety of sensors. The motivation behind creating Cog is the hypothesis that the development of intelligence for humanoids requires that they interact with the world. An early version of Cog is shown in figure 1.1j. Since that early version, the robot's head has changed, and there is now a separate project, Lazlo, concerned with developing a face for Cog. The most interesting feature about Cog is its "intelligence." The robot is designed to learn from interactions with the world, much like a child (Brooks 1997; Brooks et al. 1999; Adams et al. 2000; Breazeal and Scassellati 2001; Aryananda 2002; Fitzpatrick et al. 2003).

Cog is a test bed on which to study theories of cognitive science and artificial intelligence. As noted previously, the goal in developing Cog was to create a robot that is capable of interacting with the world—including both objects and people—in a humanlike way, thus enabling the study of human intelligence by trying to implement it. Interaction of the sort for which Cog is intended requires a rich sensory and motor apparatus. The robot has twenty-two mechanical degrees of freedom and twice as many sensors, ranging from torque sensors on motors to four cameras composing the eyes. The control and coordination of the robot's numerous degrees of freedom was developed not from mathematical models of the sensors, but from the robot's interaction with its environment and as a result of learning a predictive model of that interaction from the experience itself.

The aim of Cog's developers was to produce a relatively general system by which Cog can learn the causal relations between commands to its motors and input from its sensors, primarily vision and mechanical proprioception. This way, the robot could learn firsthand how its own movement is reflected in perceptible activity in the external world. And conversely, this ability allowed the robot to decide how to generate actions based on their intended effect. Such causal relationships are the root of the sense of kinesthesia, as well as the beginnings of what could be considered a "sense of self." By embedding knowledge of the effects of actions directly in their sensory results, one can avoid the classic symbol-grounding problem of artificial intelligence.

The current capabilities of the robot include

• Humanlike eye movements, including saccades, pursuit tracking, vergence, and the vestibular-ocular reflex.

• Head-neck orientation, allowing the robot to move its head toward a target.

• Face and eye detection, allowing the robot to detect people in its vicinity (by finding skin-toned oval-shaped objects).

• Detection of "interesting" objects in the environment and focusing on them.

• Imitation of the movement of a person's head.

• Various arm movements, including reflex withdrawal when the top of the hand is contacted, reaching for a visual target, and oscillatory arm movements. The latter enables play with a Slinky (figure 1.1j), swinging a pendulum, and turning a crank.

• Playing the drums. This is a complex behavior involving coupled neural oscillators used in the repetitive movement, as well as synchronizing to an auditory beat.

The Cog project is no longer active. Additional capabilities can be expected from new humanoids being developed at MIT.

13.5.3 Other Special-Purpose, Partial-Body Humanoids

We now consider briefly four additional special-purpose humanoids, two from Germany and two from the United States. There are numerous other humanoid projects in the world; these four have been selected somewhat arbitrarily for discussion here because they demonstrate particular applications.

13.5.3.1 The UMass Torso

Consider first the UMass Torso, a partial-body humanoid developed at the University of Massachusetts, shown in figure 13.17(a). The torso consists of two whole-arm manipulators (WAMs) from Barrett Technologies, two Barrett hands (see chapter 11), and a Bi-sight stereo head. The WAMs are 7-dof manipulators whose kinematics are roughly anthropomorphic. The individual joints are actuated through tendons and have a low gear ratio. These properties allow a natural compliance in interacting with the world and allow for the simulation of additional dynamic properties (e.g., muscle properties). The goal of the research surrounding the UMass Torso is to advance computational models of sensorimotor and cognitive development in a manner that leads to new theories for controlling intelligent robots, and provides a basis for shared meaning between humans and machines.

13.5.3.2 ISAC

ISAC (Intelligent Soft-Arm Control) is an interactive humanoid project at Vanderbilt University. The name arises from the fact that the arm is highly compliant and hence safe for working with and around people. In fact, the original goal in the development of ISAC was to aid handicapped persons (Kawamura et al. 1995); more recently it has been used for studies of human-robot interaction, with emphasis on learning and emotion. It is shown shaking hands with a human in figure 13.17(b).

ISAC has two 6-dof arms actuated by McKibben artificial muscles (see chapter 3). As described in chapter 3, these "muscles" are pneumatic actuators that shorten when pressurized; they are naturally compliant. The end effectors on the arms resemble human hands. Sensors in the system include optical encoders at the joints, six-axis force-torque sensors in the wrists, proximity sensors in the palms, and touch sensors in the fingers. There is an active vision system, infrared motion detection, speech input-output, and sonic localization, all designed to facilitate interaction with humans.

In the Intelligent Machine Architecture (IMA) employed in ISAC, human and humanoid are represented as distinct agents within a common computational framework. The human agent in the IMA maintains a model of the person with whom ISAC is interacting. The human agent receives information about the person's physical state from the robot's sensors and estimates the person's state. For this system, a person's "state" includes his movement, any sound he produces (including speech), and identifying characteristics (including face and voice). These features are sensed and processed by the robot's software. The human agent also has some ability to estimate the person's emotional state. This agent's work is supported by that of two other agents, a monitoring agent and an interactive agent. These agents are supported by others, as described in Peters et al. 2001 and Kawamura et al. 2004.

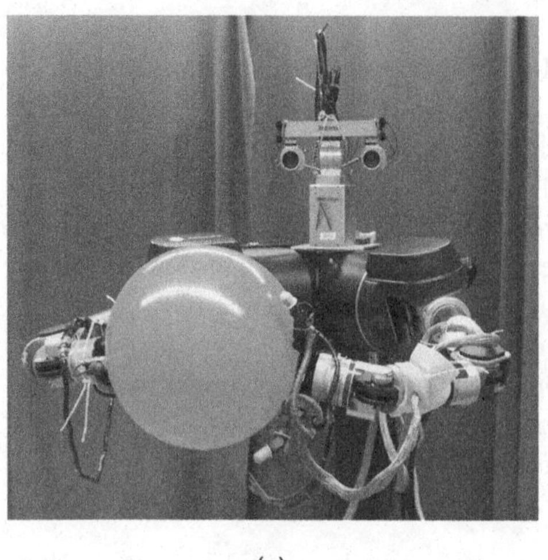

(a)

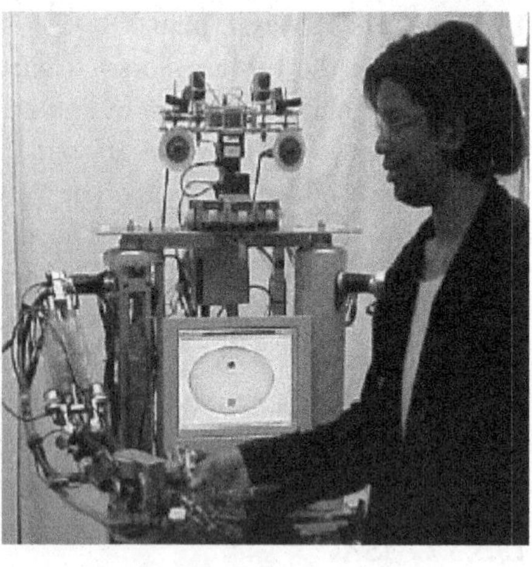

(b)

Figure 13.17
Four special-purpose humanoids: (a) UMass Torso (photograph courtesy of Andrew Fagg); (b) ISAC (photograph courtesy of Kazuhiro Kawamura); (c) HERMES (photograph courtesy of Rainer Bischoff and Volker Graefe); (d) Arnold (reproduced with permission of Thomas Bergener)

Another unique aspect of the design of ISAC (in addition to the IMA) is its memory structure. The robot has both long-term memory and short-term memory. ISAC's short-term memory was inspired by theories of the function of a portion of the brain known as the *hippocampus*, which appears to be responsible for integration of sensory inputs with motion, for short-term episodic memory, and for knowledge of the current location and orientation of an animal within its environment. These processes are embodied in an active data structure in the robot called the *sensory ego-sphere*, a concept first proposed by Albus (1991) that is also used in the computational structure of Robonaut (section 13.5.1).

It is evident that ISAC is yet another humanoid project concerned with the complex issues of human-robot interaction.

13.5.3.3 HERMES

HERMES (figure 13.18(c)) is a humanoid robot being developed by the Intelligent Robot Laboratory at the University of Munich. The goal of the project is to develop an intelligent service robot for manipulation and transportation in unknown environments and to communicate with people by voice (Bischoff and Graefe 2002). The robot is 1.85 m tall; its base is square with 0.7 m on a side. It weighs 250 kg. One of the design goals of HERMES was long-term dependability. Achievement of this goal was demonstrated by allowing HERMES to serve as a museum guide for six months, up to eighteen hours per day. During this time HERMES interacted with visitors

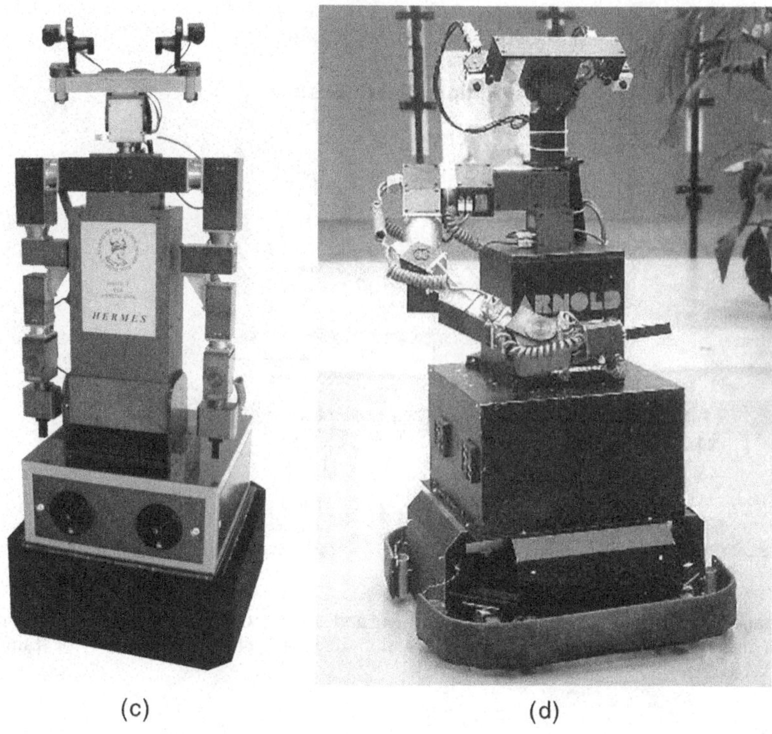

Figure 13.17
(continued)

through conversation in English, German, and French and performed requested services. During this time there were three minor failures in easily replaceable commercial components.

HERMES has an omindirectional undercarriage with four wheels, two of which are actively driven and steered. Its two arms have 6 dof each; each arm is equipped with a two-finger gripper. The robot's major sensor is a pair of video cameras mounted on independent pan-tilt units (in addition to a third pan-tilt unit that adjusts the common "head" of the robot). We have described similar vision systems in most of the other robots discussed in this chapter. A unique feature of this robot's structure is that all of its actuators are identical. There are twenty-five drive modules with the same electrical and very similar mechanical properties, controlling 22 dof. Each module contains a motor, a harmonic drive gear, a microcontroller, power electronics, a communications interface, and sensors. The structure of the modules is shown in figure 13.18, which also shows the main computer as a network of digital-signal processing (DSP) modules. Modularity certainly simplifies maintenance.

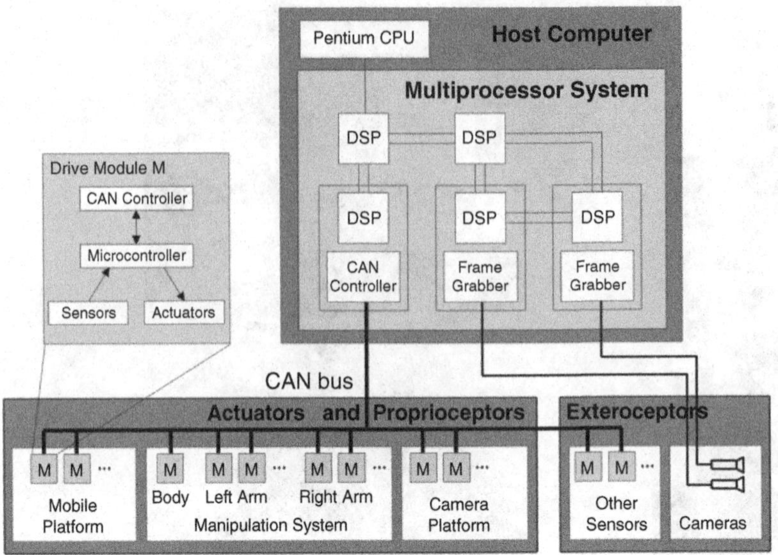

Figure 13.18
Modular architecture of HERMES for information processing and control. CAN stands for the controller area network bus and controller modules (from Bischoff and Graefe 2002, illustration courtesy of Rainer Bischoff and Volker Graefe)

It is evident that HERMES is a sophisticated robot. It successfully executes such tasks as receiving a verbal request to fetch an object, asking another person for the object, and delivering it as requested. Since interaction with it takes place in natural language, it can be operated by unskilled personnel, an essential ingredient in service robots.

13.5.3.4 Arnold and the NEUROS Project

NEUROS (Neural Robot Skills) is a project at the University of Bochum, partially funded by the German government and by a consortium of research groups and industrial partners. Its goal is to develop an autonomous robot with basic navigation and manipulation skills capable of working in unfamiliar environments (Bergener et al. 1997, 1999). Like HERMES, Arnold (figure 13.18(d)), a product of the NEUROS project, understands spoken language and is capable of locating, grasping, and handing over objects with its single 7-dof articulated arm. It is equipped with a 3-dof double stereo camera.

A similar project, also named NEUROS and also being developed at the University of Bochum, has developed a robot arm called GripSee, a gesture-controlled robot for object perception and manipulation (Becker et al. 1999). This robot's arm-and-vision system is capable of recognizing a variety of objects on a table and of being commanded, using gesture commands, to grasp one of the objects. Its hardware structure is very similar to the arm and camera system of Arnold but does not have a mobile platform. GripSee differs from Arnold in having a more complex grasp selection system and in the use of gestures to command it.

13.6 Trends in Humanoid Research

The survey of research in humanoid robots presented in this chapter makes it clear that there is a major emphasis in such research on interaction between robots and humans. Many of the projects discussed are obviously concerned with the physical appearance of the robot and its ability to walk or dance or run. However, as important as these concerns are, those involved in communication, interaction, and understanding are more fundamental to the development of a theory of humanoids (Scassellati 2002). Ultimately, the goal of humanoid research is to build robots capable of some aspect of behavior that resembles that of humans. Such robots will, as time goes on, become increasingly autonomous and able to operate in a world of humans, sharing their space and cooperating in their work. For this reason, there is increasing emphasis in humanoid research on emotional as well as cognitive interaction between humans and humanoids. As this work progresses, we believe that we will not only learn more about building intelligent robots but also learn more about ourselves.

The Web site of the highly regarded Japanese robot designer Hiraoki Kitano describes the humanoid PINO in the following words:

> In his gestation, PINO symbolically expresses not only our desires but also humankind's frail, uncertain steps towards growth and the true meaning of the word human. (Kitano 2002)

This sentence beautifully summarizes the reason for research in humanoid robots. I could not have said it better myself.

14 Localization, Navigation, and Mapping

Summary

This chapter concerns methods for enabling autonomous robots to determine where they are with respect to given coordinates or landmarks, find the best path from one point to another in a given terrain while avoiding obstacles, and map the environment. Since the measurements robots use to orient themselves are made using imprecise instruments, and since the environment in which they operate is generally varying, the resulting data are noisy. Hence, many navigation and localization algorithms are probabilistic. The explanations given in this chapter are largely intuitive, with multiple references to more precise, mathematically oriented presentations.

14.1 Overview

As robots become increasingly autonomous and able to operate in unstructured environments, they will be faced with more and more difficult problems of orientation and navigation. When a robot is moving indoors on smooth surfaces over short distances, with unobstructed visibility, these problems are not severe. In such situations the robot may have a clear view of a target location and use vision for navigation. In other cases the robot can navigate to a specified target simply by using wheel encoders and converting the distance to be traveled into wheel revolutions; this is *odometry* (discussed in section 14.3.2). This method is generally not satisfactory across long distances over uneven terrain, since different wheels will experience different amounts of slippage. Even on smooth floors, since its wheels are not identical, a robot's actual path will gradually drift away from the desired path to the goal.

The first navigation task examined in this chapter concerns the answer to the question "How do I get there?" (It is assumed here that the destination is set by a human). In general, this task will involve obstacles, uneven terrain of varying friction properties, and obstruction of the robot's view to the target. Sometimes the robot will have maps of the area to be traversed or information about known landmarks by which to navigate. The navigation task concerns getting from point *A* to point *B* along a path that avoids obstacles and is the "best" path in some sense. "Best" can be defined by such criteria as minimum fuel consumption (or battery drainage) or minimum travel time, but many other measures are possible. Note that the navigation task described here takes place in two dimensions; navigation in three dimensions, as with underwater robots or flying robots, introduces a new set of issues.

The second problem considered in this chapter concerns the question "Where am I?" This problem is known as the *localization problem*. A severe version is sometimes cast as the *kidnapped-robot problem*: A robot is picked up, blindfolded, and placed in a new location. How does it determine where it is with respect to known coordinates? What sensors does it need? How does a robot's ability to answer this question differ from that of a human or a homing pigeon?

The third problem explored in this chapter concerns the ability of a robot to prepare maps of the terrain it traverses, indoors or out. This problem has been described in terms of the answer to the question "Where have I been?" (Murphy 2000). Recently there has been growing interest in what is known as the simultaneous localization and mapping (SLAM) problem.

In the following sections of this chapter we discuss all three problems just outlined. The discussion is largely intuitive, rather than mathematical. Of course, there are some equations, as there have been throughout the book, when they help to state an issue precisely. However, the mathematical treatment of navigation, localization, and mapping is particularly difficult, because of the uncertainties associated with the sensors used in these processes and the variability of the environment. To determine where it is, a robot must rely on sensors. These sensors provide indications of distance to landmarks, distance traveled, velocity, and orientation with respect to landmarks or the coordinate directions; the robot must be able to use the sensed information to identify objects. All these measurements are noisy and imprecise. Hence, many of the algorithms used for localization and navigation in robots are probabilistic in nature and require an understanding of stochastic processes. For this reason, we attempt to provide an insight into the major issues without major reliance on mathematics. We do, however, provide extensive references to more precise treatments of these subjects. One of the best references to navigation and mapping strategies of successful robotic systems is Kortenkamp, Bonasso, and Murphy 1998.

14.2 Biological Inspiration

Many animals have highly evolved abilities to navigate and localize, including birds, fish, mammals and insects.[31] Let us consider birds first, since they are probably the champion navigators of the animal kingdom.

14.2.1 Bird Navigation

Canada geese migrate from the northern end of Canada to various locations in the United States, always returning to the same ponds and lakes. A small bird called the bobolink spends summers in Canada and the United States and winters in South America, particularly Brazil, Bolivia, and Argentina, a distance of some 7,000 km. How do these birds accomplish these remarkable navigational tasks? The answer is still not completely clear, and it varies depending on the species (Hughes 1999), but it is known that birds navigate using the earth's magnetic field, the orientation of the sun, the orientation of the stars, and near their destinations, landmarks such as mountains and lakes. Although the ability of birds to use this information is known and has been verified by ingenious experiments (such as raising birds in totally non-magnetic environments or in closed areas with an artificial sun that can be placed in an arbitrary location), the precise nature and location of the sensors is not known. It is known that birds have small deposits of magnetically stable magnetite in their heads, but the connection between these deposits and their nervous systems has not been established. Nevertheless, researchers in this field speak of birds having a *magnetic compass* for sensing the direction (and probably the gradients) in the earth's magnetic field, as well as a *sun compass* for sensing their orientation with respect to the sun (Wiltschko and Wiltschko 1995; Hughes 1999).

Homing pigeons are able to find their way home when released large distances away, even if they have been carried there in covered cages. Since they cannot see the path that has taken them from home to where they are released, they cannot store it in any way; nor can they see landmarks along the way. It has been postulated that these birds employ a *navigational map* by using naturally occurring gradients in the earth's magnetic field. In effect, this means that their magnetic compass is sensitive not to the polarity of the earth's magnetic field, but to its inclination with respect to the local gravity vector. This has been demonstrated by observation of birds flying over the equator. The pigeons combine this gradient information, which gives them a gross direction from the release point, with a memory of specific gradients or memory of landmarks near their home base. This home base information is acquired by young homing pigeons while flying in the vicinity of home (Wiltschko and Wiltschko

31. An outstanding source of information on the navigation senses (and other unique senses) of animals is Hughes 1999.

1998; Hughes 1999). Some birds are able to return home when transported to the point of release in open containers, so they can see the passing scene along the way, but not when they are moved in closed containers. This suggests that their visual system is able to record optic-flow information.

It is interesting to note that the ability to navigate using the direction of the sun, that is, by means of a sun compass, also requires that an animal have a biological clock, since the location of the sun in the sky changes during the day.

Perhaps even more amazing than a magnetic compass or a sun compass is evidence that some birds also possess a *celestial compass*. Experiments with birds raised under an artificial night sky, with a specific pattern of rotation, shows that their migratory directions can be altered. Yet the celestial compass (i.e., for birds in the Northern Hemisphere, the ability to sense the location of the North Star, Polaris) does not appear to be in the genes but is acquired from early experiences.

14.2.2 Insect Navigation

The navigation feats of Monarch butterflies are even more astounding than those of birds. These butterflies fly from northern Mexico to sites throughout the Western United States in the winter and early spring. The butterflies that return the following year are the children or grandchildren of those earlier migrants. Where is the navigational map stored in those tiny insects? How do they pass it on to later generations? Recent experiments made use of a butterfly flight simulator, in which Monarchs were tethered and subjected to a slow, laminar airflow that caused them to fly continuously for hours. Variations in the day-night cycle were used to change their circadian clocks. These experiments revealed that these butterflies possess a *time-compensated sun compass* that they use for determining their direction of migration (Mouritsen and Frost 2002). Furthermore, rotation of the magnetic field surrounding the simulator did not change the direction of flight, thus confirming that Monarch butterflies do not use a magnetic compass for navigation.

The use of a sun compass is also well documented in bees. Since the work of von Frisch (1967) it has been known that bees use a "waggle dance" to indicate the direction of a food source with respect to the sun. The dance is performed on a vertical surface of the honeycomb, apparently coded so that straight up means directly toward the sun. The workers that observe the dance and fly toward the food source do so in a straight line from the hive, no matter how convoluted the finder's flight pattern was during the search phase. Now, the most interesting part of this process is that the dancing bee can point to the proper direction with respect to the sun even on an overcast day when the sun is not visible. This feat is possible because the bee's sun sensor is sensitive to the ultraviolet radiation from the sun, not the visible light. In fact, the photoreceptors in bees' eyes can identify the polarization plane of ultraviolet light. This is a remarkable evolutionary adaptation to the need for accurate navigation even when the sun is not visible. In addition, it appears that bees also steer by visual landmarks. Their apparently circuitous routes are actually composed of path segments between landmarks.

By contrast with flying insects, most ant species use pheromones to mark their path, with one major exception. The desert ant (*Cataglyphis fortis*) navigates through a combination of path integration and visual-landmark recognition (Wehner, Michel, and Antonsen 1996; Collett et al. 1998). It appears that as they travel away from their nests, they integrate the path to obtain a *global vector* that points from the current location toward the nest, then store this vector. In experiments reported by Collett et al. (1998), ants returning from a journey were collected before reaching their nest and placed at points along a familiar route, in such a way that their global vectors were canceled. The ants then found their way home by using visual landmarks and employing local vectors to obtain the needed orientation at each path segment.

14.2.3 Navigation in Other Animals and Humans

In addition to birds, a number of other animals migrate long distances. The return of salmon to the streams where they were born, after several years of swimming in the ocean, is an amazing phenomenon. Magnetosensors have been identified in trout and perhaps salmon as well (Walker et al. 1997). Marine mammals, like whales, also migrate long distances. Some terrestrial mammals also travel long distances, and some, like caribou, do so over terrain with relatively fewer landmarks than, say, those that live in a more temperate climate. Elephants are reputed to be able to navigate to "burial grounds" from long distances.

The bottom line of this quick review is that animals have evolved a variety of specialized mechanisms to aid them in navigation, including the ability to sense the earth's magnetic field and the orientation of the sun and fixed stars, as well as visual sensing of landmarks. This naturally raises the question of comparable human abilities. A number of experiments to test the existence of a human magnetic compass have produced only negative results (Hughes 1999). Humans who are kidnapped and released in unfamiliar surroundings are simply lost. Apparently, whatever navigation senses we may have had prior to the appearance of *Homo sapiens* or as early hominids have been lost. We make up for them through intelligence and communication. Thus, a person in an unfamiliar but populated environment will ask questions of others to figure out where she is and how to get home. In unpopulated areas, lost hikers use a variety of other methods for localization and navigation. For example, they climb tall trees to sight landmarks, they follow streams, or they navigate by the sun. Of course, ideally they have compasses or, better yet, GPS receivers to assist them in localization. These and other methods are also used to assist robots in navigation.

14.3 Robot Navigation

Navigation is the process of determining and maintaining a path or trajectory to a goal destination. We have seen that animals possess a variety of remarkable abilities to assist them in navigation. To provide robots with many of the same abilities, we use sensors, such as

- *Computer vision,* to enable a robot to see and recognize landmarks (such as large rocks, trees, and beacons). Vision sensors are the eyes of the robot; object recognition software is essential for useful interpretation of the visual input these sensors provide. Other specialized sensors are used to locate specific light sources. For example, Mars rovers are equipped with *sun sensors* to enable them to determine their orientation with respect to the sun. Thus, robots can be equipped with the equivalent of a sun compass, which we saw earlier as being important for animal navigation. Similarly, a star sensor, combined with a celestial map, can be used for celestial navigation.
- *GPS,* to enable outdoor robots to determine (within centimeters) their latitude and longitude. Generally, GPS receivers do not work satisfactorily indoors because of the presence of walls and ceilings that attenuate the signals from satellites. Also, until a network of satellites for such purposes is established, GPS cannot be used for robot navigation on the moon, on Mars, or on other planets.
- *A compass,* to provide an indication of magnetic north. In some latitudes it may be necessary to correct the compass reading to obtain "true north." A compass is not useful on Mars, since it has only a very weak magnetic field.
- *A clock,* essential in connection with a sun sensor.
- *A map* of the terrain, to enable the robot to locate landmarks (and hence itself).
- *Wheel encoders,* to measure distances traveled as well as changes in orientation. Since odometric measurements made by such encoders tend to drift, they may need to be corrected using Kalman filters (see section 14.6) or periodic sightings of a fixed star or the sun.
- *Range finders,* to enable the robot to estimate its distance from objects in the environment. Ultrasonic sensors are useful primarily at short distances; laser range finders can be used at larger distances.
- *Gyroscopes,* to provide heading directions and to improve odometric readings.

Clearly, we can equip our robots with the basic functions of animal navigation sensors, even if we are not yet able to do so in comparably small, low-power packages with comparable precision. Having sensors, however, is only part of the solution. We must now look at how these sensors are used.

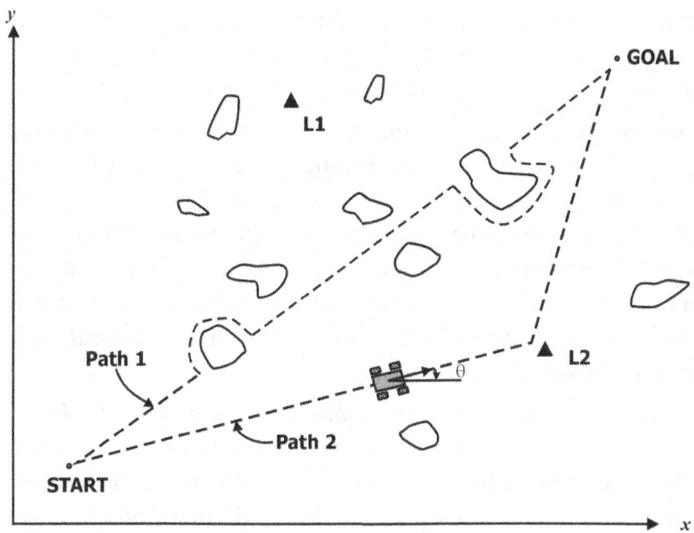

Figure 14.1
Simple navigation strategies

14.3.1 A Simplified Scenario

Consider the situation illustrated in figure 14.1. A robot is situated at the point labeled "START" in an outdoor environment. Its assignment is to travel to the point labeled "GOAL." A straight-line path is not possible, because of the presence of obstacles. The goal and two landmarks (tall pillars labeled "L1" and "L2") are visible on a clear day. Let us consider three possible scenarios:

1. *Clear day, visible goal, unknown distance to goal* Under these conditions the robot can navigate by vision and compass. It scans the environment using vision; it needs object recognition software so that it can recognize the goal. It can then travel toward the goal, keeping its image in the center of its viewfinder and moving in the same compass direction toward the goal until one of its sonars indicates the presence of an obstacle. At this point the goal is not visible, being obscured by the obstacle. The robot can make a random selection of turning right or left, and then travel around the obstacle, keeping a fixed distance from it. When the original compass direction is again sensed and the goal is visible, the robot can turn, change its heading, and repeat the previous strategy. Such a path is indicated in the figure as "Path 1." Note that the randomly selected direction was a not good choice for either obstacle.

Clearly, there are many alternate strategies for finding a path that avoids obstacles and allows the robot to reach the goal. Finding the "best" path from start to goal is termed *path planning*; it is a well studied problem in robotics (e.g., Latombe 1991).

Clearly, this is an optimization problem, requiring selection of an optimization criterion and an algorithm. Typical criteria are path length and time to reach the goal. Optimization algorithms used include various gradient methods, including A*.

2. *Goal and obstacles not visible, straight line distances known* When the goal (or even the first obstacle) is not visible, but its map location is known, the robot can travel the desired distance and in the desired direction by using GPS. If GPS is not available, the robot can use a compass, wheel encoders, and knowledge of the wheel diameters. Unfortunately, this method, known as *dead reckoning*, is useful only for short distances (see section 14.3.2). It also requires that the robot know its initial position and orientation (its *pose*) on the map. If it does not, it must first *localize* itself. Localization is discussed in section 14.6.

3. *Goal not visible from start location, landmarks visible, goal visible from landmark location* In this type of situation, the robot can navigate entirely by vision, which is known as *navigation by landmarks* (Path 2 in figure 14.1). Alternatively, if the landmarks are not visible from far away, but their coordinate or map locations are known, the robot can return to strategy 2.

These scenarios are simplified versions of some of the navigation methods discussed in the following sections.

14.3.2 Navigation by Dead Reckoning

Dead reckoning is the process of determining the change in a vehicle's position and orientation over time by integrating its velocity with respect to time. It has a long history in navigation at sea, where it was for a long time the principal method for estimating one's map position, frequently with large errors. Basically, the process consisted of recording information in the ship's log every half hour, such as the compass heading, estimates of the force and direction of the wind, the way the sails were trimmed, and an estimate of the prevailing ocean current. These data were used for calculation (*reckoning*) of the ship's new position. Early-day navigators became surprisingly good at making these calculations, and their accuracy could be surprisingly good, but they also made large errors. Clearly, estimating distance from the expression

$$\text{Distance} = \text{speed} \times \text{time},$$

and doing this every half hour, is not a good approximation to integration, but as the intervals get smaller, this approximation gets better. Estimates of position based on such approximations are clearly imprecise, and the approximation gets worse with time, unless corrected by observations of the sun or fixed stars, say, by means of a sextant (see section 14.6).

In contemporary robots we also describe as dead reckoning the process of estimating the robot's position after an elapsed time by integrating the velocity over that time. The term *odometry* describes dead reckoning whether the position estimates used are made by integrating velocity or directly from encoders mounted on a robot's wheels. Since the diameter of the robot's wheels is known, it is a simple matter to compute the distance traveled. The change in orientation is obtained by calculations based on information from encoders on the robot's steering mechanism. This method of localization is useful only for short distances. The problem is that wheels slip and terrain changes, particularly in exterior and unstructured environments, so the distance traveled and the vehicle orientation will deviate randomly from the estimated values. In addition, no vehicle is exactly symmetrical, so the distance traveled by the right side may differ from that traveled by the left side. The net effect of these factors is that the mean of the dead-reckoning estimate of position drifts further and further from the correct values as the vehicle moves. Note that odometry is a proprioceptive sense (see chapter 3), using only internal measurements. Hence, it is reasonable to assume that the addition of some exteroceptive sensors on robots can improve accuracy in navigation and path planning.

Let us define odometry more precisely. Assume that the control inputs to the robot are its linear velocity $V(t)$ and rotational velocity $\omega(t)$. Then, given the robot's starting position (x_0, y_0) and orientation θ_0 with respect to the local x-axis, we can compute the current robot *pose* $[x(t), y(t), \theta(t)]$ from the expressions

$$
\begin{aligned}
x(t) &= x_0 + \int_0^t V(t) \cos \theta(t)\, dt, \\
y(t) &= y_0 + \int_0^t V(t) \sin \theta(t)\, dt, \\
\theta(t) &= \theta_0 + \int_0^t \omega(t)\, dt.
\end{aligned}
\tag{14.1}
$$

Note that the presence of trigonometric terms makes these expressions nonlinear, and closed-form solutions are therefore possible only with very simple inputs. Furthermore, sensor errors (or, equivalently, input errors) lead to errors in the vehicle pose. These errors are frequently assumed to be Gaussian to simplify the analysis. Excellent treatments of odometric errors can be found in Borenstein and Feng 1995, Borenstein, Everett, and Feng 1996, and Kelly 2004. A method of improving odometry by combining it with the use of a gyroscope involves a technique called *gyrodometry* (Borenstein and Feng 1996).

We have already indicated that one way of dealing with dead-reckoning errors is to use additional (exteroceptive) sensors. Another way is to use statistical estimation techniques, such as Kalman filtering. The latter are discussed in section 14.6, in connection with the study of localization algorithms.

14.3.3 Inertial Navigation

Navigation in space vehicles and many military systems is accomplished by means of inertial navigation systems (INSs) (Grewal, Weill, and Andrews 2001). These systems consist of three accelerometers and three gyroscopes, along with appropriate servos and processors, in order to provide the needed sensing for 3-D navigation in an inertial coordinate system. The outputs of the INS are the coordinates $(x(t), y(t), z(t))$, at any time t, and the three angles defining the attitude of the vehicle in space (the ocean, etc.). The latter could be pitch, yaw, and roll (in the vehicle coordinate system), or they could be Euler angles defined in some global coordinate system. Linear velocity and position estimates are obtained by integrating the accelerometer outputs. The choice of coordinate system depends on the mission. Aircraft and submarines use Earth-referenced coordinates; space vehicles use inertial coordinates, perhaps referenced to fixed stars.

Note that an INS is a proprioceptive system, since it does not rely on external references (except for calibration). Hence, navigation by INS is a form of dead reckoning. However, the precision of the instruments used in INSs is much higher than that available with odometry. Accuracies on the order of 0.1% of the distance traveled are reported by Everett (1995); significantly better accuracies exist in military systems. Nevertheless, an INS must be calibrated and updated periodically, generally within hours of use. These updates were done traditionally using celestial references but currently employ GPS.

Although needed for navigation in three dimensions, INSs are not often used for terrestrial navigation by robots, mainly because of their high cost. They are employed primarily in robot aircraft and underwater vehicles (but see Barshan and Durrant-Whyte 1995).

14.3.4 Combinations of Local and Global Strategies

The use of purely proprioceptive measurements, such as those obtained from wheel encoders, is a *local* strategy. Navigation by GPS is a *global* strategy. Intuitively, it would make sense to use some global information to update and improve local information periodically. As we show subsequently, this is in fact a common approach to localization (determination of where the robot is) and hence to navigation. Strategies based on such an approach have some similarity to the navigation methods of bees and desert ants, as noted earlier. Indeed, these similarities have been used to develop robot navigation strategies (e.g., Roumeliotis, Pirjanian, and Matarić 2000).

14.4 Mapping

One of the major applications of mobile robots is to create models of the environment they traverse using sensor data; this process is known as *mapping*. Military applications of this technology are obvious. Visualize a robot that somehow enters a vacant building in hostile territory. For example, it could be thrown through an open window, crawl through drain pipes, or climb up the side of the building. Once inside, the robot can traverse the hallways and create a map showing doors, hall crossings, stairways, and other features. The robot will require sensors for this job, primarily computer vision (and associated feature recognition software), sonars, and laser range finders. The information it gathers can be transmitted to external receivers or stored on board for later capture and analysis. This type of mapping is known as *indoor mapping*; it has been studied extensively during the past twenty years. *Outdoor mapping* is clearly more difficult, since the external environment is not conveniently arranged in orthogonal corridors like those generally found in interior settings.

Localization, determination of where a robot is located, is necessary for many mapping tasks. In some situations the initial position and orientation—that is, the pose of the robot—are determined by humans. For example, if the robot is placed in a hallway with a given pose, it can proceed with its exploration and mapping task from that initial condition. On the other hand, in the example given in the previous paragraph (in which a robot is, say, thrown into a building through a window), it will need to perform localization before it can begin its task. There are also situations in which a robot will need to localize itself repeatedly, if it drifts off its course while navigating by dead reckoning. We discuss localization separately, in section 14.6.

The most commonly used approaches to mapping are termed *grid-based* or *metric mapping* and *topological mapping*. Recently hybrid methods combining these approaches have been suggested (Thrun et al. 1998). Let us look at each of these methods in turn.

14.4.1 Metric Mapping

Metric or quantitative maps, as the name implies, are based on measurements of the space they map. An indoor metric map may include the lengths of wall sections, door-opening widths, hallway widths, distances to intersections, and so forth. A typical metric navigation instruction might be: "Move 45 meters in a northerly direction, then turn 30° clockwise and move another 65 meters." Path planning in metrically mapped spaces usually includes the designation of a number of *way points* at specific (x, y) locations (such as the turning point in the navigation instruction just given), connected by straight-line segments. Paths can then be selected on the basis of some optimization criterion.

A widely used method of generating a metric map is to cover the environment to be mapped with an evenly spaced grid. Each cell in the grid is then filled with one or more values that represent the presence or absence of an obstacle (which could be another robot or a human). Grid-based mapping was first proposed in the 1980s by Elfes (Moravec and Elfes 1985; Elfes 1987, 1989). Of course, a robot traversing an area it is mapping does not have complete and accurate a priori knowledge concerning the presence of obstacles. Hence, with each cell $\{x, y\}$ is associated a probability that the cell is occupied, denoted by $P(occ_{x,y})$. This value represents the robot's belief that it can or cannot move to the center of the cell. For this reason, grid-based maps are also called *occupancy maps.*

Consider the indoor environment of figure 14.2, showing two offices (labeled A and B), an entrance from a stairway, hallways and dark areas denoting pillars (in the corners), desks, and cabinets. (Many other types of obstacles might exist in such a setting, frequently placed at inconvenient places in hallways or other areas.) The figure shows a grid superimposed on the map. The robot's goal as it traverses the

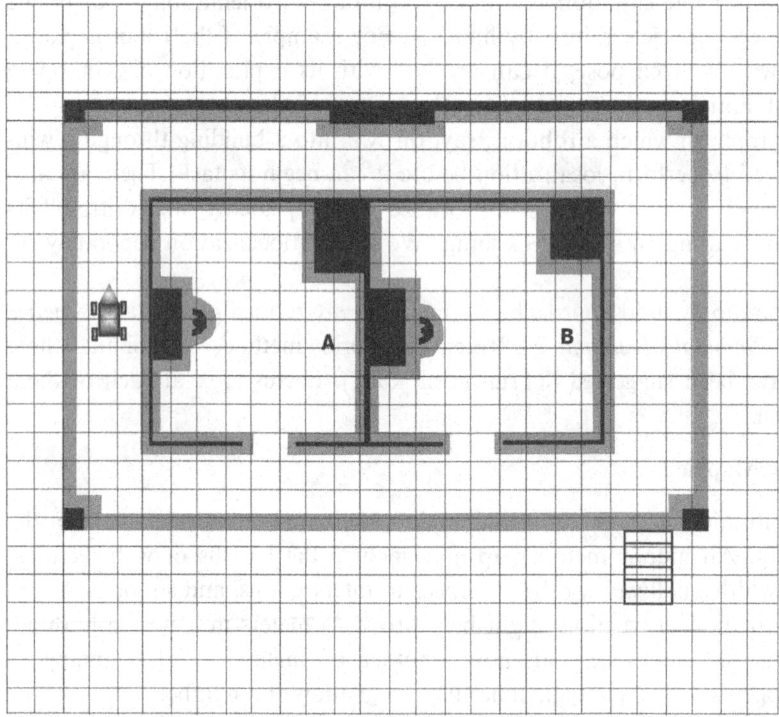

Figure 14.2
An indoor office environment

space is to estimate the probability of each square's being occupied. (Recall that when it begins the task, the robot has only the empty grid, not the information about the structure and furnishings provided in the figure to establish the task for readers.)

It is also important to note that we must be able to take the physical dimensions of the robot into account. This is usually done by performing the mapping in *configuration space*, an approach due to Lozano-Perez (Lozano-Perez and Wesley 1979; Latombe 1991). Basically, this approach replaces the physical space in which the robot travels with one in which the robot can move without colliding with any object. Since the robot has physical dimensions, one way this can be accomplished is by "growing" the boundaries by at least half the diameter of the robot. This is shown by the grey areas in figure 14.2. Then the physical robot can be replaced by a point for path-planning purposes.

The traditional way of generating occupancy grids has made use of sonars. Since sonars (like all sensors) are imprecise, the resulting occupancy grid is also imprecise. Working from an occupancy map generated from sonar readings, a system cannot say with certainty that a particular cell is occupied or empty, but it can make probabilistic statements about the cell's status.

Consider the two-dimensional representation of a sonar model projected onto an occupancy grid shown in figure 14.3. The sonar beam is actually a 3-D cone, with a maximum range indicated by R. The cone width is frequently stated as a combination of degree measurement and range measurement, as in 30° at a range of 5 m, or 15° at 3 m. (We look here only at the planar projection, disregarding the spatial dimension.) The space in front of the sonar, where range readings can be identified, is divided into four regions, labeled a, b, c, and d:

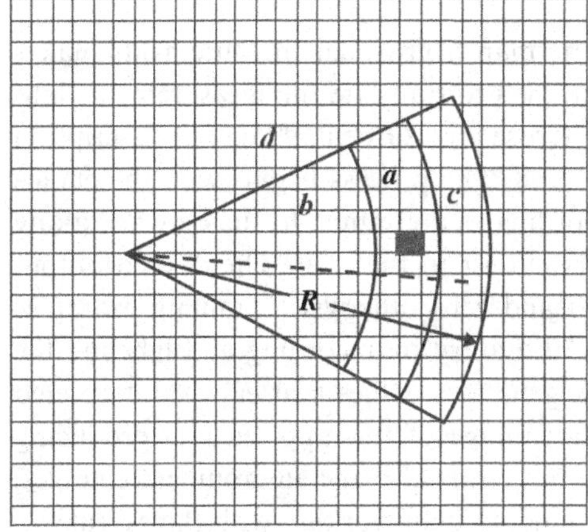

Figure 14.3
2-D sonar map projected onto an occupancy grid

- Region a (shown in the figure as containing an object) is *probably occupied.*
- Region b, in front of the object, is *probably empty*.
- Region c is in the shadow of the object, so its occupancy status is unknown.
- Region d is outside the beam and hence not of interest.

The first work on the use of sonars for mapping was completed in the 1980s (Elfes 1987, 1989; see also Borenstein, Everett, and Feng 1996 for a simpler sonar model used for high-speed computation). Interpretation of sonar data is made difficult by a variety of reflections from corners, doorways, and objects in the vicinity of the soundwaves. Nevertheless, it is possible to estimate the probability of occupancy of any grid cell given the sonar measurements. The problem is that the sensor models yield the conditional probability $P(s/H)$ (read as "the probability of s given H"), that is, the probability that the sensor output s has some specific value if the hypothesis H, that the cell is occupied, is true. What we really want is the probability that the cell is really occupied, given the sensor reading s, or $P(H/s)$. The calculation of this quantity can be made using Bayes' rule,

$$P(H|s) = \frac{P(s|H)P(H)}{P(s|H)P(H) + P(s|\neg H)P(\neg H)}, \tag{14.2}$$

where $P(H)$ and $P(\neg H)$ are a priori probabilities that H is true or not true. In dealing with occupancy grids, H represents the hypothesis that a particular cell is occupied; its negation then represents the hypothesis that the cell is empty. Bayes' rule has been used for this purpose by Moravec (1988) and Elfes (1989). Note that Bayes' rule (or other methods for performing statistical inference) are simply ways of translating sensor readings into probability-of-occupancy values, and hence a metric map of the environment.

A quite different way of building a metric map from sonar measurements makes use of an artificial neural network that is trained to translate the measurements into occupancy values (Thrun 1993; Van Dam, Kröse, and Groen 1996). The description that follows is adapted from Thrun et al. 1998. The process begins with the choice of a particular cell located at (x, y); its coordinates are inputs to the network along with values obtained from the four sensors closest to that cell. The desired output of the network is one if (x, y) is occupied and zero otherwise. Training examples can be generated in a simulator or in the laboratory by recording and labeling the sensor readings. Once trained, the network generates values that can be interpreted as probability-of-occupancy values. One advantage of this approach to producing an occupancy grid is that the network can be retrained if the environment changes, which tends to be less computationally demanding than working with Bayes' rule.

A method of obtaining occupancy information that does not rely on sonar measurements employs stereo cameras. Images of a particular area from the two cameras are sent to a high-performance computer for image processing. If vertical edges occur in both images, the disparity between them can be used to compute the proximity of the edges and from that information, an occupancy grid can be constructed (Thrun et al. 1998). Other approaches to grid-based metric mapping are described by Thrun, Burgard, and Fox (1998) and Yamauchi, Schultz, and Adams (1998).

The bottom line of grid-based metric mapping is that in order to obtain reasonable accuracy in the map it is necessary to use a small cell size. Hence, grid-based path planning tends to be expensive, both in terms of computation and in terms of storage. On the other hand, metric mapping relates directly to information on the motion of the robot, from encoders on the wheels that provide both rotation and translation information. As indicated previously, such dead-reckoning navigation tends to drift off course. However, in outdoor environments, when such measurements are corrected, say by means of GPS or periodic sighting of a fixed star, the resulting metric map may be quite accurate.

14.4.2 Topological Mapping

Topological navigation and mapping do not rely on precise measurements, but rather on landmarks. In an indoor environment such landmarks could be doors, hallway intersections, or T-junctions of hallways. A typical navigation instruction might be "Move along this hallway until you reach a hallway intersection, turn right, and then left through the first red door." Topological navigation is common in our experience, since we frequently get driving directions like "Go straight until you see a McDonald's, turn right and look for a church with two tall towers on the left, turn right again and go to a yellow house in the next block." Thus, topological mapping involves the creation of a map in which the location of landmarks is essential and not the distance between them.[32] Note that topological mapping is qualitative, whereas metric mapping is quantitative; this suggests that blends of the two approaches might have some of the advantages of both, as we discuss in section 14.4.3. Figure 14.1 indicates two pillars to be used as landmarks. These are *artificial landmarks*, placed in the landscape specifically to assist with navigation. Any of the obstacles in the landscape could be used as *natural landmarks*. In figure 14.2, the potential landmarks include doors, corners, intersections, pillars, storage cabinets, and even the desks and chairs if their locations are fixed.

Typically, topological maps are represented by graphs, in which each node is a landmark and adjacent nodes are connected by arcs. Such maps are much more

32. I frequently hike on mountain trails. When I ask for directions from other hikers, I receive metric directions from some ("Just go another half a mile, you can't miss the turnoff") and topological from others ("Look for a large log across the trail; shortly thereafter you will see a pile of rocks that marks the place to cross the creek"). Although I have a preference for landmark-based navigation in the woods, there are certainly times when precise measurements are very helpful, particularly on the highway.

compact than the corresponding grid-based maps. Thrun et al. (1998) give an example transformation of a grid-based map into a topological map. The original map had more than twenty-seven thousand occupied cells; the resulting topological graph had sixty-seven nodes. This compactness is one of the advantages of topological mapping. Another advantage is that a robot can use sensor information about the environment directly, for example, to indicate that it is at an intersection of two hallways, whereas in grid-based mapping the presence of an intersection must be inferred from the grid. On the negative side, a robot's sensors may miss a landmark, or confuse two landmarks (such as two identical doors), so artificial landmarks may be required to ensure accuracy. Basic work in topological mapping is described in Kuipers and Byun 1991, Matarić 1992b, Beccari, Caselli, and Zanichelli 1998, and Choset and Burdick 2000.

Another important advantage of the topological approach is that it can be incremental, allowing users (e.g., soldiers using a robot to map the interior hallways of a building in enemy territory) to view the most up-to-date map at any one time. As the robot continues its exploration of the environment it is mapping, it continually updates its estimates of its own position and orientation as well as the map. This approach is facilitated by a behavior-based architecture.

14.4.3 Combined Metric and Topological Mapping

As pointed out by Thrun et al. (1998) and others, metric and topological mapping have complementary strengths. Grid-based maps are easier to learn, they are easy to maintain, and they facilitate accurate localization. On the other hand, topological maps are more compact and thus lend themselves to fast mapping and path planning. Hence, a number of investigators have used combination strategies for mapping. An outstanding survey was published by Thrun (1998b). We shall encounter such combined maps in connection with simultaneous localization and mapping in section 14.7.

14.5 Case Study 14.1: Incremental Topological Mapping

This case study describes a behavior-based technique for incremental online mapping and autonomous navigation for mobile robots developed by some colleagues in the USC Robotics Research Laboratory (Dedeoglu, Matarić, and Sukhatme 1999). The experimental platform used in this study was a Pioneer AT robot, equipped with seven sonars, a drift-stabilized gyroscope, a compass, and a pan-tilt color camera, as illustrated in figure 14.4. The gripper shown in the figure was not used. The on-board processor was a PC-104 stack with a 40 MB hard disk. Although slow and small by today's standards, it was more than adequate for the task the robot

Figure 14.4
Pioneer AT mobile robot, with equipment box (containing computer and communication components), sonars, and camera (photograph courtesy of Gaurav Sukhatme)

was expected to accomplish. Communication with a remote computer (on which a graphical user interface was running) and other robots was accomplished by wireless Ethernet.

The goal of the initial phase of the study was to enable the mapping of the main topological features of the interior of a building, such as corridors, junctions, corners, and doors (both open and closed). The map was built incrementally, so a remote user could use a computer monitor to see the latest version of the map as it was updated by the robot in real time. While exploring, the robot continually estimated its position and orientation.

The control and mapping algorithm was entirely composed of behaviors, running in parallel (see chapter 5). Each behavior was responsible for a particular task such as obstacle avoidance, target following, feature extraction, or map updating. At each behavior cycle, data were passed among the behaviors as required.

The navigation strategy required the robot to traverse corridors in the building, avoiding obstacles, and turning only when the path was blocked. Right turns were preferred to left turns, and U-turns were executed only at dead ends. This strategy meant that the robot tended to penetrate deep into the building, rather than exploring rooms near its current location. The behaviors involved in the navigation strategy are shown in ovals in figure 14.5, which also shows their interconnections and the sensors on which they rely for input information.

As discussed in the previous section, a topological map is a graph in which the nodes represent landmarks and the links indicate distances and bearings between landmarks. In this study, each landmark also had associated counters that indicated how frequently it was detected and actually visited by the robot.

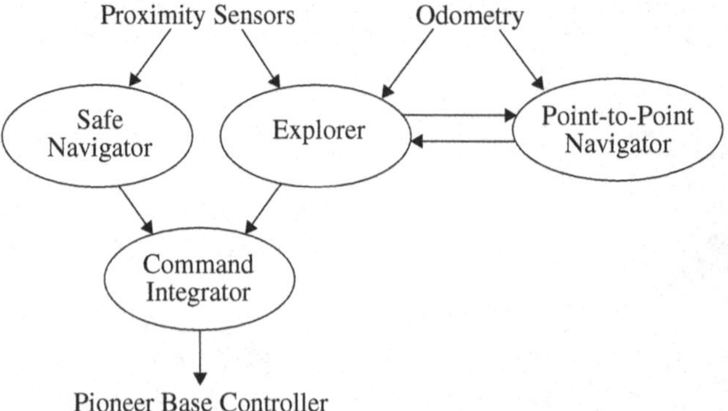

Figure 14.5
Organization of navigational behaviors (from Dedeoglu, Matarić, and Sukhatme 1999, reproduced with permission of The International Society for Optical Engineering [prior to 1999: Society of Photo-optical Instrumentation Engineers, or SPIE] and Gaurav Sukhatme)

14.5.1 Landmark Detection

The identification of topological landmarks was designed to work with sonar sensors. It consisted of feature detectors running in parallel, to identify line segments that could be associated with such features as corridors or open doors. The robot had five sonars: three in front and one on each side. Buffers were associated with each sensor, each updated with projections of the proximity sensors onto the robot's 2-D coordinate system. The detectors were enhanced by least-squares fitting algorithms and other features, as described in Dedeoglu, Matarić, and Sukhatme 1999. The specific behaviors involved in landmark detection, the sensors on which they rely, and their interconnections are shown in figure 14.6. The behaviors include an *opening detector* that looks for discontinuities in consecutive line segments in the right or left sonar buffers, thus giving an indication of a corridor or an open door. The *door detector* relies on the robot's vision system. It processes visual blob information, examines the shape of perceived color objects, and recognizes closed doors by their height and width. The outputs of all the behaviors feed into the *map builder* module.

14.5.2 Concurrent Localization and Mapping

The robot started its exploration with no a priori map. Hence, it was required to track its current position with enough accuracy to be able to distinguish nearby landmarks. It was also required to traverse corridor loops and close them with enough precision to repeat the process. ("Closing" refers to traversing a corridor loop and returning to the starting location, or close to it.) Hence, it was required to track its position and to constantly relocalize itself with respect to a local map. This meant

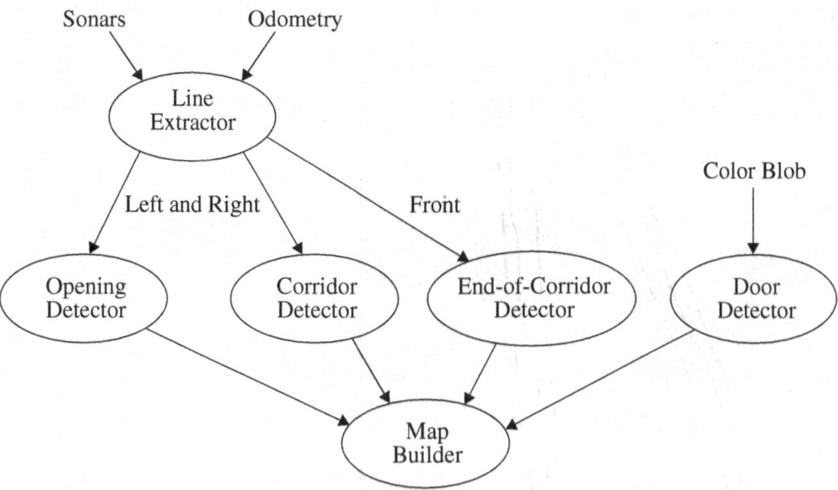

Figure 14.6
Landmark detection and map-building behaviors (from Dedeoglu, Matarić, and Sukhatme 1999, reproduced with permission of SPIE and Gaurav Sukhatme)

that odometry errors had to be bounded. Section 14.6.3 discusses how the addition of an inexpensive gyroscope and a Kalman filter were used to improve the accuracy of localization over that possible with odometry alone. Even with these tools, it was difficult to estimate the gyro drift, which resulted in mapping errors on the order of 15% of the distance traveled, as shown in figure 14.7. Note that the robot path (given by the middle line between the corridor estimates) starts at location P1 and ends up at P2. This is clearly not satisfactory. Not only is there a mismatch between P1 and P2, but the corridor estimates are neither straight nor orthogonal.

To improve the odometry, two assumptions were made: (1) that the corridors were straight and (2) that they were orthogonal to each other. An *odometry correction* behavior, shown in figure 14.8, used these two assumptions to make position corrections online. If the robot's estimated trajectory violated the assumptions, recent wall segments were rotated (and, if necessary, shifted) to match the corridor's straight heading. The complete set of behaviors involved in position tracking is shown in figure 14.8. As indicated earlier, the topological map was built incrementally, and the latest version (including the latest estimate of the landmark locations) was displayed in real time to a user, who could be remotely located. The robot made repeated passes through the hallways to improve the precision of the map. When a new landmark was detected, the mapping program searched through the current version of the topological map to see if it matched one found earlier. A matching algorithm was used to check the degree of matching. If the match was sufficiently

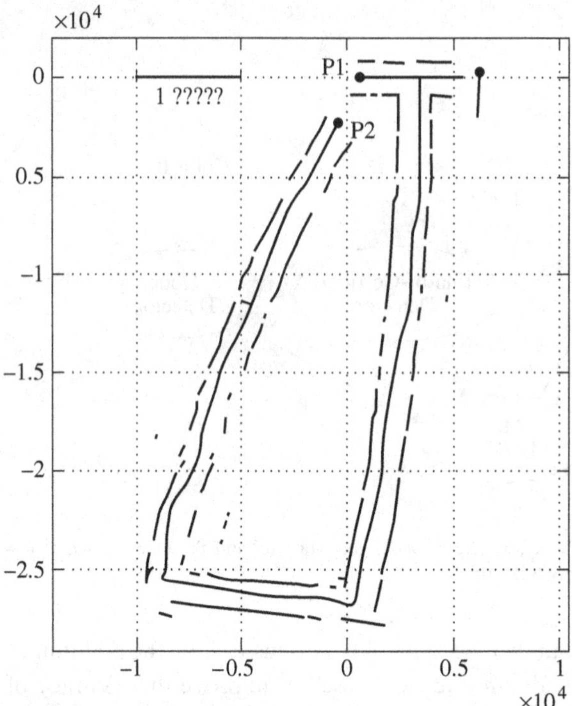

Figure 14.7
The effect of unmodeled gyroscope drift on navigation error (from Dedeoglu, Matarić, and Sukhatme 1999, reproduced with permission of SPIE and Gaurav Sukhatme)

strong to assure that the newly found landmark indeed corresponded to one detected previously, but they were not in identical locations, the program made appropriate corrections to the map. In this way, each successive pass through the hallways improved the quality of the map.

14.5.3 Experimental Evaluation

A map of the second-floor corridor in the USC Salvatori Computer Science Center was created by the system using the method described in this section and displayed on the user's computer screen. A typical screen shot is shown in figure 14.9. In the screen shot shown in the figure, the robot had completed four autonomous tours around second floor of the building, starting with no a priori knowledge, either about the environment or its location. This total run represents some 300 m traveled at an average speed of 30 cm/sec. As can be seen in the figure, not all the wall segments in the map are perfectly aligned. Such a flawless alignment was not the goal of this

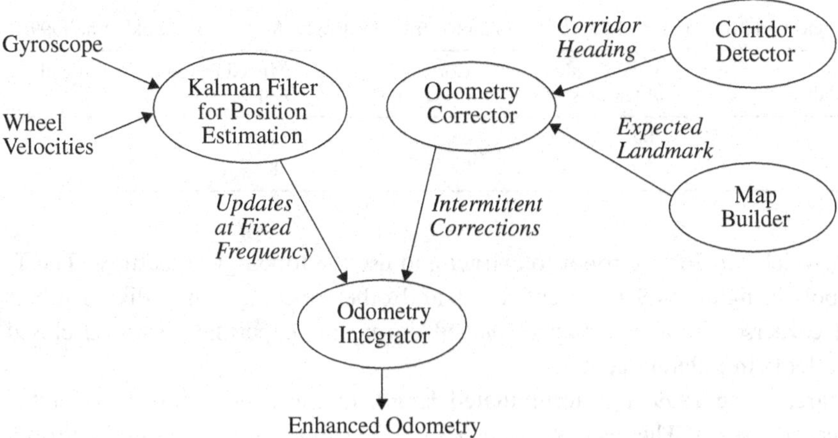

Figure 14.8
Position-tracking behaviors (from Dedeoglu, Matarić, and Sukhatme 1999, reproduced with permission of SPIE and Gaurav Sukhatme)

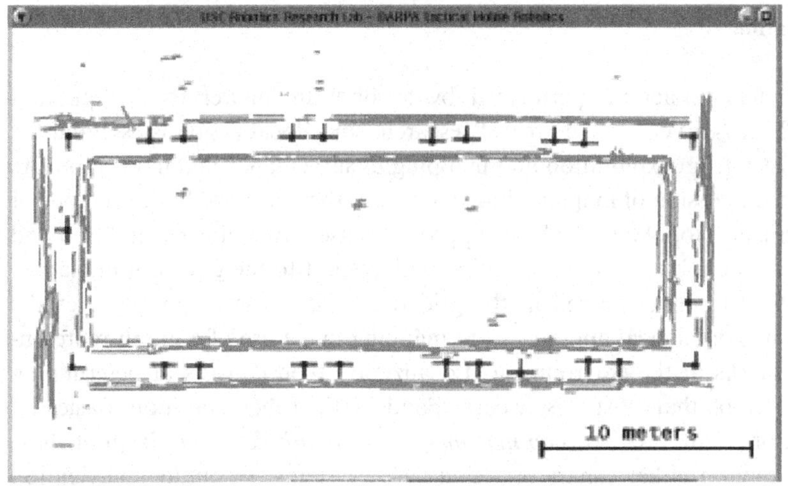

Figure 14.9
Map of the second floor of the USC Salvatori Computer Science Center, as obtained by the robot (from Dedeoglu, Matarić, and Sukhatme 1999, reproduced with permission of SPIE and Gaurav Sukhatme)

Table 14.1
Landmark detection rates in map of figure 14.9 (adapted from Dedeoglu, Matarić, and Sukhatme 1999)

Sensor modality	Total number of features	Correct detection (%)	Missed features (%)	False alarm (%)
Sonar	300	81	19	20
Color vision	180	92	8	3

mapping, which was for the robot to extract and use the topological features. The T-like symbols in figure 14.9 represent doors and other openings; the L-like symbols represent corners. The actual map displayed during the experiment showed closed and open doors in different colors.

For March–June 1999, the accumulated length of the robot's traversals of the floor exceeded 2.6 km. The resulting statistics on the robot's ability to obtain a topological map of the building are shown in table 14.1. These results indicate that the behavior-based method described in this case study is a simple and successful strategy for incremental topological mapping.

14.6 Localization

Localization refers to actions performed by a robot in answer to the question "Where am I?" It is a very active area of research, since localization is essential for many robot tasks, from exploration and mapping to surveillance and navigation. As indicated in the discussion of mapping in section 14.4, there are two basic methods of mapping: metric and topological. Metric approaches use a two-dimensional grid and attempt to place the robot on a map location with respect to the grid coordinate system, that is, to determine the cell in the grid that most closely approximates the robot's position. Topological approaches to mapping make use of a graph representation of landmarks in the environment. Localization then consists of determining the node in the graph that most closely corresponds to the robot's position. Basically, then, localization requires either *map matching* or *landmark detection*. Both of these methods may rely on either sonars or vision-based systems. In outdoor navigation robots can use GPS for very precise global localization. As shown subsequently, many localization estimates rely on Kalman filters to combine information from proprioceptive and exteroceptive sensors to improve localization accuracy. Such estimation methods generally require good models of the sensors and their uncertainties.

Landmark-based localization in indoor environments makes use of such features as doors (Thrun 1998a), corners, overhead lights, and even air diffusers in ceilings (Abe et al. 1999). Since systems for landmark-based localization are usually tailored to specific environments, they are difficult to apply to new and different situations.

A unique vision-based localization method makes use of panoramic color images (Ulrich and Nourbakhsh 2000). The algorithm is topological in nature. It uses an Omnicam panoramic camera (see chapter 3) to obtain color images of each landmark or location. The system is then trained to associate a proper label with the captured images. An image-matching system then finds which of the stored images most closely matches the current input image. This process is referred to as the system's *place recognition module*. This system was successfully employed to track a robot's position in several indoor and outdoor environments. Other successful vision-based localization systems include those discussed in Ulrich and Nourbakhsh 2000 and Kelly 2000.

14.6.1 Localization Using Landmarks (Triangulation)

For centuries before the development of GPS, mariners used sextants to localize themselves on a map. The sextant is an instrument with two mirrors that is used to measure the angle between two "landmarks," which could be, for example, the sun and the horizon. Angles obtained via sextant measurement were used in connection with tables of data to obtain latitude and longitude. Considerable calculation and knowledge were required to navigate via sextants.

Measurement of distances to landmarks by triangulation dates back to the sixteenth century. Since this method depends on the measurement of angles, it is related to sextant-based navigation. It is interesting to note that prior to the development of triangulation (made precise by Galileo), distances were measured in terms of hours of travel[33] or number of steps. Triangulation can be used to localize a robot with respect to three or more landmarks. Specifically, we consider here pose estimation given a map of the environment and the bearings of landmarks measured relative to a robot's position (Betke and Gurvits 1997).

Consider figure 14.10, which shows a robot at an unknown location p and three landmarks at unknown positions L_1, L_2, and L_3. The lengths of the vectors from the robot to the landmarks are given by z_1, z_2 and z_3. The distances between each pair of landmarks are known, and the visual angles between each pair of landmarks as seen by the robot can be measured. These angles are shown in the figure.

Then the robot's position is located on an arc of a circle that includes L_1 and L_2, and also on a circle that includes L_1 and L_3, as shown in figure 14.11, which also shows the visual angles. It can be seen that the robot must be located at the intersection of the two circles, which yields two possible locations. Generally, physical

33. Some years ago while I was trekking in Nepal with my son and daughter, we met Tibetans who had crossed the Himalayas with their sheep to sell them in Katmandu. When asked far they had come, they replied: "Three days of walking." In Los Angeles distances are also frequently given in units of travel time, like "I live about one hour from the university and a half hour from the beach." Can we localize from that?

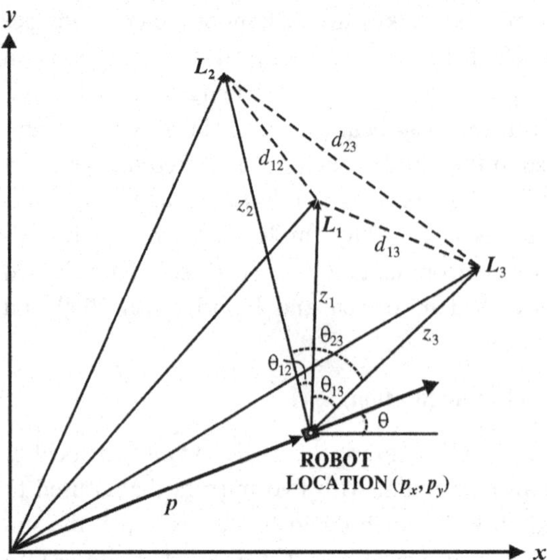

Figure 14.10
Robot location with three landmarks

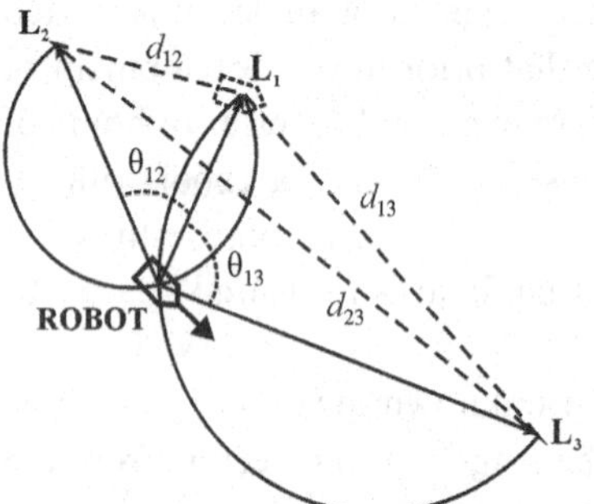

Figure 14.11
Localization by triangulation (adapted from Betke and Gurvits 1997)

reasoning based on the visual angles enables one to discard one of these, and the localization is complete, at least in principle. To obtain numerical values for the robot coordinates (p_x, p_y) we use the three visual angles and the three known distances in the law of cosines to obtain a set of equations for the three unknown distances from the robot to the landmarks.

For example, for landmarks L_1 and L_2, one can write

$$d_1^2 = z_1^2 + z_2^2 - 2z_1z_2 \cos \theta_{12}, \tag{14.3}$$

where both z_1 and z_2 are unknown. Similar equations for the other two interlandmark distances yield three equations in three unknowns. These equations are nonlinear, but they can be solved by least-squares methods. Now, given these lengths and the geometry of figure 14.10, we can calculate the robot coordinates. Once the coordinates are known, the equations can be solved for vector direction, thus specifying the orientation of the robot as well. Of course, all the angle measurements are noisy. However, using more than three landmarks and then averaging the resulting values will improve the estimate.

The method described in the foregoing is a traditional triangulation method. An alternate method, based on solution of linear equations, is described in Betke and Gurvits 1997. It is also worth noting that localization using signals from three or four satellites in a GPS system can be viewed as a form of triangulation (Grewal, Weill, and Andrews 2001).

14.6.2 Kalman Filter Basics

We have indicated previously that the measurements used to localize a robot in its environment are imprecise and include a variety of disturbances. How do we use noisy data to obtain an estimate of the true robot pose *as accurately as possible*? In view of the fact that outputs from robot sensors are noisy, it is evident that we must use some form of statistical averaging to obtain useful estimates of the values of the signals being measured by these sensors. For example, assume that we are attempting to measure a signal $x(t)$. All that we can observe, however, is the sensor output, $z(t)$, which will be affected by the addition of a disturbance or noise signal $v(t)$, to the signal we want to measure, $x(t)$, as shown in figure 14.12.

In this simple case, we can *estimate* the value of $x(t)$ simply by averaging $z(t)$ over time, that is,

$$\hat{x} = \frac{1}{t_f} \int_0^{t_f} z(t)\, dt, \tag{14.4}$$

where $\hat{x}$ is the estimated value of x, and t_f represents the end of the averaging interval. If the data are sampled, the equivalent discrete-time expression is

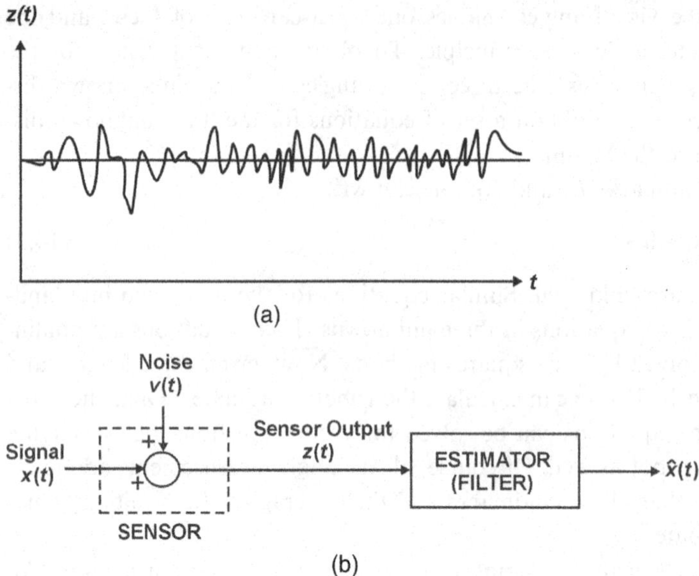

Figure 14.12
Estimating the value of a signal corrupted by noise: (a) signal, with mean value indicated; (b) estimation block diagram

$$\hat{x} = \frac{1}{N+1}\sum_{n=0}^{N} z(n\Delta t), \tag{14.5}$$

where Δt is the time interval between samples. Note that these expressions imply assumptions on the nature of the noise, namely, that it is stationary in some sense and has zero mean and a Gaussian probability distribution.

Now we must ask, "How good is this estimate?" Clearly, the longer the averaging time, the better the estimate, if the mean value of the noise remains at zero. If we average over some short interval, say one-quarter of the duration of the signal in figure 14.12, each interval will yield a different value. We can average these short-time estimates to get a new $\hat{x}$ (will it be the same as that from equations (14.4) or (14.5)?). More importantly, we can compute the variance of these estimates, which will provide an indication of their variability. Of course, if the noise mean is not constant, but, say, increasing over time, the previous expressions will not be useful. Furthermore, we have assumed that the signal $x(t)$ is constant. In practice, the sensor itself is a dynamic system, so that its output will not be constant even in the absence of the additive noise. This is where Kalman filters come in. (The remainder of this section is highly mathematical and can be skipped by readers interested only in an intuitive view of the subject.)

A Kalman filter (Kalman 1960; Gelb 1974; Grewal and Andrews 2000; Grewal, Weill, and Andrews 2001) is a procedure for processing the outputs of one or more noisy sensor outputs to enable estimation of the state of the system being measured, even when that system has uncertain dynamics. The noisy sensors could be odometers, sonars, range sensors, inertial sensors (like gyroscopes or accelerometers), or even GPS receivers. The state variables being measured by these sensors include position, velocity, acceleration, attitude, and attitude rate of a vehicle on land, in water, in the air, or in space. The system state may also have additional complexities, such as parameters that vary with time and other errors. By the system's having "uncertain dynamics," we mean that the sensor itself may be affected by such factors as wind, terrain changes, or even human activity. The Kalman filter provides estimates of the state variables as well as estimates of their variability. Since we normally deal with multiple state variables, represented by a state vector, there are also multiple uncertainties. Rather than dealing with simple scalar variances, as in the above example, we now deal with a *covariance matrix* that contains both the uncertainty of the individual state variables and their effects on one another.

Before giving an example, it is necessary to make one more observation. We referred to the filter output as an "estimate" of a state variable. In fact, it is an *optimal estimate*, but optimal in what sense? Estimates produced by Kalman filters are optimal in the *least-squares sense*. In other words, for any given state variable x_i, its estimate $\hat{x}_i$ is optimal in the sense that the quadratic cost function

$$J = f_i(x_i(t) - \hat{x}_i(t))^2 \tag{14.6}$$

is minimized. For multiple state variables, (14.6) becomes a vector-matrix expression. The use of quadratic cost functions greatly simplifies the mathematical optimization problem.

Let us illustrate the Kalman filter with a simple scalar example. Consider the system shown in figure 14.13(a), which represents the dynamics of a continuous system with a time-varying feedback term and a sensor that measures the system state $x(t)$.

The system equation can be written directly from the figure:

$$\dot{x}(t) = -a(t)x(t) + w(t). \tag{14.7}$$

The input to the system, $w(t)$, is assumed to be white noise with zero mean and variance Q; $x(t)$ is the system state; and $a(t)$ is a feedback gain term. The sensor model (also known as the *measurement equation*) in the figure is

$$z(t) = H(t)x(t) + v(t), \tag{14.8}$$

where $x(t)$ is the state variable being measured, $H(t)$ is the (possibly variable) sensor gain, and $v(t)$ is additive noise with zero mean and variance R. To simplify matters, we assume that the two noise sources (the system noise $w(t)$ and the measurement

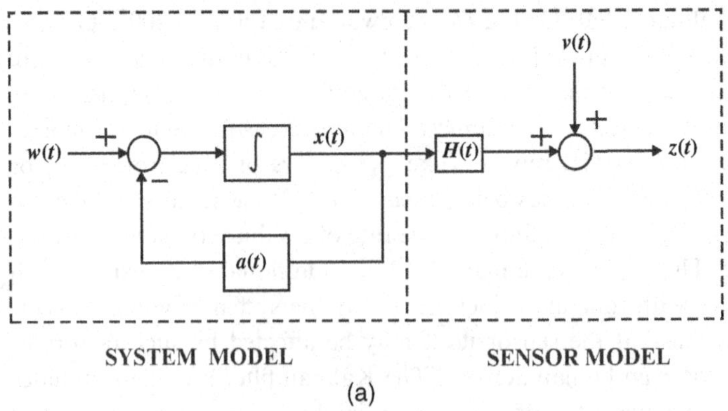

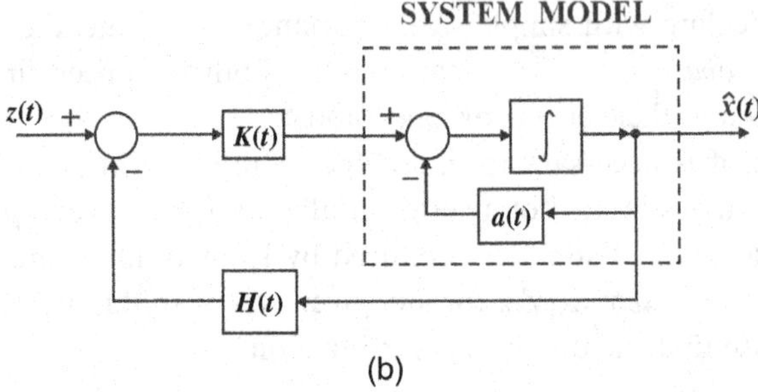

Figure 14.13
Example of state estimation in a continuous linear system (a) System and sensor models; (b) continuous Kalman filter

noise $v(t)$) are uncorrelated. Although the system equation (14.7) appears quite innocent, in fact it cannot be solved by ordinary calculus methods because the input is nondeterministic. Such equations are known as *stochastic differential equations*. Fortunately, we do not need to solve the equation to discuss the principles of filtering.

Now, since we cannot observe the state $x(t)$ directly, we have to estimate it from the measurement $z(t)$. This is accomplished by the Kalman filter shown in figure 14.13(b). Note that this figure consists of three parts:

1. a model of the system whose state we are attempting to estimate
2. the measurement or sensor gain $H(t)$
3. an additional gain term denoted by $K(t)$, known as the *Kalman gain*[34]

34. For continuous systems, like the one considered in this simple example, $K(t)$ is sometimes called the *Kalman-Bucy gain*, to distinguish it from the corresponding term in discrete-system models. See Kalman and Bucy 1961.

Item 1 in this list is important. It shows that part of the reason for a Kalman filter's ability to provide an optimum estimate of the system state is that it contains a model of the system itself. This also implies that any changes to a robot that affect its dynamics require insertion of a new model into the filter.

We need two additional equations: one for the desired estimate and one for the variability of the estimate. The state estimate equation can be written directly from figure 14.13(b):

$$\dot{\hat{x}} = -a(t)\hat{x}(t) + K(t)[z(t) - H(t)\hat{x}(t)]. \tag{14.9}$$

The solution of this equation yields the desired estimate of the state variable, but to solve it, we need the term $K(t)$. Note the term in square brackets in (14.9). It represents the difference between the measurement $z(t)$ and the *predicted measurement* $H(t)\hat{x}(t)$; in the absence of measurement noise and if our estimate were perfect, this term would be zero. The variance of the estimate, $P(t)$, is given by the equation

$$\dot{P}(t) = 2a(t)P(t) + Q. \tag{14.10}$$

This differential equation is known as the *Riccati equation.* Finally, the Kalman gain is computed from

$$K(t) = \frac{P(t)H(t)}{R(t)}. \tag{14.11}$$

Clearly, the Kalman gain depends on the variance of the estimate, the sensor gain, and the variance of the sensor noise, which could be time varying.

In general, of course, we deal with more than one state variable and systems of higher order, so these equations become vector-matrix equations. Furthermore, to solve them on a digital computer, these equations need to be replaced by their discrete-time equivalents. An excellent development of the Kalman filter and ways of computing it in practice can be found in Grewal and Andrews 2000 and Grewal, Weill, and Andrews 2001. In the discrete form, we work with values at each time step t_k, that is, the state estimate vector $\mathbf{x}(t_k)$, and the estimated covariance matrix $\mathbf{P}(t_k)$, rather than the scalar values in continuous time as in the preceding example. The resulting equations are recursive and use sequential sets of measurements. Prior knowledge of the state (expressed by the covariance matrix) is improved at each step by taking the prior state estimates and new data and using them to generate the next state estimate.

Finally, note that although the system model in the preceding example was linear, the method can be applied to nonlinear systems using the *extended Kalman filter*.

14.6.3 Kalman Filter–Based Localization

In recent years Kalman filters have been used extensively for robot localization. They have been applied for the correction of odometry errors by a number of investigators, such as Barshan and Durrant-Whyte (1995), Fuke and Krotkov (1996), and Roumeliotis, Sukhatme, and Bekey (1999b). An extended Kalman filter has been used to match beacon observations to a navigation map for sustained localization (Leonard and Durrant-Whyte 1991). Kalman filtering has also been employed for localization in both indoor and outdoor environments using a Pioneer AT robot by Roumeliotis and Bekey (1997) and Goel, Roumeliotis, and Sukhatme (1999). The wheels on the same side of this robot (illustrated in figure 14.4) were mechanically coupled, so that the encoders provided only two speeds: one for the right pair of wheels and one for the left pair. The measurements obtained were

$$\mathbf{z} = (v_L v_R \dot{\theta})^T, \tag{14.12}$$

where the first two terms are the translational velocities of the left and right side wheels, respectively, and the third term is the robot yaw rate:

$$\dot{\theta} = \frac{v_R - v_L}{l}, \tag{14.13}$$

where l is the vehicle axle length. The robot first navigates using only odometry based on the wheel encoders. The error in the final position is determined according to the relation

$$\text{Error} = \frac{\text{Actual final position} - \text{Estimated final position}}{\text{Total distance traversed}}.$$

The resulting position estimates after runs ranging from 10 m to 25 m produced final-position errors ranging from 20% to 25%. When the odometers were carefully calibrated using precise tachometers to convert wheel rotations into linear velocities, so that small differences in wheel diameter could be compensated for, the errors decreased to approximately 16%. Then, an inexpensive gyroscope (Systron-Donner QRS 14-64-109) was added to the system to measure the yaw rate. Recall that Kalman filter inputs can include signals from more than one sensor, so the gyro output was used along with the calibrated odometer outputs. The resulting errors ranged from 0.4% to 2.6%. Clearly, the addition of a sensor and the use of Kalman filters can result in a dramatic reduction in odometry errors.

In outdoor experiments the same robot was used on a course between buildings. GPS signals were available when not occluded by buildings. Hence, the robot position was estimated by GPS alone, when available, and by the Kalman filter when GPS signals were too weak to be useful. Reliance on global references can be useful in rover navigation on distant planets, such as Mars. In such situations the rover can use a sun sensor as a global reference (since GPS is not available). However, such a sensor produces a significant drain on the rover's battery. One option for preventing battery drain is to have the rover navigate most of the time by using wheel encoders, steering angle potentiometers, and gyros, augmented by Kalman filters. It can then use the sun sensor infrequently to obtain accurate localization (Roumeliotis and Bekey 1997).

We indicated in the previous section that Kalman filters require a model of the dynamic system whose state is being estimated and that this can be a problem since any change in the robot may require development of a new model for use in the filter. There is a way to avoid dynamic modeling, and that is to use a different form of the Kalman filter. The standard (or direct, total-state) formulation includes such variables as the vehicle orientation. By contrast, in the *error-state formulation* of the Kalman filter, the estimated variables are the *errors* in orientation and other state variables. This formulation has a number of advantages, in addition to avoiding the need for accurate modeling (Roumeliotis, Sukhatme, and Bekey 1999a).

14.6.4 Other Probabilistic Approaches to Localization

We hope that by this point in the chapter, the reader is convinced that robot localization, navigation, and mapping are in fact probabilistic problems.[35] It is impossible to know a robot's pose exactly, given imperfect sensors and imperfect knowledge of the robot's environment. We can only determine the probability that the robot is in location l, given a set of sensor readings $\mathbf{s}$, that is, $P(l|\mathbf{s})$. Even this information is not easy to obtain, since the sensor readings only give us $P(s|l)$. In other words, knowing something about the imperfections of the sensor, we can find the probability that its output will be some value given that the robot is indeed at location l. As we showed earlier in the chapter, Bayes' rule can be used to compute one from the other, if we know certain probabilities in advance.

Kalman filter methods are only one possible approach to the localization, navigation, and mapping problems. Others in current use and beyond the scope of this book are maximum-likelihood methods (e.g., Olson 2000; Howard, Matarić, and Sukhatme 2002b), methods based on partially observable Markov decision models (e.g., Cassandra and Kaelbling 1996; Koenig and Simmons 1998), and methods known as multiple-hypothesis tracking. We have used the last approach to combine Kalman filtering and Bayesian estimation in a unified framework for localization (Roumeliotis and Bekey 2000b). Finally, methods known as particle filtering have come into use in recent years for state estimation in mobile robotics (e.g., Fox et al. 2001; Thrun 2002; Fox 2003; Kwok, Fox, and Meilca 2004).

35. At least I hope that there is high probability that the reader is so convinced.

14.7 Simultaneous Localization and Mapping

It should be clear from the preceding discussion that a robot must know where it is while performing a mapping function. In the case study of section 14.5, it was noted that the Pioneer robot needed to perform constant relocalization to ensure that it placed landmarks it had detected and identified correctly on the map it was creating. The point is that mapping requires localization.

Both localization and mapping may be accomplished using probabilistic approaches. For example, Tomatis, Nourbakhsh, and Siegwart (2001) combine metric and topological mapping in their approach to simultaneous localization and mapping (SLAM). Local metric maps are connected by means of a global topological map. A 360° laser scanner is used to detect corners and openings for the topological maps and lines for metric localization. To allow for nonlinearities in the robot and environment models, Tomatis and colleagues use extended Kalman filters for the metric and create a *stochastic map*. With this approach, the authors combine the robot pose $(x\ y\ \theta)^T$ with a vector of landmark locations. The ith landmark is represented by $(\alpha_i\ r_i)^T$ where α is the bearing from the robot to the landmark and r is its distance from the robot. Topological navigation uses a partially observable Markov decision process for estimating the robot state. This approach has been tested in indoor environments with considerable success. In Castellanos and Devy 2000, the absolute localization is extended to local reference frames.

SLAM has become a major research topic, and recent conferences have frequently devoted entire sessions to it. A variety of approaches to SLAM are being tried, including the use of airborne cameras (Kim and Sukkarieh 2003), data fusion involving odometers and inertial sensors with extended Kalman filters (Folkesson and Christensen 2003), set theory (Di Marco et al. 2004), and genetic algorithms (Duckett 2003), among others. We expect that research interest in as well as practical applications of SLAM will continue for some time in the future.

14.8 Multirobot Localization

Increasingly, mobile robot applications require the cooperative work of many robots. All the approaches to localization discussed previously in this chapter concern only a single robot. In most early approaches to the problem of localizing a group of robots, the pose of each individual robot in the group was estimated. More recently there have been various attempts to combine the experiences of individual robots in a group in order to increase the precision of localization of the group. This is particularly interesting when the individual robots are equipped with different types of sensors. For example, consider a group of robots in which only one has a GPS receiver, and the rest have only odometry. Communication among members of the group should make it possible to improve the collective localization.

The first approach to cooperative localization of which we are aware was the *method of portable landmarks* (Kurazume, Nagata, and Hirose 1994; Kurazume et al. 1996). All the robots in the group have only odometric sensing, so they can navigate only by dead reckoning. The group is divided into two teams. At each time increment, one team moves, while the other remains stationary and acts as landmark. At the next increment, the roles are reversed. Kurazume and colleagues report average errors of 0.4% in position and 1° in orientation over distances of approximately 20 m. In a variation of this method, one robot moves, while rest of the team forms an equilateral triangle of localization beacons (Kurazume and Hirose 2000). In these and similar approaches, only one (or part of a group) moves at a time, and the two groups must maintain visual contact at all times. Still other approaches are suitable only for indoor environments or have other similar limitations.

An alternative approach, termed *collective localization*, is reported by Roumeliotis and Bekey (2002). In this approach, all the robots can move, and visual contact with one another is not required. All the robots are assumed to have proprioceptive sensors to allow for position tracking and exteroceptive sensors to monitor landmarks in the environment as well as other robots (which are considered "dynamic landmarks"). Furthermore, it is assumed that all the robots can communicate with each other. As an example of the type of improvement possible with such a system, consider figure 14.14.

This hypothetical figure shows the regions of localization uncertainty of two robots before and after they exchange information. Robot 1 has better sensors and is able to estimate its position much more accurately than robot 2. Information flow between the two robots in the form of relative pose measurements significantly reduces the localization uncertainty of robot 2 and also produces a small improvement for robot 1.

For the analysis of this approach, we consider the group as a single entity. A Kalman filter (in which each robot has knowledge of its own pose) describes the motion of the group. The estimation process is divided among all m robots, thus distributing it among m Kalman filters, each operating on a different robot. The details of the mathematical formulation and analysis of this approach are given in Roumeliotis and Bekey 2002. The results show dramatic improvements in localization as a result of the information exchange. Each time two robots meet and measure their relative poses, they exchange information not only for themselves, but for the rest of the group as well. This sharing of sensor output adds robustness to the team and makes it less sensitive to sensor failures of individual robots.

An alternative method of cooperative localization makes use of maximum-likelihood estimation rather than Kalman filtering. For example, Simmons et al.

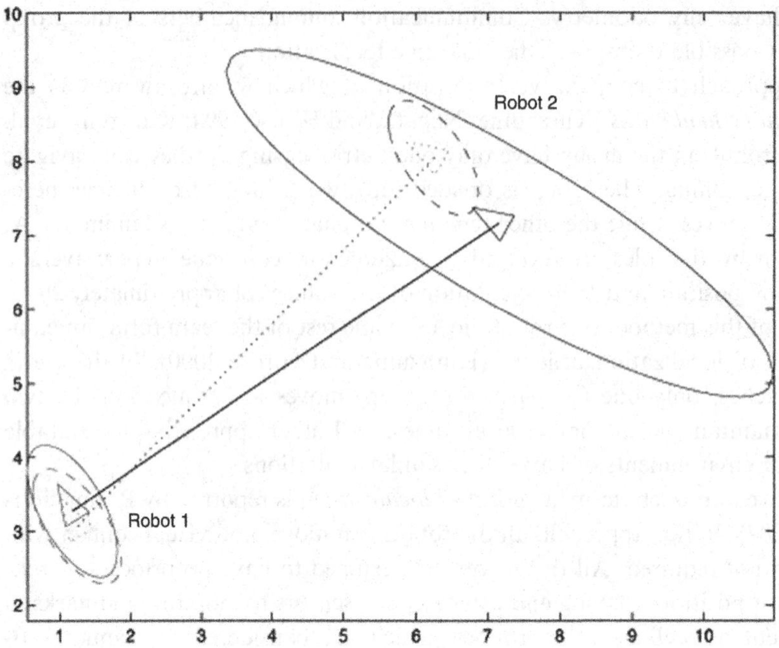

Figure 14.14
Improvement in localization of two communicating robots. The solid (dashed) lines represent 3σ regions of confidence before (after) relative pose measurements (from Roumeliotis and Bekey 2002, copyright 2002 by IEEE, reproduced with permission of IEEE and Stergios Roumeliotis)

(2000) address cooperative exploration and mapping by multiple robots. Their mapping algorithm is an online approach to likelihood maximization that finds maps that are maximally consistent with sensor data. A method for localizing the members of a robot group makes use of the robots themselves as landmarks (Howard, Matarić, and Sukhatme 2002b). Howard and colleagues assume that all the robots are equipped with sensors that allow them to measure changes in their own pose, as well as the relative pose and identity of other nearby robots. Using a combination of maximum-likelihood estimation and distributed numerical optimization for each robot, robots can estimate the relative range, bearing, and orientation of every other robot in the group. Howard and colleagues show that indeed such localization does not require external landmarks yet allows all the robots to be in motion; it is robust to changes in the environment and to poor motion sensing. Robots are able to infer the pose of other robots they have never seen.

We expect to see increasing effort in the area of multirobot localization in the future, with emphasis on methods that allow scaling to large numbers of robots.

14.9 Concluding Remarks

In this chapter we have attempted to guide the reader on a whirlwind tour through very complex and difficult material. As our robots become increasingly autonomous and able to travel in increasingly unstructured terrain, they will require highly sophisticated methods of localization, navigation, and mapping. Many of the algorithms mentioned in this chapter (and discussed in the references) require knowledge of stochastic processes as well as differential equations (to model robot dynamics) and optimization methods. As these methods grow in importance, we believe they should result in changes in the curricula for students in robotics. It should be clear that an understanding of only the deterministic aspects of artificial intelligence, or simple discrete mathematics as taught in many computer science departments, will not be sufficient for work in this field.

15 The Future of Autonomous Robots

Summary

Current trends in the development of autonomous robots are reviewed. Based on these trends, the author offers some projections to the future, including various robot applications as well as developments in group robotics, human-robot communication, and robot intelligence. The concluding section of the chapter deals with the potential danger of highly intelligent robots and offers a hopeful outlook for the future.

15.1 Introduction

In the preceding chapters of this book we have explored a number of trends in autonomous robots. We have examined a variety of mobile robots and their applications: wheeled, legged, flying, swimming, and crawling. We have studied humanoid robots that imitate some aspects of human appearance or behavior. We have looked at group robotics to understand the ways in which single robots can cooperate and communicate. In addition to reviewing complete robots and systems of robots, we have also examined some of the sensors and actuators that enable them to connect to the world and some aspects of robot intelligence, the software architecture of robots, methods of controlling them, and such application issues as localization, navigation, and grouping. It should be evident from this overview that robots at the beginning of the twenty-first century are increasingly autonomous, are able to sense and interpret a large variety of sensory stimuli, and are displaying more and more of the attributes of intelligence, such as learning, planning, and communicating, both with humans and with one another.

What, then, are the prospects for the future? Will we see just more of the same, perhaps faster and more reliable, or will there be dramatic new developments? Will the developments be gradual, or can we say, as in the title of a forthcoming book, that *The Singularity Is Near* (Kurzweil 2005)?

Autonomous robots inspire both fear and hope in the minds of many people. As we have shown in previous chapters, these robots have applications that may significantly contribute to human welfare in the years to come. At the same time, a number of researchers have warned that these contributions carry a heavy price. Improvements in robot skill and intelligence also mean that these machines may become a potential source of danger to humanity.

In the remainder of this chapter we first describe our projections for the future of autonomous robots and then discuss the possible danger to humanity from increasingly intelligent machines.

15.2 Current Trends in Robotics

This section describes our view of current trends in the field and offers some projections of likely developments within the next decade. We begin with some areas of application and then move to issues of behavior and intelligence.

15.2.1 Entertainment

We have shown the importance of autonomous robots for entertainment in various places in the book. Robots like AIBO (see chapter 8) are the beginning of a wave of autonomous, mobile artificial creatures. We expect that "dogs" like AIBO will be available in various breeds and sizes. The dogs will be supplemented by "cats" and other creatures, ranging from "monkeys" to "panda bears" and "snakes." Clearly, as these creatures become capable of larger and larger behavior repertoires, equipped with better methods of communication with their owners and increasing intelligence, their entertainment value will grow. Increasingly lifelike dolls will be available, enabling their owners to communicate with them using natural language. The current generation of humanoid robots is likely to lead to a new generation of humanlike companions and playmates, behaving like boys or girls.

15.2.2 Military Applications

One of the most rapidly growing areas of robot application is the use of robots for military purposes. At the time of this writing there are development efforts in autonomous aircraft (both fixed and rotary wing), autonomous tanks, Humvees, and other wheeled vehicles, and autonomous submarines. The current generation of such vehicles is primarily teleoperated. Unpiloted aerial vehicles are mainly drones, controlled remotely from a remote base or from a following aircraft. Unpiloted underwater vehicles may be either teleoperated or autonomous. Similarly, a wide variety of unpiloted ground vehicles is in development or in partial deployment, including tanks and a variety of trucks and armored vehicles. At the present time the primary application of such vehicles is either reconnaissance or delivery of supplies. Demining robots are also in various stages of development and implementation.

Unfortunately, it appears certain that improved sensing and reasoning on the part of military robots will lead to their use in actual combat. We expect that such robots will be equipped with weapons and will (eventually) be able to use them without human intervention or control. The only hopeful aspect of this development is a scenario in which robots from opposing camps are shooting at one another, without endangering any human lives.

15.2.3 Household and Industrial Service Robots

The author believes that the next decade will see the commercial development and application of a number of household-helper robots. Current vacuum cleaner robots from iRobot Corporation in the United States and Electrolux in Sweden (and others under development) are examples of this trend, as are lawn-mowing robots. The industrial counterparts are security robots that roam the inside of a warehouse or the perimeter of a building, mail delivery robots, and automated delivery vehicles. Developments in Germany, Japan, and other countries will lead to autonomous window-washing robots for skyscrapers, ship-cleaning robots that climb on a ship's hull while the vehicle is in dry dock, and robots that wash entire passenger aircraft. We expect that street lights will be cleaned and serviced by robots in the near future. Although this range of applications is still small, we expect it to grow significantly as the abilities of robots in such areas as navigation, manipulation, and cognition increase and it become easier to communicate with them.

15.2.4 Care of the Elderly and People with Disabilities

The population of most industrialized countries is becoming older, as the average lifespan increases and the birthrate decreases. The cost of caring for the infirm elderly is enormous and growing, which provides a strong motivation for the development of robots to assist these people in the activities of daily living. Robots for elder care are being developed in Japan, Germany, Italy, Great Britain, the United States, and other countries. The goal of this work is to produce robots capable of being commanded by means of natural speech and/or gestures to fetch items (including food from the refrigerator), turn appliances on or off, and assist with dressing, feeding, and moving to the toilet, to cite only a few of the potential applications.

Analogous applications exist in assistance to persons with a variety of disabilities. The past decade has seen the emergence of robot "seeing-eye dogs" to assist the blind. Similar developments are expected to provide "hearing-ear robots" and a variety of robotic systems to assist paraplegic persons with walking. The commercial potential in this area is enormous, but the robots to be developed must be nearly 100% fail-safe, easy to operate, affordable for a large segment of the population, and approved by health care and elder care authorities. This is a tall order indeed, but we expect major successes within the next decade.

15.2.5 Construction and Heavy Industry

The use of human-operated machinery will gradually give way to autonomous robots in a variety of construction industries, from industrial buildings to highways and bridges. Shimizu Corporation in Japan already uses robots to erect the steel framework of large buildings, spray the framework with asbestos for fireproofing, and clean up the excess from the spraying. Several companies are developing autonomous vehicles for road construction, including tractors and asphalt- or concrete-surfacing machines. There are also projects for the development of robotic machines for agricultural activities, including planting of seeds, replanting, weeding, fertilizing, and harvesting.

15.3 Human-Robot Cooperation and Interaction

It is evident from the foregoing list of applications that many of them will depend on the ability of humans and robots to cooperate, to work side by side on common projects. This is expected to be one of the major trends in the next decade. Whereas during the early days of robot manipulators in industry, it was necessary to erect barriers to keep robots from injuring humans, the new situation is the exact opposite. Future generations of robots will be designed in such a way that they cannot harm humans. Major improvements will occur in human-robot interaction (know as HRI, by analogy with HCI, human-computer interaction). Humans and robots will be able to communicate through speech using natural language, as well as through gestures and body positions. They should be able to decide jointly on task allocation, based on the capabilities and training of each of them. They should then be able to teach other humans and robots how to perform a task, using spoken and graphical instructions as well as imitation. We expect a human-robot symbiosis in which it will be natural to see cooperation between humans and robots on both simple and complex tasks. Robots will be able to learn from their human coworkers by observation and imitation and then pass on the skills they have acquired to other robots as well as apprentice humans.

15.3.1 The Three Es: Emotions, Esthetics, and Ethics

We noted in chapter 13 that one of the major research areas in humanoid robotics concerns enabling robots to understand, react to, and display emotions. Thus, HRI will involve not only factual communication, but the ability of robots to discern and react to emotions displayed by their human coworkers. This implies the ability to sense and understand body language, facial expressions, tone of voice, gestures, and other manifestations of emotional states. Significant work at Waseda University in Japan, at MIT in the United States, and at other institutions is beginning to address these issues. In other words, emotions are no longer outside of the realm of discussion in robotics. Clearly, the expression and understanding of emotional states are important parts of human-robot communication.

Now, if a robot can display and sense emotion, can it also develop an esthetic sense, and indicate that it thinks a proposed project is ugly but could be beautified? We believe that such developments are not only possible, but likely. Furthermore, robots will require an ethical awareness as well, to ensure that in pursuit of some goal they do not perform actions that would be harmful to people or other robots or unethical in some way. The three Es of emotion, esthetics, and ethics represent a formidable barrier to robot design at the present time, but we expect that this barrier will be overcome within the next decade or two.

15.4 Multirobot Systems

We discussed multiple-robot systems in chapter 12, including some examples and some of the design principles for such systems. We anticipate much progress on this front in the next decade. We expect to see an increased emphasis on scaling to much larger numbers of cooperative robots. Such activities as construction of a solar power station in earth orbit, habitats on the moon or Mars, or undersea facilities may require the cooperative actions of hundreds or thousands of robots. Although individually autonomous, these robots will require some form of organization and task assignment. It may be desirable that the robots be taskable (i.e., capable of performing different tasks, either by assignment from a supervisor robot, or perhaps by reconfiguring to make them more suitable for a new task). There will be a need for diagnostic software and the ability to perform repairs and part replacements. We expect significant progress in scaling robot teams to much larger numbers. A greater challenge is the development and implementation of multirobot learning, which may require a second decade to be completed. A similar challenge is the development of autonomous, self-sustaining colonies, capable of generating their own energy supplies.

15.5 Micro- and Nanorobots

Within the next ten or twenty years, we will see the development of a variety of very small robots. Even today it is possible to design a 1 cm^3 robot that contains a processor, a camera and one or more other sensors, a motor and controller, a radio, and a battery. Hence, such a robot can be fully autonomous, albeit with limited capability. Rapid development in MEMS (micro electromechanical systems) and

NEMS (nano electromechanical systems) will lead to amazing new applications. We envision the development of molecular-sized nanorobots that can be injected into the bloodstream of a patient. These robots should be able to locate and destroy tumors, repair aneurisms, and perform a variety of other surgical procedures. We predicted such applications many years ago (Lewis and Bekey 1992), and now they appear to be on the horizon.

Ideas for tiny robots have appeared in the science fiction literature for a long time. One of the stories in Ray Bradbury's *Martian Chronicles* (Bradbury 1945) describes a highly automated home on earth, where each day microscopic robots appear from under the carpet; each captures a grain of dust and carries it away. Of course, nanorobots are a much more recent concept. Ray Kurzweil (2001, 153), a pioneer in artificial intelligence, forecasts a time when *nanobots*, "miniature robots the size of blood cells," will "travel through the capillaries of our brains and communicate with biological neurons." Clearly, such applications are possible and will occur. On the other hand, we maintain some skepticism concerning Kurzweil's claim that these nanobots "will work with neurotransmitters in our brains to vastly extend our mental abilities" (151).

15.6 Reconfigurability

Reconfigurable robots were discussed in chapter 7. There are clearly advantages in building robots capable of changing their configuration to meet particular needs. As we have seen, the CONRO robot developed by Shen and Will (Castano, Shen, and Will 2000; Shen, Salemi, and Will 2002) can morph from a snake into a legged machine or roll like a wheel. We believe there will be significant new developments in such self-reconfigurable robots. However, it is not clear whether these developments will be long-lasting. There is a polarity between multipurpose systems and numerous single-purpose systems. In the past, single-purpose systems tended to prevail, because they are less expensive to build. The reconfiguration mechanism and their associated sensors and processors make multipurpose systems expensive. Nature has a tendency to favor specialized, single-purpose organisms. Thus social insects are specialized into gatherers, fighters, or nursemaids. On the other hand, humans are highly multipurpose systems. Reconfigurability will be one of the interesting developments in robotics to watch in the future.

15.7 The Implications of Computer Power

Recent increases in robot capability and intelligence would not have been possible without major increases in the speed and capability of their on-board processors. Indeed, these increases have been dramatic. From about 1980 to the present time, the speed and processing capacity of computers has increased by a factor of two every eighteen months. This increase was predicted by Gordon Moore, one of the founders of Intel Corporation, and is known as Moore's law. Some researchers had predicted that Moore's law would not apply after the year 2000, but the increases in computing power and speed continue and are now expected to continue for at least another decade. Clearly, the ability to increase the computing power of the robot's "brain" has major implications for the future.

Kurzweil (1999), whom we quoted earlier, has used Moore's law to forecast the following:

- By the year 2010 $1,000 desktop computers will be able to perform about a trillion calculations per second and enable communication with simulated people (avatars) using natural language. These avatars will be our primary interface with machine intelligence.
- By 2020 a $1,000 computing device will have a processing capability approximately equal to that of the human brain.
- By 2030 a $1,000 computing device will have a processing capability of approximately one thousand human brains.

The increase of a factor of one thousand in ten years forecast by Kurzweil assumes a doubling every year, since $2^{10} = 1{,}024$, and this rate exceeds the commonly quoted figure of a doubling every eighteen months. Others have suggested that Moore revised his law in 1975 to indicate a doubling of speed *and* a doubling of the number of transistors on a chip every twenty-four months, which for some applications could be a factor of four, or a doubling every year. The precise numbers do not matter; what matters is that the growth in computer speed and capacity has been exponential and promises to continue exponentially for the near future. The implications of such growth are staggering, even if one takes Kurzweil's projections with a grain of salt. Exponential growth in computing capacity for a robot's on-board computer implies that within twenty or thirty years, humanoid robots will be able to interact with people on human terms; that robots may be not only physical, but intellectual, partners of humans. This is the logical extension of the current trends discussed in previous sections of this chapter.

15.8 Self-Organization, Self-Repair, Autonomous Evolution, and Self-Replication

Eventually all forecasts about the future of robotics must face the question of self-replication; that is, will there come a day when robots can make copies of themselves? This has been the stuff of many science fiction stories, beginning with the play *RUR* (Capek 1921). It is also the concern of serious scientists who view self-reproduction as the eventual goal of intelligent machines. Certainly, at least in principle, robots could operate a factory that produces robots. However, at the present state of technology, humans would have to design and build the factory. We believe that true self-reproducing robots are still not on the horizon. (Perhaps they should never exist, as we show in the next section.)

Self-repair, at least for a certain level of damage, is a much less controversial issue and much more feasible. We believe that within the next ten to twenty years, robots will be able to repair their own failures (up to a point) and that "master-mechanic" robots will be able to perform more complex jobs of maintenance and upgrading.

We have already indicated that very large colonies of robots will require some form of organization. This is certainly true of human societies in which complete anarchy is seldom a wise choice for the inhabitants. We visualize the development of "social software" that will make it possible for groups of robots to organize themselves into operating units most appropriate to a given task. Under certain conditions, the robots may choose a single leader (a "king" or "dictator") to lead them. In other circumstances, a highly democratic structure or a hierarchical structure may be more appropriate. Robots will be able to select and implement whatever organizational unit suits the situation.

Finally, in this enumeration of less probable scenarios, we add autonomous evolution. We have shown that evolutionary or genetic algorithms are very useful optimization tools that can be used to select appropriate robot parameters to accomplish particular goals. In chapter 6 we presented a case study in which evolutionary algorithms were used to enable a six-legged robot to select the sequence of leg movements that enabled it to move most rapidly. If the goals in such an algorithm were to be set by the robots themselves, or by humans interested in developing superrobots, the results would be unpredictable.

15.9 The Potential Dangers of Robotics

It is evident that Moore's law will lead to the continuation of the exponential growth in computer power we have seen in recent years, at least for some time. Earlier we cited Kurzweil's prediction that by the year 2030, a desktop computer could have the processing power of a thousand human brains. If such powerful computers are used as on-board processors in robots, the consequences are unpredictable, and they could be dangerous for humanity. We know that increasing complexity may give rise to emergent behaviors not foreseen in the design of a machine. Kurzweil (1999) predicts that by 2030, machines will claim to be conscious, and that these claims will be largely accepted. In other words, the consequences of robots' achieving human levels of intelligence are unknown. It is not at all clear that robots with such advanced levels of consciousness will continue to obey their human creators.

Recall that Kurzweil also foresees the development and deployment of huge numbers of nanobots. He visualizes the use of some form of replication to create billions or trillions of nanobots, since they may be useful only in very large numbers. But then the question arises:

> And who will control the nanobots? Organizations (governments or extremist groups) or just a clever individual could put trillions of undetectable nanobots in the water or food supply of an entire population. These "spy" nanobots could then monitor, influence, or even take over our thoughts and action. (Kurzweil 2001, 153)

The concerns articulated by Kurzweil were stated even more clearly in a remarkable article by Bill Joy (2000) entitled "Why the Future Doesn't Need Us." Joy is the cofounder and former chief scientist of Sun Microsystems. In this article he expresses his concern about the effects of technology, especially robotics, genetic engineering, and nanotechnology, on human lives in the future. His concerns about robotics are perhaps even more serious than those of Kurzweil. Basically, he believes that intelligent robots will be a new and superior species, perhaps capable of self-reproduction, and certainly capable of producing mass death and destruction if controlled by individuals with malevolent intent. Further, if robots can self-replicate, he sees a terrible potential ahead. In his words:

> Thus we have the possibility not just of weapons of mass destruction but knowledge-enabled mass destruction (KMD), this destructiveness hugely amplified by the power of self-replication. (Joy 2000, 242)

Hans Moravec (1999), a distinguished computer scientist at Carnegie Mellon University, sees intelligent robots eventually succeeding humans and believes that humans, as a species, face extinction. Thus, he views intelligent robots as a new species to replace *Homo sapiens*. However, for Moravec, this is not a bleak vision, since he sees robots as the next step in the evolution of humanity.

All these projections are just that, projections. However, they come from people with a great deal of knowledge of the field and should be taken seriously. One possible consequence of the growth of robot intelligence might a popular uprising against technology. Kurzweil foresees the emergence of neo-Luddite[36] movements, intent on destroying technology "before it destroys us," but he believes that they will fail to stop the growth of robotic power.

36. The term "Luddite" refers to the followers of Ned Ludd, a quasi-mythical populist leader in England at the beginning of the Industrial Revolution who rallied people to the cause of destroying knitting machines, since their introduction resulted in the loss of jobs for people in the textile mills. Needless to say, the original Luddites failed, but Ludd's name is still associated with antitechnology movements at the present time.

15.10 Concluding Remarks

Robots are here to stay. They may create fear in some people and great hope for the future in others, but it is clear that in some form, robotic systems will become an increasingly integrated part of the human landscape as time goes on. In the near future we are likely to see robotic systems in homes, factories, schools, hospitals, and other institutions. Rodney Brooks (2004) of MIT believes that robots today are where personal computers were in 1978. That is the year when PCs began to appear in large numbers; fifteen years later they became ubiquitous. The same thing will happen, Brooks believes, with robots. In many cases they may not look like robots (at least as most of us think of them): Their processors will be embedded within other systems, and the systems themselves will display various levels of autonomy. As robots become increasingly intelligent and autonomous, our predictions about them become increasingly uncertain. Clearly, the "doomsday" predictions of some scientists could come to pass, but we agree with Kurzweil (2001) in believing that the constructive and creative application of robotic technology will ultimately prevail. However, this will be possible only if those of us who design, build, and apply robots are aware of the possible dangers that may arise. This is our challenge and our opportunity. Autonomous robots are here; let us use them for the benefit of humanity.

Appendix: Introduction to Linear Feedback Control Systems

A.1 Linear Control Systems in the Frequency Domain

The purpose of this appendix is to introduce the analysis of linear continuous-feedback control systems using frequency domain methods. As indicated in chapter 4, such systems contain dynamic elements and are described by differential equations; hence they are also termed *dynamical systems.* (Note that discrete-time or sampled systems are described by difference equations, rather than differential equations. Since methods for continuous and discrete-time systems are comparable, we will not discuss difference equations here.) Of course, in these few pages we can only touch on the highlights of the control system field, and interested readers are referred to current textbooks in the field, such as D'Azzo and Houpis 1995 and Kuo 1995.

Figure A.1 illustrates the basic feedback system described in chapter 4. We concentrate initially on the mathematical description of the plant or controlled system.

In the control system example discussed at the beginning of chapter 4, the plant (furnace) properties were described simply by a constant gain K_p. In practice, physical systems do not respond instantaneously to changes in input. The furnace may be described by a differential equation such as

$$\frac{dy(t)}{dt} + ay(t) = K_p u(t). \tag{A.1}$$

Let us assume that the constant $a > 0$. Let us also assume that the initial value of the output $y(t)$ is zero and the input changes suddenly (a step change) from zero to one. Now, since $y(t) = \int [dy(t)/dt]\,dt$, it is evident that the output $y(t)$ cannot change instantaneously. Hence, from equation (A.1), the derivative will jump to $dy/dt = K_p$, and the output will now change exponentially:

$$y(t) = \frac{K_p}{a}(1 - e^{-at}). \tag{A.2}$$

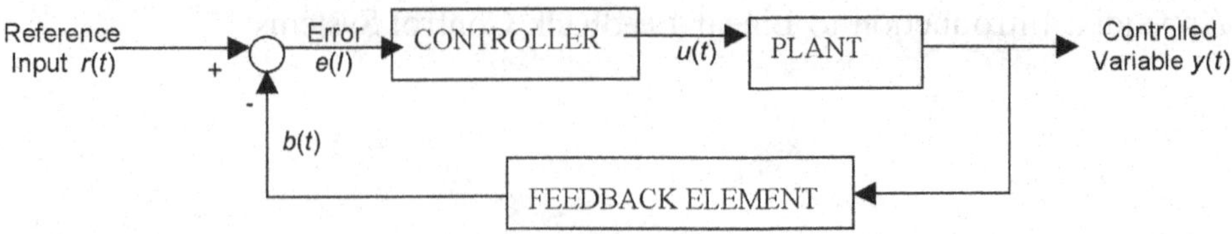

Figure A.1
Feedback control system

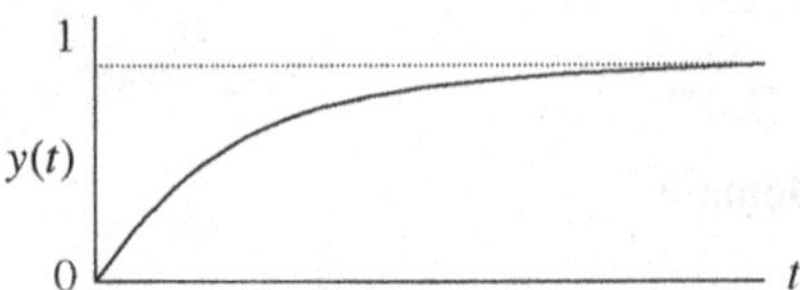

Figure A.2
Response of first-order system of equation (A.1) to a step input

This response is shown graphically in figure A.2, in which it has been assumed that $K_p/a = 1$.

Since equation (A.1) contains only the first derivative of the output $y(t)$, the system it describes is known as a *first-order system*. Many physical systems can be approximated using equations of this form.

To obtain the controlled variable $y(t)$ in figure A.1, it was necessary to solve a differential equation. However, equation (A.1) is linear[37] and has constant coefficients; hence we can take advantage of the fact that sinusoidal inputs to linear systems produce sinusoidal responses of the same frequency, but changed in amplitude and shifted in phase. To illustrate this fact, let the plant input in equation (A.1) be

$$u(t) = A \sin \omega t = A \sin 2\pi f t, \tag{A.3}$$

where ω is the frequency in radians per second. Then the output will be given by

$$y(t) = B \sin(\omega t + \varphi), \tag{A.4}$$

where B represents the amplitude of the output sine wave and φ is the phase shift of the output with respect to the input, as illustrated in figure A.3.

Thus, linear systems with constant coefficients can be described simply by their effect on the amplitude and phase of input sine waves, that is, the parameters B/A and φ. Since any periodic input can be represented as a sum of sinusoids, these parame-

37. A system is linear if it obeys the principle of superposition: If the response to input u is y, then the response to a composite input $a_1 u_1 + a_2 u_2$ will be $a_1 y_1 + a_2 y_2$, where a_1 and a_2 are constants.

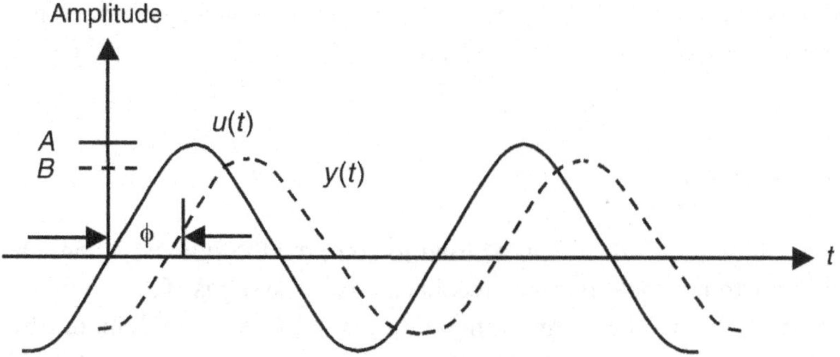

Figure A.3
Sinusoidal input (solid line) and output (broken line) waveforms of system of equation (A.1)

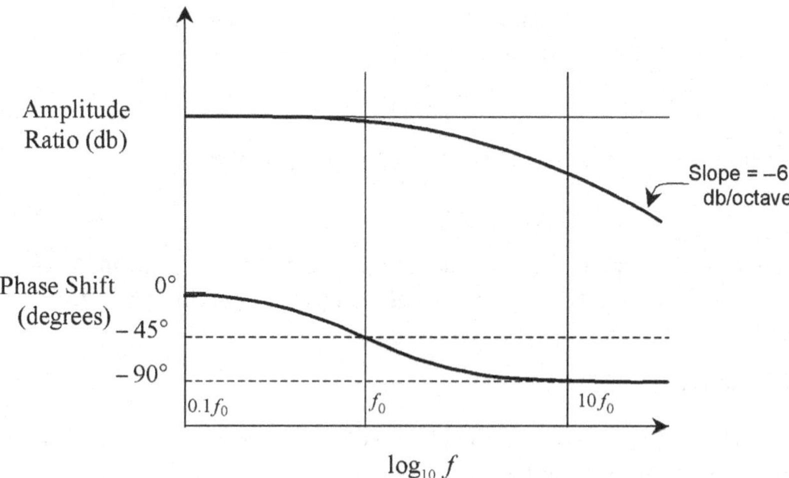

Figure A.4
Gain and phase of first-order system as a function of frequency

ters can be used to characterize the system for an arbitrary input $u(t)$. (Note that if the input is not periodic, but of finite duration, we can still represent it by means of a Fourier series in which the fundamental period is equal to the signal duration.)

To illustrate this fact for the system of equation (A.1), we let $u(t)$ be a sine wave, and we record the amplitude change and phase shift for various frequencies. The results are shown in figure A.4. For simplicity, it has been assumed that the gain factor $K_p/a = 1$. It is evident from the figure that the gain of the system decreases as a function of frequency. The phase shift also changes as a function of frequency, from a value of 0° at zero frequency to a maximum of 90° as the frequency becomes large.

The curves of figure A.4 are known as the *frequency response* of the system. When plotted on a logarithmic frequency scale, this type of figure is also known as a *Bode diagram*.

A.2 The Transfer Function

Now we need to ask how these effects can be used to compute the system response to a given input and how to use them to study feedback control systems. To accomplish these goals we need to change the system representation from the time domain (differential equations) to the frequency domain. This is accomplished using Laplace or Fourier transforms. Readers not familiar with these transforms are referred to the works cited in the first paragraph of the appendix. Here we simply indicate that the Laplace transform of a function $y(t)$ will be indicated by $Y(s)$, where s is a complex frequency variable, $s = \sigma + j\omega$. Letting $s = j\omega = j2\pi f$, where ω is the frequency in radians per second, we denote the Laplace transform of a variable $y(t)$ by

$$L[y(t)] = Y(s). \tag{A.5}$$

The Laplace transform of the time derivative of $y(t)$ is given by

$$L[dy(t)/dt] = sY(s), \tag{A.6}$$

which shows that in the Laplace domain, differentiation is converted to multiplication by the variable s. We can use this result and transform equation (A.1) term by term to yield

$$sY(s) + aY(s) = KU(s), \tag{A.7}$$

where we have omitted the subscript on the plant gain for simplicity. Note that this is now an algebraic equation that can be solved for $Y(s)$:

$$Y(s) = \frac{K}{s+a}U(s). \tag{A.8}$$

The ratio of output to input transforms is known as the *transfer function* of the system (or simply as the *system function*), since it represents the system properties in a compact way, that is:

$$G(s) = \frac{Y(s)}{U(s)} = \frac{K}{s+a}. \tag{A.9}$$

To visualize the frequency domain behavior of this function, we let $s = j\omega = j2\pi f$, where ω is the frequency in radians per second and f is the frequency in Hertz. With this substitution, the transfer function can be written as

$$G(j\omega) = \frac{K}{a + j\omega}. \tag{A.10}$$

This is a complex function of frequency. We can clarify its significance by recalling that any complex number can be represented by its magnitude and phase, that is,

$$G(j\omega) = \mathrm{Re}(j\omega) + j\ \mathrm{Im}(j\omega) = |G(j\omega)|e^{-j\phi(\omega)}.$$

To obtain the real and imaginary parts of (A.10), we multiply both numerator and denominator by the complex conjugate term $(a - j\omega)$ to obtain

$$G(\omega) = \frac{a - j\omega}{a^2 + \omega^2}. \tag{A.11}$$

Plotting the magnitude and phase of this expression yields the curves of figure A.4.

Thus, the transfer function contains all the frequency domain behavior of the linear system of equation (A.1), and the two can be considered equivalent representations. Now, to obtain the output for a given input, we simply use equation (A.9) and write

$$Y(s) = G(s)U(s). \tag{A.12}$$

Actually, this expression yields the Laplace transform of the output, which must then be converted back to the time domain.

To illustrate this process, consider the transfer function of (A.9). If the input to the system is a unit step $u_o(t)$, we can look up its Laplace transform in a table such as table A.1 and write

$$U(s) = \frac{1}{s}. \tag{A.13}$$

Then the output transform becomes

$$Y(s) = G(s)U(s) = \left(\frac{K}{s + a}\right)\left(\frac{1}{s}\right) = \frac{K}{s(s + a)}. \tag{A.14}$$

There are two common ways of inverting this expression to obtain $y(t)$, the output in the time domain. The easiest way is to find the expression of (A.14) in a table of Laplace transforms. Unfortunately, a small table such as table A.1 does not contain this expression, so we must expand it using *partial fractions*:

$$G(s) = \frac{K}{s(s + a)} = \frac{A}{s} + \frac{B}{s + a}, \tag{A.15}$$

where the terms A and B can be obtained from (A.15) by eliminating one term at a time:

Table A.1
A short table of Laplace transforms

$f(t)$	$F(s)$
Functions	
$u(t)$	$\frac{1}{s}$
t	$\frac{1}{s^2}$
e^{-at}	$\frac{1}{s+a}$
te^{-at}	$\frac{1}{(s+a)^2}$
$\sin \beta t$	$\frac{\beta}{s^2+\beta^2}$
Operations	
$\frac{df}{dt}$	$sF(s) - f(0)$
$\int_0^t f(t)\,dt$	$\frac{F(s)}{s}$

$$A = sG(s)|_{s=0} = \frac{K}{a}, \tag{A.16}$$

$$B = (s+a)G(s)|_{s=-a} = -\frac{K}{a}. \tag{A.17}$$

Substituting these expressions in equation (A.15), we obtain

$$Y(s) = \frac{K}{a}\left(\frac{1}{s} - \frac{1}{s+a}\right). \tag{A.18}$$

We can now obtain the inverse transform for each of these two terms from table A.1, which yields the time domain expression

$$y(t) = \frac{K}{a}(1 - e^{-at}). \tag{A.19}$$

Not surprisingly, this equation is identical to equation (A.2).

We have demonstrated that Laplace transforms can be used to convert the differential equations describing dynamical systems into algebraic equations in the frequency domain. Solution of these equations and inversion of the resulting transforms yields the output in the time domain. To apply these ideas to closed-loop (feedback) control systems, we return to figure A.1. The controller, plant, and feedback elements will be represented by their transfer functions $G_c(s)$, $G_p(s)$, and $G_f(s)$, respectively. The controller output is given by

$$U(s) = G_c(s)E(s) = G_c(s)[R(s) - B(s)], \tag{A.20}$$

where the capital letters represent Laplace transforms of the respective time functions $u(t), e(t)$, and $b(t)$. The feedback signal is related to the output by

$$B(s) = G_f(s)Y(s). \tag{A.21}$$

The feedback element transfer function $G_f(s)$ may represent the properties of the sensor used to measure the output variable and the change of units, say, from degrees to volts. We now note that the output is given as the product of the control input $U(s)$ and plant transfer function:

$$Y(s) = G_p(s)U(s). \tag{A.22}$$

We can eliminate the feedback signal variable $B(s)$ by substituting (A.21) in (A.20) and can eliminate the controller output $U(s)$ by combining (A.20) and (A.22) to obtain:

$$Y(s) = G_p(s)G_c(s)[R(s) - G_f(s)Y(s)]. \tag{A.23}$$

Solving for the ratio of the output or controlled variable $Y(s)$ to the reference input transform $R(s)$ yields

$$\frac{Y(s)}{R(s)} = \frac{G_c(s)G_p(s)}{1 + G_c(s)G_p(s)G_f(s)}. \tag{A.24}$$

Thus, the *closed-loop transfer function* of the feedback control system in figure A.1 is given by

$$G(s) = \frac{G_c(s)G_p(s)}{1 + G_c(s)G_p(s)G_f(s)}. \tag{A.25}$$

Note that the numerator is the product of the forward-path transfer functions in figure A.1 and the denominator is one plus the product of all the transfer functions in the loop. As we have seen earlier, this is an algebraic expression in the complex variable s, so it contains the frequency domain behavior of the complete closed-loop system. The subscripted quantities represent the individual transfer functions of the controller, plant, and feedback elements, respectively, all of which are functions of frequency. Equation (A.25) can be used to obtain the frequency domain characteristics of the closed-loop system and to compute its response to an arbitrary input, just as in the foregoing open-loop case.

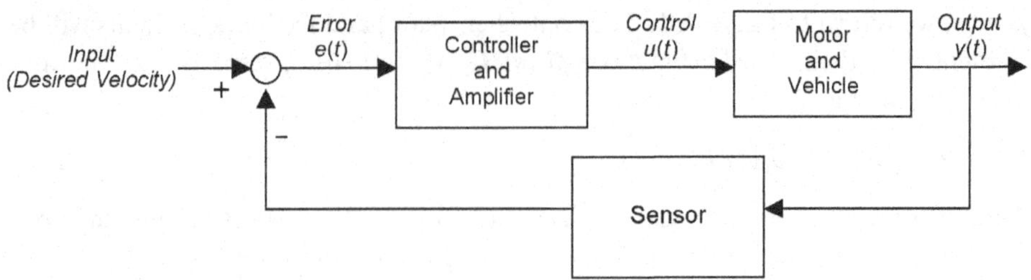

Figure A.5
Feedback control system for vehicle velocity

A.3 Stability

Consider again the control system of figure 4.2, which is reproduced here as figure A.5 for convenience. We have shown (in chapter 4) that the addition of feedback reduces the sensitivity of the system to parameter variations. What is the price paid for this improvement in performance? Perhaps the heaviest price arises from the fact that feedback control systems may break into uncontrolled oscillations. How is this possible? There are numerous ways of studying the stability of closed-loop control systems, some graphical and some analytical. Here we examine two relatively simple approaches to checking the stability of a control system.

A.3.1 Stability from Frequency Response Curves

Observe that the system of figure A.5 uses negative feedback; that is, the output of the sensor recording the vehicle velocity is subtracted from the desired input value. (If the feedback signal is positive, it adds to the input.) Assume that the input signal is positive. Then, if the feedback signal adds to this value, it will produce an even larger positive value at the output of the comparator. The controller will respond to this signal by sending a large negative signal to the motor. It is evident that successively larger signals of opposite polarities are possible; the effect will be an oscillation of the velocity that eventually will either saturate the motor signals or cause physical harm to the system. Now, consider the amplitude and phase curves of figure A.3. Note that if the output sine wave were shifted by 180°, it would have the opposite sign of the input; that is,

$$y(t) = -B \sin \omega t. \tag{A.26}$$

Now, if this signal is compared to the input, the "error" will be

$$e(t) = A \sin \omega t + B \sin \omega t = (A + B) \sin \omega t, \tag{A.27}$$

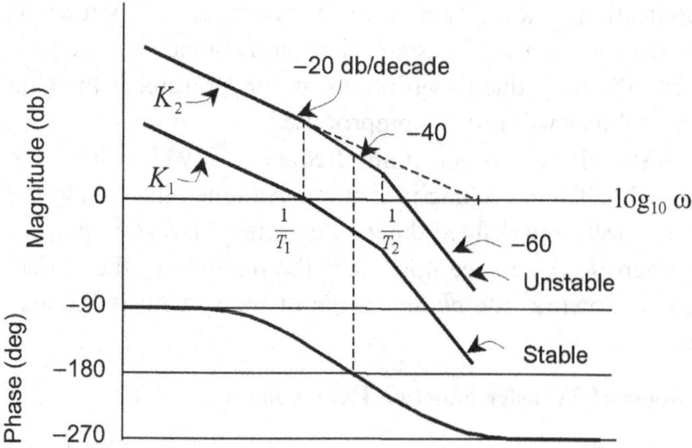

Figure A.6
Frequency properties of a loop transfer function

which is larger than the input, and we have the potential for instability if the loop gain is greater than one. As figure A.4 shows, the first-order system of equation (A.1) has a maximum phase shift of 90° at infinite frequency. It can be shown that each unit increase in the order of the system results in an additional 90° of potential phase shift. Hence, second- and higher-order systems may break into oscillation. In fact, one of the design criteria for control systems is based on this reasoning.

Consider figure A.6, which represents the gain and phase of a loop transfer function,

$$G(s) = G_c(s)G_p(s)G_f(s) = \frac{K}{s(1 + T_1 s)(1 + T_2 s)}. \tag{A.28}$$

Like figure A.4, figure A.6 is a Bode diagram. It represents the frequency response of $G(s)$ for two different values of the gain K, denoted by K_1 and K_2. As previously indicated, it is possible for a system to become unstable if the transfer function in (A.28) has a gain greater than one at the frequency where the total phase shift is $-180°$. It can be seen from the figure that at the frequency where the phase shift is $-180°$ and the system gain $K = K_2$,

$$\log_{10}|G(j\omega)| > 0. \tag{A.29}$$

Thus, for this gain value, the magnitude curve is greater than one, and the system will be unstable. On the other hand, when $K = K_1$, the magnitude curve is below zero at this frequency, so the system is stable. The range of possible increase values in the gain up until the point at which the magnitude of the transfer function is exactly one is clearly an indication of a safety margin for system design, known as the *gain margin*. The gain margin is a useful design criterion. Assume that the gain of the system can change by 10%, say, due to variations in temperature. If the gain margin is greater than 10%, stability will not be compromised.

As indicated previously, instability can occur at any frequency at which the phase shift is 180° and the gain is unity. The preceding paragraph examined the gain at the frequency where the phase is −180° and defined the gain margin. It is also important to examine the frequency where $K = 1$ to see how close the phase is to the critical value of −180°. This value is known as the *phase margin* of the system; it can also be obtained from figure A.6.

A.3.2 Stability from the Roots of Transfer Function Denominator

Given the loop transfer function $G(s)$ of a control system (A.25), it is possible to determine stability conditions from its denominator. The detailed analysis of control system stability is treated in numerous textbooks (e.g., D'Azzo and Houpis 1995; Kuo 1995) and is beyond the scope of this book. However, we can indicate some of the principles here. The denominator of the transfer function $G(s)$ can be factored and then expanded in partial fractions. Consider the typical transfer function

$$G(s) = \frac{N(s)}{D(s)} = \frac{N(s)}{s(s+a)(s+b)(s^2+c^2)}, \tag{A.30}$$

where $N(s)$ and $D(s)$ represent the numerator and denominator polynomials in s. The denominator has been factored in (A.30). (Note that this is only an illustrative example. In general any of the terms of the denominator might be raised to higher powers than in this example, and there could also be many more terms than in this example. However, this expression is sufficient to illustrate the method.) We first expand the transfer function as follows:

$$G(s) = \frac{N(s)}{s(s+a)(s+b)(s^2+c^2)} = \frac{A_1}{s} + \frac{A_2}{s+a} + \frac{A_3}{s+b} + \frac{A_4}{s^2+c^2}. \tag{A.31}$$

Now, using table A.1, we can obtain the inverse Laplace transform of (A.31) as

$$g(t) = A_1 u_0(t) + A_2 e^{-at} + A_3 e^{-bt} + A_4 \sin ct. \tag{A.32}$$

This time domain expression is the response of the system of (A.30) to a unit impulse. It includes a step function, two exponentials, and a sine wave. (Clearly, other functions are possible, but this example illustrates the principles involved.) Consider the second and third terms in this time domain expression. It is evident that if either a or b is negative, these functions will grow exponentially without bound. The step function is bounded, as is the sine wave. Hence, a stability condition is that the roots of the denominator of the closed-loop expression (A.30) must lie in the left half of the

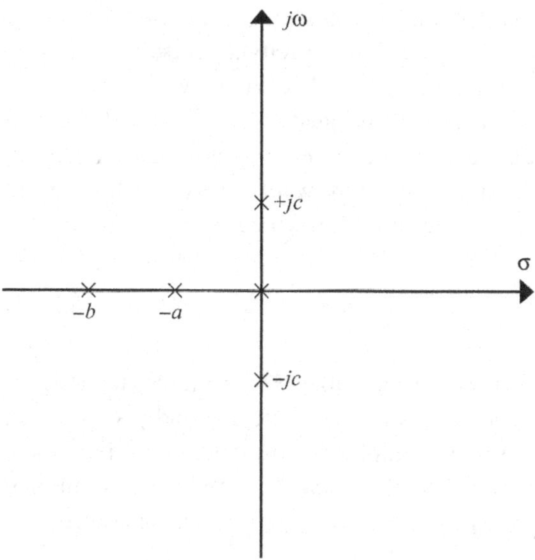

Figure A.7
Location of poles of transfer function of equation (A.30)

complex s-plane, or along the imaginary axis. As indicated previously, the variable s is a complex quantity, $s = \sigma + j\omega$, so the location of the roots of the denominator of (A.30) can be shown as in figure A.7. The roots of the denominator of the transfer function are known as the *poles* of the system, and the roots of the numerator are known as *zeros*.

It can be seen in figure A.7 that all the poles of (A.30) are indeed located on the left half of the complex plane, or on the imaginary axis. As indicated by equation (A.32), poles located on the negative real axis give rise to decaying exponentials, poles on the imaginary axis give rise to sinusoidal waves, and a root on the origin corresponds to a unit step. Intuitively it should be obvious that a pair of complex poles located in the left half plane, but not on the $-\sigma$ axis, will represent sine waves modulated by a decaying exponential. Poles in the right half plane represent growing time functions (i.e., instability).

A.4 Control System Design

The previous section has shown that the choice of loop gain may influence the stability of an entire system. It is not difficult to imagine that controller gain may be selected in such a way as to provide for a desired gain margin. Further, if the controller introduces some positive phase shift (also known as *phase lead*, in contrast with negative phase shift, or *phase lag*) it may be possible to increase the phase margin of a system to a desired value, say 30°. Thus, specifying these two margins is one way of specifying design criteria for a feedback control system. However, they are not the only possible criteria that can be specified.

Ideally, one would like to design a system to meet a desired *performance* criterion, as well as a stability criterion. For example, we may want a system whose mean square error is bounded. Then the performance criterion would be of the form

$$J = \int_0^T e^2\,dt. \tag{A.33}$$

Clearly, many other performance criteria are possible. For example, to design a control system in such a way that it reaches a particular state as rapidly as possible, we use a *minimum time* criterion. When more complex performance criteria are used, the resulting controllers will also be more complex than the simple and traditional PID controllers. Interested readers are urged to consult the literature on control system design.

In practical situations, the form of the controller may be specified in advance. As indicated in chapter 4, many robot motion control systems use proportional-integral-derivative controllers. The gains associated with the three components of these controllers can be selected to meet various criteria, including stability and steady-state error.

References

Abe, Y., M. Shikano, T. Fukuda, F. Arai, and Y. Tanaka. (1999). Vision-based navigation system for autonomous mobile robot with global matching. *IEEE International Conference on Robotics and Automation*, 1299–1304, Detroit, Mich.

Adams, B., C. Breazeal, R. Brooks, and B. Scassellati. (2000). Humanoid robots: A new kind of tool. *IEEE Intelligent Systems* 15(4): 25–31.

Agah, A. (1994). Sociorobotics: Learning for coordination in robot colonies. Ph.D. dissertation, Computer Science Department, University of Southern California, Los Angeles.

Agah, A. (1996). Robot teams, human workgroups and animal sociobiology: A review of research on natural and artificial multi-agent autonomous systems. *Advanced Robotics* 10: 523–545.

Agah, A., and G. A. Bekey. (1997a). Cognitive architecture for robust adaptive control of robots in a team. *Journal of Intelligent and Robotic Systems* 20: 251–273.

Agah, A., and G. A. Bekey. (1997b). Emergence and effectivess of communications interface in a system of distributed intelligent agents. *Journal of Robotics and Mechatronics* 9(2): 146–151.

Agah, A., and G. A. Bekey. (1997c). Phylogenetic and ontogenetic learning in a colony of interacting robots. *Autonomous Robots* 4(1): 85–100.

Agah, A., and G. A. Bekey. (1997d). Tropism-based cognition: A novel software architecture for agents in colonies. *Journal of Experimental and Theoretical Artificial Intelligence* 9(2–3): 393–404.

Aiken, T. (2004). MOS quality sensors make vehicle cabins safer. *Sensors* 20(2): 40–42.

Albergoni, V., C. Cobelli, and G. Francini. (1974). *Biological systems: An engineering approach to physiology*. Bologna: Pitagora Editrice.

Albus, J. S. (1971). A theory of cerebellar function. *Mathematical Biosciences* 10: 25–61.

Albus, J. S. (1981). *Brains, behavior and robotics.* Peterborough, N.H.: BYTE.

Albus, J. S. (1991). Outline for a theory of intelligence. *IEEE Transactions on Systems, Man and Cybernetics* 21(3): 473–509.

Albus, J. S., R. Bostelman, and N. Dagalakis. (1992). The NIST Robocrane. *Journal of Robotic Systems* 10(5): 27–34.

Albus, J. S., H. McCain, and R. Lumia. (1987). NASA/NBS standard reference model for telerobot control system architecture (NASREM). Technical Report, Robots Systems Division, National Bureau of Standards, Gaithersburg, Md.

Alexander, R. M. (1984). Walking and running. *American Scientist* 72: 348–354.

Altendorfer, R., N. Moore, H. Komsuoglu, M. Buehler, H. B. Brown Jr., D. McMordie, U. Saranli, R. J. Full, and D. E. Koditschek. (2001). RHex: A biologically inspired hexapod runner. *Autonomous Robots* 11(3): 207–213.

Ambrose, R. O., H. Aldridge, R. S. Askew, R. Burridge, W. Bluethmann, M. A. Diftler, C. Lovchik, D. Magruder, and F. Rehnmark. (2000). ROBONAUT: NASA's space humanoid. *IEEE Intelligent Systems* 15(4): 57–63.

Ambrose, R. O., R. S. Askew, W. Bluethmann, and M. A. Diftler. (2000). Humanoids designed to do work. *First IEEE International Conference on Humanoid Robots*, Cambridge, Mass.

Amidi, O. (1996). An autonomous vision-guided helicopter. Ph.D. dissertation, Robotics Institute, Carnegie Mellon University, Pittsburgh.

Amit, R., and M. Matarić. (2002). Learning movement sequences from demonstration. *Proceedings of the International Conference on Development and Learning*, 302–306. Cambridge, Mass.

Apostolopoulos, D., and J. E. Bares. (1995). Locomotion configuration for a robust rappeling robot. *International Conference on Intelligent Robots and Systems*, 3: 280–284, Pittsburgh, Pa.

Arbib, M. A., ed. (1995). *The handbook of brain theory and neural networks.* Cambridge, Mass.: MIT Press.

Arbib, M. A. (2000). The mirror system, imitation and the evolution of language. In *Imitation in natural and artificial systems*, edited by K. Dautenhahn and C. Nehaniv. Cambridge, Mass.: MIT Press.

Arbib, M. A., T. Iberall, and D. M. Lyons. (1985). Coordinated control programs for movements of the hand. In *Hand function and the neocortex*, edited by A. W. Goodwin and I. Darian-Smith, 111–129. Berlin: Springer-Verlag.

Arbib, M. A., and G. Rizzolatti. (1996). Neural expectations: A possible evolutionary path from manual skills to language. *Communication and Cognition* 29: 393–424.

Arikawa, K., and S. Hirose. (1996). Development of the quadruped robot TITAN-VIII. *International Conference on Intelligent Robots and Systems (IROS)*, 208–214, Osaka, Japan.

Arkin, R. C. (1986). Path planning for a vision-based mobile robot. *Proceedings of the SPIE Conference on Mobile Robots*, 240–249, Cambridge, Mass.

Arkin, R. C. (1987). Motor schema-based navigation for a mobile robot: An approach to programming by behavior. *IEEE International Conference on Robotics and Automation*, 264–271, Raleigh, N.C.

Arkin, R. C. (1993). Cooperation without communication: Multi-agent schema based robot navigation. *Journal of Robotic Systems* 9(3): 351–364.

Arkin, R. C. (1998). *Behavior-based robotics.* Cambridge, Mass.: MIT Press.

Arkin, R. C., and T. Balch. (1998). Cooperative multiagent robotic systems. In *Artificial intelligence and mobile robots*, edited by D. Kortenkamp, R. P. Bonasso, and R. Murphy, 277–295. Menlo Park, Calif., and Cambridge, Mass.: AAAI Press/MIT Press.

Arkin, R. C., T. Balch, and E. Nitz. (1993). Communication of behavioral state in multi-agent retrieval tasks. *IEEE International Conference on Robotics and Automation*, 3: 588–594, Atlanta, Ga.

Arkin, R. C., and J. D. Hobbs. (1992). Dimensions of communication and social organization in multi-agent robotic systems. In *Second International Conference on Simulation of Adaptive Behavior*, 486–493. Cambridge, Mass.: MIT Press.

Arkin, R. C., and D. C. Mackenzie. (1994). Temporal coordination of perceptual algorithms for mobile robot navigation. *IEEE Transactions on Robotics and Automation* 10(3): 276–286.

Armada, M., P. Gonzalez de Santos, M. Prieto, and J. C. Grieco. (1998). REST: A six-legged climbing robot. In *European Mechanics Colloquium (Euromech 375), Biology and Technology of Walking*, 159–164.

Aryananda, L. (2002). Recognizing and remembering individuals: Online and unsupervised face recognition for humanoid robot. *IEEE/RSJ International Conference on Intelligent Robots and Systems*, 2: 1202–1207, Lausanne, Switzerland.

Asada, H., T. Kanade, and I. Takeyama. (1982). Control of a direct drive arm. *Proceedings of ASME Annual Meeting (Robotics Research and Advanced Applications)*, 63–72.

Asada, M., and H. Kitano, eds. (1999a). *RoboCup-98: Robot Soccer World Cup II.* Berlin: Springer-Verlag.

Asada, M., and H. Kitano. (1999b). The RoboCup challenge. *Robotics and Autonomous Systems* 29(1): 3–12.

Asada, M., H. Kitano, I. Noda, and M. Veloso. (1999). RoboCup: Today and tomorrow—What we have learned. *Artificial Intelligence* 110: 193–214.

Atkeson, C. G., J. G. Hale, F. Pollick, M. Riley, S. Kotosaka, S. Schaal, T. Shibata, G. Tevatia, and A. Ude. (2000). Using humanoid robots to study human behavior. *IEEE Intelligent Systems* 15: 46–56.

Atkeson, C. G., A. W. Moore, and S. Schaal. (1997). Locally weighted learning. *Artificial Intelligence Review* 11: 11–73.

Atkeson, C. G., and S. Schaal. (1997a). Learning tasks from a single demonstration. *IEEE International Conference on Robotics and Automation*, 1706–1712, Albuquerque, N.M.

Atkeson, C. G., and S. Schaal. (1997b). Robot learning from demonstration. In *Machine Learning: Proceedings of the Fourteenth International Conference*, edited by D. Fisher, 12–20. San Francisco: Morgan Kaufmann.

Autumn, K., Y. A. Liang, S. T. Hsieh, W. Zesch, W. P. Chan, T. W. Kenny, R. Fearing, and R. J. Full. (2000). Adhesive force of a single gecko foot-hair. *Nature* 405: 681–685.

Ayers, J. (2001). Building a robotic lobster. In *Artificial ethology*, edited by O. Holland and D. MacFarland, 139–155. Oxford: Oxford University Press.

Bahr, B., Y. Li, and M. Najafi. (1996). Design and suction cup analysis of a climbing robot. *Journal of Computers and Electrical Engineering* 22(3): 193–209.

Bailey, S. A., J. G. Cham, M. R. Cutkosky, and R. J. Full. (2000). Comparing the locomotion dynamics of the cockroach and a shape deposition manufactured biomimetic hexapod. *International Symposium on Experimental Robotics (ISER)*, 239–248.

Baker, S., and S. K. Nayar. (2001). Single viewpoint catadioptric cameras. In *Panoramic vision: Sensors, theory, applications*, edited by R. Benosman and S. B. Kang. New York: Springer-Verlag.

Balch, T., and R. C. Arkin. (1994). Communication in reactive multiagent robotic systems. *Autonomous Robots* 1(1): 27–52.

Balch, T., and R. C. Arkin. (1998). Behavior-based formation control for multiagent robot teams. *IEEE Transactions on Robotics and Automation* 14(6): 926–993.

Balch, T., and L. Parker, eds. (2002). *Robot Teams: From Diversity to Polymorphism.* Wellesley, Mass.: AK Peters.

Bar-Cohen, Y. (2000). Electroactive polymers as artificial muscles—Capabilities, potentials and challenges. Paper presented at Robotics 2000, Conference, Feb. 28, Albuquerque, N.M.

Bar-Cohen, Y., ed. (2001). *Electroactive polymer (EAP) actuators as artificial muscles—Reality, potential and challenges.* Bellingham, Wash.: SPIE Press.

Bar-Cohen, Y., and C. Breazeal, eds. (2003). *Biologically inspired intelligent robots.* Bellingham, Wash.: SPIE Press.

Bares, J. E., and D. Wettergreen. (1999). Dante II: Technical description, results and lessons learned. *International Journal of Robotics Research* 18(7): 621–649.

Bares, J. E., and W. L. Whittaker. (1993). Configuration of autonomous walkers for extreme terrain. *International Journal of Robotics Research* 12(6): 535–559.

Barshan, B., and H. F. Durrant-Whyte. (1995). Inertial navigation systems for mobile robots. *IEEE Transactions on Robotics and Automation* 11(3): 328–342.

Barto, A. G. (1992). Reinforcement learning and adaptive critic methods. In *Handbook of intelligent control*, edited by D. A. White and D. A. Sofge, 469–491. New York: Van Nostrand-Reinhold.

Bassler, U. (1977). Sensory control of leg movement in the stick insect Carausius morosus. *Biological Cybernetics* 25(1): 61–72.

Bassler, U. (1983). *Neural basis of elementary behavior in stick insects.* Berlin: Springer-Verlag.

Bayliss, L. E. (1966). *Living control systems.* London: English University Press.

Beard, R. W., J. Lawton, and F. Y. Hadaegh. (2000). A feedback architecture for formation control. *Proceedings of the American Control Conference*, 4087–4091, Chicago, Ill.

Beard, R. W., J. Lawton, and F. Y. Hadaegh. (2001). A coordination architecture for spacecraft formation flying. *IEEE Transactions on Control Systems Technology* 9(6): 777–790.

Beccari, G., S. Caselli, and F. Zanichelli. (1998). Qualitative spatial representations from task-oriented perception and exploratory behavior. *Robotics and Autonomous Systems* 25(3–4): 147–157.

Becker, M., E. Kefalea, E. Mael, C. von der Malsburg, M. Pagel, J. Triesch, J. C. Vorbruggen, R. P. Wurtz, and S. Zadel. (1999). GripSee: A gesture-controlled robot for object perception and manipulation. *Autonomous Robots* 6(2): 203–221.

Beer, R. D. (1990). *Intelligence as adaptive behavior: An experiment in computational neuroethology*. New York: Academic Press.

Beer, R. D., and H. J. Chiel. (1995). Locomotion, invertebrate. In *The Handbook of Brain Theory and Neural Networks*, edited by M. A. Arbib, 553–556. Cambridge, Mass.: The MIT Press.

Beer, R. D., H. J. Chiel, and L. S. Sterling. (1989). Heterogeneous neural networks for adaptive behavior in dynamic environments. In *Advances in neural information processing systems*, vol. 1, edited by D. S. Touretzky, 577–585. San Mateo, Calif.: Morgan Kaufmann.

Beer, R. D., R. E. Ritzmann, and T. McKenna, eds. (1993). *Biological neural networks in invertebrate neuroethology and robotics*. San Diego: Academic Press.

Bekey, G. A., and W. J. Karplus. (1968). *Hybrid Computation*. New York: Wiley.

Bekey, G. A., J. J. Kim, J. K. Gronley, E. L. Borrtrager, and J. Perry. (1992). GAIT-ER-AID: An expert system for diagnosis of human gait. *Artificial Intelligence in Medicine* 4: 293–308.

Bekey, G. A., H. Liu, R. Tomović, and W. J. Karplus. (1993). Knowledge-based control of grasping in robot hands using heuristics from human motor skills. *IEEE Transactions on Robotics and Automation* 9(6): 709–722.

Bekey, G. A., and R. Tomović. (1986). Robot control by reflex actions. *IEEE International Conference on Robotics and Automation*, 240–247, San Francisco.

Bekey, G. A., R. Tomović, and I. Zeljković. (1990). Control architecture for the Belgrade-USC hand. In *Dextrous robot hands*, edited by S. T. Venkataraman and T. Iberall, 136–149. New York: Springer-Verlag.

Beni, G. (1988). The concept of cellular robotic systems. *IEEE International Symposium on Intelligent Control*, 57–62.

Beni, G., and S. Hackwood. (1992). Stationary waves in cyclic swarms. *IEEE International Symposium on Intelligent Control*, 234–242.

Beni, G., and J. Wang. (1989). Swarm intelligence. *Proceedings of the Seventh Annual Meeting of the Robotics Society of Japan*, 425–428, Tokyo.

Beni, G., and J. Wang. (1991). Theoretical problems for the realization of distributed robotic systems. *IEEE International Conference on Robotics and Automation*, 1914–1920, Sacramento, Calif.

Bergener, T., C. Bruckhoff, P. Dahm, H. Janssen, F. Joublin, and R. Menzner. (1997). Arnold: An anthropomorphic autonomous robot for human environments. In *Proceedings of Workshop SOAVE 97*, 25–34. Ilmenau, Germany: VDI Verlag.

Bergener, T., C. Bruckhoff, P. Dahm, H. Janssen, F. Joublin, R. Menzner, A. Steinhage, and W. von Seelen. (1999). Complex behavior by means of dynamical systems for an anthropomorphic robot. *Neural Networks* 12(7–8): 1087–1099.

Bernard, C. (1865). *Introduction to the study of experimental medicine*. Reprint, New York: Macmillan, 1957.

Berns, K. (1994). Neural nets for controlling the running machine LAURON. *Artificial Intelligence* 8(3): 27–32.

Betke, M., and L. Gurvits. (1997). Mobile robot localization using landmarks. *IEEE Transactions on Robotics and Automation* 13(2): 251–258.

Bicchi, A., and V. Kumar. (2001). Robotic grasping and manipulation. In *Ramsete: Articulated and mobile robots for services and technology*, edited by S. Nicosia, B. Siciliano, A. Bicchi, and P. Valigi, Lecture Notes in Control and Information series, 270: 55–74. Berlin: Springer-Verlag.

Biederman, I. (1987). Recognition-by-components: A theory of human image understanding. *Psychological Reviews* 94: 115–147.

Billock, J. (1996). Clinical evaluation and assessment: Principles in orthotics and prosthetics. *Journal of Prosthetics and Orthotics* 8(2): 41–44.

Bischoff, R., and V. Graefe. (2002). HERMES—An intelligent humanoid robot, designed and tested for dependability. In *Eighth International Symposium on Experimental Robotics (ISER02)*, edited by B. Siciliano and P. Dario. Heidelberg: Springer-Verlag.

Blickhan, R., and R. J. Full. (1993). Similarity in multilegged locomotion: Bouncing like a monopod. *Journal of Comparative Physiology* A173: 509–517.

Bluethmann, W., H. Aldridge, M. A. Diftler, and R. O. Ambrose. (2000). Control architecture for the ROBONAUT space humanoid. *First IEEE International Conference on Humanoid Robots*, Cambridge, Mass.

Bluethmann, W., R. O. Ambrose, M. A. Diftler, R. S. Askew, E. Huber, M. Goza, F. Rehnmark, C. Lovchik, and D. Magruder. (2003). Robonaut: A robot designed to work with humans in space. *Autonomous Robots* 14(2–3): 179–197.

Bonabeau, E., M. Dorigo, and G. Theraulaz. (1999). *Swarm intelligence: From natural to artificial systems.* New York: Oxford University Press.

Bonasso, R. P., R. J. Firby, E. Gat, D. Kortenkamp, D. Miller, and M. Slack. (1997). Experiences with an architecture for intelligent, reactive agents. *Journal of Experimental and Theoretical Artificial Intelligence* 9(2): 237–256.

Borenstein, J., H. R. Everett, and L. Feng. (1996). *Navigating mobile robots: Systems and techniques.* Wellesley, Mass.: Peters.

Borenstein, J., and L. Feng. (1995). Correction of systematic odometry errors in mobile robots. *International Conference on Intelligent Robotic Systems (IROS95)*, 569–574, Pittsburgh.

Borenstein, J., and L. Feng. (1996). Gyrodometry: A new method for combining data from gyros and odometry in mobile robots. *IEEE International Conference on Robotics and Automation*, 423–428, Minneapolis.

Borenstein, J., and Y. Koren. (1987). Motion control analysis of a mobile robot. *ASME Journal of Dynamic Systems, Measurement and Control* 109(2): 73–79.

Bradbury, R. (1945). *The martian chronicles.* New York: Doubleday.

Breazeal, C. L. (2002). *Designing sociable robots.* Cambridge, Mass.: MIT Press.

Breazeal, C. L., and B. Scassellati. (2001). Challenges in building robots that imitate people. In *Imitation in animals and artifacts*, edited by K. Dautenhahn and C. Nehaniv. Cambridge, Mass.: MIT Press.

Breder, C. M. (1926). The locomotion of fishes. *Zoologica* 4: 159–256.

Brooks, R. (1986). A robust layered control system for a mobile robot. *IEEE Journal of Robotics and Automation* 2(1): 14–23.

Brooks, R. (1989). A robot that walks: Emergent behaviors from a carefully evolved network. *Neural Computation* 1(2): 253–262; reprinted in R. Brooks, *Cambrian Intelligence: The Early History of the New AI* (Cambridge, Mass.: MIT Press), chap. 2.

Brooks, R. (1997). The Cog project. *Journal of the Robotic Society of Japan* 15(7).

Brooks, R. (1999). *Cambrian intelligence: The early history of the new AI.* Cambridge, Mass.: MIT Press.

Brooks, R. (2004). The robots are here. *Technology Review*, April 2, 30.

Brooks, R., C. Breazeal, M. Marjanovic, and B. Scassellati. (1999). *The Cog project: Building a humanoid robot.* MIT Lecture Notes in Computer Science, 1562. Cambridge, Mass.

Brooks, R., and M. Matarić. (1993). Real robots, real learning problems. In *Robot learning*, edited by J. H. Connell and S. Mahadevan, 193–213. Boston: Kluwer Academic.

Brooks, R., and L. Stein. (1994). Building brains for bodies. *Autonomous Robots* 1(1): 7–25.

Brutzman, D. P., Y. Kanayama, and M. J. Zyda. (1992). Integrated simulation for rapid development of autonomous underwater vehicles. *Proceedings of the IEEE Conference on Autonomous Underwater Vehicles (AUV 92)*, Washington, D.C.

Bryson, A. E., and Y.-C. Ho. (1969). *Applied optimal control.* New York: Blaisdell.

Burdick, J., J. Radford, and G. S. Chirikjian. (1993). A sidewinding locomotion gait for hyper-redundant robots. *IEEE International Conference on Robotics and Automation*, 101–106, Atlanta.

Buskey, G., G. F. Wyeth, and J. Roberts. (2001). Autonomous helicopter hover using an artificial neural network. *IEEE International Conference on Robotics and Automation (ICRA 2001)*, 1635–1640, Seoul.

Butterfass, J., M. Grebenstein, H. Liu, and G. Hirzinger. (2001). DLR-Hand II: Next generation of a dextrous robot hand. *IEEE International Conference on Robotics and Automation*, 109–114, Seoul.

Caccia, M., G. Bruzzone, and G. Veruggio. (2003). Bottom-following for remotely operated vehicles: Algorithms and experiments. *Autonomous Robots* 14(1): 17–32.

Caccia, M., G. Indiveri, and G. Veruggio. (2000). Modeling and identification of open-frame variable configuration unmanned underwater vehicles. *IEEE Journal of Oceanic Engineering* 25(2): 227–240.

Campbell, D., and M. Buehler. (2003). Stair descent in the simple hexapod RHex. *IEEE International Conference on Robotics and Automation*, 1380–1385, Taipei, Taiwan.

Cannon, W. B. (1939). *The wisdom of the body*. New York: Norton.

Cao, Y. N., A. S. Fukunaga, and A. B. Kahng. (1997). Cooperative mobile robotics: Antecedents and directions. *Autonomous Robots* 4(1): 7–28.

Capek, K. (1921). *RUR: Rossum's universal robots*. Prague.

Cassandra, A. R., and L. P. Kaelbling. (1996). Acting under uncertainty: Discrete Bayesian models for mobile robot navigation. *IEEE/RSJ International Conference on Intelligent Robots and Systems*, Osaka.

Castano, A., A. P. Behar, and P. Will. (2002). The CONRO modules for self-reconfigurable robots. *IEEE Transactions on Mechatronics* 7(4): 403–409.

Castano, A., W.-M. Shen, and P. Will. (2000). CONRO: Towards deployable robots with inter-robot metamorphic capabilities. *Autonomous Robots* 8(3): 309–324.

Castellanos, J. A., M. Devy, and J. D. Tardós. (2000). Simultaneous localization and map building for mobile robots: A landmark-based approach. *IEEE International Conference on Robotics and Automation* (Workshop on mobile robot navigation and mapping), San Francisco.

Chang, K.-S., R. Holmberg, and O. Khatib. (2000). The Augmented Object Model: Cooperative manipulation and parallel mechanism dynamics. *IEEE International Conference on Robotics and Automation*, San Francisco, 470–475.

Chiel, H. J., R. D. Beer, R. D. Quinn, and K. Espenschied. (1992). A distributed neural network architecture for hexapod robot locomotion. *IEEE Transactions on Robotics and Automation* 8(3): 293–303.

Chirikjian, G. S., and J. Burdick. (1994). A modal approach to hyper-redundant manipulator kinematics. *IEEE Transactions on Robotics and Automation* 10(3): 343–354.

Chirikjian, G. S., and J. Burdick. (1995). The kinematics of hyper-redundant robot locomotion. *IEEE Transactions on Robotics and Automation* 11(6): 781–793.

Choi, S. K., and J. Yuh. (1993). Design of advanced underwater robotic vehicle and graphic workstation. *IEEE International Conference on Robotics and Automation*, 99–105, Atlanta.

Choset, H., and J. Burdick. (2000). Sensor-based motion planning: The hierarchical generalized Voronoi graph. *International Journal of Robotics Research* 19(2): 96–125.

Chun, W., and T. Jochem. (1994). Unmanned ground vehicle Demo II. *Unmanned Systems* 12(1): 14–20.

Clark, J. E., J. G. Cham, S. A. Bailey, E. M. Froehlich, P. K. Nahata, R. J. Full, and M. Cutkosky. (2001). Biomimetic design and fabrication of a hexapedal running robot. *IEEE International Conference on Robotics and Automation*, 3643–3649, Seoul, Korea.

CNN (Cable News Network). (1999). Silicone sea bream lures Japanese anglers. Atlanta: CNN, Feb. 25, 1999.

Cohen, A. H. (1992). The role of heterarchical control in the evolution of central pattern generators. *Brain Behavior and Evolution* 40: 112–124.

Cohen, A. H., and D. L. Boothe. (1999). Sensorimotor interactions during locomotion: Principles derived from biological systems. *Autonomous Robots* 7(3): 239–245.

Collett, M., P. Collett, S. Bisch, and R. Wehner. (1998). Local and global vectors in desert ant navigation. *Nature* 394: 269–272.

Collins, J. J., and S. A. Richmond. (1994). Hard-wired central pattern generators for quadrupedal locomotion. *Biological Cybernetics* 71(5): 375–385.

Connell, J. H., and S. Mahadevan. (1993a). Introduction to robot learning. In *Robot learning*, edited by J. H. Connell and S. Mahadevan, 1–17. Boston: Kluwer Academic.

Connell, J. H., and S. Mahadevan. (1993b). Rapid task learning for real robots. In *Robot learning*, edited by J. H. Connell and S. Mahadevan, 105–140. Boston: Kluwer Academic.

Connell, J. H., and S. Mahadevan, eds. (1993c). *Robot learning*. Boston: Kluwer Academic.

Conrad, J. M., and J. W. Mills. (1997). *Stiquito: Advanced experiments with a simple and inexpensive robot*. Los Alamitos, Calif.: IEEE Press.

Conrad, J., and J. Mills. (2003). *Stiquito controlled! Making a truly autonomous robot*. Los Alamitos, Calif.: IEEE Computer Society Press.

Craig, J. J. (1989). *Introduction to robotics: Mechanics and control*, 2nd ed. Reading, Mass.: Addison-Wesley.

Cruse, H. (1976). The control of body position in the stick insect (Carausius morosus), when walking over uneven surfaces. *Biological Cybernetics* 24: 25–33.

Cruse, H. (1979). A new model describing the coordination pattern of the legs of a walking stick insect. *Biological Cybernetics* 32: 107–113.

Cruse, H. (1990). What mechanisms coordinate leg movement in walking arthropods? *Trends in Neuroscience* 13(1): 15–21.

Cruse, H., C. Bartling, M. Dreifert, J. Schmitz, D. E. Brunn, J. Dean, and T. Kindermann. (1995). Walking: A complex behavior controlled by simple networks. *Adaptive Behavior* 3(4): 385–418.

Cutkosky, M. R. (1985). *Robotic grasping and fine manipulation*. Boston: Kluwer Academic.

D'Azzo, J. J., and C. Houpis. (1995). *Linear control system analysis and design: Conventional and modern*. New York: McGraw-Hill.

Dean, J., T. Kindermann, J. Schmitz, M. Schumm, and H. Cruse. (1999). Control of walking in the stick insect: From behavior and physiology to modeling. *Autonomous Robots* 7(3): 271–288.

Dean, J., and G. Wendler. (1983). Stick insect locomotion on a walking wheel: Interleg coordination of leg position. *Journal of Experimental Biology* 103: 75–94.

Dedeoglu, G., M. Matarić, and G. Sukhatme. (1999). Incremental, on-line topological map building with a mobile robot. *Proceedings of the SPIE Conference on Mobile Robots XIV*, 129–139, Cambridge, Mass.

DeLaurier, J. D. (1999). The development and testing of a full-scale piloted ornithopter. *Canadian Aeronautics and Space Journal* 45(2): 72–82.

Delcomyn, F. (1989). Walking in the american cockroach: The timing of motor activity in the legs during straight walking. *Biological Cybernetics* 60(5): 373–384.

Desai, J. P., J. Ostrowski, and V. Kumar. (1998). Controlling formations of multiple mobile robots. *IEEE International Conference on Robotics and Automation (ICRA-98)*, 2864–2869, Leuven, Belgium.

Desai, J. P., M. Zefran, and V. Kumar. (1998). Two-arm manipulation with friction-assisted grasping. *Advanced Robotics* 13(4): 485–507.

Devjanin, E. A., V. S. Gurfinkel, E. V. Gurfinkel, V. A. Kartashev, A. V. Lensky, A. Y. Shneider, and A. G. Shtilman. (1983). The six legged walking robot capable of terrain adaptation. *Mechanism and Machine Theory* 18(4): 257–260.

Dickmanns, E. D. (1998). Vehicles capable of dynamic vision—A new breed of technical beings? *Artificial Intelligence* 103(1–2): 49–76.

Diftler, M. A., C. J. Culbert, R. O. Ambrose, R. Platt, and W. Bluethmann. (2003). Evolution of the NASA/DARPA robonaut control system. *IEEE International Conference on Robotics and Automation*, 2543–2548, Taipei, Taiwan.

Di Marco, M., A. Garulli, A. Giannitrapani, and A. Vicino. (2004). A set theoretic approach to dynamic robot localization and mapping. *Autonomous Robots* 16(1): 5–23.

Doherty, L., B. A. Warneke, B. E. Boser, and K. S. J. Pister. (2001). Energy and performance considerations for smart dust. *International Journal of Parallel and Distributed Systems and Networks* 4(3): 121–133.

Dowling, K. (1997). Limbless locomotion: Learning to crawl with a snake robot. Ph.D. dissertation, Robotics Institute, Carnegie Mellon University, Pittsburgh.

Dubowsky, S., C. Sunada, and C. Mavroidis. (1999). Coordinated motion and force control of multi-limbed robotic systems. *Autonomous Robots* 6(1): 7–20.

Duckett, T. (2003). A genetic algorithm for simultaneous localization and mapping. *IEEE International Conference on Robotics and Automation*, 434–440, Taipei, Taiwan.

Dudek, G., and M. Jenkin. (2000). *Computational principles of mobile robotics.* Cambridge: Cambridge University Press.

Elfes, A. (1987). Sonar-based real-world mapping and navigation. *IEEE Transactions on Robotics and Automation* 3(3): 249–265.

Elfes, A. (1989). Using occupancy grids for mobile robot perception and navigation. *IEEE Computer* 22(6): 46–57.

Engelberger, J. F. (1980). *Robotics in practice.* New York: American Management Association.

Erwin-Wright, S., D. Sanders, and S. Chen. (2003). Predicting terrain contours using a feedforward neural network. *Engineering Applications of Artificial Intelligence* 16: 465–472.

Everett, H. R. (1995). *Sensors for mobile robots.* Wellesley, Mass.: Peters.

Fagg, A. H., M. A. Lewis, and J. F. Montgomery. (1993). The USC autonomous flying vehicle: An experiment in real-time behavior-based control. *IEEE/RSJ International Conference on Intelligent Robots and Systems* 2: 1173–1187, Yokohama, Japan.

Fagg, A. H., D. Lotspeich, and G. A. Bekey. (1994). A reinforcement learning approach to reactive control policy design for autonomous robots. *IEEE International Conference on Robotics and Automation*, 39–44, San Diego.

Fearing, R. S. (1986). Simplified grasping and manipulation with dextrous robot hands. *IEEE Journal of Robotics and Automation* 2(4): 188–195.

Fielding, M. R., C. J. Damaren, and R. Dunlop. (2001). Hamlet: Force/position controlled hexapod walker—Design and systems. Paper presented at IEEE Conference on Control Applications, Sept. 5–7, Mexico City.

Fikes, R. E., and N. J. Nilsson. (1971). STRIPS: A new approach to the application of theorem proving to problem solving. *Artificial Intelligence* 2(3–4): 189–208.

Fiorini, P., and J. Burdick. (2003). The development of hopping capabilities for small robots. *Autonomous Robots* 14(2–3): 239–254.

Fitzpatrick, P., G. Metta, L. Natale, S. Rao, and G. Sandini. (2003). What am I doing? Initial steps towards artificial cognition. *IEEE International Conference on Robotics and Automation*, 3140–3145, Taipei, Taiwan.

Folkesson, J., and H. I. Christensen. (2003). Outdoor exploration and SLAM using a compressed filter. *IEEE International Conference on Robotics and Automation*, 419–426, Taipei, Taiwan.

Fontan, M. S., and M. J. Matarić. (1998). Territorial multi-robot task division. *IEEE Transactions on Robotics and Automation* 14(5): 815–822.

Fox, D. (2003). Adapting the sample size in particle filters through KLD-sampling. *International Journal of Robotics Research (IJRR)* 22(12): 985–1003.

Fox, D., S. Thrun, F. Dellaert, and W. Burgard. (2001). Particle filters for mobile robot localization. In *Sequential Monte Carlo Methods in Practice*, edited by A. Doucet, N. de Freitas, and N. Gordon, 401–428. New York: Springer-Verlag.

Francois, C., and C. Samson. (1998). A new approach to the control of the planar one-legged hopper. *International Journal of Robotics Research* 17(11): 1150–1166.

Frank, A. A. (1968). Automatic control synthesis for legged locomotion machines. Ph.D. dissertation, Department of Electrical Engineering, University of Southern California, Los Angeles.

Frank, A. A. (1970). An approach to the dynamic analysis and synthesis of biped locomotion machines. *Medical and Biological Engineering* 8: 465–476.

Frank, A. A., and R. B. McGhee. (1969). Some considerations relating to the design of autopilots for legged vehicles. *Journal of Terramechanics* 6(1): 23–35.

Fratantoni, D. M., D. A. Glickson, D. C. Webb, C. P. Jones, and T. K. Campbell. (2000). The Slocum autonomous glider. Eos: Transactions, *American Geophysical Union*. 80(49): 43.

Fredslund, J., and M. Matarić. (2002a). A general algorithm for robot formations using local sensing and minimal communication. *IEEE Transactions on Robotics and Automation* 18(5): 837–846.

Fredslund, J., and M. Matarić. (2002b). Huey, Dewey, Louie and GUI—Commanding robot formations. *IEEE International Conference on Robotics and Automation (ICRA)*, 175–180, Washington, DC.

Fredslund, J., and M. Matarić. (2002c). Robots in formation using local information. *Seventh International Conference on Intelligent Autonomous Systems (IAS-7)*, 100–107, Marina del Rey, Calif.

Fujita, M., and K. Kageyama. (1997). An open architecture for robot entertainment. *First International Conference on Autonomous Agents. (Agents '97)*, 435–442, Santa Monica, Calif.

Fujita, M., and H. Kitano. (1998). Development of an autonomous quadruped robot for robot entertainment. *Autonomous Robots* 5(1): 7–20.

Fuke, Y., and E. Krotkov. (1996). Dead reckoning for a lunar rover on uneven terrrain. *IEEE International Conference on Robotics and Automation*, 411–416, Minneapolis.

Fukuda, T., and Y. Kawauchi. (1990). Cellular robotic system (CEBOT) as one realization of a self-organizing intelligent universal manipulator. *IEEE International Conference on Robotics and Automation*, 662–667, Cincinnati.

Fukuda, T., Y. Kawauchi, and F. Hara. (1994). A study on a dynamically reconfigurable robotic system. *Japan Society of Mechanical Engineers International Journal* 37(1): 202–208.

Fukuda, T., and S. Nakagawa. (1987). A dynamically reconfigurable robotic system. *International Conference on Industrial Electronics, Control and Instrumentation*, 588–595.

Fukuda, T., T. Ueyama, and F. Arai. (1992). Control strategies for cellular robotic network. In *Distributed intelligence systems*, edited by A. H. Levis and H. E. Stephanou. Oxford: Pergamon.

Fukuoka, Y., H. Kimura, and A. H. Cohen. (2003). Adaptive dynamic walking of a quadruped robot on irregular terrain based on biological concepts. *International Journal of Robotics Research* 22(3–4): 187–202.

Full, R. J., K. Autumn, J. I. Chung, and A. Ahn. (1998). Rapid negotiation of rough terrain by the death-head cockroach. *American Zoologist* 38: 81A.

Full, R. J., and M. S. Tu. (1991). Mechanics of a rapid running insect: Two-, four- and six-legged locomotion. *Journal of Experimental Biology* 156(1): 215–231.

Gaßmann, B., K.-U. Scholl, and K. Berns. (2001). Behavior control of LAURON III for walking in unstructured terrain. *International Conference on Climbing and Walking Robots (CLAWAR 2001)*, 651–658, Karlsruhe, Germany.

Gat, E. (1992). Integrating planning and reacting in a heterogeneous asynchronous architecture for controlling real world mobile robots. *National Conference on Artificial Intelligence*, 809–815, San Jose, Calif.

Gat, E. (1995). Towards principled experimental study of autonomous mobile robots. *Autonomous Robots* 2(3): 179–190.

Gelb, A., ed. (1974). *Applied optimal estimation*. Cambridge, Mass.: MIT Press.

Gerkey, B. P., and M. J. Matarić. (2000). Principled communication for dynamic multi-robot task allocation. *Proceedings of the International Symposium on Experimental Robotics (ISER)*, Honolulu, 341–352.

Gerkey, B. P., and M. Matarić. (2001). Principled communication for dynamic multi-robot task allocation. In *Experimental robotics*, vol. 7, edited by D. Rus and S. Singh, 353–362. Berlin: Springer-Verlag.

Godjevac, J. (1997). *Neuro-fuzzy controllers: Design and application*. Lausanne, Switzerland: Presses Polytechniques et Universitaires Romandes.

Goel, P., G. Dedeoglu, S. Roumeliotis, and G. Sukhatme. (2000). Fault detection and identification in a mobile robot using multiple model estimation and neural network. *IEEE International Conference on Robotics and Automation*, 2302–2309, San Francisco.

Goel, P., S. I. Roumeliotis, and G. Sukhatme. (1999). Robust localization using relative and absolute position estimates. *Intelligent Robotic Systems Conference (IROS 99)*, 1134–1140, Kyongju, Korea.

Goldberg, D. E. (2002). *The design of innovation: Lessons from and for competent genetic algorithms.* Genetic Algorithms and Evolutionary Computation 7. Boston: Kluwer Academic.

Golden, J., and Y. Zheng. (1990). Gait synthesis for the SD-2 biped robot to climb stairs. *International Journal of Robotics and Automation* 5(4): 149–159.

Gomi, T., ed. (2000). *Evolutionary robotics: From intelligent robots to artificial life.* Ottawa: AAI Books.

Gonzalez de Santos, P., J. A. Galvez, J. Estremera, and E. Garcia. (2003). SILO4: A true walking robot for the comparative study of walking machine techniques. *IEEE Robotics and Automation Magazine* 10: 23–32.

Gopalkrishnan, R., M. S. Triantafyllou, G. S. Triantafyllou, and D. Barrett. (1994). Active vorticity control in a shear flow using a flapping foil. *Journal of Fluid Mechanics* 274: 1–21.

Grasmeyer, J. M., and M. T. Keennon. (2001). Development of the Black Widow micro air vehicle. AIAA paper no. 2001-0217, presented at American Institute of Aeronautics and Astronautics (AIAA) Annual Conference, Reston, Va.

Graver, J. G., R. Bachmayer, N. E. Leonard, and D. M. Fratantoni. (2003). Underwater glider model parameter identification. Paper presented at thirteenth International Symposium on Unmanned, Untethered Submersible Technology (UUST), Aug. 24, 2003, Durham, N.H.

Grefenstette, J. J. (1987). *A user's guide to GENESIS.* Naval Center for Applied Research. (This version is no longer available. The current version is GENESIS v. 5 (1990), available at http://www.aic.nrl.navy.mil.)

Grewal, M. S., and A. P. Andrews. (2000). *Kalman filtering: Theory and practice.* New York: Wiley.

Grewal, M. S., L. R. Weill, and A. P. Andrews. (2001). *Global positioning systems, inertial navigation, and integration.* New York: Wiley.

Grillner, S. (1981). Control of locomotion in bipeds, tetrapeds, and fish. In *Handbook of physiology*, vol. 1, edited by V. B. Brooks, 1179–1236, Bethesda, Md.: American Physiological Society.

Grillner, S., and R. Dubuc. (1988). Control of locomotion in vertebrates: Spinal and supraspinal mechanisms. *Advanced Neurology* 47(1–2): 425–453.

Grodins, F. (1963). *Control theory and biological systems.* New York: Columbia University Press.

Hackman, J. R. (1990). *Groups that work (and those that don't).* San Francisco: Jossey-Bass.

Hackwood, S., and J. Wang. (1988). The engineering of cellular robotic systems. *Proceedings of the Third IEEE International Symposium on Intelligent Control*, 70–75, Arlington, Va.

Hasegawa, Y., Y. Ito, and T. Fukuda. (2000). Behavior coordination and its modification on brachiation-type mobile robot. *IEEE International Conference on Robotics and Automation*, 3983–3988, San Francisco.

Hashimoto, S., S. Narita, H. Kasahara, K. Shirai, and T. Kobagashi. (2000). Humanoid robots in Waseda University—Hadaly-2 and WABIAN. *First IEEE International Conference on Humanoid Robots*, Cambridge, Mass.

Hayati, S. (1986). Hybrid position/force control of multi-arm cooperating robots. *IEEE International Conference on Robotics and Automation*, 82–89, San Francisco.

Healey, A. J., and D. Lienard. (1993). Multivariable sliding mode control for autonomous diving and steering of unmanned underwater vehicles. *IEEE Journal of Oceanic Engineering* 18(3): 327–339.

Hebert, M. H., C. Thorpe, and A. Stentz, eds. (1997). *Intelligent unmanned ground vehicles: Autonomous navigation research at Carnegie Mellon.* Boston: Kluwer Academic.

Heinsohn, R., and C. Packer. (1995). Complex cooperative strategies in group-territorial African lions. *Science* 269: 1260–1262.

Hemami, H., and R. L. Farnsworth. (1977). Postural and gait stability of a planar five-link biped by simulation. *IEEE Transactions on Automatic Control* 22(3): 452–458.

Hemami, H., and B. F. Wyman. (1979). Modeling and control of constrained dynamic systems with application to biped locomotion in the frontal plane. *IEEE Transactions on Automatic Control* 24(4): 526–535.

Hildebrand, M. (1965). Symmetrical gaits of horses. *Science* 150: 701–708.

Hill, A. V. (1938). The heat of shortening and the dynamic constants of muscle. *Proceedings of the Royal Society, section B* 126: 136–195.

Hirose, S. (1984). A study of design and control of a quadruped walking vehicle. *International Journal of Robotics Research* 3(2): 113–133.

Hirose, S. (1993). *Biologically inspired robots: Snake-like locomotion and manipulation.* Oxford: Oxford University Press.

Hirose, S., and O. Kunieda. (1991). Generalized standard foot trajectory for a quadruped walking vehicle. *International Journal of Robotics Research* 10(1): 3–12.

Hirose, S., A. Nagakubo, and R. Toyama. (1991). Machine that can walk and climb on floors, walls and ceilings. *Fifth International Conference on Advanced Robotics (ICAR)*, 753–758, Pisa, Italy.

Hirose, S., and Y. Umetani. (1976). Kinematic control of active cord mechanism with tactile sensors. *Second International CISM-IFT Symposium on Theory and Practice of Robots and Manipulators* 241–252.

Hirose, S., and Y. Umetani. (1978). Some considerations on a feasible walking mechanism as a terrain vehicle. *Third RoManSy Symposium*, 357–378, Udine, Italy.

Hirose, S., and Y. Umetani. (1981). An active cord mechanism with oblique swivel joints and its control. *Fourth RoManSy Symposium*, 327–340, Warsaw, Poland.

Hirose, S., K. Yoneda, K. Arai, and T. Ibe. (1991). Design of prismatic quadruped walking vehicle TITAN VI. *Fifth International Conference on Advanced Robotics (ICAR)*, 723–728, Pisa, Italy.

Hirose, S., K. Yoneda, and H. Tsukagoshi. (1997). TITAN VII: Quadruped walking and manipulating robot on a steep slope. *IEEE International Conference on Robotics and Automation*, 494–500, Albuquerque, N.M.

Hodgins, J., and M. H. Raibert. (1991). Adjusting step length for rough terrain locomotion. *IEEE Transactions on Robotics and Automation* 7(3): 289–298.

Holland, J. H. (1975). *Adaptation in natural and artificial systems.* Ann Arbor: University of Michigan Press.

Holldobler, B., and E. Wilson. (1990). *The ants.* Cambridge, Mass.: Belknap Press of Harvard University Press.

Hopgood, A. A. (2000). *Intelligent systems for engineers and scientists.* Boca Raton, Fla.: CRC Press.

Houk, J. C., P. E. Crago, and W. Z. Rymer. (1980). Functional properties of the Golgi tendon organs. In *Spinal and supraspinal mechanisms of voluntary motor control and locomotion*, edited by J. E. Desmedt, 33–43. Basel, Switzerland: Karger.

Houk, J. C., W. Z. Rymer, and P. E. Crago. (1991). Responses of muscle spindle receptors to transitions in stretch velocity. In *Proceedings of the Symposium: Muscle Afferents and Spinal Control of Movement*, Paris 53–61. New York: Pergamon.

Howard, A., M. J. Matarić, and G. S. Sukhatme. (2002a). An incremental self-deployment algorithm for mobile sensor networks. *Autonomous Robots* 13(2): 113–126.

Howard, A., M. J. Matarić, and G. S. Sukhatme. (2002b). Localization for mobile robot teams: A distributed MLE approach. In *Experimental robotics*, vol. 8, edited by B. Siciliano and P. Dario, 146–155. Berlin: Springer-Verlag.

Hu, J., and G. Pratt. (2000). Nonlinear switching control of bipedal walking robots with provable stability. *First IEEE International Conference on Humanoid Robots*, Cambridge, Mass.

Hu, J., J. Pratt, and G. Pratt. (1998). Adaptive dynamic control of a bipedal walking robot with radial basis function neural networks. *International Conference on Intelligent Robots and Systems (IROS)*, 400–405. Victoria, British Columbia, Canada.

Hughes, H. C. (1999). *Sensory exotica: A world beyond human experience.* Cambridge, Mass.: MIT Press.

Hyon, S. H., S. Kamijo, and T. Mita. (2002). "Kenken"—A biologically inspired one-legged running robot. *Journal of the Robotic Society of Japan* 20(4): 103–112.

Iagnemma, K., A. Rzepniewski, S. Dubowsky, and P. Schenker. (2003). Control of robotic vehicles with actively articulated suspension in rough terrain. *Autonomous Robots* 14(1): 5–16.

Iberall, T. (1987). The nature of human prehension: Three dextrous hands in one. *IEEE International Conference on Robotics and Automation*, 396–401, Raleigh, N.C.

Iberall, T., and C. L. MacKenzie. (1990). Opposition space and human prehension. In *Dextrous robot hands*, edited by S. T. Venkataraman and T. Iberall, 32–54. New York: Springer-Verlag.

Iida, F., and R. Pfeifer. (2004). "Cheap" rapid locomotion of a quadruped robot: Self-stabilization of bounding gait. *Intelligent Autonomous Systems* 8: 642–649.

Ikeda, H., and N. Takanashi. (1987). Joint assembly moveable like a human arm. *US Patent 4,683,406*, NEC Corporation, issued July 1987.

Ilg, W., and K. Berns. (1995). A learning architecture based on reinforcement learning for adaptive control of the walking machine LAURON. *Robotics and Autonomous Systems* 15: 321–334.

Ilg, W., K. Berns, T. Muhlfriedel, and R. Dillmann. (1997). Hybrid learning concepts based on self-organizing neural networks for adaptive control of walking machines. *Robotics and Autonomous Systems* 22(3): 317–327.

Ingvast, J., C. Ridderström, F. Hardarson, and J. Wikander. (1993). Warp1: Towards walking in rough terrain—Control of walking. *International Conference on Climbing and Walking Robots (CLAWAR)*, Catania, Italy.

Inoue, H., K. Tachi, K. Tanie, S. Yokoi, and H. Hirai. (2000). HRP: Humanoid Robotics Project of MIT. *First IEEE-RAS International Conference on Humanoid Robots*, Massachusetts Institute of Technology, Cambridge, Mass.

Ishii, K., T. Fujii, and T. Ura. (1998). Neural network system for online controller adaptation and its application to underwater robot. *IEEE International Conference on Robotics and Automation*, 756–761, Leuven, Belgium.

Isidori, A. (1997). *Nonlinear control systems*. Berlin: Springer-Verlag.

Isidori, A. (1999). *Nonlinear control systems*, vol. 2. Berlin: Springer-Verlag.

Jacobsen, S. C., E. K. Iversen, D. F. Knutti, R. T. Johnson, and K. B. Biggers. (1986). Design of the Utah/MIT Dextrous Hand. *IEEE International Conference on Robotics and Automation*, 1520–1532, San Francisco.

Jalics, L., H. Hemami, and B. Clymer. (1997). A control strategy for terrain adaptive bipedal locomotion. *Autonomous Robots* 4(3): 243–257.

James, H. M., N. B. Nichols, and R. S. Phillips. (1947). *Theory of servomechanisms*. New York: McGraw-Hill.

Jeannerod, M. (1981). Intersegmental coordination during reaching at natural visual objects. In *Attention and performance*, vol. 9, edited by J. Long and A. Baddeley, 153–169. Hillsdale, N.J.: Erlbaum.

Jeannerod, M. (1989). *The neural and behavioural organization of goal-directed movements*. New York: Oxford University Press.

Jones, H. L., E. W. Frew, B. Woodley, and S. M. Rock. (1998). Human-robot interaction for field operation of an autonomous helicopter. In *SPIE, Mobile Robots XII and Intelligent Transportation Systems*, 244–252, Cambridge, Mass.

Jordan, M. I., and D. E. Rummelhart. (1992). Forward models: Supervised learning with a distal teacher. *Cognitive Science* 16: 307–354.

Joy, W. (2000). Why the future doesn't need us. *Wired* 9: 238–262.

Jung, B., and G. S. Sukhatme. (2001). Cooperative tracking using multiple robots and environment-embedded, networked sensors. *Proceedings of the IEEE International Symposium on Computational Intelligence in Robotics and Automation (CIRA)*, 206–211, Banff, Alberta.

Kahn, J. M., R. H. Katz, and K. S. J. Pister. (1999). Mobile networking for "smart dust." Paper presented at ACM/IEEE International Conference on Mobile Computing and Networking (MobiCom 99), Aug. 17, 1999, Seattle.

Kalman, R. E. (1960). A new approach to linear filtering and prediction problems. *ASME Transactions (Journal of Basic Engineering)* 82: 35–45.

Kalman, R. E., and R. Bucy. (1961). New results in linear filtering and prediction. *ASME Transactions (Journal of Basic Engineering)* 83: 95–108.

Kapandji, L. A. (1982). *The Physiology of the Joints, Volume 1: Upper Limb*. 5th ed. Edinburgh: Churchill Livingstone.

Kato, K., and S. Hirose. (2001). Development of quadruped walking robot, TITAN-IX—Mechanical design concept and application for the humanitarian demining robot. *Advanced Robotics* 15(2): 191–204.

Kato, K., A. Takanishi, H. Jishikawa, and I. Kato. (1983). The realization of quasi-dynamic walking by the biped walking machine. In *Fourth Symposium on Theory and Practice of Robots and Manipulators*, edited by A. Morecki, G. Bianchi, and K. Kedzior, 341–351. Warsaw: Polish Scientific.

Kato, N. (1995). Application of fuzzy algorithm to guidance and control of underwater vehicles. In *Underwater robotic vehicles: Design and control*, edited by J. Yuh. Albuquerque, N.M.: TSI Press.

Kawamura, K., R. A. Peters II, R. E. Bodenheimer, et al. (2004). A parallel distributed cognitive system for a humanoid robot. *International Journal of Humanoid Robotics* 1: 65–93.

Kawamura, K., R. A. Peters II, S. Bagchi, M. Iskarous, and M. Bishay. (1995). Intelligent robotic systems in service of the disabled. *IEEE Transactions on Rehabilitation Engineering* 1(3): 14–21.

Kawato, M. (1999). Internal models for motor control and trajectory planning. *Current Opinion in Neurobiology* 9: 718–727.

Kawauchi, Y., M. Inaba, and T. Fukuda. (1992). Self-organizing intelligence for cellular robotic system CEBOT with genetic knowledge production algorithm. *IEEE International Conference on Robotics and Automation*, 813–818, Nice, France.

Kawauchi, Y., M. Inaba, and T. Fukuda. (1993). A principle of distributed decision making of cellular robotic system (CEBOT). *IEEE International Conference on Robotics and Automation*, 833–838, Atlanta.

Kelly, A. (2000). Mobile robot localization from large scale appearance mosaics. *International Journal of Robotics Research* 19(11): 1–20.

Kelly, A. (2004). Linearized error propagation in odometry. *International Journal of Robotics Research* 23(2): 179–218.

Khalil, H. K. (2001). *Nonlinear systems*. Englewood Cliffs, N.J.: Prentice Hall.

Khoo, M. C. K. (2000). *Physiological control systems*. Piscataway, N.J.: IEEE Press.

Khoshnevis, B., and G. A. Bekey. (1998). Centralized sensing and control of multiple mobile robots. *Computers in Industrial Engineering* 35(3–4): 503–506.

Kim, H. J., D. H. Shim, and S. Sastry. (2002). Flying robots: Modeling, control and decision making. *IEEE International Conference on Robotics and Automation*, 66–71, Washington, DC.

Kim, J.-H., and S. Sukkarieh. (2003). Airborne simultaneous localization and map building. *IEEE International Conference on Robotics and Automation*, 406–411, Taipei, Taiwan.

Kimura, H., S. Akiyama, and K. Sakurama. (1999). Realization of dynamic walking and running of the quadruped using neural oscillator. *Autonomous Robots* 7(3): 247–258.

Kirsner, S. (2003). The robot air corps. *Wired* 11: 46–47.

Kitano, H. (2002). The design of the humanoid robot Pino. Available at http://www.symbio.jst.go.jp/%7Etmatsui/pinodesign.htm.

Klafter, R. D., T. A. Chmielewski, and M. Negin. (1989). *Robotic engineering: An integrated approach*. Englewood Cliffs, N.J.: Prentice-Hall.

Klute, G. K., and B. Hannaford. (2000). Accounting for elastic energy storage in McKibben artificial muscle actuators. *ASME Journal of Dynamic Systems, Measurements and Control* 122(2): 386–388.

Kobayashi, T., and S. Haruyama. (1997). Partly-hidden Markov models and their application to gesture recognition. *International Conference on Acoustics, Speech and Signal Processing (ICASSP97)* 3081–3084.

Koditschek, D. E., and M. Buehler. (1991). Analysis of a simplified hopping robot. *International Journal of Robotics Research* 10(6): 269–281.

Koenig, P., and G. A. Bekey. (1993). Generation and control of gait patterns in a simulated horse. *IEEE International Conference on Robotics and Automation*, 359–366, Atlanta.

Koenig, S., and R. Simmons. (1998). Xavier: A robot navigation architecture based on partially-observable Markov decision process models. In *Artificial intelligence and mobile robots*, edited by D. Kortenkamp, R. P. Bonasso, and R. R. Murphy, 91–122. Menlo Park, Calif., AAAI Press.

Koivo, A. J., and G. A. Bekey. (1987). *Report of NSF Workshop on Coordinated Multiple Robot Manipulators: Planning, Control and Applications*. Lafayette/Los Angeles: Purdue University/University of Southern California.

Konolige, K., and K. Myers. (1996). The Saphira architecture for autonomous mobile robots. In *AI-based mobile robots: Case studies of successful robot systems*, edited by D. Kortenkamp, R. P. Bonasso, and R. R. Murphy, 211–242. Cambridge, Mass.: MIT Press.

Kortenkamp, D., R. P. Bonasso, and R. R. Murphy, eds. (1998). *Artificial intelligence and mobile robots*. Menlo Park, Calif.: AAAI Press.

Kosko, B. (1994). *Fuzzy thinking: The new science of fuzzy logic*. New York: Hyperion Books.

Kotay, K., and D. Rus. (1996). Navigating 3D steel web structures with an inchworm robot. *International Conference on Intelligent Robots and Systems (IROS)*, Munich, Germany.

Kotay, K., and D. Rus. (2000). The inchworm robot: A multi-functional system. *Autonomous Robots* 8(8): 53–69.

Krotkov, E., R. Simmons, and W. L. Whittaker. (1995). Ambler: Performance of a six-legged planetary rover. *Acta Astronautica* 35(1): 75–81.

Krylow, A. M., T. G. Sandercock, and W. Z. Rymer. (1995). Muscle models. In *The handbook of brain theory and neural networks*, edited by M. A. Arbib, 609–613. Cambridge, Mass.: MIT Press.

Kube, C. R., and N. H. Zhang. (1994). Collective robotics: From social insects to robots. *Adaptive Behavior* 2(2): 189–218.

Kuipers, B., and Y.-T. Byun. (1991). A robot exploration and mapping strategy based on a semantic hierarchy of spatial representations. *Robotics and Autonomous Systems* 8: 47–63.

Kuniyoshi, Y., M. Inaba, and H. Inoue. (1994). Learning by watching: Extracting reusable task knowledge from visual observation of human performance. *IEEE Transactions on Robotics and Automation* 10(6): 799–822.

Kuniyoshi, Y., Y. Yorozu, M. Inaba, and H. Inoue. (2003). From visuo-motor self-learning to early imitation—A neural architecture for humanoid learning. *IEEE International Conference on Robotics and Automation (ICRA 2003)*, 3132–3139, Taipei, Taiwan.

Kuo, B. C. (1995). *Automatic control*. New York: Wiley.

Kurazume, R., and S. Hirose. (2000). An experimental study of a cooperative positioning system. *Autonomous Robots* 8(1): 43–52.

Kurazume, R., S. Hirose, S. Nagata, and N. Sashida. (1996). Study on cooperative positioning systems (Basic principle and measurement experiment). *IEEE International Conference on Robotics and Automation*, 1421–1426, Minneapolis.

Kurazume, R., S. Nagata, and S. Hirose. (1994). Cooperative positioning with multiple robots. In *IEEE International Conference on Robotics and Automation*, 1250–1257.

Kuroki, Y., T. Ishida, J. Yamaguchi, M. Fujita, and T. Doi. (2001). A small biped entertainment robot. *Proceedings, International Conference on Humanoid Robots*, 181–186, Tokyo, Japan.

Kurzweil, R. (1999). *The age of spiritual machines*. New York: Viking Penguin.

Kurzweil, R. (2001). Accelerated living. *PC Magazine* 20: 151–152.

Kurzweil, R. (2005). *The singularity is near: When humans transcend biology*. New York: Viking/Penguin Books.

Kwok, C. T., D. Fox, and M. Meilca. (2004). Real-time particle filters. *Proceedings of the IEEE* 92(2): 469–484.

Kyberd, P. J., C. Light, P. H. Chappell, J. M. Nightingale, D. Whatley, and M. Evans. (2001). The design of anthropomorphic prosthetic hands: A study of the Southampton Hand. *Robotica* 19: 593–600.

Latombe, J. C. (1991). *Robot motion planning*. Boston: Kluwer Academic.

Lawton, J., R. W. Beard, and F. Y. Hadaegh. (1999). An adaptive control approach to satellite formation flying with relative distance constraints. *Proceedings of the American Control Conference*, 1545–1549, San Diego.

LeDoux, J. E., and J.-M. Fellous. (1995). Emotion and computational neuroscience. In *The handbook of brain theory and neural networks*, edited by M. A. Arbib, 356–359. Cambridge, Mass.: MIT Press.

Leonard, J. J., and H. F. Durrant-Whyte. (1991). Mobile robot localization by tracking geometric beacons. *IEEE Transactions on Robotics and Automation* 7(3): 376–382.

Lettvin, J. Y., H. R. Maturana, W. S. McCulloch, and W. H. Pitts. (1959). What the frog's eye tells the frog's brain. *Proceedings of the IRE (Institute of Radio Engineers)* 47(11): 1940–1951.

Lewis, M. A., and G. A. Bekey. (1992). The behavioral self-organization of nanorobots using local rules. *IEEE/RSJ International Conference on Intelligent Robots and Systems (IROS)*, 1333–1338, Raleigh, N.C.

Lewis, M. A., and G. A. Bekey. (2002). Gait adaptation in a quadruped robot. *Autonomous Robots* 12(3): 301–312.

Lewis, M. A., A. H. Fagg, and G. A. Bekey. (1994). Genetic algorithms for gait synthesis in a hexapod robot. In *Recent trends in mobile robots*, edited by Y. F. Zheng, 317–331. Singapore: World Publishing.

Lewis, M. A., A. H. Fagg, and A. Solidum. (1992). Genetic programming approach to the construction of a neural network for control of a walking robot. *IEEE International Conference on Robotics and Automation*, 2618–2623, Nice, France.

Lewis, M. A., and K.-H. Tan. (1997). High precision formation control of mobile robots using virtual structures. *Autonomous Robots* 4(4): 387–403.

Li, Z., and S. Sastry. (1988). Task oriented optimal grasping by multifingered robot hands. *IEEE Journal of Robotics and Automation* 4: 32–44.

Lindström, M., H. Orebäck, and H. Christensen. (2000). BERRA: A research architecture for service robots. *IEEE International Conference on Robotics and Automation* 4: 3278–3283, San Francisco.

Liston, R. A., and R. S. Mosher. (1968). A versatile walking truck. *Proceedings of the Transportation Engineering Conference*, Institution of Civil Engineers, London.

Liu, H., T. Iberall, and G. A. Bekey. (1988). Reasoning about grasping from task descriptions. *Proceedings of the SPIE Conference on Intelligent Robots* 1002: 642–651, Cambridge, Mass.

Liu, H., T. Iberall, and G. A. Bekey. (1989). Neural network architecture for robot hand control. *IEEE Control Systems Magazine* 9(3): 38–43.

Lozano-Perez, T., and M. A. Wesley. (1979). An algorithm for planning collision-free paths among polyhedral obstacles. *Communications of the ACM* 22(10): 560–570.

MacKenzie, C. L., and T. Iberall. (1994). *The grasping hand.* Amsterdam: North-Holland.

Mackenzie, D. (2003). Shape shifters tread a daunting path toward reality. *Science* 301: 754–756.

Mackenzie, D. C., R. C. Arkin, and J. M. Cameron. (1997). Specification and execution of multiagent missions. *Autonomous Robots* 4(1): 29–52.

Maes, P., and R. Brooks. (1990). Learning to coordinate behaviors. *Proceedings of the Eighth National Conference on Artificial Intelligence (AAAI-90)*, 796–802. San Matco, CA: Morgan Kaufmann.

Maharaj, D. Y. (1994). The application of nonlinear control theory to robust helicopter control. Ph.D. dissertation, Aeronautics Department, Imperial College, London.

Majchrzak, A., and L. Gasser. (1992). Toward a conceptual framework for specifying manufacturing workgroups congruent with technological change. *International Journal of Computer Integrated Manufacturing* 5: 118–131.

Mange, D., and M. Tomassini, eds. (1998). *Bio-inspired computing machines—Towards novel computational architectures.* Lausanne, Switzerland: Presses Polytechniques et Universitaires Romandes.

Martinoli, A., and F. Mondada. (1995). Collective and cooperative group behaviors: Biologically inspired experiments in robotics. *Proceedings of the Fourth International Symposium on Experimental Robotics*, 3–10. Berlin: Springer-Verlag.

Mason, M., and J. K. Salisbury. (1985). *Robot hands and the mechanics of manipulation.* Cambridge, Mass.: MIT Press.

Matarić, M. (1992a). Behavior-based control: Main properties and implications. *Workshop on Intelligent Control Systems; IEEE International Conference on Robotics and Automation*, 46–54, Nice, France.

Matarić, M. (1992b). Integration of representation in goal-driven behavior-based robots. *IEEE Transactions on Robotics and Automation* 8(3): 304–312.

Matarić, M. (1992c). Minimizing complexity in controlling a mobile robot population. *IEEE International Conference on Robotics and Automation*, 830–835, Nice, France.

Matarić, M. (1993). Designing emergent behaviors: From local interactions to collective intelligence. *Second International Conference on Simulation of Adaptive Behavior (From Animals to Animats 2)*, 432–441.

Matarić, M. (1994). Interaction and intelligent behavior. Ph.D. dissertation, Department of Electrical Engineering and Computer Science, Massachusetts Institute of Technology, Cambridge, Mass.

Matarić, M. (2001). Visuo-motor primitives as a basis for learning by imitation. In *Imitation in animals and artifacts*, edited by K. Dautenhahn and C. Nehaniv, 391–422. Cambridge, Mass.: MIT Press.

Matarić, M. (2002). Situated robotics. In *Encyclopedia of cognitive science*, edited by L. Nadel, London: Nature Publishing Group.

Matarić, M., G. Sukhatme, and E. Østergaard. (2003). Multi-robot task allocation in uncertain environments. *Autonomous Robots* 14(2–3): 253–261.

Matsuoka, Y. (1997). The mechanisms in a humanoid robot hand. *Autonomous Robots* 4(2): 199–209.

McCulloch, W. S., and W. Pitts. (1943). A logical calculus of ideas immanent in nervous activity. *Bulletin of Mathematical Biophysics* 5: 115–133.

McGeer, T. (1990). Passive dynamic walking. *International Journal of Robotics Research* 9(2): 62–82.

McGhee, R. B. (1967a). Finite state control of quadruped locomotion. *Simulation* 9(September): 135–140.

McGhee, R. B. (1967b). Some finite state aspects of legged locomotion. *Mathematical Biosciences* 2: 67–84.

McGhee, R. B., and A. A. Frank. (1968). On the stability properties of quadruped creeping gaits. *Mathematical Biosciences* 3: 331–351.

McGhee, R. B., and G. I. Iswandhi. (1979). Adaptive locomotion of a multilegged robot over rough terrain. *IEEE Transactions on Systems, Man and Cybernetics* 9(4): 176–182.

McGhee, R. B., D. E. Orin, D. R. Pugh, and M. R. Patterson. (1984). A hierarchically-structured system for computer control of a hexapod walking machine. *Fifth IFToMM Symposium on Robots and Manipulator Systems (RoManSy 1984)*, Udine, Italy; reprinted in *Theory and Practice of Robot Manipulators*, edited by A. Morecki (London: Hermes Publishing, 1985), 368–381.

McHenry, M. C., and G. A. Bekey. (1995). Foot placement for legged robots. *Proceedings, First ECPD International Conference on Advanced Robotics and Intelligent Automation*, 302–306, Athens, Greece.

McMahon, T. (1984). *Muscles, reflexes and locomotion*. Princeton, N.J.: Princeton University Press.

Meltzoff, A. N., and M. K. Moore. (1977). Imitation of facial and manual gestures by human neonates. *Science* 198: 75–78.

Mendel, J. M., and K. S. Fu, eds. (1970). *Adaptive, learning and pattern recognition systems: Theory and applications*. New York: Academic Press.

Meystel, A. (1986). Planning in a hierarchical nested controller for autonomous robots. *IEEE 25th Conference on Decision and Control*, Athens, Greece, 1237–1249.

Miki, N., and I. Shimoyama. (1999). Study on micro-flying robots. *Advanced Robotics* 13(3): 245–246.

Milhorn, H. T. (1966). *The application of control theory to physiological systems*. Philadelphia: Saunders.

Miller, G. (2000). Research on biomimetic snakes. In *Neurotechology for biomimetic robots*, edited by J. Ayers, J. L. Davis, and A. Rudolph. Cambridge, Mass.: MIT Press.

Miller, G. (2002). Snake robots for search and rescue. In *Neurotechnology for biomimetic robots*, edited by J. Ayers, J. L. Davis, and A. Rudolph, 271–284. Cambridge, Mass.: MIT Press.

Miller, J. R., O. Amidi, C. Thorpe, and T. Kanade. (1999). Precision 3-D modeling for autonomous helicopter flight. *Proceedings of the International Symposium on Robotics Research (ISRR)*.

Miller, W. T., R. P. Hewes, F. H. Glanz, and L. G. Kraft. (1990). Real-time dynamic control of an industrial manipulator using a neural network–based learning controller. *IEEE Transactions on Robotics and Automation* 6(1): 1–9.

Milsum, J. H. (1966). *Biological control systems analysis*. New York: McGraw-Hill.

Minsky, M. (1961). Steps toward artificial intelligence. *Proceedings of the IRE (Institute of Radio Engineers)* 49(1): 8–30; reprinted in *Computers and thought*, edited by E. Feigenbaum and J. Feldman (New York: McGraw-Hill, 1963).

Minsky, M., and S. Papert. (1969). *Perceptrons: An introduction to computational geometry*. Cambridge, Mass.: MIT Press.

Miura, H., and I. Shimoyama. (1984). Dynamic walk of a biped. *International Journal of Robotics Research* 3(2): 60–74.

Mondada, F., G. C. Pettinaro, I. W. Kwee, A. Guignard, L. M. Gambardella, D. Floreano, S. Nolfi, J. L. Deneubourg, and M. Dorigo. (2002). SWARM-BOT: A swarm of autonomous mobile robots with self-assembling capabilities. In *Proceedings of the International Workshop on Self-Organization and Evolution of Social Behavior*, Ascona, Switzerland, edited by C. K. Hemelrijk and E. Bonabeau, 307–312. Zurich: University of Zurich.

Montgomery, J. F. (1999). Learning helicopter control through "teaching by showing." Ph.D. dissertation, Department of Computer Science, University of Southern California, Los Angeles.

Montgomery, J. F., and G. A. Bekey. (1998). Learning of fuzzy-neural behavior-based control through "teaching by showing." *Proceedings of the IEEE Conference on Computational Intelligence in Robotics and Automation (CIRA)*, 1–6.

Montgomery, J. F., A. H. Fagg, and G. A. Bekey. (1995). The USC AFV-1—A behavior-based entry in the 1994 International Aerial Robotics Competition. *IEEE Expert* 10(2): 16–22.

Moore, E. Z., D. Campbell, F. Grimminger, and M. Buehler. (2002). Reliable stair climbing in the simple hexapod RHex. *IEEE International Conference on Robotics and Automation*, 2222–2227, Washington, D.C.

Moravec, H. (1988). Sensor fusion in certainty grids for mobile robots. *AI Magazine* 9(2): 61–74.

Moravec, H. (1999). *Robot: Mere machine to transcendent mind*. New York: Oxford University Press.

Moravec, H., and A. Elfes. (1985). High-resolution maps from wide-angle sonar. *IEEE International Conference on Robotics and Automation*, 116–121.

Morgansen, K. A., V. Duindam, R. J. Mason, J. W. Burdick, and R. M. Murray. (2001). Nonlinear control methods for planar carangiform robot fish locomotion. *IEEE International Conference on Robotics and Automation*, 427–434, Seoul, Korea.

Morgansen, K. A., P. A. Vela, and J. Burdick. (2002). Trajectory stabilization for a planar carangiform robot fish. *IEEE International Conference on Robotics and Automation*, 756–762, Washington, D.C.

Morita, T., H. Iwata, and S. Sugano. (1999). Development of human symbiotic robot: WENDY. *IEEE International Conference on Robotics and Automation*, 3183–3188, Detroit.

Morita, T., H. Iwata, and S. Sugano. (2000). Human symbiotic robot design based on division and unification of functional requirements. *IEEE International Conference on Robotics and Automation*, 2229–2234, San Francisco.

Mouritsen, H., and B. J. Frost. (2002). Virtual migration in tethered flying monarch butterflies reveals their orientation mechanism. *Proceedings of the National Academy of Sciences USA* 99(15): 10162–10166.

Murphy, R. R. (2000). *Introduction to AI robotics*. Cambridge, Mass.: MIT Press.

Murphy, R. R. (2002). Marsupial robots. In *Robot teams: From diversity to polymorphism*, edited by T. Balch and L. E. Parker. Wellesley, Mass.: A. K. Peters.

Murray, I. M. (1969). *Human anatomy made simple*. Garden City, N.Y.: Doubleday.

Mussa-Ivaldi, F., A. Giszter, and E. Bizzi. (1994). Linear combination of primitives in vertebrate motor control. *Proceedings of the National Academy of Sciences USA* 91(16): 7354–7358.

Muybridge, E. (1899). *Animals in motion*. London: Chapman & Hall. Reprint, New York: Dover, 1957.

Näder, M., and H. G. Näder, eds. (2002). *Otto Bock prosthetic compedium (Lower Leg Prostheses)*. Berlin: Schiele & Schon.

Naffin, D., and G. Sukhatme. (2002). A test bed for autonomous formation flying. University of Southern California Institute for Robotics and Intelligent Systems Technical Report IRIS02-412.

Naffin, D., and G. Sukhatme. (2004). Negotiated formations. *Proceeding of the International Conference on Intelligent Autonomous Systems*, 181–190.

Nagakubo, A., and S. Hirose. (1994). Walking and running of the quadruped wall-climbing robot. *IEEE International Conference on Robotics and Automation*, 1005–1012, San Diego.

Namiki, A., Y. Imai, M. Ishikawa, and M. Kanek. (2003). Development of a high-speed multifingered hand system and its application to catching. *IEEE/RSJ International Conference on Intelligent Robots and Systems (IROS)*, Las Vegas, Nev., 2666–2671.

Napier, J. R. (1956). The prehensile movements of the human hand. *Journal of Bone and Joint Surgery* 38B: 902–913.

Narasinham, S., D. M. Siegel, and J. M. Hollerbach. (1990). CONDOR: A computational architecture for robots. In *Dextrous robot hands*, edited by S. T. Venkataraman and T. Iberall, 117–135. Berlin: Springer-Verlag.

Netter, F. H., and J. H. Hansen. (2003). *Atlas of human anatomy*, 3rd ed. Teterboro, N.J.: ICON Learning Systems.

Niku, S. B. (2001). *Introduction to robotics: Analysis, systems, applications.* Upper Saddle River, N.J.: Prentice-Hall.

Nilsson, M. (1998). Why snake robots need torsion-free joints and how to design them. *IEEE International Conference on Robotics and Automation*, 412–417, Leuven, Belgium.

Nilsson, N. J. (1969). A mobile automaton: An application of artificial intelligence techniques. *International Joint Conference on Artificial Intelligence (IJCAI)*, 509–520, Washington, D.C.

Nolfi, S., and D. Floreano. (2000). *Evolutionary robotics: The biology, intelligence and technology of self-organizing machines.* Cambridge, Mass.: MIT Press.

Noreils, F. R. (1993). Toward a robot architecture integrating cooperation between mobile robots: Application to indoor environment. *International Journal of Robotics Research* 12(1): 79–98.

Okhotsimski, D. E., and A. K. Platonov. (1973). Control algorithms of legged vehicle capable of mastering obstacles. *Proceedings of the Fifth IFAC Symposium on Automatic Control in Space*, Geneva, Switzerland.

Olson, C. F. (2000). Probabilistic self-localization for mobile robots. *IEEE Transactions on Robotics and Automation* 16(1): 55–66.

Orebäck, A., and H. Christensen. (2003). Evaluation of architectures for mobile robotics. *Autonomous Robots* 14(1): 33–49.

Ostrowski, J., and J. Burdick. (1996). Gait kinematics of a serpentine robot. *IEEE International Conference on Robotics and Automation*, Minneapolis, 1294–1299.

Ostrowski, J., and J. Burdick. (1998). The geometric mechanics of undulatory robotic locomotion. *International Journal of Robotics Research* 17(7): 683–702.

Otis, A. B., W. Fenn, and H. Rahn. (1950). Mechanics of breathing in man. *Journal of Applied Physiology* 2: 592–607.

Paap, K. L., T. Christaller, and F. Kirchner. (2000). A robot snake to inspect broken buildings. *IEEE/RSJ International Conference on Intelligent Robots and Systems (IROS2000)*, 2079–2082.

Paap, K. L., M. Dehlwisch, and B. Klaassen. (1996). GMD-Snake: A semi-autonomous snake-like robot. *Third International Symposium on Distributed Autonomous Robot Systems (DARS 96)*, Saitama, Japan, 71–77.

Paluska, D., J. Pratt, D. Robinson, and G. Pratt. (2000). Design of a humanoid biped robot for walking research. *First IEEE-RAS International Conference on Humanoid Robots*, Massachusetts Institute of Technology Cambridge, Mass.

Pamecha, A., I. Ebert-Uphoff, and G. S. Chirikjian. (1997). Useful metrics for modular robot motion planning. *IEEE Transactions on Robotics and Automation* 13(4): 531–545.

Parker, L. E. (1994). ALLIANCE: An architecture for fault-tolerant, cooperative control of heterogeneous mobile robots. *International Conference on Intelligent Robots and Systems (IROS)*, 776–783.

Parker, L. E. (1997). Cooperative motion control for multi-target observations. *IEEE/RSJ International Conference on Intelligent Robots and Systems*, 1591–1598.

Parker, L. E. (1998). ALLIANCE: An architecture for fault tolerant multirobot cooperation. *IEEE Transactions on Robotics and Automation* 14(2): 220–240.

Parker, L. E. (1999). Cooperative robotics for multi-target observation. *Intelligent Automation and Soft Computing* 5(1): 5–19.

Parker, L. E., G. A. Bekey, and J. Barhen, eds. (2000). *Distributed autonomous robotic systems*, vol. 4. Tokyo: Springer-Verlag.

Parker, L. E., and B. Emmons. (1997). Cooperative multi-robot observation of multiple moving targets. *IEEE International Conference on Robotics and Automation*, Albuquerque, N.M., 2082–2089.

Payton, D., M. Daily, R. Estkowski, M. Howard, and C. Lee. (2001). Pheromone robotics. *Autonomous Robots* 11(3): 319–324.

Payton, D., R. Estkowski, and M. Howard. (2002). Progress in pheromone robotics. *Proceedings of the Seventh International Conference on Intelligent Autonomous Systems*, Marina del Rey, Calif.

Pearson, K. G. (1973). Central programming and reflex control of walking in cockroach. *Journal of Experimental Biology* 56: 173–193.

Pearson, K. G. (1976). The control of walking. *Scientific American* 276: 72–82.

Perrier, M., and C. Canudas de Wit. (1996). Experimental comparison of PID vs. PID plus nonlinear controller for subsea robots. *Autonomous Robots* 3(2–3): 195–212.

Perry, J. (1992). *Gait analysis: Normal and pathological function.* Thorofare, N.J.: SLACK, Inc.

Perry, J., and J. K. Gronley. (1989). *Observational gait analysis handbook.* Downey, Calif.: Rancho Los Amigos Medical Center.

Perry, J., and H. J. Hislop. (1967). *Principles of lower-extremity bracing.* Washington, D.C.: American Physical Therapy Association.

Peters II, R. A., K. Kawamura, D. M. Wilkes, K. A. Hambuchen, T. E. Rogers, and A. Alford. (2001). ISAC humanoid: An architecture for learning and emotion. *IEEE-RAS International Conference on Humanoid Robots*, Massachusetts Institute of Technology, Cambridge, Mass.

Petit, C. W. (2003). Beautiful swimmers. *US News & World Report*, October 27, 56–58.

Pfeiffer, F., J. Eltze, and H.-J. Weidemann. (1995). The TUM walking machine. *Intelligent Automation and Soft Computing* 1(3): 307–323.

Pfeiffer, F., H.-J. Weidemann, and J. Eltze. (1993). Leg design based on biological principles. *IEEE International Conference on Robotics and Automation (ICRA)*, Atlanta, 352–358.

Pfeiffer, F., H.-J. Weidemann, and J. Eltze. (1990). Dynamics of the walking stick insect. *IEEE International Conference on Robotics and Automation (ICRA)*, Cincinnati, 1458–1463.

Piaget, J. (1962). *Play, dreams and imitation in children.* New York: Norton.

Poole, D., A. Mackworth, and R. Goebel. (1998). *Computational intelligence: A logical approach.* Oxford: Oxford University Press.

Popović, D., and L. Schwirtlich. (1993). Design and evaluation of the self-fitting modular orthosis (SFMO). *IEEE Transactions on Rehabilitation Engineering* 1(3): 165–174.

Popović, D., R. Tomović, and L. Schwirtlich. (1989). Hybrid assistive system—The motor neuro prosthesis. *IEEE Transactions on Biomedical Engineering* 36(7): 729–738.

Porsin-Sirirak, T. N., Y.-C. Tai, H. Nassef, C.-M. Ho, J. Grasmeyer, and M. Keenon. (2000). Microbat: A palm-sized battery-powered ornithopter. Paper presented at NASA/JPL Workshop on Biomorphic Robotics, Pasadena, Calif, August 14.

Poulakakis, I., J. A. Smith, and M. Buehler. (2003). On the dynamics of bounding and extensions toward the half-bound and the gallop gaits. *Proceedings of the Second International Symposium on Adaptive Motion of Animals and Machines*, Kyoto, Japan.

Pratt, G. A. (2000). Legged robots: What's new since Raibert. *IEEE Robotics and Automation Magazine* 7: 15–19.

Pratt, G. A., and M. M. Williamson. (1995). Series elastic actuators. *International Conference on Robots and Systems*, 399–406, Pittsburgh.

Premvuti, S., and S. Yuta. (1991). Consideration on the cooperation of multiple autonomous mobile robots. In *Autonomous mobile robots: Control, planning and architecture*, edited by S. S. Iyengar and A. Elfes, 219–223. Los Alamitos, Calif.: IEEE Computer Society Press.

Prior, S. D. (1990). An electric wheelchair mounted robotic arm—A survey of potential users. *Journal of Medical Engineering and Technology* 14(4): 143–154.

Quinn, R. D., and K. Espenschied. (1993). Control of a hexapod robot using a biologically inspired neural network. In *Biological neural networks in invertebrate neuroethology and robotics*, edited by R. D. Beer, R. E. Ritzmann, and T. McKenna, 365–381. San Diego, Calif.: Academic Press.

Quinn, R. D., and R. E. Ritzmann. (1998). Construction of a hexapod robot with cockroach kinematics benefits both robotics and biology. *Connection Science* 10: 239–254.

Raibert, M. H. (1986). *Legged robots that balance*. Cambridge, Mass.: MIT Press.

Raibert, M. H., H. B. Brown, and M. Chepponis. (1984). Experiments in balance with a 3D one-legged hopping machine. *International Journal of Robotics Research* 3(1): 75–92.

Rao, K., G. Medioni, H. Liu, and G. A. Bekey. (1988). Robot hand-eye coordination: Shape description and grasping. *IEEE International Conference on Robotics and Automation* 1: 407–411, Philadelphia.

Riggs, D. S. (1970). *Control theory and physiological feedback mechanisms*. Baltimore: Williams & Wilkins.

Ritzmann, R. E., R. D. Quinn, J. T. Watson, and S. N. Zill. (2000). Insect walking and biorobotics: A relationship with mutual benefits. *Bioscience* 50(1): 23–33.

Rizzolatti, G., L. Gadiga, V. Gallese, and L. Fogassi. (1996). Premotor cortex and the recognition of motor actions. *Cognitive Brain Research* 3: 131–141.

Rock, S. M., E. W. Frew, H. L. Jones, E. LeMaster, and B. Woodley. (1998). Combined CDGPS and vision-based control of a small autonomous helicopter. *American Control Conference* 2: 694–698, Philadelphia.

Rosenblatt, F. (1957). *The perceptron: A perceiving and recognizing automaton*. Ithaca, N.Y.: Cornell Aeronautical Laboratory.

Rosenblatt, F. (1962). *Principles of neurodynamics: Perceptrons and the theory of brain mechanisms*. Washington, D.C.: Spartan.

Rosheim, M. (1989). *Robot wrist actuators*. New York: Wiley.

Rosheim, M. (1994). *Robot evolution: The development of anthrobotics*. New York: Wiley.

Roumeliotis, S. I., and G. A. Bekey. (1997). An extended Kalman filter for frequent local and infrequent global sensor data fusion. *SPIE Conference on Sensor Fusion and Decentralized Control in Autonomous Robotic Systems*, 11–22, Pittsburgh.

Roumeliotis, S. I., and G. A. Bekey. (2000a). Collective localization: A distributed Kalman filter approach to localization of groups of mobile robots. *IEEE International Conference on Robotics and Automation (ICRA)*, 2958–2965, San Francisco.

Roumeliotis, S. I., and G. A. Bekey. (2000b). Bayesian estimation and Kalman filtering: A unified framework for mobile robot localization. *IEEE International Conference on Robotics and Automation*, 2895–2992, San Francisco.

Roumeliotis, S. I., and G. A. Bekey. (2002). Distributed multirobot localization. *IEEE Transactions on Robotics and Automation* 18(5): 781–795.

Roumeliotis, S. I., P. Pirjanian, and M. Matarić. (2000). Ant-inspired navigation in unknown environments. *AAAI International Conference on Autonomous Agents*, 25–26, Barcelona.

Roumeliotis, S. I., G. S. Sukhatme, and G. A. Bekey. (1999a). Circumventing dynamic modeling: Evaluation of the error-state Kalman filter applied to mobile robot localization. *IEEE International Conference on Robotics and Automation*, 1656–1663, Detroit.

Roumeliotis, S. I., G. S. Sukhatme, and G. A. Bekey. (1999b). Smoother-based 3-D attitude estimation for mobile robot localization. *IEEE International Conference on Robotics and Automation*, 1979–1986, Detroit.

Ruiz de Angulo, V., and C. Torras. (1997). Self-calibration of a space robot. *IEEE Transactions on Neural Networks* 8(4): 951–963.

Rus, D., Z. Butler, K. Kotay, and M. Vona. (2002). Self-reconfiguring robots. *Communications of the ACM* 45(3): 39–45.

Rus, D., and G. S. Chirikjian, eds. (2001). Self-reconfigurable robots. *Autonomous Robots* 10(1): 5–6.

Rus, D., B. R. Donald, and J. Jennings. (1995). Moving furniture with teams of autonomous mobile robots. *International Conference on Intelligent Robots and Systems (IROS)*, 235–242, Pittsburgh.

Rus, D., and M. Vona. (2001). Crystalline robots: Self-reconfiguration with compressible unit modules. *Autonomous Robots* 10(1): 107–124.

Russel, M. (1983). ODEX 1: The first functionoid. *Robotics Age* 5: 12–18.

Russell, S., and P. Norvig. (2002). *Artificial intelligence: A modern approach*. Upper Saddle River, N.J.: Prentice-Hall.

Rybski, P., D. Krantz, M. Stoeter, M. Gini, R. Voyles, D. Hougen, B. Yesin, B. Nelson, and M. Erickson. (2000). Rangers and Scouts: A team of robots for reconnaissance and surveillance. *IEEE Robotics and Automation Magazine* 7: 14–24.

Rybski, P., S. Stoeter, M. Gini, D. Hougen, and N. P. Papanikolopoulos. (2002). Performance of a distributed robotic system using shared communications channels: A framework for the operation and coordination of multiple miniature robots. *IEEE Transactions on Robotics and Automation* 18(5): 713–727.

Safak, K. K., and G. G. Adams. (2002). Dynamic modeling and hydrodynamic performance of biomimetic underwater robot locomotion. *Autonomous Robots* 13(3): 223–240.

Saito, F., T. Fukuda, and F. Arai. (1994). Swing and locomotion control for a two-link brachiation robot. *IEEE Control Systems Magazine* 14(1): 5–12.

Saranli, U., M. Buehler, and D. E. Koditschek. (2001). RHex: A simple and highly mobile hexapod robot. *International Journal of Robotics Research* 20(7): 616–631.

Saripalli, S., J. F. Montgomery, and G. Sukhatme. (2003). Visually guided landing of an unmanned aerial vehicle. *IEEE Transactions on Robotics and Automation* 19(3): 371–380.

Sato, A. (2003). A strategic approach by Yamaha Motor Co. for its lineup of unmanned helicopters. Paper presented at Association of Unmanned Vehicle System International (AUVSI), Baltimore, Md., July 15–17.

Scassellati, B. (2002). Theory of mind for a humanoid robot. *Autonomous Robots* 12(1): 13–24.

Schaal, S. (2002). Arm and hand movement control. In *The Handbook of Brain Theory and Neural Networks,* 2nd ed., edited by M. A. Arbib, 110–113. Cambridge, Mass.: MIT Press.

Schmiedeler, J. P., and K. J. Waldron. (1999). The mechanics of quadrupedal galloping and the future of legged vehicles. *International Journal of Robotics Research* 18(12): 1224–1234.

Schneider, A., and U. Schmucker. (1999). Force control of six-legged robot for service operations. *Second International Conference on Climbing and Walking Robots (CLAWAR)*, 231–237, Portsmouth, U.K.

Schultz, A. C., and L. E. Parker, eds. (2002). *Multi-robot systems: From swarms to intelligent automata (Proceedings of the 2002 NRL Workshop on Multi-Robot Systems)*. Dordrecht, Netherlands: Kluwer Academic.

Sekmen, A., A. Koku, and S. Sabatto. (2001). Multi-robot cooperation based on vocal communication. *Proceedings of the IASTED International Conference on Robotics and Applications*, 124–129, Tampa, Fla.

Shahinpoor, M., and K. J. Kim. (2001). Ionic polymer-metal composites: I. Fundamentals. *Smart Materials and Structures* 10: 819–833.

Sharp, C. S., O. Shakernia, and S. Sastry. (2001). A vision system for landing an unmanned aerial vehicle. *IEEE International Conference on Robotics and Automation*, 1720–1728, Seoul, Korea.

Shen, W.-M., B. Salemi, and P. Will. (2002). Hormone-inspired adaptive communication and distributed control for CONRO self-reconfigurable robots. *IEEE Transactions on Robotics and Automation* 18(5): 1–12.

Shih, C., and W. Gruver. (1992). Control of a biped robot in the double-support phase. *IEEE Transactions on Systems, Man and Cybernetics* 22(4): 729–735.

Shih, C., W. Gruver, and T.-T. Lee. (1993). Inverse kinematics and inverse dynamics for control of a biped walking machine. *Journal of Robotic Systems* 10(June): 531–555.

Shim, H., T. J. Koo, F. Hoffman, and S. Sastry. (1998). A comprehensive study on control design of autonomous helicopters. *Proceedings of the IEEE Conference on Decision and Control*, Orlando, Fla.

Shurr, D., and J. Michael. (2002). *Prosthetics and orthotics.* Upper Saddle River, N.J.: Prentice-Hall.

Simmons, R. S. (1994). Structured control for autonomous robots. *IEEE Transactions on Robotics and Automation* 10(1): 34–43.

Simmons, R. S., and D. Apfelbaum. (1998). A task description language for robot control. *International Conference on Intelligent Robots and Systems (IROS)* 3: 1931–1937, Vancouver, British Columbia, Canada.

Simmons, R. S., D. Apfelbaum, W. Burgard, D. Fox, M. Moors, S. Thrun, and H. Younes. (2000a). Coordination for multi-robot exploration and mapping. *National Conference on Artificial Intelligence*, 852–858, Austin, Texas.

Simmons, R. S., S. Singh, D. Hershberger, J. Ramos, and T. Smith. (2000b). First results in the coordination of heterogeneous robots for large-scale assembly. *Seventh International Symposium on Experimental Robotics (ISER 00)*, Honolulu, Hawaii.

Song, S. M., and K. J. Waldron. (1989). *Machines that walk: The adaptive suspension vehicle.* Cambridge, Mass.: MIT Press.

Specht, D. F. (1991). A general regression neural network. *IEEE Transactions on Neural Networks* 2(6): 568–576.

Spong, M. W. (1995). The swing-up control problem for the acrobot. *IEEE Control Systems Magazine* 15: 49–55.

Spong, M. W. (1996). Motion control of robot manipulators. In *Handbook of control*, edited by W. Levine, 1339–1350. Boca Raton, Fla.: CRC Press.

Stansfield, S. (1991). Robotic grasping of unknown objects: A knowledge-based approach. *International Journal of Robotics Research* 10(4): 314–326.

Sternad, D., M. Duarte, H. Katsumata, and S. Schaal. (2000). Dynamics of a bouncing ball in human performance. *Physical Review E* 63: 1–8.

Stone, H., ed. (1980). *Introduction to computer architecture.* Chicago: Science Research Associates (SRA), a division of McGraw-Hill.

Stone, P., M. Asada, T. Balch, R. D'Andrea, M. Fujita, B. Hengst, G. Kraetzschmar, et al. (2001). RoboCup–2000: The fourth robotic soccer world championships. *AI Magazine* 22(1): 11–38.

Sukhatme, G. S. (1997). The design and control of a prototype quadruped microrover. *Autonomous Robots* 4(2): 211–220.

Sukhatme, G. S., and G. A. Bekey. (1995). An evaluation methodology for autonomous mobile robots for planetary exploration. *First ECPD International Conference on Advanced Robotics and Intelligent Automation*, 558–563, Athens, Greece.

Sukhatme, G. S., and M. Matarić. (2000). Embedding robots into the Internet. *Communications of the ACM* 43(5): 67–73.

Sukhatme, G. S., J. F. Montgomery, and M. Matarić. (1999). Design and implementation of a mechanically heterogenous robot group. *Proceedings of Mobile Robots XIV Conference (SPIE 99)*, Boston, 111–122.

Sukhatme, G. S., J. F. Montgomery, and R. T. Vaughan. (2001). Experiments with aerial-ground robots. In *Robot teams: From diversity to polymorphism*, edited by T. Balch and L. E. Parker. Wellesley, Mass.: A. K. Peters.

Sutton, R. S. (1988). Learning to predict by the method of temporal differences. *Machine Learning* 3: 9–44.

Sutton, R. S. (1990). Integrated architectures for learning, planning and reacting based on approximating dynamic programming. *Seventh International Conference on Machine Learning*, 216–224. Menlo Park, Calif.: Morgan Kaufmann.

Swinson, M. L., and D. J. Bruemer. (2000). Expanding frontiers of humanoid robotics. *Intelligent Systems* 15(4): 12–16.

Takanishi, A., M. Ishida, Y. Yamazaki, and I. Kato. (1985). The realization of dynamic walking by the biped walking robot WL-10RD. *International Conference on Advanced Robotics (ICAR'85)*, 459–466.

Taylor, C. L., and R. J. Schwartz. (1955). The anatomy and mechanics of the human hand. *Artificial Limbs* 2(2): 22–35.

Takayama, H., and S. Hirose. (2001). Development of HELIX: A hermetic 3D active cord with novel spiral swimming motion. Paper presented at *TITech COE/Super Mechano-Systems Symposium*, Session D-3, Nov. 20.

Terada, Y. (2000). A trial for animatronic systems including aquatic robots. *Journal of the Robotic Society of Japan* 18(2): 37–39.

Thorpe, C., T. Jochem, and D. Pomerleau. (1997). The 1997 Automated Highway Free Agent Demonstration. *Proceedings of IEEE Conference on Intelligent Transportation Systems*, 496–501.

Thrun, S. (1993). Exploration and model building in mobile robot domains. In *Proceedings of the International Conference on Neural Networks (ICNN-93)*, edited by E. Ruspini, 175–180. Piscataway, N.J.: IEEE.

Thrun, S. (1998a). Finding landmarks for mobile robot navigation. *IEEE International Conference on Robotics and Automation*, 958–963.

Thrun, S. (1998b). Learning metric-topological maps for indoor mobile robot navigation. *Artificial Intelligence* 99(1): 21–71.

Thrun, S. (2002). Particle filters in robotics. In *Proceedings of the 18th Conference in Uncertainty in Artificial Intelligence*, 511–518. San Francisco, Calif.: Morgan Kaufmann.

Thrun, S., M. Beetz, M. Bennewitz, W. Burgard, A. B. Cremers, F. Dellaert, D. Fox, et al. (2000). Probabilistic algorithms and the interactive museum tour-guide robot Minerva. *International Journal of Robotics Research* 19(11): 972–979.

Thrun, S., A. Bucken, W. Burgard, D. Fox, T. Fröhlinghaus, D. Henning, T. Hofmann, M. Krell, and T. Schmidt. (1998). Map learning and high-speed navigation in RHINO. In *Artificial intelligence and mobile robots*, edited by D. Kortenkamp, R. P. Bonasso, and R. R. Murphy, 21–52. Menlo Park, Calif.: AAAI Press/MIT Press.

Thrun, S., W. Burgard, and D. Fox. (1998). A probabilistic approach to concurrent mapping and localization for mobile robots. *Autonomous Robots* 5(3–4): 253–271.

Tinbergen, A. (1996). *Social behavior in animals*. London: Methuen.

Tomatis, N., I. Nourbakhsh, and R. Siegwart. (2001). Simultaneous localization and map building: A global topological model with local metric maps. *IEEE/RSJ International Conference on Intelligent Robots and Systems (IROS 2001)*, Maui, Hawaii.

Tomović, R. (1961). A general theoretical model of creeping displacement. *Cybernetica* 4.

Tomović, R., and G. Boni. (1962). An adaptive artificial hand. *IRE Transactions on Automatic Control* 7: 3–10.

Tomović, R., and R. B. McGhee. (1966). A finite state approach to the synthesis of bioengineering control systems. *IEEE Transactions on Human Factors in Electronics* 7(2): 65–69.

Torras, C. (1995). Robot adaptivity. *Robotics and Autonomous Systems* 15(1–2): 11–23.

Torras, C. (2002). Robot arm control. In *Handbook of brain theory and neural networks*, 2nd ed., edited by M. A. Arbib, 979–983. Cambridge, Mass.: MIT Press.

Townsend, W. T. (2000). The Barrett Hand grasper—programmably flexible part handling and assembly. *Industrial Robot* 27(3): 181–188.

Ulrich, I., and I. Nourbakhsh. (2000). Appearance-based place recognition for topological localization. In *IEEE International Conference on Robotics and Automation*, 1023–1029, San Francisco.

Valavanis, K., D. Gracanin, M. Matijasevic, R. Kolluru, and G. A. Demetriou. (1997). Control architectures for autonomous underwater vehicles. *IEEE Control Systems Magazine* 17(6): 48–64.

Van Dam, J. W. M., B. J. A. Kröse, and F. C. A. Groen. (1996). Neural network applications in sensor fusion for an autonomous mobile robot. In *Reasoning with uncertainty in robotics*, edited by L. Dorst, M. van Lambalgen, and F. Voorbraak, 263–277. Berlin: Springer-Verlag.

van der Pol, B., and J. van der Mark. (1928). The heartbeat considered as a relaxation oscillator, and an electrical model of the heart. *Philosophical Magazine* 6: 763–775.

Veloso, M., T. Balch, P. Stone, and H. Kitano. (2002). RoboCup-2001: The fifth robotic soccer world championships. *AI Magazine* 23(1): 55–68.

Venkataraman, S. T., and T. Iberall, eds. (1990). *Dextrous robot hands*. New York: Springer-Verlag.

von Frisch, K. (1967). *The dance language and orientation of bees*. Cambridge, Mass.: Belknap Press of Harvard University Press.

Voth, D. (2002). Nature's guide to robot design. *IEEE Intelligent Systems* 17(6): 4–6.

Voyles, R. (2000). TerminatorBot: A robot with dual-use arms for manipulation and locomotion. *IEEE International Conference on Robotics and Automation*, 61–66, San Francisco.

Voyles, R., A. Agah, P. K. Khosla, and G. A. Bekey. (1997). Tropism-based cognition for the interpretation of context-dependent gestures. *IEEE International Conference on Robotics and Automation*, 3481–3486, Albuquerque, N.M.

Voyles, R., J. D. Morrow, P. K. Khosla. (1999). Gesture-based programming for robotics: Human-augmented software adaptation. *IEEE Intelligent Systems* 14(6): 22–29.

Vukobratović, M., B. Borovac, D. Surla, and D. Stokić. (1989). *Scientific fundamentals of robotics*, vol. 7, *Biped locomotion: dynamics, stability, control and applications*. Berlin: Springer-Verlag.

Vukobratović, M., D. Hristić, and Z. Stojiljković. (1974). Development of active anthropomorphic skeletons. *Medical and Biological Engineering* 12(1): 66–80.

Vukobratović, M., and D. Juricić. (1968). Contributions to the synthesis of the biped gait. *Engineering in Biology and Medicine* 8: 188–196.

Waldron, K. J., and R. B. McGhee. (1986). The adaptive suspension vehicle. *IEEE Control Systems Magazine* 6: 7–12.

Waldron, K. J., V. J. Vohnout, A. Pery, and R. B. McGhee. (1984). Configuration design of the adaptive suspension vehicle. *International Journal of Robotics Research* 3(2): 37–48.

Walker, M., C. Diebel, C. Haugh, P. Pankhurst, J. Montgomery, and C. Green. (1997). Structure and function of the vertebrate magnetic sense. *Nature* 390: 371–376.

Walter, W. G. (1950). An imitation of life. *Scientific American* 182: 42–45.

Walter, W. G. (1953). *The living brain*. New York: Norton.

Wang, H. H., S. M. Rock, and M. J. Lee. (1996). OTTER: The design and development of an intelligent underwater robot. *Autonomous Robots* 3(2–3): 297–320.

Wang, L.-X. (1994). *Adaptive fuzzy systems and control*. Englewood Cliffs, N.J.: Prentice-Hall.

Wang, P. K. C. (1991). Navigation strategies for multiple autonomous mobile robots moving in formation. *Journal of Robotic Systems* 8(2): 177–195.

Wang, P. K. C., F. Y. Hadaegh, and K. Lau. (1999). Synchronized formation rotation and attitude control of multiple free-flying spacecraft. *AIAA Journal of Guidance, Control and Dynamics* 22: 28–35.

Watkins, C., and P. Dayan. (1992). Q-learning. *Machine Learning* 8: 279–292.

Webb, B., and T. R. Consi, eds. (2001). *Biorobotics: Methods and applications*. Menlo Park, Calif.: AAAI Press/MIT Press.

Webb, P. W. (1984). Form and function in fish swimming. *Scientific American* 251: 58–68.

Weber, S., O. C. Jenkins, and M. Matarić. (2000). Imitation using perceptual and motor promitives. *Proceedings, Fourth International Conference on Autonomous Agents*, 136–137, Barcelona.

Wehner, R., B. Michel, and P. Antonsen. (1996). Visual navigation in insects: Coupling of egocentric and geocentric information. *Journal of Experimental Biology* 199(1): 129–140.

Weimerskirch, H., J. Martin, Y. Clerquin, P. Alexandre, and S. Jiraskova. (2001). Energy saving in flight formations. *Nature* 413: 697–698.

Weitzenfeld, A. (1995). NSL: The Neural Simulation Language. In *The Handbook of Brain Theory and Neural Networks*, edited by M. A. Arbib, 654–658. Cambridge, Mass.: MIT Press.

Werbos, P. (1974). *Beyond regression: New tools for prediction and analysis in the behavioral sciences.* Cambridge, Mass.: Harvard University Press.

Werbos, P. (1995). Backpropagation: Basics and new developments. In *The Handbook of Brain Theory and Neural Networks,* edited by M. A. Arbib, 134–139. Cambridge, Mass.: MIT Press.

Wettergreen, D., C. Thorpe, and W. L. Whittaker. (1993). Exploring Mount Erebus by walking robot. *International Conference on Intelligent Autonomous Systems (IAS),* 72–81.

Wiener, N. (1961). *Cybernetics: Control and communication in the animal and the machine.* New York: Wiley.

Wilson, A. B. (1998). *A primer on limb prosthetics,* Springfield, Ill.: C. C. Thomas.

Wilson, E. O. (1971). *The insect societies.* Cambridge, Mass.: Belknap Press of Harvard University Press.

Wilson, E. O. (2000). *Sociobiology: The new synthesis.* 25th anniversary Ed. Cambridge, Mass.: Belknap Press of Harvard University Press.

Wiltschko, W., and R. Wiltschko. (1995). *Magnetic orientation in animals.* London: Springer-Verlag.

Wiltschko, W., and R. Wiltschko. (1998). The navigation system in birds and its development. In *Animal cognition in nature,* edited by R. P. Balda, I. M. Pepperberg, and A. C. Kamil, 155–199. New York: Academic Press.

Winston, P. H. (1992). *Artificial intelligence,* Reading, Mass.: Addison-Wesley.

Winter, D. A. (1984). Biomechanics of human movement with applications to the study of human locomotion. *Critical Review of Biomedical Engineering* 9(4): 287–314.

Winter, D. A. (1987). *The biomechanics and motor control of human gait.* Waterloo, Ontario, Canada: University of Waterloo Press.

Wolf, A., H. Ben Brown Jr., R. Casciola, A. Costa, M. Schwerin, E. Shammas, and H. Choset. (2003). A mobile hyper-redundant mechanism for search and rescue tasks. In *IEEE/RSJ International Conference on Intelligent Robots and Systems (IROS 2003),* Las Vegas, Nev., 3: 2889–2995.

Yamaguchi, H., and J. Burdick. (1998). Asymptotic stabilization of multiple nonholonomic mobile robots forming group formations. In *IEEE International Conference on Robotics and Automation,* Leuven, Belgium, 3573–3580.

Yamaguchi, H., T. Arai, and G. Beni. (2001). A distributed control scheme for multiple robotic vehicles to make group formation. *Robotics and Autonomous Systems* 36: 125–147.

Yamamoto, I., and Y. Terada. (1999). Development of oscillating fin propulsion system and its application to ships and artificial fish. *Mitsubishi Heavy Industries Technical Review* 36(3).

Yamasaki, F., T. Matsui, T. Miyashita, and H. Kitano. (2000). PINO, the humanoid that walks. *First IEEE Conference on Humanoid Robots,* Cambridge, Mass.

Yamashiro, S., and F. Grodins. (1971). Optimal regulation of respiratory airflow. *Journal of Applied Physiology* 30: 597–602.

Yamauchi, B., A. C. Schultz, and W. Adams. (1998). Mobile robot exploration and map-building with continuous localization. *IEEE International Conference on Robotics and Automation (ICRA-98),* 3715–3720, Leuven, Belgium.

Yates, F. E. (1988). *Self-organizing systems: The emergence of order.* Boston: Plenum.

Yeung, D.-Y., and G. A. Bekey. (1989). Using a context-sensitive learning network for robot arm control. In *IEEE International Conference on Robotics and Automation,* 1441–1447, Seottsdale, Ariz.

Yim, M., K. Roufas, D. Duff, Y. Zhang, C. Eldershaw, and S. Homans. (2003). Modular reconfigurable robots in space applications. *Autonomous Robots* 14(2–3): 225–234.

Yim, M., Y. Zhang, and D. Duff. (2002). Modular robots. *IEEE Spectrum* 39(2): 30–34.

Yuh, J., ed. (1995). *Underwater robotic vehicles: Design and control.* Albuquerque: TSI Press.

Yuh, J. (2000). Design and control of autonomous underwater vehicles: A survey. *Autonomous Robots* 8(1): 7–24.

Yuh, J., T. Ura, and G. A. Bekey, eds. (1997). *Autonomous underwater robots.* Boston: Kluwer Academic.

Yun, X. (1993). Object handling using two arms without grasping. *International Journal of Robotics Research* 12(1): 99–106.

Zadeh, L. (1965). Fuzzy sets. *Information and Control* 8: 338–353.

Zeglin, G., and B. Brown. (1998). Control of a bowleg hopping robot. *IEEE International Conference on Robotics and Automation*, 793–798, Leuven, Belgium.

Zhang, Y., K. Roufas, and M. Yim. (2001). Software architecture for modular self-reconfigurable robots. *IEEE/RSJ International Conference on Intelligent Robots and Systems (IROS)*, 2355–2360, Maui, Hawaii.

Zheng, Y. F., and J. Y. S. Luh. (1985). Control of two coordinated robots in motion. *Proceedings of the IEEE Conference on Decision and Control*, 1761–1765.

Author Index

Subject Index

Printed in the United States
By Bookmasters